Principles of
plant biology for the tropics

Frontispiece The Baobab (*Adansonia digitata* L.) casts its spell on all who love the tropics. (Zambesi valley, Central Africa).

Principles of plant biology for the tropics

A.R. Loveless

Department of Biological Sciences,
Portsmouth Polytechnic
(formerly at the University of the West Indies, Jamaica
and the University of Rhodesia,
now University of Zimbabwe)

Longman
London and New York

Longman Group Limited
Longman House
Burnt Mill, Harlow
Essex CM20 2JE, England
Associated companies throughout the world

*Published in the United States of America
by Longman Inc., New York*

First published 1983

British Library Cataloguing in Publication Data

Loveless, A. R.
   Principles of plant biology for the tropics.
   1. Tropical plants
   I. Title
   581.909′3   QK936

   ISBN 0-582-44757-7

Library of Congress Cataloging in Publication Data
Loveless, A. R. (Arthur Raymond), 1922–
   Principles of plant biology for the tropics.

   Includes bibliographies and index.
   1. Botany.   2. Botany–Tropics.   I. Title.
QK47.L796   581.909′13   81-20876
                        AACR2

---

*Printed in Singapore by
The Print House (Pte) Ltd*

Contents

Preface

It has been my privilege to be among the first members of the teaching staff to be appointed at both the University College (now the University) of the West Indies in Jamaica, and the University College of Rhodesia and Nyasaland (now the University of Zimbabwe). There are, or at any rate were, two sides to being a botanist at a newly established institution in the tropics. On the one side there is the thrill of being something of a pioneer, but on the other there is the frustration which comes from a lack of information about the local vegetation, which must provide the specimens to illustrate the teaching if the subject is to make any impact on the students. Biologists in the universities of the Northern Hemisphere who have never left their home country, where the standard teaching types have been established for a long time, rarely appreciate the time and effort that has to be put into the sifting of local material for teaching purposes. It was, and still remains, part of the duty of staff to mould their teaching around the flora of the country in which they find themselves. Although botanists who have not worked in the tropics may not appreciate it, the classical route of studying the basic structure, physiology, classification, ecology and genetics of the local flora must be followed, as it always has been elsewhere in the past. The microtome, the light microscope and the plant press should rightly be very much more in evidence than the electron microscope and the computer. The satisfaction that comes from tackling a problem that is crying out for a solution is much greater than competing feverishly with a multitude of other researchers in the hope of discovering today what one of them will discover tomorrow in any case. To a biologist with an interest in the wider aspects of the subject, the study of biology in the tropics offers a very exhilarating, and at times humiliating, challenge.

These few remarks will help to explain the aim and content of this book. It is an accident of history that, until recently, the centres of research and publication of biological works have all been located in temperate regions, and as far as plant biology is concerned, this has led to the propagation of 'laws' and 'rules' that are almost exclusively based on the specialized and often depauperate vegetation of these regions. Some have refused to admit the myopia of their approach, whereas others, in their enthusiasm, have tended to go to the other extreme and extolled 'tropical biology' as if it were a completely new subject. The author takes the view that an exclusively tropical biology is just as myopic as an exclusively temperate, or classical, biology. In a world where travel between tropical and temperate regions is becoming commonplace, the aim should be the formulation of general principles applicable to the world as a whole. This is a book of principles; it is a textbook of plant biology for use in the tropics, and not a textbook of tropical botany.

As will be appreciated, to present some kind of logical and balanced treatment of the principles of plant biology between the covers of one book involves problems of inclusion and exclusion. A glance through the chapter headings can hardly fail to provoke the comment, 'Where are the accounts of bacteria, fungi and the genetics of micro-organisms?' My immediate answer, admittedly lame, would be that the book would be too big if they were included, but my more fundamental answer would be that it has been my experience that a student's ultimate satisfaction is best served by not offering too much too soon. The confidence that comes from a thorough understanding of a subject at one stage is the best preparation for the next. This book attempts to give a first-year university student a taste of several educational experiences, implicit throughout the subject but especially evident in certain areas. A taste of critical observation and three-dimensional interpretation is offered in Part I which approaches plant structure from a developmental point of view. A taste of the experimental and biochemical aspects of the subject is the theme of Part II, which deals with plant physiology at the level of the whole plant, although the insights of biochemistry into respiration and photosynthesis are not neglected. A taste of the 'aesthetic' pleasure inherent in the diversity of plants is intended to flavour the accounts of the plant kingdom and ecology; and, lastly, a taste of the statistical aspect of all modern biology is introduced in a very brief account of Mendelian genetics, in which problem solving by deductive reasoning is the keynote. Within the scope of this menu it is impossible to cover all the firmly established aspects of plant biology, let alone to take extensive excursions into the growing points of the subject. My justification for omitting even a short account of the fungi, which are in fact my specialist interest, is that they are not plants. But this pedantic reason aside, the real reason is that the importance of fungi as plant pathogens warrants an entirely different approach from that appropriate to the plant kingdom, and this cannot be achieved satisfactorily in a few pages. To include a few fungal life cycles, such as those of a pin-mould and a mushroom, would probably have done more harm than good, by encouraging a rote learning of a few life cycles. Regretfully I exclude my favourite organisms, leaving them to speak for themselves at a later stage.

Many will think that it is foolhardy nowadays for an author to attempt a textbook single-handed. Today, it is sometimes argued, is the age of attractively presented pamphlets and booklets which are 'disposable', to be replaced when necessary by revisions with ever more colourful illustrations. The hard fact remains, however, that the acquisition of a proper scientific attitude is achieved not by flitting from one bright idea to the next, but by a disciplined study of a hard core of original observations and an appraisal of their interpretation. To achieve this end it is my belief that a student in his first year at university needs a single book which will present him with an integrated bird's eye view of the subject. This is probably best achieved by one or two authors rather than a panel of several authors. Pamphlets and readers by different authors can lead to fragmentation of information rather than to its synthesis. I am not arguing against the use of pamphlets and short studies, but in my view they have their greatest educational value when the student can fit their contents into an already existing framework. This book is my attempt to provide such a framework, to which molecular biology, microbial genetics, plant pathology and other specialist topics can be added at a later stage.

Such a hope as I have outlined has meant that I have inevitably had to tread in areas where I have not been a participant at the laboratory bench. Nevertheless my tropical experience has inevitably forced me, more than most, into becoming something of a 'Jack of all trades and a . . .' However, I have been able to enlist the criticism and advice of a number of friends who are experts in the fields where I have least knowledge. I am grateful to these colleagues not only for their sometimes deservedly blunt criticism, but also, perhaps more surprisingly, for their encouragement to persevere in my task. They did not adopt the attitude that I was poaching on their territory. With their kind help many glaring errors were removed from earlier drafts, but it is inevitable that in a book of this type some errors of fact and interpretation still remain. I should greatly appreciate receiving notice of errors, and suggested improvements.

Nobody writes a textbook of plant biology 'out of the blue'. His mind has been shaped and his outlook fashioned by all sorts of outside influences, some of which he will recall and admit, and others which, because he has assimilated them years ago, he may have long since forgotten. These latter influences have been just as formative as the former, but patient research might be required to discover what they were. There are, however, two colleagues of whose influence I am very much aware, and whom it would be an injustice not to acknowledge. The first is Professor Geoffrey Asprey, my first 'Prof', who instilled and then fostered my interest in tropical botany. More recently my friend, Dr Brian Strophair, has read and re-read drafts of all the chapters of Part II which is his specialist field. In addition he has volunteered his help as a guinea-pig for much of the rest of the book. Many of his suggestions I was only too willing to accept, but there were some issues on which I was resolved to follow my own inclination in an attempt to achieve the balance I was seeking. Errors must therefore be laid at my door! I must also thank Mr Eddie Hawton for his invaluable help in preparing many of the illustrations for reproduction in this book.

Finally I have to acknowledge an unrepayable debt to two members of my family who have had to endure the affliction of 'the book' for more years than I care to recall. Nevertheless it gives me great pleasure to record the part played by my youngest daughter, Wendy, who has converted many of my 'roughs' into beautiful line diagrams bearing no trace of the scruffiness of the originals. If the value of this book is enhanced by its diagrams, then much of the credit is due to the unstinted efforts of my daughter. In conclusion, the vital role played by my wife can only be fully appreciated by myself. The wife plays a specially important part in the life of a tropical botanist, and there is some basis for the facetious remark that wives should be interviewed before staff are appointed in the tropics. To Mary, 'botanist by marriage', I dedicate this book for her part in making it a reality.

Part 1

Structure of the flowering plant

Chapter 1

The plant body

A flowering plant may be said to have a body with as much justification as an animal has a body. The idea that a plant has a body emphasizes that it is the entire plant which is the living unit, the different parts or organs merely performing their respective functions in the life of the whole. Each plant organ is, as a rule, dependent for its livelihood upon the existence and proper working of the other organs, and so cannot be considered in isolation from them. Thus, when studying the structure and functions of any particular organ, it must always be considered in the context of the entire plant. Only in this way will the study of each separate organ assume its full meaning.

Root and shoot systems

The body of nearly all flowering plants can be divided into two systems, the root system which grows below ground and the shoot system which grows above it. The typical root system anchors the plant firmly in the soil, and absorbs water and various dissolved raw materials from it. The typical shoot system consists of a main stem upon which are borne leaves, branch shoots, and sometimes flowers. It should be emphasized that the words stem and shoot are not synonymous, the word 'shoot' being a collective term for both stem and leaves. The leaves are the chief food-producing organs of the plant, making complex foods in the presence of light by a process called photosynthesis. The stem supports the leaves, displaying them in the light needed for photosynthesis, and also acts as the main channel of communication between the various organs of the plant. Water and dissolved minerals absorbed by the roots are carried in the stem to the leaves and flowers, and foods synthesized in the leaves are conducted away through the stem to other regions of the plant, both above and below ground, where they are either used for growth or else stored for future use. Flowers are organs that are concerned with the reproduction of the species by seeds.

Although, generally speaking, it is true that the shoot and root systems of a plant consist of the above-ground and under-ground parts respectively, the shoots of some plants grow below ground while the roots of others grow above ground. Hence to distinguish botanically between a root and shoot system, a more precise basis of characterization is needed. For this purpose, the external structure of a typical herbaceous plant will be considered in more detail (Fig. 1.1). The plant consists of a main axis, partly above and partly below ground, which bears lateral members. The axis is constantly

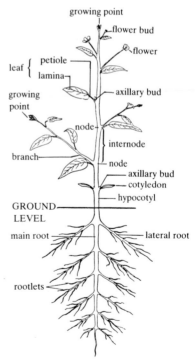

Fig. 1.1 The root and shoot systems of a typical herbaceous flowering plant.

elongating as a result of growth at both ends. The descending portion of the axis, the main root, bears lateral members similar to itself. These lateral roots are produced some distance behind the apex in what is called acropetal succession, i.e. the youngest lateral members are nearest the apex, the older ones progressively further back. An important character concerning the origin of lateral roots is that they arise internally and push their way outwards, causing a small slit to appear in the outer tissues of the parent root at the points at which they emerge. Lateral roots are therefore said to be endogenous in origin. The extreme tip of every root, main or lateral, is covered by a small cap termed the root cap. The ascending portion of the axis, the main stem, bears its lateral members in pairs, one on top of the other. The lower member of each pair is a leaf, the upper member is a bud. A bud is a compact, undeveloped shoot, consisting of a short stem bearing crowded, overlapping, immature leaves. A bud which is produced as the upper member of a pair of laterals is called an axillary bud, because the angle between the upper face of a leaf and the stem is called the leaf axil. An axillary bud may later develop into a branch or lateral shoot, which is a replica of the main shoot. Leaves and axillary buds, like lateral roots, also develop in acropetal succession, but they differ from them in two respects. Firstly, they originate at the apex and not in the older portion of the axis, and secondly, they arise as superficial bulges of tissue and not from deep-seated tissues. Leaves and axillary buds are therefore said to be exogenous in origin. At the tip of the main stem the undeveloped leaves are packed closely together to form a bud, which is called the terminal bud to distinguish it from the axillary buds.

Buds and branching

As has already been mentioned, a bud is a condensed shoot. Its stem is very short and its leaves are so close together that they overlap, each one wrapping round the next above it. The inner leaves are crinkled and folded, since a large surface area is packed into a small space (Fig. 1.2b). Most buds are too small for their structure to be easily examined, but the European cabbage (*Brassica oleracea*), a vegetable sometimes grown in tropical highlands, is an extremely enlarged terminal bud which shows very clearly the main features of bud structure.

Trees which shed their leaves during the dry season (deciduous) form resting buds in which the outer leaves are modified into bud scales. If a terminal bud of a deciduous species of *Ficus* (some species are evergreen) is examined during the dry season, it will be seen that the bud scales are represented by a number of conical sheaths one inside the other (Fig. 1.2a). Inside the sheaths are foliage leaves which, although very small, are perfectly formed. When the dry season is over, the stem in the centre of the resting bud begins to elongate and, in doing so, splits open the outer sheaths which soon fall off. The stem continues to increase in length and thereby spaces out the leaves, which unfold and spread out their surface. When the terminal bud sprouts the sheaths are not spaced out by the elongating stem, so that when they fall off they leave scars which are close together (Fig. 1.2c, d). These bud scars, or girdle scars, mark the position of each terminal bud, and therefore the length of stem between two sites of bud scars represents one season's growth.

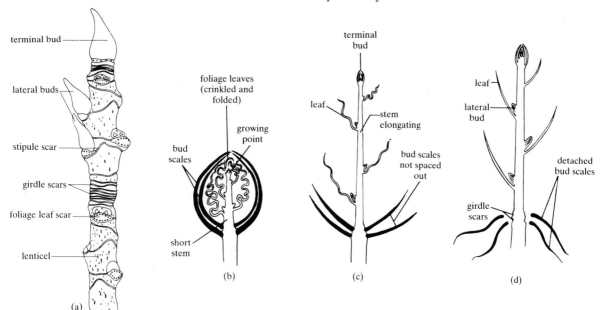

Fig. 1.2 Buds. (a) Twig of *Ficus* collected from tree in leafless condition. (b)–(d) Stages in the opening and growth of a resting bud. (a from Hall 1970, p. 96; b–d from D. G. Mackean (1965), *Introduction to Biology*, p. 31, John Murray, London)

Modes of branching

In some trees the terminal bud of the seedling axis continues to grow year after year, so that the tree has one main stem, or trunk, with its branches formed from axillary buds. This mode of growth in which the original terminal bud carries on the main axis indefinitely, is called *monopodial*, and the stem is said to be unlimited or indefinite in its growth. A tree produced by monopodial growth tends to be cone-shaped; a good example is *Casuarina*.

If a terminal bud ceases to grow for any reason, such as the production of a flowering shoot, further increase in length of the axis is achieved by the growth of a lateral bud which becomes, at least for the present, the new terminal bud. This increase in length by successive development of lateral buds is called *sympodial* growth and can happen in several ways. For example, each twig of the almond tree (*Terminalia catappa*), after growing sideways for a time, turns upwards at its end and bears a cluster of leaves. This twig grows no further but, from one of the leaf axils several centimetres behind, a lateral bud grows out to form another twig which continues the main growth of the branch. This second twig then behaves like the first, and so the process is repeated. Sympodial growth is also shown by plants, notably grasses, which have horizontal underground stems. At intervals the end of the stem bends upwards and becomes an aerial leafy shoot, but an axillary bud at the base of the erect portion becomes active and grows forward in the soil, giving the false impression of being a continuation of the original axis.

Arrangement of leaves on the stem

During the growth of a bud, the stem apex grows forward with the result that the lateral outgrowths are stretched further and further apart until, when the region of the stem on which they are borne stops growing, they show the arrangement typical of the adult stem. The leaves may then be

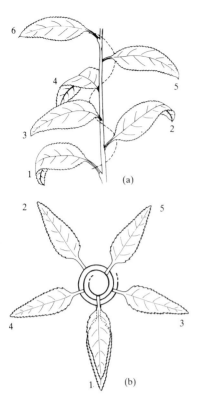

Fig. 1.3 Portion of a mature shoot with $\frac{2}{5}$ phyllotaxis, as seen from the side (a) and above (b). The broken line, which passes twice around the stem in passing from leaf 1 to leaf 6, indicates the phyllotaxis.

seen to occur singly, in pairs, or in whorls of three or more at definite points on the stem. These points of leaf insertion are called *nodes*, and the stem regions between them are called *internodes*. On the adult stem the leaves are borne in a remarkably regular and orderly way. Usually successive leaves are inserted singly at the nodes, and arranged spirally around the stem in such a manner that all the leaves fall into a number of vertical rows or *orthostichies*. The pattern in which leaves are borne on a stem is called *phyllotaxis* and is expressed by means of a fraction. The denominator of the fraction is the number of orthostichies, and the numerator is the number of times that a line connecting the leaf bases of successive leaves spirals around the stem in passing from one leaf of an orthostichy to the leaf inserted directly above it on the same orthostichy. Thus on the plant illustrated (Fig. 1.3) the phyllotaxis is $\frac{2}{5}$ because the dotted line passes twice around the stem in passing from leaf 1 to leaf 6 which lies vertically above it. The appearance, as seen from above, of the five orthostichies of a $\frac{2}{5}$ phyllotaxis is shown in Fig. 1.3b. The fraction characterizing a particular phyllotaxis also measures the angular divergence between any two successive leaves (i.e. the fraction of the circumference which separates successive leaves), and this in the plant under consideration is $\frac{2}{5}$ of 360 °, or 144 °. On the same basis a plant with distichous leaves (i.e. regularly arranged one above another in two opposite rows, one on each side of the stem) has a phyllotaxis of $\frac{1}{2}$ because the angular divergence is 180°.

Experience shows that the commonly occurring systems of spiral phyllotaxis fall into the series $\frac{1}{2}$, $\frac{1}{3}$, $\frac{2}{5}$, $\frac{3}{8}$, $\frac{5}{13}$.... As can be seen the numerator and denominator of each fraction are given by the sum of the numerators and denominators respectively of the two preceding fractions. Thus both the numerators and denominators follow the mathematical series known as the Fibonacci series. As the series is followed into the higher fractions, $\frac{5}{13}$, $\frac{8}{21}$, $\frac{13}{34}$, ... etc., the numbers rise very rapidly and clearly the vertical rows will become more and more difficult to recognize until it becomes impossible in practice to find a leaf which one can be sure lies vertically above another. Phyllotaxis has been the subject of extensive research, but as yet there is no satisfactory explanation of why the Fibonacci series should occur in phyllotactic systems. It has been suggested that the angular divergences represented by the fractions of the phyllotactic series give the spatial arrangements which expose the foliage leaves to the light with the minimum overlap consistent with a uniformly balanced plant body. It is difficult, however, to see how leaf arrangement, which is determined in the bud, can be controlled by the ultimate distribution of the adult leaves in the light. It seems more reasonable that the explanation depends on factors operating during early development. Irrespective of the underlying causes, the facts of phyllotaxis emphasize that the plant body is a highly organized system which is amenable to investigation by precise scientific techniques.

Root systems

The root system of a given species of plant has a characteristic form, although environmental conditions, such as the nature of the soil and amount of moisture available, profoundly affect both the shape of the root system and the extent to which it develops. The root which is formed directly from the axis of the embryo plant is the primary root, and roots arising from this are called lateral roots. Root systems vary in structure according to the relative development of the primary and lateral roots, but two main types are recognized. In the *tap root* system either the primary root or one or more lateral roots which replace the primary root at a very early stage in the development of the seedling, grow more rapidly and become stouter than the other roots, so that one or several main roots are formed. The roots of most beans and peppers provide good examples of a tap root system with one main axis. Tap roots may become swollen as a result of the storage of large amounts of food material in them, e.g. water-leaf (*Talinum triangulare*) and monkey gun or many roots (*Ruellia tuberosa*). In the *fibrous root* system the primary root is replaced by many roots all of which grow to approximately the same length and diameter. Root systems of this type are often easily uprooted because they tend not to grow to any great depth in the soil. Fibrous root systems are found in all grasses, including cereal crops and sugar cane (*Saccharum officinarum*).

Adventitious and modified organs

The above description of the external features of flowering plants applies to most plants belonging to this group, but there are many organs which do not agree with such an ideal type. Such atypical organs result from two causes: (1) they may not arise in the normal position; and (2) they may be modified to perform some specialized function such as storage, or additional support.

Organs which do not arise in the normal position are said to be *adventitious*. Lateral roots frequently arise on stems, especially those of climbing and creeping plants (Fig. 1.4). Likewise lateral buds, which normally arise in the axils of leaves, may arise from the leaf edges of the resurrection plant or leaf of life (*Bryophyllum pinnatum*), or from the root tubers of sweet potato (*Ipomoea batatas*) which themselves are swollen adventitious roots.

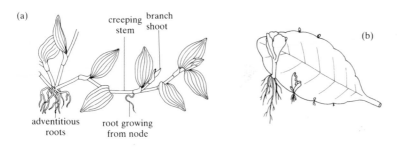

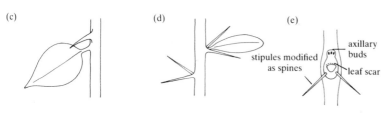

Fig. 1.4 Adventitious and modified organs. (a) Adventitious roots of *Commelina*. (b) Adventitious buds, which give rise to shoots, arising in the notches of a leaf of the leaf of life, or resurrection plant (*Bryophyllum pinnatum*). (c) Branch spines of *Bougainvillea*. (d) Stipular spines of crown of thorns (*Euphorbia milii*). (e) A single pair of stipular spines, seen from the front.

Stems and leaves, and to a lesser extent roots, may be so modified that it is sometimes difficult to recognize them for what they are. However, by studying their position in relation to other structures it is usually possible to determine their morphological nature. For example, woody spines occur on the shoots of *Bougainvillea* spp. and crown of thorns (*Euphorbia milii*), but from their relative positions it can be readily established that the spines of *Bougainvillea* are modified branches, whereas those of crown of thorns are modified stipules. (N.B. Stipules are small lateral outgrowths, usually leaf-like, which occur at the base of the leaf stalks of many plants, e.g. pigeon pea (*Cajanus cajan*).)

Factors affecting the shape of plant organs

No two plants are exactly alike. Some plants, for example certain members of the Cucurbitaceae, the family which includes melons, squashes and other sorts of gourds, have leaves which vary widely in shape, whereas other plants are relatively uniform in appearance. The variation found in plants is the outcome of the interaction between genetic, environmental and developmental effects.

Genetic variation is due to differences in the hereditary make-up of different plants. Consistent differences between groups of plants which occur in every generation must be due to the genetic differences between them. For instance, the grasses (Gramineae) all have very characteristic leaves with (a) a blade portion in which a number of veins, approximately equal in size, run parallel with one another from the base to the tip of the leaf blade, (b) a long tubular sheath which surrounds a cylindrical stem (usually hollow except at the nodes, but sometimes pithy) and is usually split lengthwise on the side opposite the leaf blade, and (c) a small outgrowth, the *ligule*, either membranous or represented by a fringe of hairs, which stands out from the upper surface of the leaf at the junction of the blade and sheath (see Fig. 33.5a). The sedges (Cyperaceae) superficially resemble grasses, but differ genetically from them in the following ways. In Cyperaceae the leaf sheath forms a continuous sleeve around a solid triangular stem, and there is no ligule present. Genetic effects extend, of course, to chemical as well as structural features. Thus one of the easiest means of distinguishing a columnar cactus (member of the Cactaceae) from a columnar *Euphorbia* (member of the Euphorbiaceae), which vegetatively may be very similar to each other, is to make a small slit in the stem. If a white latex exudes from the slit the plant is a *Euphorbia*, whereas if no exudation occurs or a colourless juice exudes the plant is a cactus.

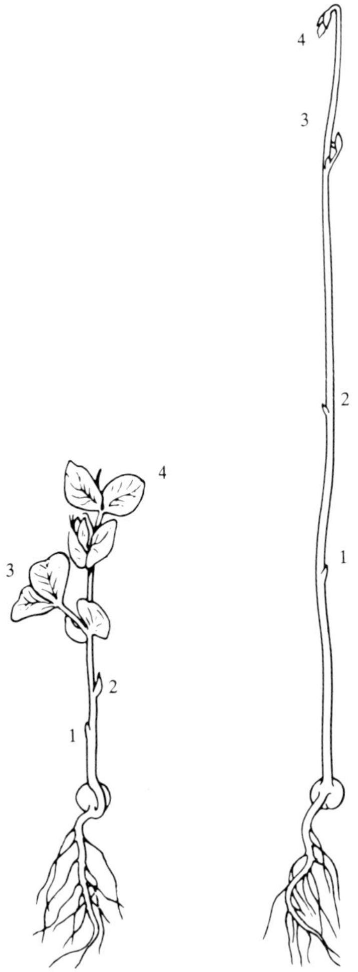

Environmental variation is the direct result of the action of the environment on individual plants. The influence of the environment on plant form can be demonstrated by growing plants with a similar or identical genetic make-up in different environments. Nowhere is this more easily and strikingly shown than by the effect of light on the growth of seedlings. In the dark the green colour does not develop, the leaves do not expand but the internodes elongate many times more than normal. Such a seedling, which is long and straggly, and yellow in colour, is said to be *etiolated* (Fig. 1.5). This response is of great biological value to the plant because it enables a buried shoot to elongate rapidly, unhindered by bulky leaves, until it reaches the light when the normal pattern of development takes over. The environmental effect on growth is also seen in the lopsided shapes of trees growing near windswept coasts. These shapes are due to the inhibition of growth on the seaward side of the plant resulting from excessive unilateral evaporation and the toxic effects of salt spray.

Fig. 1.5 Seedlings of *Pisum* grown in light (a) and dark (b). Note the relative positions and sizes of the leaves at nodes 1, 2, 3 and 4. The hook at the end of the stem of the etiolated seedling is characteristic.

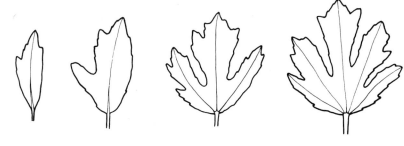

Fig. 1.6 Variation in shape of successive leaves on one plant of *Urena lobata*.

In the case of etiolation and wind-pruned trees it is the external environment which is responsible for producing the abnormal pattern of growth, but the influences of the conditions inside the plant are also continuously being exerted. In the course of its development a plant passes through a series of stages, the most obvious of which are the vegetative and reproductive. The vegetative phase does not, however, consist in the production of a series of uniform organs. This may be verified by examining almost any young plant and noting how the leaves borne at different positions (i.e. produced at different stages of development) show a progressive increase in the complexity of lobing or toothing (Fig. 1.6). Sometimes the transition from juvenile to adult leaf shape is not gradual but sudden. Various species of *Eucalyptus*, for example, have juvenile leaves that are ovate but mature leaves that are sickle-shaped. There are numerous other examples of such sudden phase changes, but little is known about the physiological reasons for their occurrence.

The classification and naming of plants

Plants are so many and varied in their form and structure that they must be classified in some way before it is feasible to study them as a group of organisms rather than as isolated individuals. When a number of individual plants are examined carefully, it immediately becomes apparent that there are degrees of likeness among them. Some resemble each other so closely that they are difficult to distinguish, while others are clearly very different. The recognition of this fact leads logically to an attempt to arrange them in groups, the members of any one of which resemble each other more closely than they resemble the plants in any other group. The difficulty, of course, lies in sorting out the groups by criteria which are fundamental and not superficial. In practice, however, when the whole structure and life history of many different kinds of plants are compared, there are about ten basic patterns of organization underlying the bewildering diversity. Thus grasses, palms, melons and cacti, although very different at first sight, all have typical or modified roots and stems, a type of leaf (macrophyll) which has originated in a particular way, and all reproduce by seeds. In these fundamental features of their organization and reproduction they are all alike, each being a modification of a common pattern. Together with the cone-bearing trees, ferns, clubmosses and several other types of plants they are all placed in one of the ten large *divisions* of the plant kingdom, the *Tracheophyta*, which includes all plants possessing a specialized conducting tissue, called xylem, used for transport of water and solutes. Similarly the mosses and liverworts have certain fundamental features in common and are grouped together in another division, the *Bryophyta*.

In exactly the same way that the plant kingdom can be split up into divisions, so each of the divisions can be further divided into smaller groups. Grasses, palms, melons and cacti are all placed in the class *Angiospermae* which, mainly on account of its large size (about 250 000 different kinds of plants), is subdivided into two subclasses. Grasses and palms both have only one seed leaf or cotyledon in their seeds, a feature which separates them as members of one subclass, the *Monocotyledons*, from the other subclass, the *Dicotyledons*, whose members have two cotyledons and include melons and cacti. Classes are divided into *orders* (grasses and palms belong to separate orders, as also do melons and cacti), the orders into *families*, the families into *genera* (singular genus), and finally the genera into *species*. The species is the smallest classificatory unit in general use. Despite its importance on this account, it is difficult, if not impossible, to define a species rigidly but, broadly speaking, it consists of a group of individuals which can interbreed and which resemble one another so closely that for all normal purposes they can be regarded as being of the same kind. The application of this system of classification to one species, the common melon (not the water melon), is set out in Table 1.1.

Table 1.1 Classification of the melon

Kingdom	Plantae
Division	Tracheophyta
Class	Angiospermae
Subclass	Dicotyledonae
Order	Cucurbitales
Family	Cucurbitaceae
Genus	*Cucumis*
Species	*melo*

Table 1.1 also serves to demonstrate the standard way in which plants are named for purposes of designation, so that botanists can specify very precisely the plant to which they are referring. Vernacular names vary from language to language and even from place to place within a country, and so are useless for this purpose. *Yam* is a good example of an ambiguous vernacular name. Strictly speaking, a yam is 'any of the commercially useful plants of the botanical genus *Dioscorea*, or the tubers or rhizomes of these plants'. However, in the United States the word yam is commonly understood to mean the sweet potato (*Ipomoea batatas*). Yams are also often confused with several of the edible aroids, *Colocasia, Alocasia* and *Xanthosoma* spp., which are more properly referred to as cocoyams, taros, dasheens, eddoes, tanias or yautias. In fact almost any edible starchy root, tuber or rhizome that is grown within the tropics, has at one time or another been described as a yam. For scientific purposes every species of living and fossil plant is given a two-word name, a binomial, written in Latin. The first word denotes the genus to which the plant belongs, and is called the *generic name*. All generic names begin with a capital letter. The second name distinguishes the particular species and is called the specific name, or *specific epithet*. The specific epithet of all plants should be written with a small initial letter, e.g. *Bunchosia swartziana*, although formerly the practice was to write the first letter as a capital when the epithet is named after a person, usually the discoverer of the plant, e.g. *Bunchosia Swartziana*, which was named after Swartz. Because sometimes the same species has been given different names (synonyms) by separate botanists, and more rarely the same name (homonym) has been accidently given to different species, the bino-

mial name is followed by the name of the person or persons responsible for its publication. This procedure gives a fixed point of reference for a name, because the original description, usually accompanied by pressed and dried specimens, is permanently available for study. The genus to which the melon and cucumber belong is *Cucumis*, but whereas the melon is *Cucumis melo* L., the cucumber is *Cucumis sativus* L. where L. stands for Linnaeus. Similarly the full citation for the choyote, chocho or christophine is *Sechium edule* (Jacq.) Swartz, where the enclosure of the name Jacq. (short for Jacquin) in brackets means that Jacquin originally placed the christophine in one genus, *Sicyos*, but that Swartz subsequently showed that it was better placed in another genus, *Sechium*, and published the new binomial name. His reasons for changing the name were accepted, and so *Sechium edule* became the correct name. In such cases of a change in taxonomic position the name of the original author is cited in brackets in the full citation of the revised binomial. In practice it is usually not necessary, when giving the name of a plant, to add the authority except when there is the possibility of ambiguity, or where absolute precision is required as in international scientific journals.

Dicotyledons and Monocotyledons

This book is largely concerned with flowering plants, and reference will continually be made in the following pages to the two subclasses Dicotyledons (often colloquially called dicots) and Monocotyledons (or monocots). To conclude this first chapter, therefore, it is convenient to mention the outstanding features of these two well-defined types of flowering plants. At an early stage in their evolution the flowering plants diverged into two stocks which gave rise respectively to the Dicotyledons and Monocotyledons. The members of one stock have seeds containing embryos with two seed-leaves or cotyledons, and those of the other have seeds containing embryos with only one cotyledon. The two stocks are named after the number of cotyledons in the embryo, because this feature is so constant. The 'bifurcation' into Dicotyledons and Monocotyledons was a very fundamental one, and reflected many differences in the structure and physiology of the two stocks. In other words, the number of cotyledons is correlated with many other more obvious characters. The reason why one of these was not chosen as the basis for naming the two stocks is that all of them have more exceptions than the character for number of cotyledons. In practice, however, a flowering plant can usually be recognized straightaway from these more obvious characters as either a Dicotyledon or a Monocotyledon. Palms, lilies, orchids and grasses (which include bamboos) are typical Monocotyledons, most other flowering plants are Dicotyledons. Table 1.2 summarizes the main distinguishing features of the two groups.

Table 1.2 The main distinguishing features of the Dicotyledons and the Monocotyledons

Dicotyledons	Monocotyledons
1. Seeds with two (rarely more) cotyledons	1. Seeds with one cotyledon only
2. Leaves usually with reticulate venation	2. Leaves usually having parallel veins with cross-connections
3. Root system usually with one main root bearing laterals	3. Root system usually fibrous
4. Floral parts in fours or fives, or multiples of these numbers, or a large indefinite number	4. Floral parts usually in threes or multiples of three
5. Vascular bundles arranged regularly near the circumference of the stem	5. Vascular bundles irregularly scattered throughout the stem

Chapter 2

The plant cell

It is a simple matter of observation that houses are often built of constructional units called bricks. In a similar sort of way the bodies of most plants are made up of units called *cells* which are either alive or have once been alive. This piece of information, though now common knowledge, is not self-evident because cells, unlike bricks, cannot be seen with the naked eye. Indeed 200 years ago it would have been more reasonable to think that the trunk of a tree consisted of a solid mass of hard material (wood) rather than of millions of tiny units stuck together. Perhaps the most direct way of demonstrating the cellular construction of plants is to soak any plant tissue in a type of softening fluid called a *macerating agent*. When this is done, even with a piece of lignum vitae (*Guaiacum officinale*) wood which is so hard that it was formerly used for making cog wheels, it is found that the tissue disintegrates into its component cellular units. The process can be compared to the way in which the bricks of a wall come apart if the mortar disintegrates.

The results of maceration and other techniques leave no doubt that cells are the constructional units of plants, but this must not be taken to mean that cells are indivisible. In fact cells can be broken up by appropriate means into smaller units which can then be separated into fractions and studied in isolation. However, it is found that, although some cellular fragments (e.g. mitochondria) continue to function more or less as they did in the intact cell for a day or two, no subcellular unit is capable of a continued independent existence. A second conclusion may therefore be drawn, namely that the cell is the functional as well as the structural unit of living matter, i.e. it is the smallest unit that can sustain life.

The concept that the cell is the basic unit of living matter has been recognized since about 1830 but this should not blind us to the fact that it is one of the foundations of modern biology. The concept is important because, by defining the fundamental unit of life, it focused attention on the structure that had to be investigated and understood if biology was to advance. Indeed it would not be an overstatement to say that we understand life only in so far as we understand the structure and functioning of cells.

The simplest plants consist of only one cell, and in these the cell and the plant are identical. Most plants, however, consist of a large and ever increasing number of cells, the upper limit being determined only by the size the plant attains at maturity. Some appreciation of the numbers involved can be gained when it is realized that an average sized leaf may contain anything up to 50 million cells. It is difficult enough to accept a startling fact like this, but even more difficult to answer the question 'Why should plants (or animals for that matter) consist of cells?' or, to frame the question in another way, 'Why should living matter be parcelled up in little packages?' Biologists do not have a complete answer, but one important factor is the relationship between area of cell surface and volume of cell contents, i.e. the surface/volume ratio. Every living part of a cell requires the supply of food material and the removal of waste products, but it is only through the surface that substances can enter or leave. However, as a cell gets bigger the surface does not increase as rapidly as the volume. Consider the case of a small spherical cell with a radius of one unit. When such a cell increases in size, the volume increases as the cube of the radius (the volume of a sphere $= \frac{4}{3}\pi r^3$) but the surface area increases only as the square of the radius (the surface area of a sphere $= 4\pi r^2$). Table 2.1 shows how the surface/volume ratio of a sphere decreases as its size increases. Because it is only through the surface of a cell that materials can enter or leave, the bigger a cell grows the less efficient it becomes in exchanging substances with its surroundings. It follows therefore that if a cell were to grow beyond a certain size the central parts would not be able to function properly. In view of this limitation imposed on a cell by the surface/volume ratio, it is not surprising that plants get bigger by increasing the number rather than the size of their cells.

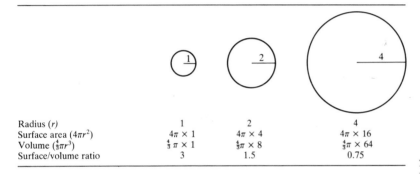

Radius (r)	1	2	4
Surface area ($4\pi r^2$)	$4\pi \times 1$	$4\pi \times 4$	$4\pi \times 16$
Volume ($\frac{4}{3}\pi r^3$)	$\frac{4}{3}\pi \times 1$	$\frac{4}{3}\pi \times 8$	$\frac{4}{3}\pi \times 64$
Surface/volume ratio	3	1.5	0.75

Table 2.1 Comparison of the surface/volume ratio of spheres with radius of 1, 2, and 4 units of length

The structure of plant cells

Mature plant cells differ from one another in size, shape, structure and function. Nevertheless all plant cells have certain features in common, and it is possible to envisage a hypothetical type in which the basic features are present in a relatively unmodified form. This generalized cell (Fig. 2.1) consists of three parts: (1) a *cell wall* on the outside; (2) a lining layer of protoplasm, the *protoplast*; and (3) a large cavity, called the *central vacuole*, which occupies the bulk of the interior of the cell. The features of each of these parts will be considered in turn.

The cell wall

The recognition that a plant cell consists of a protoplast encased within a wall means that the partitions between adjoining cells of a tissue are double walls. This fact is often not easy to detect under the light microscope, but by differential staining and maceration techniques it can be shown that the walls of cells which are in contact with one another are always held together by an intercellular substance, or *middle lamella*. When a cell divides into two, this middle lamella is laid down as the boundary layer and upon it the

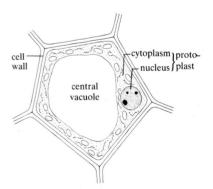

Fig. 2.1 The three regions of a hypothetical plant cell.

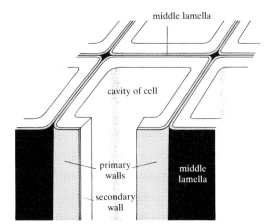

Fig. 2.2 Diagram showing the three layers of a cell wall. The middle lamella is shown in black, followed by the primary wall (stippled) and the secondary wall (white) which surrounds the cavity of the cell. (Modified from Jane 1970, p. 26)

true wall of each daughter cell is deposited. The way in which the middle lamella is formed during cell division is described in Chapter 38.

The cell walls of plants have a complex structure but three fundamental parts may be distinguished: the *middle lamella*, the *primary wall* and the *secondary wall* (Fig. 2.2). All cells have a middle lamella and a primary wall, but a secondary wall is present only in certain types of cell.

The *middle lamella* is a layer of intercellular 'cement' which binds together the primary walls of two adjacent cells. It is composed mainly of water and colloidal pectic compounds which have the important property of being plastic (i.e. easily moulded). The plasticity of the pectic compounds makes possible the intercellular movements and adjustments which are necessary before cells can assume their mature size and shape.

The *primary wall* is the first true wall to be formed by a new cell, and consists mainly of cellulose and hemicelluloses, although water, pectic compounds, and proteins are also present. It starts to form on each side of the middle lamella immediately after a cell has divided into two, and continues to grow as long as the two daughter cells increase in size. This is made possible by the continuous addition of new material to the first-formed wall, which increases both in surface area and in thickness. A primary wall is deposited during the development of all cells, and in those which have living contents at functional maturity it is often the only wall. Mature cells which have only primary walls can revert to the embryonic condition (i.e. can resume the ability to divide), and when this happens the wall becomes thinner. Thus the criteria of a primary wall are that it is capable, firstly, of increasing in surface area, and, secondly, of undergoing reversible changes in thickness.

When a cell has ceased enlarging, wall material sometimes continues to be deposited on the inner surface of the primary wall so that the wall as a whole is conspicuously thickened, and in extreme cases may become so thick that it occupies most of the inside of the cell. This additional wall material is known as a *secondary wall*, even though it may be laid down in two or more (often three) separate layers. The deposition of a secondary wall causes irreversible changes in wall structure, and cells with secondary walls cannot increase in volume by surface growth or revert to the embryonic condition. When first laid down the secondary wall consists mainly of cellulose (with hemicelluloses relatively less important than in a primary wall), but subsequently it usually becomes impregnated with other substances, notably lignin. Cells in which a secondary wall is formed are almost invariably non-living at maturity. Because the presence of a secondary wall gives mechanical strength to a cell, it is not surprising to find that secondary walls are typical of cells which have either a mechanical or a conducting function, or both. In cells which have thick secondary walls it is impossible under the light microscope to identify separately the middle lamella and the two primary walls on either side of it. These three layers are seen as a single line, and should be represented as such in high-power drawings. The term *compound middle lamella* is a convenient term to denote this thin 'three-ply' layer which is sandwiched between two thick secondary walls.

Plasmodesmata

Although a plant cell is surrounded by a wall, the protoplast inside is not completely isolated. Under the electron microscope it can be seen that at certain places on the primary wall fine protoplasmic strands, known as *plasmodesmata* (singular, plasmodesma), pass through minute pores in the primary walls and middle lamella between adjacent cells so that the two protoplasts are connected (Fig. 2.3a). Plasmodesmata facilitate the passage of materials from one cell to the next without the necessity of having to pass through living membranes. The presence of plasmodesmata provides a physical basis for the fact that a plant behaves as a single organism rather than as an assemblage of independent cellular units.

Pits

In the formation of the primary wall of cells the wall substance is rarely laid down evenly over the entire middle lamella with the result that depressions (called either *primordial pits* or *primary pit fields*) occur in its surface (Fig. 2.3b). Such depressions usually, but not invariably, develop in the region of groups of plasmodesmata, and they vary in depth according to the thickness of the primary wall. In cells in which a secondary wall is deposited, the primordial pit is not covered by secondary wall material, so that a clearly defined cavity, called a *pit*, develops in the secondary wall. A pit in the wall of a given cell nearly always occurs opposite a complementary pit in the wall of the adjacent cell so that two pits are combined into a paired structure, the *pit-pair* (Fig. 2.3c). The two cavities of a pit-pair are separated only by the pit membrane, consisting of two thin layers of primary wall and the middle lamella between them (i.e. the compound middle lamella).

There are two kinds of pit: simple pits and bordered pits. In a simple pit the pit cavity (as seen from the side) is uniform in width from the pit membrane at the bottom to the aperture facing the inside of the cell at the top. The pit cavity may be circular or slit-like in cross-section, so that a simple pit appears in surface view as a circular or slit-like area which is more transparent than the surrounding wall (Fig. 2.3d). In a bordered pit the secondary wall projects over the pit cavity to form an overarching roof, called the pit border (hence the adjective bordered), so that the pit cavity varies in width along its length. By having a pit border arching over an uncovered area of primary wall, the bordered pit combines maximum area of pit membrane with minimum loss of rigidity. In essence, therefore, a bordered pit consists of four parts: a pit membrane, a pit cavity, a pit border and a pit aperture (Fig. 2.3e).

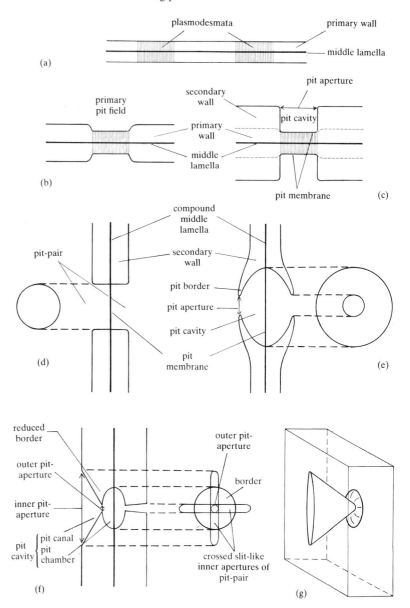

Fig. 2.3 Structure of pits. (a)–(c) Stages in the formation of a simple pit. (d) Simple pit-pair in surface view and L.S. (e) Bordered pit-pair in L.S. and surface view. (f) Reduced bordered pit-pair with the two inner apertures at right angles to each other, as seen in L.S. and surface view. (g) Three-dimensional diagram of reduced bordered pit.

In the simplest types of bordered pit the pit cavity consists of a single space, and the pit aperture is round, elliptic or linear according to the shape of the pit membrane. When the pit apertures are linear they are arranged on the wall in vertical ladder-like rows, and such bordered pits are given the special name *scalariform pits*. More complicated types of bordered pit, known as *reduced bordered pits*, occur in cells with very thick secondary walls. Here the pit cavity is divided into two parts, an outer pit chamber (towards the pit membrane) and an inner pit canal (towards the inside of the cell). As a result of the division of the pit cavity into two parts, there are two pit apertures, an *outer aperture* between the pit chamber and the pit canal and an *inner aperture* facing the inside of the cell. Although the outer aperture, like the pit membrane, is usually circular, the shape of the inner aperture tends to become long and narrow (often longer than the diameter of the pit membrane) because, as the cell wall continues to thicken, the pit canal assumes the form of a flattened funnel (Fig. 2.3f). When this happens, the two slit-like inner apertures of a pit-pair are often arranged at right angles to each other, giving the 'cross pits' which are diagnostic of xylary fibres, one of the commonest of the cell types composing wood.

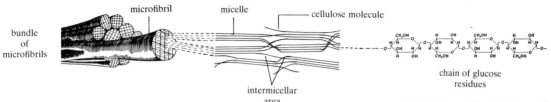

Fig. 2.4 The relationship between the thread-like molecules of cellulose and the microfibrils which make up the framework of plant cell walls. The microfibrils are large enough to be seen under the electron microscope.

The chemical composition of cell walls

The main structural component of all cell walls is cellulose, a single molecule of which consists of a long straight unbranched chain of glucose units joined end to end. Hundreds of these chain-like molecules, each of which contains an indefinite number (sometimes over 10 000) of glucose units, are combined into larger threads, called *microfibrils*, which are still so thin that they can be seen only with the electron microscope (Fig. 2.4). Although the cellulose chains always follow the general direction of the microfibril they vary in their relationship to one another in different regions along its length. In some regions the chains run strictly parallel to one another to form regular bundles which have the properties of a crystalline lattice; these regions are called *micelles* or *crystalline areas*. Elsewhere (i.e. in the intervals between forming part of one micelle and part of another) the chains are arranged less regularly and form non-crystalline regions, the *intermicellar areas*. This arrangement whereby hundreds of cellulose molecules are neatly packed into microfibrils may be compared roughly to a steel cable.

At a higher level of organization the microfibrils themselves are interwoven (and sometimes even grouped into larger bundles) to form a three-dimensional network, which is the structural framework of the cell wall. Water and any wall substances other than cellulose are present in the spaces within and between the microfibrils of the cellulose framework. The addition of substances to the framework is called *encrustation*, a word which strictly means deposition *on* but is here used to imply deposition *between*. Encrusting substances can be removed from the cell wall leaving the cellulose framework intact. From photographs taken under the electron microscope (EM) it has been observed that cell shape is related to the arrangement of the microfibrils within the wall (Fig. 2.5). When the microfibrils are orientated at random the cell tends to grow equally in all directions and hence to develop into a spherical cell. When, on the other hand, most of the microfibrils are laid down in one particular direction, the cell grows mainly at right angles to the general direction of the microfibrils and assumes a cylindrical shape. The situation is more complicated than this, however, because the microfibrils may be differently orientated in the various wall layers of the same cell. Cellulose microfibrils are well suited to play the essential role in wall structure because they form a layer which is not only freely permeable to water and dissolved solutes, but also extremely strong (a cellulose microfibril being nearly as strong as a steel wire of comparable width).

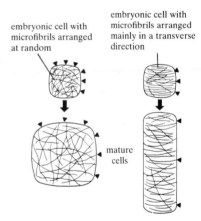

Fig. 2.5 Diagram showing how orientation of the cellulose microfibrils in the wall of an embryonic cell determines the shape of the mature cell into which it develops. Note that the microfibrils are represented as more widely spaced in the mature cells to indicate that the entire wall expands during growth, as shown by the displacement of markers (black triangles) on the surface of the embryonic cells. In fact, new microfibrils are continuously laid down during cell enlargement so that the density of microfibrils in the wall remains approximately constant throughout growth. (After Ray 1972, p. 143)

Unlike starch (which is also a polymer of glucose), cellulose does not react with an aqueous or alcoholic solution of iodine, but it is stained blue by chlor–zinc–iodine solution (alternatively called Schultze's solution) which consists of iodine dissolved in an aqueous solution of zinc chloride and potassium iodide.

Although cellulose is the fundamental constituent of all cell walls, the physical and chemical properties of the walls of cells specialized for particular functions may be modified as a result of the deposition of such encrusting substances as lignin, cutin and suberin.

Lignification   The markedly thickened walls of cells which provide mechanical strength are usually impregnated with lignin, a complex polymer consisting of phenylpropanoid units which are apparently linked together in various ways to form highly branched and complex three-dimensional molecules of variable composition. The deposition of lignin in a cell starts in the middle lamella and then spreads through the primary and secondary walls. This modification in the structure of the walls nearly always results in the ultimate death of the protoplast inside. Lignification does not greatly interfere with the permeability of the wall to water and solutes, but it alters both the physical and chemical properties of the wall. Lignified walls are much harder and resist compression better than cellulose walls. In this respect lignification may be compared with the principle of reinforced concrete used in buildings; the cellulose provides rods of high tensile strength and the lignin is a substance that is resistant to pressures. The properties of wood are due mainly to the high percentage of lignified cells in it. Lignified walls can be distinguished chemically from cellulose walls by the following staining reactions: (1) Schultze's solution stains cellulose walls blue but lignified walls yellow; (2) an alcoholic solution of phloroglucinol followed by a few drops of strong hydrochloric acid stains lignified walls a bright red colour; (3) a solution of any aniline salt (usually the chloride or sulphate) stains lignified walls yellow.

A lignin test should always be made when examining plant tissues to determine which cell types are present, but it must be remembered that a cell type which is lignified at maturity may not give a positive reaction to a lignin stain during its early development.

Cutinization   The epidermal cells of leaves and young stems have their outer walls impregnated with, and covered by, a waxy material called cutin. The superficial non-cellular layer of cutin is known as a cuticle. In sections of leaves and young stems the cuticle looks like a layer of varnish that has been applied over the surface. Cutin is stained red with the fat-soluble dye Sudan IV· The presence of cutin in and on the outer walls of epidermal cells renders them almost impermeable to water and so reduces evaporation from their exposed surfaces.

Suberization   In older stems and roots the original surface layer (epidermis) is usually replaced by a protective layer of phellem. The walls of phellem cells are impregnated with a waxy material, suberin, which is similar in many of its properties to cutin (e.g. stains red with Sudan IV) and is likewise relatively impervious to water. The suberin is usually deposited as a distinct layer on the inner surface of the primary cellulose wall. In thick-walled phellem cells a layer of cellulose or lignified cellulose may occur on the inside of the suberin layer, which then becomes sandwiched between two cellulose layers.

The protoplast

The protoplasmic layer lining the inside of the cell wall is differentiated into two main regions: (1) the nucleus, which is usually seen as a distinct granular structure; and (2) the cytoplasm, which is all the protoplasm outside the nucleus (Fig. 2.6).

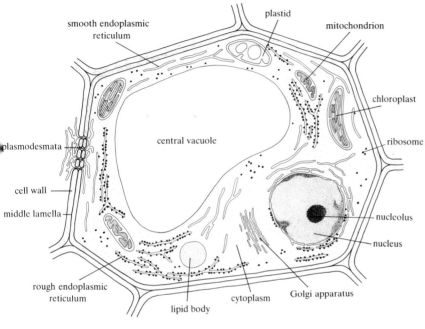

Fig. 2.6 Diagram of a hypothetical plant cell as it might appear under an electron microscope.

The nucleus

In cells which are dividing the nucleus is not constant in its appearance because it is involved in a cycle of changes associated with its division into two daughter nuclei. For most of the time, however, the nucleus can be readily identified as a spherical or ellipsoidal body separated from the rest of the protoplasm by a definite *nuclear envelope*. The term nuclear envelope is preferable to the older one of nuclear membrane, because under the EM the boundary layer is seen to consist of two membranes. In the living state the nucleus appears homogeneous except for one or more small dense bodies, the *nucleoli* (singular nucleolus), which are concerned with the synthesis of ribonucleic acid (RNA) although their precise role in this function is not understood. After fixation and staining, however, the nucleus is seen to contain an irregular network of darkly-staining threads, called *chromonemata*, and a clear fluid, the *nuclear sap* (Fig. 2.7). At the beginning of cell division the chromonemata resolve themselves into a definite number of short thick rods, the *chromosomes*, which are responsible for the transmission of hereditary characters from parent to offspring (see Chapter 38). The names chromonemata (literally, coloured threads) and chromosomes (literally, coloured bodies) were given to these structures because they are readily stained by certain dyes.

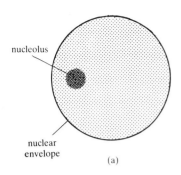

nucleolus

nuclear
envelope

(a)

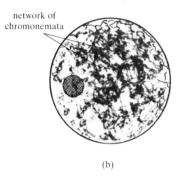

network of
chromonemata

(b)

Fig. 2.7 Appearance of a non-dividing nucleus as seen (a) in the living state, and (b) after fixation and staining. In neither case can any detailed structure be discerned.

The nucleus has two important properties. First, it is the centre which controls the activity of the cell. This property can be deduced from the behaviour of cytoplasm deprived of its nucleus. When, for instance, certain unicellular organisms are cut into two parts, one with and the other without a nucleus, only the nucleate part will survive and regenerate. Similar circumstantial evidence for the controlling function of the nucleus is provided by the observation that the nucleus frequently migrates to the region of the cell where activity is greatest. In an actively growing root hair, for example, the nucleus is usually present at the growing tip.

The second important property of the nucleus, which is actually another but more far-reaching aspect of its controlling function, is that it contains the material determining the hereditary characters of an organism. This property can be demonstrated rather strikingly by reciprocal grafting experiments with two species of *Acetabularia*, a genus of unicellular green algae. All species of *Acetabularia* have a unicellular body which consists of three parts, a hat, a stem, and a foot containing the nucleus (Fig. 2.8). In one species, *A. mediterranea*, the hat is rather like an inverted umbrella, consisting of numerous radiating ribs webbed together. In another species, *A. crenulata*, the hat consists of finger-like outgrowths, which are loose and not joined together. If the hat of either species is cut off a new hat will grow. It is also possible, however, to graft the stem of one species on to the foot of the other species. When this is done a new hat also grows, but in this case the hat always has the shape characteristic of the species contributing the foot and therefore the nucleus. If two feet, one belonging to each species, are grafted on to a single stem, the hat that regenerates has an intermediate form reflecting the influence of each nucleus. These experiments demonstrate that the type of hat, which is an inherited character of each species, is determined by the nucleus in the foot and not by the cytoplasm of the immediately adjacent stem. In addition to the results of reciprocal grafting experiments there is now irrefutable evidence from the study of viruses that the nucleus contains the molecules which form the chemical basis of heredity (see Chapter 39).

The cytoplasm

In the past the cytoplasm was regarded as a ground substance in which various droplets and granules were suspended. The ground substance was thought to lack any organization above the colloidal level because, for example, a punctured cell collapses as its contents ooze out. This concept of the cytoplasm as a structureless matrix is now known to be completely wrong because, under the EM, the cytoplasm is seen to be permeated by an extensive and complex system of interconnecting fluid-filled spaces which are enclosed by a double membrane. By reconstructing electron micrographs in three dimensions the spaces can be shown to vary in shape; in some areas they are tube-like or ribbon-like, whereas in others they are distended into vacuoles. The whole system of membranes and enclosed spaces is called the *endoplasmic reticulum*, usually abbreviated as ER (Fig. 2.9). Much remains to be discovered about the ER, but it seems to be concerned with the storage of proteins, lipids and other substances, and their transport between the various regions of the cell. In addition, the folds and channels of the membrane system provide an extensive surface for the siting of organized sequences of enzymes, particularly those concerned with protein and lipid synthesis. Where the walls of adjoining cells are pierced by plas-

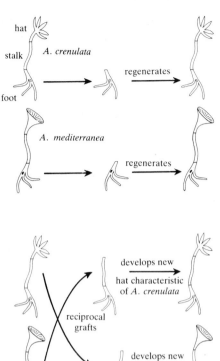

hat
stalk   *A. crenulata*
foot
regenerates

A. mediterranea
regenerates

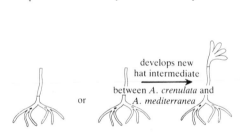

develops new
hat characteristic
of *A. crenulata*

reciprocal
grafts

develops new
hat characteristic
of *A. mediterranea*

or

develops new
hat intermediate
between *A. crenulata* and
A. mediterranea

Fig. 2.8 The role of the nucleus demonstrated by reciprocal grafting of the unicellular algae, *Acetabularia crenulata* and *A. mediterranea*.

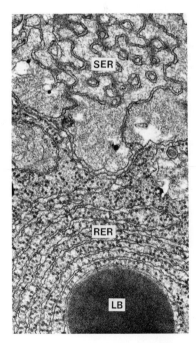

SER

RER

LB

Fig. 2.9 The endoplasmic reticulum (ER) as seen under an electron microscope. The ER is called *rough* when covered with ribosomes, and *smooth* when ribosomes are absent. The regular parallel arrangement of the rough ER in this electron micrograph is due to the presence of a large lipid body. LB = lipid body; RER = rough endoplasmic reticulum; SER = smooth endoplasmic reticulum. × 40 000 (Courtesy of Dr Stephen Moss, Portsmouth Polytechnic)

modesmata it is likely that there are connections between the endoplasmic reticulum of the two protoplasts. The cytoplasm is bounded on its outer surface by a membrane known as the *plasmalemma* (or plasma membrane) and on its inner surface, where it borders on the central vacuole, by another membrane known as the *tonoplast* (or vacuolar membrane). It appears likely that the plasmalemma, the nuclear envelope, and the ER are interconnected to form a single system of membranes which permeates throughout the cytoplasm (Fig. 2.10). All biological membranes are differentially permeable, allowing some substances to pass through them quickly, others more slowly, and still others not at all. The plasmalemma and tonoplast are vitally important in the physiology of cells because they largely control the exchange of materials between the cytoplasm and the extracytoplasmic space outside the cell and inside the vacuole. Unlike the nuclear envelope and the ER, the plasmalemma and tonoplast are single membranes.

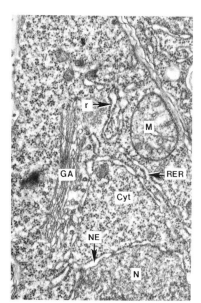

Fig. 2.10 Electron micrograph of inner region of cytoplasm to show the continuity of the components of the protoplast. The rough ER at the bottom right appears to be converging on the nuclear envelope. The Golgi apparatus (GA) is an organelle which, in plant cells, is thought to be concerned in the synthesis of wall polysaccharides which are then transported to the plasmalemma. The alignment of ribosomes (r) on the surface of the ER is clearly seen. Cyt = cytoplasm; GA = Golgi apparatus; M = mitochondrion; N = nucleus; NE = nuclear envelope; r = ribosomes; RER = rough endoplasmic reticulum. × 21 000. (Courtesy of Dr Stephen Moss, Portsmouth Polytechnic)

Structure of membranes   Membrane structure has long been a matter of controversy because it is difficult to conceive of a surface which is both differentially permeable and yet allows only the one-way passage of certain substances, which is exactly what many biological membranes do. Although various types of 'sieves' have been suggested as molecular models of cell membranes, it seems that many of the properties of membranes result from their chemical nature rather than their physical porosity. For example, it has been known since the turn of the century that some membranes are more permeable to substances which are readily soluble in lipids than to those which are more soluble in water. On the basis of the evidence available at the time Danielli suggested in 1938 a 'sandwich' model, in which a continuous inner layer of lipids, two molecules thick, is sandwiched between a layer of protein on each side (Fig. 2.11a).

Membranes are extremely thin layers, about 6 to 10 nm thick (1 nm = 10^{-9} m), and their structure can be seen only with the aid of the EM. Under the EM, membranes are seen to have a three-ply structure, a clear (electron-transparent) inner region bounded on each side by a dark (electron-dense) band. The central electron-transparent region is composed largely of fatty material, and the outer electron-dense layers are mainly protein. Electron micrography of membranes thus provides direct evidence for the essential correctness of Danielli's model which was proposed entirely on circumstantial evidence. EM studies of many types of cells have shown that the appearance of biological membranes is basically the same not only for the surface membranes of all cells but also for the membranes of the internal organelles (literally, small organs) as well. This constant appearance has led to the three-ply structure of a membrane being called a *unit membrane* (Fig. 2.11b). It should nevertheless be stressed that, despite the uniformity of their appearance, membranes differ widely in function and this is reflected by variation in their detailed structure and chemical composition.

Plastids   Broadly speaking, the cytoplasm can be regarded as the 'workshop' of the cell because in it are carried out most of the chemical activities necessary for the maintenance of life. Associated with the performance of these activities the cytoplasm contains a number of organelles, ranging in size from those readily visible under the low power of the light microscope to those visible only under the EM. Organelles are localized areas of living cytoplasm which are specialized for particular functions and, like the ER, are surrounded by two unit membranes. In addition to organelles the cytoplasm also contains other structures which are non-living; these are collectively referred to as ergastic substances (the adjective ergastic is derived from the Greek word meaning 'to work') because they are products of the working or metabolism of the cell.

Plastids are the largest cytoplasmic organelles of plant cells and are clearly visible under the light microscope. They vary considerably in size and shape, but in the cells of flowering plants they commonly assume the form of small biconvex discs. Although the different types of plastids are associated with definite physiological functions, it is customary to classify them according to colour into leucoplasts (colourless), chloroplasts (green), and chromoplasts (red, orange or yellow).

Leucoplasts occur in cells that are not normally exposed to light, e.g. in the deep-seated tissue of aerial parts and in the tissue of underground parts.

Leucoplasts are centres for the synthesis and deposition of reserve food materials such as starch.

Chloroplasts contain chlorophyll, the mixture of pigments to which plants owe their green colour and without which photosynthesis cannot take place. They occur in large numbers in the cells of leaves and young stems; as many as 100 or more may be present in a single cell. Electron microscopy has shown that chloroplasts have an elaborate internal structure (Fig. 2.12). The chloroplast of a higher plant consists of a colourless proteinaceous matrix, the *stroma*, enclosed by two unit membranes. Embedded throughout the stroma is an elaborate system of paired membranes, called *lamellae*, which are joined at their edges and thus enclose an internal space. At regular intervals the lamellae widen to form flattened vesicles, called *thylakoids*, which are stacked on top of one another to form structures resembling piles of coins. Such stacks of thylakoids are called *grana*. The paired membranes between the grana are called *intergranal lamellae*. The chlorophyll is located in the membranes of the thylakoids and intergranal lamellae, which collectively present an enormous surface area for the trapping of light energy for photosynthesis.

Chromoplasts contain pigments other than chlorophyll and are responsible for most of the red, orange and yellow colours of plants parts. They are common in the petals of many flowers, and in some fruits such as red pepper (*Capsicum annuum*). The function of chromoplasts is obscure. They generally do not develop in a cell if chloroplasts are present. During the ripening of certain fruits such as red pepper, the change in colour from the green of the immature fruit to the red of the mature fruit is correlated with a decline in the number of chloroplasts and a simultaneous development of chromoplasts.

Plastids develop from relatively simple structures, called *proplastids*, which gradually enlarge and assume the structure of mature plastids. Proplastids are present in embryonic cells and increase in number by division. When an embryonic cell divides, the proplastids are distributed more or less at random between the two daughter cells.

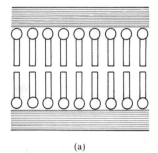

(a)

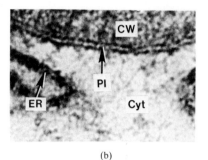

(b)

Fig. 2.11 The unit membrane. (a) The Danielli 'sandwich' model of biological membranes. Lipid molecules are represented by the club-shaped structures, and protein material by the parallel lines covering the surface of the lipid bilayer. (b) Single unit membrane of plasmalemma showing central electron-transparent layer bounded on each side by an electron-dense band. CW = Cell wall: Cyt = cytoplasm: ER = endoplasmic reticulum; Pl = plasmalemma. × 300 000. (Electron micrograph by courtesy of Dr Stephen Moss, Portsmouth Polytechnic)

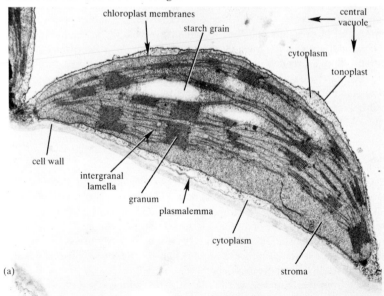

(a)

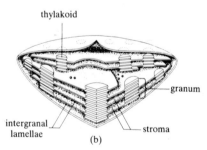

(b)

Fig. 2.12 Internal structure of a chloroplast. (a) Section of a single chloroplast of *Chrysanthemum segetum* showing the limiting double membrane, the grana and intergranal lamellae, and the stroma which in this preparation is densely granular and contains starch grains. Magnification × 16 500 (Courtesy of Professor John Dodge, Royal Holloway College, University of London) (b) Model of a chloroplast, based on electron micrographs of serial sections through a chloroplast. (From Adams *et al.* 1970, p. 166)

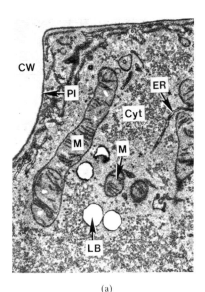

(a)

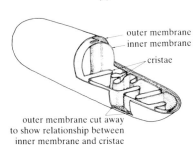

outer membrane cut away
to show relationship between
inner membrane and cristae

(b)

Fig. 2.13 Mitochondria. (a) Electron micrograph of mitochondria in longitudinal and transverse section. The cristae greatly increase the working surface of the organelle. M = mitochondrion; CW = cell wall; Cyt = cytoplasm; ER = endoplasmic reticulum; LB = lipid body; Pl = plasmalemma. × 13 500. (Courtesy of Dr Stephen Moss, Portsmouth Polytechnic) (b) Stereogram of a generalized mitochondrion, showing how the surface area is increased by inward folds called cristae. (From Hurry 1965, p. 24).

Mitochondria   In addition to plastids, the cytoplasm also contains hundreds of minute spherical or sausage-shaped bodies called mitochondria. These are difficult to distinguish from other small cytoplasmic inclusions except by the use of special stains. The EM has revealed that mitochondria, despite their minute size, have a complex internal organization with numerous intuckings which collectively present a large internal surface area (Fig. 2.13).

Like plastids, mitochondria are self-reproducing structures. At cell division all the mitochondria divide, half going to one daughter cell and half to the other. Although the existence of mitochondria was established long before their function was discovered, they are now known to contain the enzymes necessary for cellular respiration.

Ribosomes   Ribosomes are even smaller particles than mitochondria and can be seen only under the EM. They are commonly, but not invariably, found attached to the outside of the endoplasmic reticulum. Ribosomes, although very small, are extremely important because they are the sites of protein synthesis in the cytoplasm. Cells that are active in protein synthesis have a well-developed ER associated with an increased number of ribosomes. Protein synthesis has been achieved *in vitro* using isolated ribosomes.

Ergastic substances   Being products of the chemical activity of the cell, ergastic substances are regarded as either reserve food materials or waste products according to whether or not they are known to be useful to the plant at some future time. Depending on their physicochemical properties, ergastic substances are present either in solution, as granular or crystalline bodies, or suspended as droplets.

One of the most common ergastic substances in the cytoplasm is starch which occurs as grains in the leucoplasts and chloroplasts. Starch grains vary in size and shape in different plants but commonly show conspicuous layering around a point, the hilum, which is at the centre of the grain in some plants but towards one side of the grain in others (Fig. 2.14). One or more starch grains may develop in a single plastid. Starch is the principal reserve material of most plants and next to cellulose, which is the chief component of plant cell walls, is the most abundant carbohydrate in the plant kingdom. Starch grains are widely distributed in the plant body, but are particularly numerous in storage organs (such as fleshy roots, tubers, rhizomes and corms) where they may almost completely fill the cells containing them.

Other common cytoplasmic inclusions are droplets of oil, and various types of crystals composed in most cases of calcium oxalate.

The central vacuole

The large cavity in the interior of a cell was originally called a vacuole because it was thought to be empty. In fact, this space is filled with vacuolar sap, an aqueous solution of various inorganic and organic substances which are mostly reserve food materials or metabolic by-products, i.e. they are mostly ergastic substances. The vacuole is commonly colourless but may be pigmented due to the presence of soluble bluish or reddish pigments belonging to the anthocyanin group of chemicals.

Apart from being merely a 'tank' for the deposition of storage or excretory substances, the central vacuole can be regarded as an answer on the part of the individual plant cell to the problem of the surface/volume ratio. By pushing the cytoplasm to the outside of a cell, the presence of the vacuole ensures that no part of the protoplast is far from the surface and thus promotes more efficient exchange of materials between a cell and its surroundings. A vacuolate cell is thus capable of attaining a larger size than would otherwise be possible. The central vacuole also has an important skeletal function because normally the volume of its fluid contents is sufficiently large to cause the wall on the outside of the cell to be slightly stretched. The imposition of an inward pressure on the vacuolar sap by the stretched wall confers rigidity on the wall and hence on the cell as a whole. The situation can be compared to the rigidity of an inflated football, but here the leather casing (which by itself is flimsy but very resistant to extension) is held rigid by offering resistance to air under pressure instead of water under pressure. In small herbaceous plants the rigidity of the individual cells is largely responsible for the rigidity of the whole plant, and if water is lost from the vacuoles faster than it can be replaced the plant *wilts*, i.e. the leaves droop and the stem falls over. Provided the wilting is not too severe or too prolonged, the plant recovers when the vacuoles are 're-inflated' as a result of the roots absorbing water faster than it is being lost from the rest of the plant. This relationship between rigidity and water supply is often shown by the alternate wilting of seedlings on hot dry days and their recovery during the cooler nights.

The dynamic nature of plant cells

Unless qualified, the above account might leave the impression that the plant cell is a more or less static structure, whereas in fact nothing could be further from the truth. The nucleus, as already mentioned, undergoes a cycle of changes in connection with cell division. The ER also is a labile structure (i.e. capable of being broken down and re-formed) because at certain times it is much better developed than at others, and at cell division it even fragments. More spectacular evidence for the dynamic state of protoplasm is provided by those cells which exhibit the phenomenon of protoplasmic streaming, or cyclosis. In many plant cells (those forming the hairs on the stamens of *Tradescantia* are very suitable material for showing cyclosis) the entire cytoplasm circulates around the cell, sweeping the various organelles along with it. Although the mechanism of protoplasmic streaming is not understood, it is probable that the movement results in more efficient exchange of materials between the various organelles of a single cell and between the protoplasts of neighbouring cells.

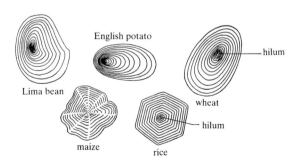

Fig. 2.14 Starch grains from different plants. (From Bonner, J. and Galston, A. W. (1952), *Principles of plant physiology*, p. 195, published by W. H. Freeman & Co.)

Chapter 3

The tissues of the plant body

Meristems

As a general rule, one of the obvious differences between plants and animals is the pattern of growth by which they attain their adult forms. Animals grow until they reach physical maturity, and during this time growth occurs throughout the body. Plants, however, continue to grow until they die but, as soon as the embryonic stage is passed, growth is restricted to cer-

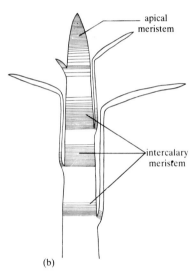

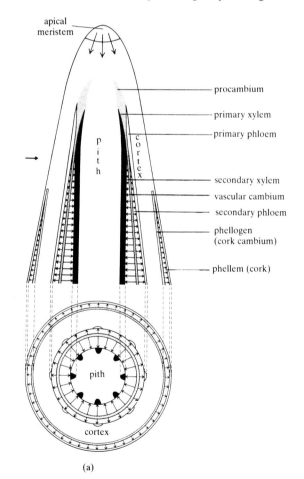

Fig. 3.1 Meristems. (a) Diagram to show the position and pattern of growth of the apical and lateral meristems in the stem axis of a woody dicotyledon. The bold arrow on the left indicates the distance from the apex where primary growth stops and secondary growth begins. (b) Diagram to show the mode of origin and subsequent position of intercalary meristems. The closely lined areas are the youngest; the unlined areas are mature or growing only very slowly. (b modified from Eames and MacDaniels 1947, p. 64)

tain regions called meristems (Fig. 3.1). Here the cells remain embryonic and continue to divide, whereas in the rest of the plant they reach maturity and assume a permanent form. One type of meristem is always present at the tip of every root and stem. The activity of such *apical meristems* is responsible for the increase in length of the plant body and, in the case of shoot meristems, for the production of lateral branches, leaves and flowers. The growth initiated by apical meristems is known as *primary growth*, and all tissues formed from apical meristems are called primary tissues. In some plants, notably grasses, the increase in length of the stem is also due to the division of meristematic cells located at the base of each node and leaf sheath. These meristematic regions are called *intercalary meristems* (Fig. 3.1b) because they are inserted or intercalated between mature primary tissues both above and below them. They are really portions of the apical meristem which become separated from the main body of the meristem and are left behind as the apex grows forward. Intercalary meristems may remain active long after the cells of the internodes above them have fully matured. Growth of the cells produced by intercalary meristems is responsible for the rapid elongation of the stem which often occurs just· before flowering. The tissues formed from intercalary meristems are similar to the adjacent tissues derived from the apical meristem and are therefore classified as primary tissues.

Some plants, especially monocotyledons, complete their life cycle by primary growth. In other plants, including most dicotyledons, the stem and root increase in thickness by means of a process called *secondary growth* which is initiated by *lateral meristems* or *cambia* (singular, cambium). These develop within the already existing primary tissues of the root and stem, and form secondary tissues in planes parallel with the surface of these organs (Fig. 3.1a). There are two cambia that may develop in a plant showing secondary growth, the vascular cambium and the phellogen (or cork cambium). The vascular cambium is responsible for most of the increase in thickness during secondary growth, whereas the phellogen produces a protective layer of periderm (or cork). This is formed in the outer region of the expanding root or stem when the primary surface layer (epidermis) is ruptured by the increase in thickness due to the activity of the vascular cambium.

Maturation of meristematic cells

Since all the cells of the adult plant body result from the activity of meristems, the question arises as to how meristematic cells are converted into mature cells. The essential characteristic of a meristematic region is that it consists of actively dividing cells which have the dual property of maintaining the meristem as a distinct region and, at the same time, of adding new cells to the rest of the plant body. In every meristem there are certain cells which divide in such a way that at each division one of the daughter cells (the *initial*) remains in the meristem, whereas the other daughter cell (the *derivative*) gradually passes out of the meristem and eventually becomes one or more cells within the main body of the plant. The stages by which a newly formed derivative reaches maturity can more easily be followed by reference to apical meristems rather than to lateral meristems.

The meristematic cells at the apices of roots and stems (but *not* those of lateral meristems) have the characteristic appearance shown in Fig. 3.2.

Although they appear approximately square in section, their basic shape is that of a 14-faced polyhedron, each face pentagonal. This is the shape assumed when a number of similar elastic spheres are subjected to pressure equally from all directions, until there are no air spaces between them. The cells have thin walls, the cell cavity is filled with dense cytoplasm and a large nucleus (up to two-thirds the diameter of the cell), and visible vacuoles are absent. During the process of becoming transformed into a mature cell, a derivative passes through three distinct but overlapping phases. The ability to divide is not confined to the initial cells but extends also to their immediate derivatives, which usually divide several times before starting to mature. The first phase in the growth of a derivative is therefore *cell division*, but this soon passes into the second phase, namely *cell enlargement*. Because the apical meristem constantly grows forward as a result of cell division, the newly formed derivatives come to occupy the region just behind the apex. It is in this subapical region that they begin to enlarge, and they continue to do so until by the time they reach maturity they are commonly many times larger than the meristematic cells from which they were derived (Fig. 3.2).

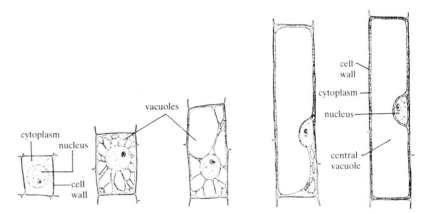

Fig. 3.2 Successive stages in the enlargement of cells at the tip of a root. The process of cell enlargement is accompanied by the formation of the central vacuole.

As a cell enters the phase of enlargement, droplets of cell sap form in the cytoplasm and these, after further increase in size, fuse to form several small vacuoles. The nucleus meanwhile remains suspended in the centre of the cell by strands of cytoplasm. The vacuoles continue to enlarge by uptake of water and finally coalesce to form a single central vacuole. As a consequence, the nucleus is moved to a peripheral position in the thin layer of cytoplasm lining the cell. While the vacuoles are expanding the whole cell increases in size but, despite the increase in surface area, the cell wall does not decrease in thickness because new wall material is continually being added to the original wall. Thus cell enlargement is not merely the inflation of a cell due to vacuolation, but it is an active process in which the amount of plant substance is increased as a result of intense biochemical activity by the cell.

In the process of assuming their mature form, cells not only increase in size but also become structurally modified to fulfil particular physiological functions in the adult plant. This modification for specialized functions is called *differentiation* (Fig. 3.3) and forms the third and last phase in the maturation of a meristematic cell. Differentiation starts while cells are still enlarging but is never complete until after they have ceased to enlarge. It

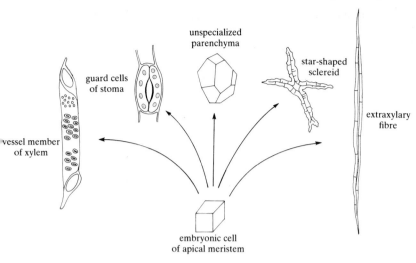

unspecialized
parenchyma

star-shaped
sclereid

guard cells
of stoma

vessel member
of xylem

extraxylary
fibre

embryonic cell
of apical meristem

Fig. 3.3 The concept of differentiation. The diagram shows some of the modifications which take place in derivatives of an apical cell as they assume their particular functions in the adult plant. (The different cells are not drawn at the same magnification.)

should be regarded as the process by which cells become different not only from their meristematic precursors but also from their immediate neighbours. The extent to which a cell becomes differentiated depends on its final function. Some cells differ only slightly from meristematic cells (e.g. packing cells) while others become markedly different (e.g. elongate conducting cells). There are many ways in which plant cells can become specialized to serve particular functions, but most of them involve modifications of the cell wall. These modifications include such features as the deposition of a thick secondary wall, and the development of various types of pits.

Types of plant tissues

It should now be clear that the different types of cells in the adult plant are the product of the three overlapping processes of cell division, cell enlargement and cell differentiation. The resulting mature cells are not arranged at random but are associated in various ways to form recognizable groups called *tissues*. There are various ways of classifying tissues according to whether they are considered from a structural or functional point of view. Perhaps the simplest classification, and the one that will be followed here, is to divide tissues into two categories, simple and complex, on the basis of whether they consist of only one or more than one type of cell.

Simple tissues

There are three simple tissues, parenchyma, collenchyma and sclerenchyma. These terms are also applied to individual cells showing the characters of these tissues. Thus a parenchyma cell is not necessarily a unit of the simple tissue parenchyma, but it may also be a component of a complex tissue.

Parenchyma

This is the simplest type of mature tissue, being less modified from meristematic cells than any other tissue. Parenchyma cells are living cells which are sufficiently unspecialized to be capable of reverting to the meristematic condition. It is often difficult to draw sharp lines between tissues because cell types sometimes merge into one another. Parenchyma, in particular,

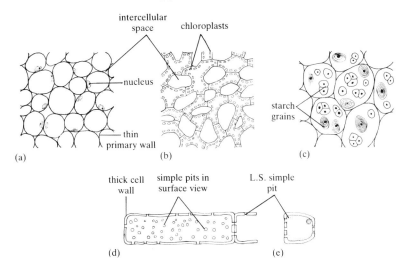

Fig. 3.4 Parenchyma. (a) Spherical (isodiametric) parenchyma typical of the pith of many plants. (b) 'Armed' parenchyma from spongy mesophyll of *Ficus* leaf. Note abundance of chloroplasts and large intercellular spaces. (c) Parenchyma from storage root of sweet potato (*Ipomoea batatas*). The cellular inclusions are starch grains, some of which are compound grains. (d) and (e) Xylem parenchyma in L.S. and T.S. (the protoplast omitted in L.S.). Note simple pits.

has very ill-defined limits and covers a range of cells which differ widely in structure and function. It is thus impossible to say that a parenchyma cell will inevitably have certain features apart from the very general one of being alive at functional maturity. However, 'typical' parenchyma cells (Fig. 3.4) can be expected to be more or less isodiametric in shape, with or without intercellular spaces between them, and to have thin cellulose walls; pits, if present, are always of the simple type. Parenchyma, as its Greek derivation (literally 'poured beside') indicates, forms the basic packing tissue in which more specialized tissues appear to be embedded. Besides its packing function, parenchyma may also be concerned with photosynthesis and with the storage of starch and other substances. In relation to these two functions parenchyma cells vary considerably both in shape and living contents. Thus the parenchyma cells of the photosynthetic tissue of a leaf are rich in chloroplasts and may be considerably elongated or lobed, while those found in food-storing structures such as a bean seed (*Phaseolus*) or a sweet potato (*Ipomoea batatas*) are packed with starch grains. In addition to forming a simple tissue, parenchyma cells are a regular component of the two complex tissues, xylem and phloem, where they commonly serve for the storage of various substances, particularly starch.

Collenchyma

This is a living tissue which has many of the characteristics of parenchyma, and may indeed be interpreted as a form of parenchyma structurally specialized as a supporting tissue in young organs. When collenchyma and parenchyma lie next to each other they frequently merge into one another through transitional types. The resemblance to parenchyma is also indicated by the common occurrence of chloroplasts in collenchyma, and by the ability of this tissue to resume meristematic activity. Collenchyma occurs immediately beneath or near the surface of young stems and petioles, and along the main veins of foliage leaves; it is rarely found in roots. The cells of collenchyma (Fig. 3.5) are elongated in the direction of the long axis of the organ in which they occur, and are characterized by the presence of thick, non-lignified, primary walls. The wall thickening, however, is not uniformly deposited around the inside of the cell but is thickest in the corners

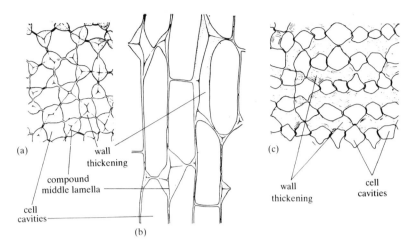

(a)

wall
thickening

compound
middle lamella

cell
cavities

(b)

(c)

wall
thickening

cell
cavities

Fig. 3.5 Collenchyma. (a) and (b) Angular collenchyma as seen in transverse and longitudinal section. (c) Lamellar collenchyma as seen in transverse section. (c from Fahn 1974, p. 96)

of the cell (*angular* collenchyma), although the two tangential walls may also be thickened (*lamellar* collenchyma). In longitudinal section therefore collenchyma shows thin and thick portions depending on the plane of the cut (Fig. 3.5b).

The thick cellulose walls of collenchyma combine considerable tensile strength with plasticity. Hence collenchyma is particularly suitable for the support of actively growing organs because its cells can extend to keep pace with the elongation of the organ and yet retain their strength.

Sclerenchyma

Whereas collenchyma is the main supporting tissue of actively growing organs, sclerenchyma performs a similar function in mature plant parts. Sclerenchyma cells have thick secondary walls, usually lignified, and their protoplasts are dead, or at any rate inactive, at maturity. Sclerenchyma is a very variable tissue but two broad categories of it can be recognized, fibres and sclereids; in general fibres are very much longer than sclereids.

Fibres   Fibres are long narrow cells with tapering pointed ends (Fig. 3.6). They are usually massed together in long strands, their tapering ends overlapping one another and fitting tightly together. While fibres are young and actively growing, their end walls tend to slide over one another to produce the pointed state of the mature fibre. The increase in length of a fibre by the intrusion of its growing tips between the walls of neighbouring cells is called *intrusive growth*. After a fibre has ceased to elongate, its walls continue to thicken until at maturity the cell cavity is very much reduced, and sometimes almost obliterated. Fibres occur in most parts of the plant body and, according to their position, they may be classified into two types: *xylary* fibres which occur in the complex tissue xylem, and *extraxylary* fibres which occur in any tissue other than xylem. Xylary fibres have reduced bordered pits, the inner apertures being slit-like and those of a pit-pair often crossed with each other. Extraxylary fibres have very narrow, simple pits. Although the thick secondary walls of fibres are usually lignified, some fibres have unmodified cellulose walls (e.g. those of the flax plant (*Linum usitatissimum*), from which linen is made). Xylary fibres are a major component of wood,

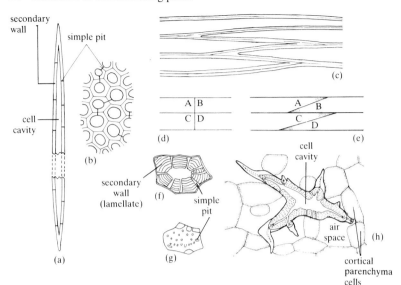

Fig. 3.6 Sclerenchyma. (a) L.S. part of an extraxylary fibre showing thick secondary wall with simple pits. (b) T.S. group of extraxylary fibres. (c) L.S. group of fibres showing dovetailing of pointed ends. (d) and (e) Diagrams to illustrate intrusive growth; the dotted areas in (e) indicate the growing tips. (f) and (g) Sectional and surface views of sclereids from stem cortex of *Hoya*, wax plant. (h) Irregularly branched sclereid from petiole of *Trochodendron* (an evergreen of Japan) as seen in sectional view. (h from E. M. Gifford Jr (1974), *Plant diversity: cells and tissues*, Biocore Unit xi, p. 12, published by McGraw-Hill, USA)

and on account of their heavily lignified walls, are responsible for the hardness and rigidity of this material. Extraxylary fibres, some lignified and others not, are common sources of rope, sacking and clothing textiles.

Sclereids   These are very variable in shape, ranging from approximately isodiametric to very irregular, and they may even be branched (Fig. 3.6 f–h). They have very thick, lignified walls, often showing concentric layering, which are pierced by numerous simple pits. Frequently the pits become branched as a result of the fusion of two or more pit cavities during the increase in thickness of the secondary wall. Sclereids may occur either singly or as small clusters among other cells in, for example, the gritty specks in the flesh of a guava (*Psidium guajava*), or as continuous masses as in the hard shell of a coconut (*Cocos nucifera*).

Complex tissues

Although a complex tissue is composed of several different types of cells it merits being recognized as a single tissue because it has certain features peculiar to itself, occurs in definite positions in the plant body, and is associated with definite functions. There are two complex tissues, the xylem which is concerned primarily with the conduction of water and inorganic solutes, and the phloem which is concerned primarily with the conduction of food manufactured in the leaves. These two complex tissues always occur side by side, and together constitute the *vascular* or conducting system which extends throughout the plant.

Xylem

The xylem is composed of parenchyma and fibres, which have already been described, and tracheids and vessel members, which are characteristic of this tissue. These four elements occur in different proportions and combinations, and all four are not necessarily present in the xylem of any given plant.

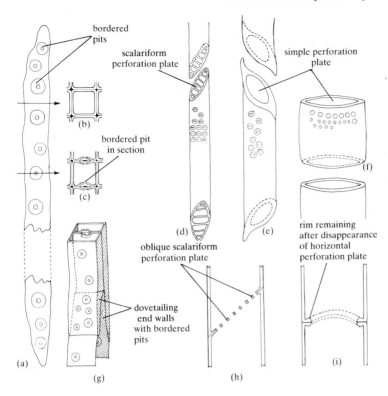

Fig. 3.7 Tissue elements of xylem. (a) Tracheid from *Pinus*, about one-third of cell shown; note the prominent bordered pits characteristic of tracheids in this genus. (b) and (c) Transverse sections of (a) at levels indicated by the arrows; note that pits are present only on opposite walls of the tracheid. (d), (e) and (f) Different types of vessel members ranging from the narrow type with scalariform perforation plates (d) to wide type with simple perforation plates (f); pitting is only shown on part of these cells but it extends over all the walls. (g), (h) and (i) Diagrams showing how tracheids (g) and vessel members (h and i) are connected longitudinally.

Tracheids Tracheids are more or less elongated cells, which are angular in cross-section and have oblique or tapering end walls (Fig. 3.7). Tracheids are dead at functional maturity when they consist only of lignified cell walls. All tracheids have secondary walls but these are deposited in various patterns which are related to the state of maturity of the region in which the tracheid is being formed (Fig. 3.8). In the tracheids formed nearest to the apical meristem the secondary walls are deposited as a series of horizontal rings (*annular* thickening) and in those further back as one or more spirals (*helical* thickening). These are succeeded by tracheids in which the secondary wall forms a net-like pattern (*reticulate* thickening) over the primary wall, and finally by tracheids in which the secondary wall is continuous ex-

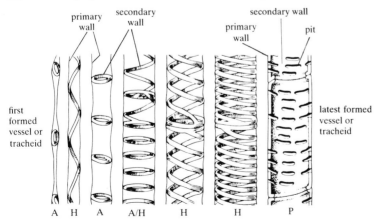

Fig. 3.8 Different patterns of wall thickening in the conducting elements of primary xylem: A, annular; H, helical; A/H, transitional between annular and helical; P, pitted. The drawings to the left of centre illustrate the stretching of the earlier formed conducting elements due to the continued elongation of the surrounding tissues. (From Ray 1972, p. 129)

cept for elongate or circular pits (*pitted* walls). These different patterns of secondary wall thickening form a series in which progressively more of the primary wall is covered, but the different types often merge and not all types are necessarily represented in any given plant. All four types of thickening give support to the tracheids in which they occur, so that the central cavity or lumen is kept open despite pressure from neighbouring cells. It is interesting to note, however, that tracheids with annular and helical thickenings are formed while the surrounding tissues are still elongating, and they continue to be stretched even after they are structurally mature. The occurrence of stretching is shown by the fact that, when several helical tracheids are present end to end in a longitudinal series within the region of elongation, the spirals of the older tracheids are more drawn out than those of the younger ones. Tracheids with more extensive, and therefore more rigid, wall thickening are not readily stretched, and such tracheids differentiate in plant organs which have stopped elongating.

Water moving through tracheids passes from one cell into the next through the areas, whether extensive or restricted to pits, where the primary wall only is present. Tracheids perform the double function of a supporting and water-conducting element, and in such vascular plants as ferns and conifers, which are more primitive than angiosperms, they are the only cell type present in the xylem apart from parenchyma. During the course of evolution the tracheid diverged along two routes, one in the direction of greater mechanical efficiency which resulted in the xylary fibre (already described), and the other in the direction of greater water-conducting efficiency which resulted in the vessel member. Tracheids, however, were not eliminated when fibres and vessel members were evolved.

Vessel members   These are cylindrical cells, non-living when mature, which are joined end to end to form multicellular water-conducting tubes, called vessels. The end walls (and sometimes also the side walls) of vessel members become perforated by one or more holes through which water can pass freely from cell to cell (Fig. 3.7d–f). Hence, instead of a tier of superimposed cells, there is formed a tube rather like a drainpipe in structure, which is the vessel. Vessel members are commonly shorter and much wider than tracheids, but they develop lignified secondary walls in a similar way and show the same patterns of wall thickening (annular, helical, etc.) as tracheids.

Vessel members begin life as separate cells with a continuous primary wall, but at a certain stage in development the middle lamella swells in the areas which are to become perforated (Fig. 3.9). Subsequently a secondary wall develops, but this is not deposited on the areas (e.g. primordial pits) which persist as thin primary wall or on the areas of future perforation. As the vessel members mature, the primary walls and middle lamella in the perforation areas break down, so leaving open holes. The perforated areas (called *perforation plates*) may be simple, with one large perforation, or multiperforate, with more than one perforation (Fig. 3.7d–f, h, i). In a mature vessel with simple perforation plates the previous position of each cross-wall is often indicated by the presence of a distinct rim of thickening around the lumen of the vessel. Vessels are not of indefinite length, although they frequently extend for several metres. Where a vessel ends, the terminal vessel member is perforated at its proximal end but not at its distal end. Therefore the passage of water from vessel to vessel takes place

(a)   (b)   (c)   (d)   (e)

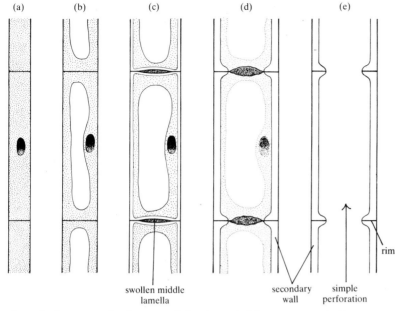

swollen middle
lamella

secondary   simple
wall   perforation

rim

Fig. 3.9 Stages in the development of a vessel member. (a) Meristematic cell. (b) Cell widens and becomes vacuolate. (c) Middle lamella swells in the areas of future perforation. (d) Lignified secondary wall laid down except over future perforation areas, and protoplast begins to disintegrate. (e) Mature vessel member in which the protoplast has disappeared, and the cross-walls have ruptured leaving a rim where they were originally attached.

through the areas of primary wall (such as pits) in a manner similar to conduction in tracheids.

In transverse section vessels can usually be recognized by their large, almost circular outlines. Whereas the widest tracheids scarcely exceed 100 μm in diameter and most are much narrower, vessels are commonly 300 μm across and some, especially those of climbers, reach a width of 700 μm. The wide diameter of vessels, coupled with the perforations in their end walls, enables water to be conducted more efficiently in these elements than in tracheids.

Phloem

The characteristic components of the phloem are sieve-tube members and companion cells, but parenchyma and, in some cases, fibres also occur.

Sieve-tube members   These are elongated cells joined end to end into sieve tubes (Fig. 3.10) in a manner analogous to the joining of vessel members to form vessels in the xylem. Unlike vessel members, however, the cell walls of sieve-tube members are not lignified and are classed as primary walls, even though they are usually thickened. Also unlike vessel members, the cross-walls of sieve-tube members, although much modified, do not break down completely. Instead, the cross-walls become perforated by open pores through which the protoplasts of two superimposed members are continuous. These connecting strands of cytoplasm are comparable with plasmodesmata but very much wider. The perforated cross-walls are called *sieve plates* because in surface view they resemble sieves (Fig. 3.11). It is from the presence of sieve plates that this characteristic cell type derives its name. Usually the sieve plate covers the entire cross-wall, but sometimes it is broken up into several perforated sieve areas, thus forming a compound sieve plate.

In the sieve plate each cytoplasmic strand is enclosed in a cylinder of a special kind of carbohydrate called callose. The thickness of this callose

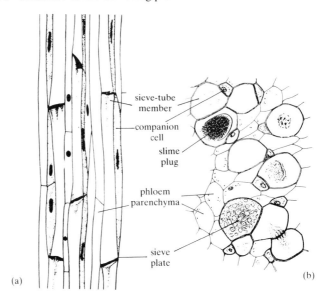

Fig. 3.10 Phloem. (a) Longitudinal section of phloem from *Ecballium* stem. (b) Transverse section of phloem from *Cucurbita* stem. (a from Coult 1973, p. 31; b from Priestley, Scott and Harrison 1964, p. 310)

sieve-tube member

companion cell

slime plug

phloem parenchyma

sieve plate

(a)   (b)

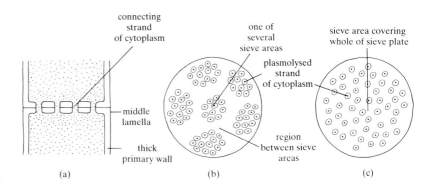

connecting strand of cytoplasm

one of several sieve areas

sieve area covering whole of sieve plate

plasmolysed strand of cytoplasm

middle lamella

thick primary wall

region between sieve areas

Fig. 3.11 The sieve plate. (a) L.S. through sieve plate showing the pores in the cross-walls. (b) Surface view of compound sieve plate. (c) Surface view of simple sieve plate.

(a)   (b)   (c)

cylinder obviously affects the size of the sieve pore. When the plant is dormant thick callose pads completely block the pores, but at the onset of the growing season much of the callose is removed and new connecting strands of cytoplasm appear. If the plant is injured callose is also deposited, often within a few seconds of the injury, thus blocking the pores and minimizing loss of food material. One consequence of this instantaneous reaction is that in most microscopic sections the pores are completely blocked. This is the result of the treatment of the tissue for examination, and does not represent the situation in an actively growing plant.

Sieve-tube members are living cells and only function as long as they are alive. A remarkable characteristic of sieve-tube members is that during development the nucleus that was originally present breaks down and completely disappears at maturity. They are the only known example of plant cells which are living but lack a nucleus. Investigation with the EM has indicated that sieve-tube members are also remarkable in possessing very few mitochondria and almost no other cytoplasmic organelles. Furthermore, the normal distinction between vacuole and cytoplasm seems to be lacking, the main volume of the cell being occupied by a fibrillar protein, called *P-protein*, whose exact structure and function is currently the subject of active

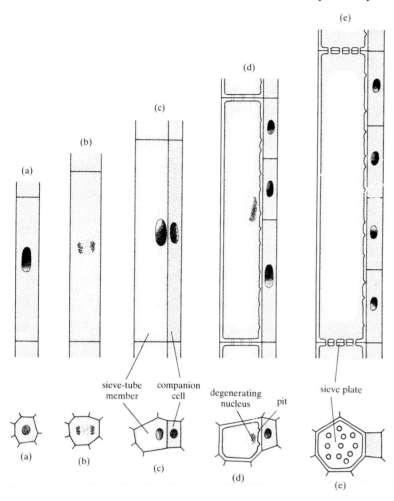

Fig. 3.12 Stages in the development of primary phloem. (a) Meristematic cell. (b) and (c) Meristematic cell during and after division into sieve-tube member and precursor of companion cells. (d) Thick wall laid down in sieve-tube member, and nucleus degenerating; precursor of companion cells has formed vertical file of three companion cells. (e) Mature phloem with sieve plates on end walls of sieve-tube member, and pits in lateral wall facing companion cells; nucleus now absent from sieve-tube member.

debate among plant physiologists. This proteinaceous material readily takes up staining reagents, and in stained longitudinal sections usually appears attached to the sieve plates as so-called *slime plugs*. The presence of slime plugs provides one of the most useful criteria for identifying the position of phloem in longitudinal sections (Fig. 3.10a).

Companion cells   Each sieve-tube member is closely associated with one or more slender parenchyma cells, called companion cells, which have dense cytoplasm and conspicuous nuclei. Companion cells arise by an unequal longitudinal division of the same mother cell as the sieve-tube member (Fig. 3.12). After being cut off from the sieve-tube member the companion cell does not necessarily remain as a single cell but may divide horizontally into a vertical file of companion cells. Companion cells always lie alongside a sieve-tube member, and the cytoplasm of the two partners is in direct contact through plasmodesmata in the pit membranes on the side walls. The two types of cells form a single physiological unit because the companion cells die when their associated sieve-tube members cease to function. The nature of the interdependence is not known, but it can be assumed that the sieve-tube member, which has no nucleus of its own, depends on the nucleus of the companion cell to support those activities for which a nucleus is normally essential.

Tissue systems

Just as individual cells are arranged into different types of tissue, so individual tissues in their turn are arranged in definite patterns throughout the plant body. Thus the tissues concerned with the conduction of water and food form a continuous system extending throughout the entire plant. These so-called vascular tissues connect places of water intake and food synthesis with regions where water and food are used for such functions as growth or storage. The non-vascular tissues are similarly continuous, which again indicates the interdependence of leaves, stems and roots. The individual tissues (parenchyma, sclerenchyma, xylem, phloem, etc.) are thus organized into larger units called *tissue systems*. In this book the main tissues of the plant are classified on the basis of their spatial continuity into three tissue systems, the dermal, the vascular and the ground (or fundamental) systems.

The *dermal system*, as its name denotes, forms the outer covering of the plant. It includes both the epidermis (i.e. the continuous layer which is formed over the surface of the primary tissues of the plant body) and also the periderm (i.e. the protective tissue which replaces the epidermis near the surface of stems and roots which undergo secondary thickening). Dermal tissues have special characteristics, such as the impregnation of their walls with cutin or suberin, which are related to their superficial position. The *vascular system* is concerned with the conduction of water and food throughout the plant, and contains two kinds of conducting tissues, the xylem (conducting water) and the phloem (conducting food). Because of the presence of fibres in them, the vascular tissues, particularly the xylem, are also concerned with support. The *ground system* includes the tissues that can be regarded as forming the ground substance in which the vascular tissues are embedded. The main ground tissues are the three simple tissues, parenchyma, collenchyma and sclerenchyma.

Within the plant body the three tissue systems are distributed in characteristic patterns depending on the organ in which they occur, on the taxonomic group to which the plant belongs, or on both (Fig. 3.13). Since the dermal system is always superficial, the principal differences in pattern depend on the relative distribution of the vascular and ground systems. In a dicotyledonous root, for example, the vascular tissue usually forms a central rod embedded in the ground tissue (*cortex*). In a dicotyledonous stem, however, the vascular tissue typically forms a hollow, perforated cylinder with some ground tissue (*pith*) enclosed within the vascular cylinder and some located between the vascular cylinder and the dermal tissue. The latter region of ground tissue is called cortex because, like the ground tissue in the root, it occurs outside the vascular tissue. This method of interpreting the anatomical pattern of an organ is not only descriptively useful, but also reflects the basic unity underlying the organization of the plant as a whole.

Whatever the arrangement of tissues in a mature organ may be, the pattern is established immediately behind the apical meristem from which it is derived. In this subapical region, where the cells are still actively dividing, it is thus possible to recognize three types of meristematic tissue, the *protoderm, ground meristem* and *procambium*. These are sometimes called the three primary meristems because they are the precursors of the dermal, ground and vascular tissue systems respectively. Although leaves originate on the flanks of the shoot apex, these three types of meristematic tissue are also present in a developing leaf.

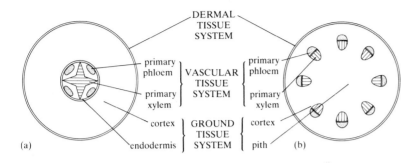

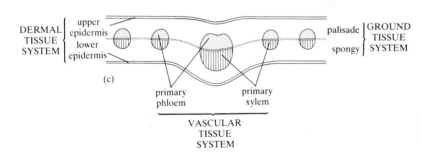

Fig. 3.13 The concept of tissue systems applied to the anatomical patterns of (a) the primary root, (b) the primary stem, and (c) the foliage leaf of a dicotyledonous plant.

Chapter 4

The primary root

The roots of a plant usually grow underground, where they anchor the aerial portion of the plant in a fixed position and absorb water and inorganic salts from the soil solution. This is, however, not always the case because the roots of plants such as epiphytic orchids grow in the air, and the stems of plants such as *Canna* grow underground. Although roots cannot reliably be distinguished from stems by whether they are growing below or above soil level, there are certain basic differences between these two types of organ. Externally, a root differs from a stem mainly in that it has no lateral appendages comparable to leaves, and is therefore a simple axis and not a segmented one as is a stem. The tip of every root is also covered by a thimble-like structure, the root-cap, which has no counterpart in stems. Internally, a root differs from a stem in the development and arrangement of its vascular tissue system.

External regions of the primary root

Every root, like every stem, grows from an apical meristem but in a root this meristem does not occur at the extreme tip but just behind it. The reason for this subapical position is that, in addition to producing new cells which are responsible for the increase in length of the root axis, the meristem also produces towards the apex a limited number of cells which are responsible for the formation of the *root cap*. It is commonly assumed that the function of the root cap is to protect the underlying meristem, and to lubricate the root in its passage through the soil by reducing the friction between the growing tip and the soil particles. The root cap is, however, a great deal more than a mere protecting and lubricating organ, because it is a source of substances which regulate the pattern of growth of the rest of the root.

Behind the root cap the surface of the root is smooth for a short distance, usually about a centimetre (Fig. 4.1). This region normally appears translucent when held up to the light owing to the absence of any differentiated tissues. It is known as the *region of elongation* because it is here that the cells, newly formed from the meristem, grow tremendously in size and, as a result, the root increases in length. It is possible to demonstrate this elongation by marking the apical centimetre of an actively growing root with lines drawn 1 mm apart, and then leaving the root to grow vertically downwards in a dark moist chamber (Fig. 4.1b). After 24 to 48 hours it will be seen that the spaces between the 2nd, 3rd and 4th millimetre lines from the apex have increased to several times their original length. The spaces further back show progressively smaller increments until at about 1 cm from the apex no elongation has occurred. It is thus clear that maximum elongation occurs in the few millimetres just behind the root tip.

From the point where elongation is more or less complete, the surface of the root is covered for a distance of one to a few centimetres with thin white hairs, which may be so numerous as to form a dense felt-like covering. This is called the *root-hair region*. Where plenty of space is available, as when seedlings are grown in moist air, the root hairs assume a simple tubular form, but in soil they take up the shape of the pore spaces between the soil particles (Fig. 4.1c). The outer surface of a root hair is mucilaginous and comes into such close contact with soil particles that the latter cannot easily be removed without breaking off the hair. This is one of the reasons why extreme care should be taken when transplanting seedlings. As the root increases in length new root hairs are continually formed behind the region of elongation. However, the life of individual root hairs is short so that the extent and relative position of the root-hair region remains more or less constant.

Behind the root-hair region the root is again bare, and here *lateral roots* make their appearance. They develop in acropetal succession (i.e. with the youngest nearest the tip) and are arranged in vertical rows along the length of the parent root. For example, in the root of *Phaseolus* they usually arise in four rows. Lateral roots originate as internal growths which push their way outwards until they emerge from small vertical splits which they make in the outer tissues of the parent root. Because of their deep-seated origin, lateral roots are said to be endogenous (i.e. arising from within). The continued production of new lateral roots by the main axis and by the laterals themselves results in a branching root system which, as it enlarges, comes into contact with an increasing volume of soil.

Internal regions of the primary root

The external division of a root into root cap, region of elongation, root-hair region, and region of lateral roots is merely the visible expression of an internal sequence of cellular activities which are ultimately responsible for the development of the mature root. The most important of these activities, apart from those involved in the formation of lateral roots, can be seen by examining a median longitudinal section through the terminal portion of a root (Fig. 4.2). At the extreme tip of the root is the root cap which appears as a more or less conical mass of cells, several to many layers thick at the apex of the cone but decreasing in thickness laterally. The cells of the root cap vacuolate almost as soon as they are formed and their walls become mucilaginous. As the root penetrates the soil, the outer cells of the root cap are continually rubbed off but they are as constantly replaced from within by the meristem. Immediately behind the root cap is the root meristem, from which all the cells of the mature root are derived. It consists of columns of cells which converge on a central area of cells situated at the junction of the meristem with the root cap. A high percentage of the cells in these columns can be shown to contain mitotic figures, which indicates that, in the living state, these cells are actively dividing. Since the cells within the meristem do not get progressively smaller even though they are continually dividing, it follows that rapid synthesis of new protoplasm and new wall material must be taking place. Cells within the root meristem have the

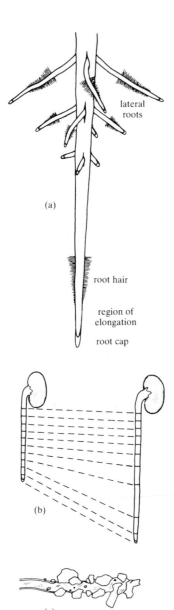

(a)

lateral roots

root hair

region of elongation

root cap

(b)

(c)

Fig. 4.1 External regions of the primary root. (a) Root system of seedling showing production of root hairs and of lateral roots. (b) Marking experiment to determine the distribution of growth in a root. (c) Tip of root hair growing in soil, with soil particles firmly adherent to the surface.

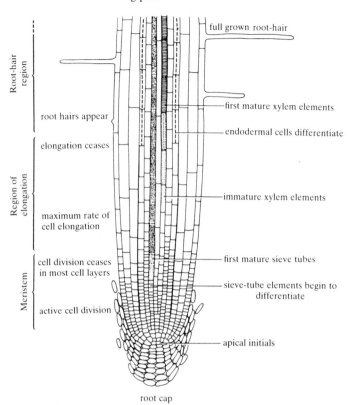

Root-hair region

full grown root-hair

root hairs appear

first mature xylem elements

endodermal cells differentiate

Region of elongation

elongation ceases

maximum rate of cell elongation

immature xylem elements

Meristem

cell division ceases in most cell layers

first mature sieve tubes

sieve-tube elements begin to differentiate

active cell division

apical initials

root cap

Fig. 4.2 Diagrammatic L.S. of the terminal portion of a root, showing the pattern of primary growth. To make this diagram as simple as possible, the number of cells actually present in roots has been very much reduced. (From Ray 1972, p. 125)

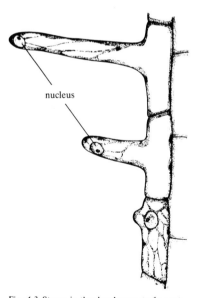

nucleus

Fig. 4.3 Stages in the development of a root hair. The nucleus always occurs at the point of root-hair formation, and in the mature root hair commonly occurs at the apex.

appearance characteristic of embryonic cells (i.e. thin walls, dense cytoplasm, conspicuous nuclei, etc.).

At about the point along the root where the columns of meristematic cells straighten out and lead into the older part of the root, the cells begin to enlarge considerably. From this point onwards cell division practically ceases, but the rate of growth nevertheless increases to a maximum because the cells are now enlarging. This enlargement is mainly in a longitudinal direction, and it is the growth of these cells which is the main cause of elongation of the root. It is relatively easy to verify by microscopic examination that this region of cell enlargement coincides with the region of maximum elongation shown by the marking experiment. Although the water which causes the developing cells to enlarge is absorbed by osmosis, it is now known that the onset of extension is determined by an increase in the plasticity of the young cell walls rather than by an increase in the osmotic potential of the cell contents. This change in plasticity is probably associated with the action of a growth–regulating substance diffusing back from the root apex.

While the cells in these columns are enlarging they are also undergoing structural differentiation for specialized functions. Differentiation becomes more and more pronounced as the cells increase in size, but is not complete until after elongation has ceased. It is behind the region of elongation, therefore, that cells assume their mature structure in relation to their position in the root. On the surface, many of the epidermal cells develop narrow tubular outgrowths which are the root hairs (Fig. 4.3). Each root hair has a thin wall and a large central vacuole extending throughout its length. A root hair is formed as a small papilla which continues to grow at its tip where the wall is thinner and more delicate than along the sides. The nucleus is almost invariably found close behind the growing tip. Inside the root, at about the level at which the root hairs first develop, the most conspicuous feature is the differentiation of the xylem. Since the root hairs are the main entry points of water into the plant, the newly formed conducting elements of the xylem complete their development at precisely the distance from the apex where they are needed to conduct the water to the rest of the plant. This correlation between the position of the root hairs and that of the differentiating xylem can readily be demonstrated by crushing the end of a thin root between two glass slides and examining it under the microscope.

It is important to realize that tissue differentiation involves not only the functional and structural modification of cells, but also the formation of an anatomical pattern. In the development of roots this pattern begins to form so early that even just behind the tip of the meristem the cells which are destined to become the three tissue systems of the mature root can already be recognized from differences in size and shape of the cells (Fig. 4.4).

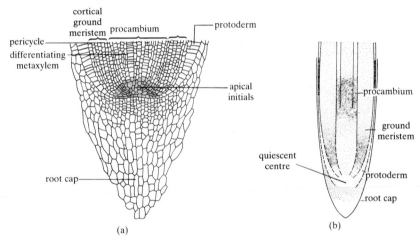

Fig. 4.4 The root tip. (a) Median L.S. of root apex of *Zea mays* showing the three tissue systems beginning to differentiate within the root meristem. The quiescent centre is dotted. (b) Diagram of L.S. root tip of *Allium cepa* (onion) showing, by means of shading, the distribution of mitotic activity; frequency of mitosis is indicated by density of shading. (a from F. A. L. Clowes (1959), *Biological Reviews*, **34**, 501–28; b from Jensen, W. A. and Kavaljian, L. G. (1958), *American Journal of Botany*, **45**, 365–72)

It will be apparent, from what has already been said, that the distance of any cell from the tip of the meristem is roughly a measure of its age. The only cells that remain a permanent part of the meristem are those located at the very tip where the columns of meristematic cells converge. When one of these tipmost cells divides, one of the daughter cells remains within the tip but the other, after having itself grown and divided many times, ends up as a mature cell in some tissue remote from the meristem. The overall effect is that cells appear to move backwards through the meristem until they become part of a mature tissue. Only the tipmost cells in the root meristem are thus the true initials. The cells of the root cap also arise by the division of apical initials, which may be the same as those producing the root meristem, or they may form an adjacent group of initials which divide independently to give rise to the root cap.

Although the apical initials are the source of all the cells of the root, it should be noted that they do not divide as rapidly as the cells a short distance behind them. Recent studies have shown that the frequency of cell di-

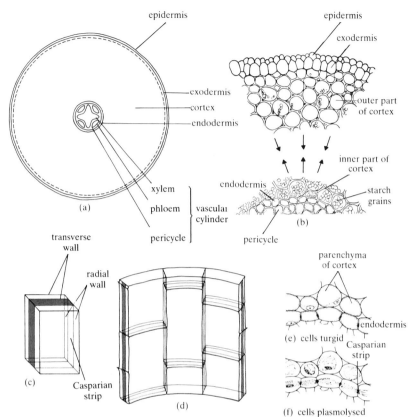

Fig. 4.5 Structure of the primary root. (a) Diagrammatic T.S. of root of dicotyledonous plant to show distribution of tissue systems. (b) T.S. cortex of primary root. (c) Stylized representation of an isolated endodermal cell to show the Casparian strip in the radial and transverse walls. (d) Three-dimensional diagram of small part of endodermal cylinder showing the Casparian strip continuous around each cell. (e) and (f) Cells of endodermis and of cortical parenchyma before (e) and after (f) plasmolysis; in the plasmolysed cells of the endodermis the protoplasts remain attached to the Casparian strip.

vision in the middle part of the meristem is considerably greater than in the extreme tip where the initials occur (Fig. 4.4b). The region of apical initials is therefore sometimes called the *quiescent centre* to distinguish it from the rest of the root meristem.

Structure of primary roots

The structure of the mature primary root can conveniently be described by considering the appearance of a transverse section across a young dicotyledonous root in the region of the older root hairs. At this distance from the root tip the tissue systems characteristic of the mature primary root have been laid down. Such a transverse section (Fig. 4.5a) clearly shows three regions: (1) an outer covering layer, the epidermis; (2) a wide cortex; and (3) a small central vascular cylinder. This basic pattern characterizes most young dicotyledonous roots, although the details of the pattern vary somewhat according to the particular root being studied. From the following description it should be possible to interpret the structure of almost any young dicotyledonous root, provided it is not definitely specialized for some particular function such as support (e.g. the prop roots of mangroves of the genus *Rhizophora*) or storage (e.g. sweet potato, *Ipomoea batatas*).

Epidermis

The epidermis consists of a single layer of closely packed, thin-walled cells which have a cuticle, but this is so thin that its presence has only recently been discovered. Some or most of the epidermal cells form root hairs by a lateral extension of their outer walls. The root hairs greatly extend the absorbing surface of the root, although the function of absorption is not restricted to them.

Cortex

Inside the epidermis is the cortex which occupies most of the cross-section of the root. It consists mainly of loosely packed, thin-walled parenchyma cells which appear circular in cross-section but which are elongated in the longitudinal direction, so that their actual shape approximates more to cylinders than to spheres. The cortex is thus permeated by a system of intercellular air spaces running mainly along the length of the root. The cortical cells frequently contain abundant starch grains (Fig. 4.5b). Sometimes one to several layers of cells immediately beneath the epidermis develop suberin in their walls and become morphologically distinct from the underlying cortical cells. Such a specialized layer, if present, is called an *exodermis*. The innermost layer of the cortex, the *endodermis*, consists of match-box shaped cells which fit tightly together with no intercellular spaces. Collectively they form a cylinder (Fig.4.5d) enclosing the vascular tissues. The endodermal cells are characterized by a peculiar band of suberized thickening on their radial and transverse walls. This band is known as the *Casparian strip* (Fig. 4.5c) after Caspary who discovered it. The Casparian strip is not merely a localized thickening of the wall due to the addition of secondary wall material, but it is a deposit of suberin (or suberin and lignin) laid down across the full thickness of the adjoining walls including the middle lamella. The protoplast is firmly attached to the Casparian strip with the result that, when endodermal cells are plasmolysed, the cytoplasm shrinks from the tangential walls but remains attached to the radial and transverse walls. As seen in transverse section this gives a very characteristic plasmolysis figure, in which the protoplast stretches across the cell as an unbroken strip (Fig. 4.5f). Since the Casparian strip is impervious to water it would seem that water and solutes cannot pass across the endodermis except through the living protoplasm of the endodermal cells, which must therefore have a controlling influence upon such movement. There can be no doubt that the endodermis is an important 'physiological barrier' between the cortex and the vascular cylinder, but its precise function remains unknown.

Fig. 4.6 The vascular cylinder of the root of a dicotyledon (*Ranunculus repens*) in transverse (a) and longitudinal (b) section. (a from Fritsch, F. E. and Salisbury, E. J. (1948), *Plant form and function*, p. 101, G. Bell & Sons, London; b from Priestley, Scott and Harrison 1964, p. 381)

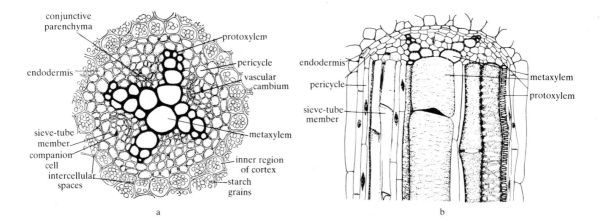

Vascular cylinder

The vascular cylinder (Fig. 4.6) comprises the pericycle and the vascular tissues. The pericycle occurs immediately within the endodermis and commonly consists of a single layer of parenchyma cells; it is the layer in which lateral roots arise. The xylem and phloem, which are the actual conducting tissues of the vascular cylinder, are arranged upon alternating radial planes. In transverse section the xylem is seen as a number of wedges which meet in the centre to form a star-shaped figure, with the points of the star extending outwards to the pericycle. The phloem is less easy to observe than the xylem but occurs in distinct strands between the arms of the star-shaped xylem. The phloem strands abut on the pericycle and are separated from the xylem by parenchyma which is continuous with the pericycle. This layer of parenchyma is sometimes referred to as *conjunctive parenchyma*, the adjective conjunctive being applied to ground tissue which occurs between areas of vascular tissue. The number of xylem wedges varies between different roots but is roughly constant for any given species of plant. According to the number of xylem wedges, roots are described as diarch (if they have two wedges), triarch (three wedges), tetrarch (four wedges) or polyarch (many wedges). The outermost conducting cells in each xylem wedge are the first to differentiate and are also the narrowest; they constitute the *protoxylem*. The conducting cells closer to the centre, collectively called the *metaxylem*, mature progressively later and are increasingly wider. This sequence of development of the xylem from outside inwards is called *centripetal*, and the xylem wedges formed as a result are said to be *exarch*. Although the protoxylem and metaxylem are not sharply delimited from each other as seen in a transverse section of a root, there are nevertheless structural differences between them. These differences are connected with the fact that differentiation of the primary xylem begins within the region of elongation but is not complete until after elongation ceases. The first xylem elements to differentiate, the protoxylem, do so while elongation is still occurring. They have the minimal amount of secondary wall thickening, laid down either as rings (annular thickening) or spirals (helical thickening). The remainder of the cell surface consists merely of the thin primary wall which is able to stretch as the tissue grows. As elongation continues the rings become pulled apart and the spirals become uncoiled. On the other hand, the later formed elements, the metaxylem, differentiate after extension growth has ceased. They normally develop pitted secondary walls in which most of the primary wall area is covered by secondary wall material. The pitted thickening of the metaxylem is stronger than the annular or helical thickening of the protoxylem but, since it covers practically the whole surface of the walls, it cannot be stretched without being torn.

The phloem strands also develop in a centripetal direction but, since the earlier and later formed parts are not easily distinguishable, it is common practice to regard a phloem strand as a single area of tissue without trying to divide it into protophloem and metaphloem.

The way in which the vascular tissues differentiate can best be appreciated by reference to a three-dimensional diagram (Fig. 4.7). The first vascular cells to become structurally mature are the outermost sieve tubes of the phloem which reach maturity within the meristem itself. The newly formed sieve tubes are connected in the longitudinal direction with already existing ones in the older phloem, and so provide food-conducting channels to supply the actively dividing cells of the meristem. Considerably further back,

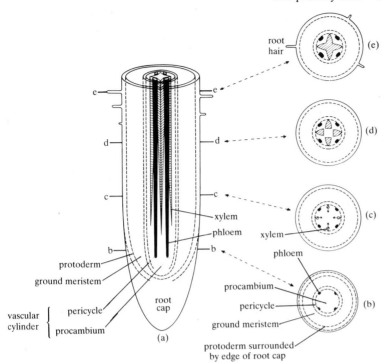

Fig. 4.7 Differentiation of vascular tissues in roots. (a) Root tip of a tetrarch root as it would appear if all the tissues were made transparent except the xylem (shaded with lines) and the phloem (black). The vascular cylinder is delimited from the cortex close to the apex as a result of the very early differentiation of the pericycle. Within the vascular cylinder the phloem begins to differentiate before the xylem. (b)–(e) Transverse sections of the same root taken at levels b–b, c–c, d–d, and e–e in Diagram (a). Both the xylem and phloem develop centripetally with increasing distance from the root apex.

usually within the region of elongation, the first tracheids or vessels of the protoxylem begin to differentiate. These also develop in continuity with the existing protoxylem elements of the older part of the root. Beyond the region of elongation, i.e. in the root-hair region, further differentiation of both phloem and xylem cells occurs progressively inwards in the areas destined to become phloem and xylem respectively. Differentiation is usually completed in the region of the older root hairs, by which time the xylem has become a fluted column and the phloem has been blocked out as a number of solid strands. A transverse section at this level of the root will show the mature arrangement not only of the vascular tissues but also of the other specialized tissues, such as the endodermis, which were also differentiating as their cells enlarged in the region of elongation.

Features of monocotyledonous roots

The basic arrangement of tissues in primary roots (i.e. central vascular cylinder, xylem and phloem on alternating radial planes, exarch protoxylem, endodermis with Casparian strips, etc.) is essentially the same for all angiosperms, but the roots of monocotyledons tend to differ from those of dicotyledons in three main respects (Fig. 4.8).

1. Whereas dicotyledonous roots generally have from two to five protoxylem groups (i.e. they are diarch to pentarch), monocotyledonous roots generally have somewhere in the order of ten or more protoxylem groups (i.e. they are polyarch). Roots with between five and ten protoxylem groups are very rare.
2. Whereas in dicotyledonous roots the developing strands of primary xylem meet in the centre of the vascular cylinder to form a fluted column of mature xylem, in monocotyledonous roots differentiation of the pri-

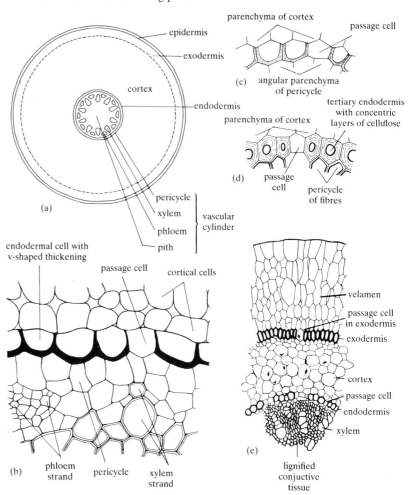

Fig. 4.8 Roots of monocotyledonous plants.
(a) T.S. of root with typical anatomical
pattern; note the vascular cylinder with
central pith and a large number of xylem
strands. (b) T.S. of small part of root of
banana (*Musa paradisiaca*) through region
where vascular cylinder adjoins cortex. (c)
Endodermis with 'U' shaped thickening
common in many monocotyledonous roots,
e.g. *Musa paradisiaca* (illustrated) and *Zea
mays*. (d) Endodermis with 'O' type
thickening from aerial root of tropical orchid,
Epidendron sp. (e) T.S. part of root of an
epiphytic orchid (*Dendrobium*) showing the
velamen and lignification of the ground tissue
within the vascular cylinder. (b from Eames
and MacDaniels 1947, p. l60; e from Fritsch,
F. E. and Salisbury, E. J. (1948), *Plant form
and function*, p. 112, G. Bell & Sons,
London)

mary xylem stops before a solid core is formed. The mature primary xylem
of monocotyledonous roots is therefore composed of separate strands,
and a parenchymatous pith results in the centre of the vascular cylinder.

3. In monocotyledonous roots the endodermis often develops thick, secon-
    dary walls (Fig. 4.8b). In suitably stained transverse sections of mono-
    cotyledonous roots the endodermis is readily observed as a distinct
    boundary layer, whereas in dicotyledonous roots it may be difficult
    to recognize. In the endodermis of dicotyledonous roots, suberin is
    deposited in the radial and transverse walls as the Casparian strip,
    and the viability of the cells in which this deposition occurs does not
    seem to be impaired. In the endodermis of monocotyledons, however, a
    continuous layer of suberin is deposited on the inner side of the entire
    primary wall, including the Casparian strip. This suberin lamella is then
    covered by layers of lignified cellulose, which may reach a considerable
    thickness. The thickening may be uneven or even, giving rise to the so-
    called 'U' type or 'O' type endodermis respectively. In the 'U' type, which
    is more common, the lignified secondary wall is thin on the outer peri-
    clinal wall but thick on the other walls, forming the characteristic 'U'
    shape in transverse section (Fig. 4.8c). In the 'O' type, the secondary-

wall material is equally thick on all walls, so that the cell cavity has a characteristic 'O' shape in transverse section (Fig. 4.8d). In contrast to the thin-walled endodermis of dicotyledonous roots, the 'U' and 'O' types of endodermis soon die because the walls of their cells become almost completely impermeable. In these thickened types of endodermis, some of the endodermal cells which are opposite the xylem strands remain thin-walled and retain their living protoplasts. Such cells are called *passage cells* because they are thought to provide passage for water and solutes between the cortex and vascular cylinder.

Some monocotyledons, notably tropical orchids and aroids which grow as epiphytes in the forks of trees, produce tufts of *aerial roots* projecting into the surrounding air. Such roots, while showing the anatomical features characteristic of monocotyledonous roots, exhibit an interesting modification of structure which corresponds with their special function of absorbing water from air instead of from soil. In these aerial roots the protoderm divides periclinally to form a multiple epidermis, several to many cells thick, called the *velamen* (Fig. 4.8e). The cells of the velamen, which are dead at functional maturity, have holes in their walls, and soak up by capillary action any water falling or condensing on the root surface. A thickened exodermis, superficially resembling an 'O' type endodermis, separates the velamen from the cortex. When the air is dry the cells of the velamen are filled with air, but when water is deposited on an aerial root it is immediately absorbed into the velamen, and then gradually passes through passage cells in the exodermis into the cortex, and from there through the passage cells of the endodermis into the vascular cylinder. Although the velamen is usually regarded as an adaptation of epiphytes, the roots of some terrestrial orchids, e.g. *Eulophia*, also possess it.

The origin of lateral roots

Lateral roots are endogenous, originating in the pericycle of the parent root. This site of origin is significant in that a lateral root develops in close proximity to the vascular tissues of the parent root, with which the vascular tissues of the lateral root eventually establish contact. Lateral roots arise singly at points along the vertical edges of the protoxylem strands, and this is why they are arranged on the surface of the parent root in vertical rows.

Fig. 4.9 Origin of lateral roots. (a) Very early stages in the development of a lateral root in *Descurainia pinnata*. a, pericycle; b, endodermis; c, primordium of protoderm and root cap; d, primordial ground meristem; e, primordial procambium. (b) Connecting the vascular tissues of the lateral root to those of the parent root. (a from Dittmer, H. J. and Spensley, R. D. (1947). *University of New Mexico Publications in Biology No. 3*, University of New Mexico Press, Albuquerque)

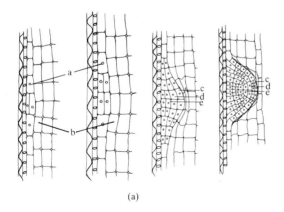

(a)

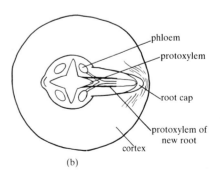

(b)

When a lateral root begins to form, several adjacent pericyclic cells divide by tangential walls so that the pericycle layer is doubled (Fig. 4.9). Further divisions, both tangential and radial, form a small protrusion, the root primordium, which presses outwards into the cortex. As a result of further growth, the primordium elongates and mechanically forces its way through the cortex until it emerges on the surface. The endodermal cells often divide to keep pace with the growth of the primordium, and it is possible, but doubtful, that they even assist its growth by secreting enzymes which break down the cortical cells in front of the advancing tip. Before the young root primordium reaches the surface of the parent root a definite root cap and apical meristem have been formed, and the cells behind the apical meristem have begun to differentiate into xylem and phloem elements. These later connect up with their counterparts in the parent root by the pericyclic cells at the proximal end of the newly formed lateral root differentiating into vascular elements. The ruptured cortex and epidermis, however, do not connect up with the corresponding tissues of the young root protruding through them.

Chapter 5

The primary stem

The term stem usually implies the branching axis of the aerial portion of a plant but, from a botanical point of view, the most distinctive feature of a stem is not that it grows in the air but that it bears leaves, and is therefore divisible into nodes (the points of leaf attachment) and internodes (the regions between the nodes). The explanation for this characteristic division of the stem into successive joints is to be found in the structure and pattern of growth of the stem apex.

The stem apex

The external morphology of the shoot, as the composite system of stems and leaves is called, has already been described in Chapter 1 and illustrated in Fig. 1.1. As was mentioned in that chapter, a stem always terminates in a bud which is composed of a number of small leaves surrounding and overarching the central portion of the bud, the stem apex. If a median longitudinal section of a terminal bud is examined under the microscope it will be seen that the tip of the stem apex is occupied by a dome-shaped mass of meristematic cells (Fig. 5.1). On the flanks of this apical meristem are small bumps of tissue which are the beginnings, or primordia, of young leaves; two such leaf primordia are evident just behind the apical meristem in Fig. 5.1. The surface of the apical meristem and leaf primordia is covered by a thin cuticle which is continuous with the cuticle covering the rest of the primary shoot.

Although the apical meristem is a region of active cell division, it is not just a homogeneous mass of embryonic cells but a highly organized structure. According to the most widely accepted modern view the apical meristem consists of two main regions which differ from each other in their mode of cell division. On the outside there is a layer, known as the *tunica*, which is one or a few cells thick. The cells of this layer divide by forming walls at right angles to the surface (anticlinally) with the result that the tunica grows in area but not in thickness. Enclosed within the tunica is the *corpus*, a mass of cells which divide in various planes including periclinally (i.e. by walls parallel with the surface), so that the whole mass grows in volume.

Leaf primordia are formed on the surface of the meristem at points where some of the corpus cells divide to form a small mound of tissue underlying the tunica. The tunica cells at such places divide to keep pace with the growth of the mound so that the tunica is maintained as a continuous layer over the developing leaf primordium. In some plants the primordium is

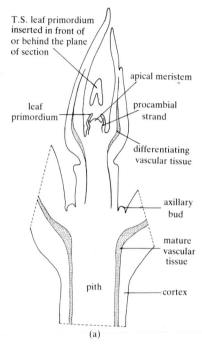

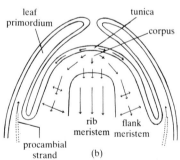

Fig. 5.1 The shoot apex. (a) Median L.S. of shoot tip of *Coleus* (which has opposite leaves, with each pair at right angles to the one above it and the one below it) showing apical meristem and successively older leaf primordia. (b) Diagram of median L.S. of stem apex of an angiosperm to illustrate the tunica-corpus theory of apical organization in shoots. The arrows indicate direction of growth.

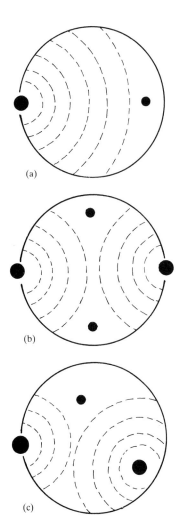

(a)

(b)

(c)

Fig. 5.2 The field theory of phyllotaxis. Each diagram represents a shoot apex as seen from above. The leaf primordia are shown as black dots, which increase in size as the primordia they represent grow larger. The presumed inhibitory fields are indicated by dotted lines. (a) Only one primordium inhibits at a time, resulting in a phyllotaxis of ½, i.e. the distichous arrangement. (b) Two opposite primordia of equal age, and therefore exerting equal inhibitory forces, result in the decussate arrangement. (c) A spiral phyllotaxis results when two or more primordia of unequal age cause an unbalanced distribution of inhibitory forces; here only the two main inhibitions from the two most recently formed primordia are shown as determining the presumptive position of the next primordium.

finger-like in shape, in which case it develops into a leaf with a narrow attachment to the stem. In other plants the primordium is a ridge that partially or completely encircles the stem, and in this case the primordium develops into a leaf with a broad base which may even ensheath the stem, as in many monocotyledons. Whatever its shape, the leaf primordium elongates rapidly so that it arches over the stem apex to form part of the terminal bud. During the early stages of this elongation the leaf assumes its typically flattened form as a result of a faster rate of cell division along the two edges of the primordium.

Arrangement of the leaf primordia

According to the species of plant, leaf primordia may develop on the stem apex in spirals, in whorls or in opposite pairs. The way in which they develop largely determines the arrangement of the leaves on the mature shoot, i.e. its phyllotaxis. Botanists have long been interested in the mechanisms underlying phyllotaxis, and two main types of explanation have been proposed. The first ('field' theory) depends upon the idea that each existing leaf primordium liberates a chemical substance, possibly a growth inhibitor, which prevents the development of further primordia in its immediate vicinity. The inhibitory effect is thought to diminish as the substance diffuses away from the primordium. A new primordium therefore arises where the sum of the inhibitions due to all the existing primordia is at a minimum (Fig. 5.2). The second type of explanation ('free space' theory) postulates that a new primordium cannot begin to form until a certain minimum space becomes available above the pre-existing primordia. The extent of this free space might depend on the minimum number of uncommitted cells necessary for a primordium to form.

To decide which of these possibilities is correct, many experiments involving micro-surgery and chemical treatments have been made. In support of the field theory, C. W. Wardlaw, using fern apices, showed that a leaf primordium isolated by surgical cuts grew more rapidly than normal, suggesting that on an uncut apex its growth is inhibited by substances diffusing from adjacent primordia. However, in support of the free space theory, R. and M. Snow have show that, if an area of a shoot apex where a leaf primordium can be expected to arise is isolated by two cuts radiating from the tip of the stem apex, no primordium will develop if the angle between the cuts is small, even though in other respects the isolated sector continues to grow apparently normally. This suggests that the physical size of the space between the two cuts is important in the mechanism of primordium initiation, i.e. it is a matter purely of geometry. Unfortunately the results of any micro-surgical experiments are open to severe criticism because wound hormones are almost invariably formed and abnormal metabolism can therefore be expected. Equally severe limitations also apply to the chemical evidence which has been advanced in support of one or other of the two theories. As a result, it is not possible at present to arrive at a firm conclusion regarding the factors which determine the arrangement of the leaf primordia on the stem apex.

Development of the shoot

When the developing leaf is still minute and forms part of the apical bud, another bulge of meristematic cells arises in the axil between each leaf and

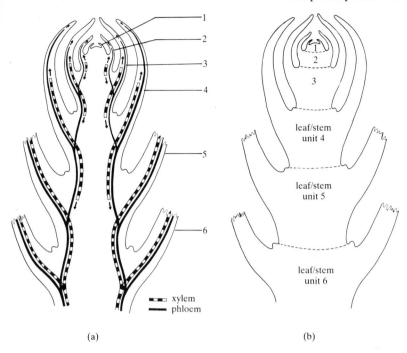

(a)   (b)

Fig. 5.3 Development of the vascular system in shoots. In order to show how the vascular strands connect at the nodes, alternate leaf pairs shown in this figure have been turned through 180°. Compare this figure with *Fig. 5.* (a) Diagrammatic L.S. of a shoot apex, showing the longitudinal differentiation of the vascular strands. Leaf pair 1, procambium only; leaf pair 2, some mature phloem continuous with the phloem of older part of stem, and an isolated length of immature xylem at the base of the leaf primordium; leaf pairs 3 and 4, phloem continues to extend into developing leaf, and xylem spreads upwards into leaf and downwards into stem; leaf pairs 5 and 6, mature phloem and mature xylem, the latter now connected to older xylem of internode below. (b) The division of a shoot into successive leaf/stem units as a consequence of the pattern of vascularization shown in (a). The outline of the two diagrams is the same.

the stem. This bulge is the rudiment of an axillary bud and when fully developed shows the same structure as the terminal bud. The fate of an axillary bud depends upon a number of internal and external factors; it may become dormant at an early stage of development or it may grow into a branch shoot with a terminal bud of its own.

In the region below the corpus the cells divide more slowly, begin to vacuolate, and develop thicker walls. This change is usually first shown in the centre of the axis where the cells are arranged in very regular, parallel rows. When the cells in this central region divide the new walls are formed transversely to the long axis so that vertical files of cells are produced, each file originating from a single cell near the apex. The cells in these files, which are collectively known as the *rib meristem*, elongate and cause the surrounding tissues to be stretched with the result that the leaf primordia are pulled apart from one another.

The cells of the rib meristem lose their meristematic ability sooner than those surrounding them, which form the region called the *flank meristem*. If the cells of the flank meristem continue to divide and elongate for a considerable time after the cells of the rib meristem have stopped enlarging, the central tissue is torn apart and a hollow stem results. When the flank meristem eventually ceases to grow in length the leaves occupy their final position on the adult stem.

While the young stem is elongating and the leaf primordia are developing, differentiation of the vascular tissues is taking place (Fig. 5.3). Procambium, which is the meristematic precursor of the vascular system, arises as longitudinal strands of tissue between the rib and flank meristems at positions related to the points of attachment of the leaf primordia. The procambial strands of the youngest leaf primordia develop as extensions of those already formed lower down the stem, so that the vascular tissues are laid down as a continuous system. As each leaf primordium develops, one or more procambial strands in the stem grow into it with the result that the vascular tissue of a mature leaf can be regarded as merely an outward extension of the vascular tissue in the stem. As the procambial strands extend into the leaf primordia, their cells differentiate into the various elements of the mature vascular strands. During this process the procambial cells grow mainly in one direction and produce the elongated cells characteristic of conducting tissue. Differentiation of the phloem within a procambial strand is acropetal, proceeding up the stem (from where it is already continuous with more mature phloem at a lower level in the stem) and then extending outwards into the leaf. Differentiation of the xylem, however, begins at the base of each leaf primordium and then proceeds in both directions, acropetally into the leaf and basipetally (from above towards the base) into the internode below until it joins up with the xylem of the older strands lower down the stem. This origin of the vascular strands in units divided between a leaf and the internode below emphasizes that, from the point of view of its development, the shoot system can be regarded as a series of leaf–stem units successively formed from the stem apex.

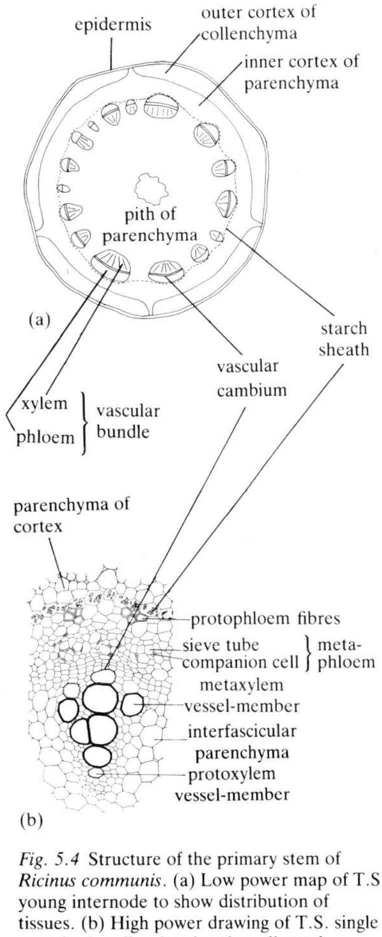

Structure of primary stems

The internal organization of the primary stem is more complex than that of the primary root because a stem is not a simple axis but is divided into nodes and internodes, with one or more leaves attached at each node. This close association of the stem with the leaves results in differences in structure between nodes and internodes, whereas in the primary root the arrangement of tissues shows little variation between different levels. As in the root, however, there is a clear separation into the three tissue systems, the dermal, the ground and the vascular. The distribution of the ground and vascular tissues in stems differs so radically in dicotyledons and monocotyledons that the stem structure of each is best treated separately.

Primary stem structure of dicotyledons

A transverse section, supplemented by longitudinal sections, through a young internode which has ceased to elongate gives the best idea of the structure of a dicotyledonous stem in the primary state of growth. Seedlings of the castor oil plant (*Ricinus communis*) are suitable material for sectioning provided they have been grown under conditions which favour rapid growth and hence the production of a main stem with long internodes. In such a section (Fig. 5.4) the vascular system appears as a number of distinct units, called *vascular bundles*, which are arranged in a ring between two regions of ground tissue, a wide central region (the pith) and a narrow peripheral region (the cortex). The vascular bundles are separated laterally by panels of parenchyma, the interfascicular parenchyma, which is also part of the ground tissue. This parenchyma is called interfascicular because it occurs between the fascicles, a name formerly used for the vascular bundles. A panel of interfascicular parenchyma is sometimes called a pith ray, because it connects the pith with the ground tissue of the cortex.

Fig. 5.4 Structure of the primary stem of *Ricinus communis*. (a) Low power map of T.S. young internode to show distribution of tissues. (b) High power drawing of T.S. single vascular bundle, showing the collateral arrangement of primary xylem and phloem. (b from Brown 1935, p. 131)

Epidermis

The epidermis consists of a single layer of cells which have, as their most characteristic feature, a cuticle on their outer surface. Most of the epidermis is occupied by relatively unspecialized cells, the epidermal cells proper, which fit tightly together as a skin. At intervals, however, the epidermis is pierced by openings, called *stomata* (singular, stoma), which are bounded by two specialized epidermal cells, the guard cells. Below each stoma there is a space which is connected with a system of intercellular air spaces among the underlying cells of the cortex. Stomata constitute a more prominent component of the leaf epidermis than of the stem epidermis, and therefore consideration of the structure and operation of guard cells will be deferred until Chapter 8 dealing with the leaf.

Cortex and pith

From their positions in the stem it will be readily appreciated that the pith originates from the rib meristem of the shoot apex and the cortex from the flank meristem. The cortex is usually divided into two regions. The outer cortex, which is not always present, consists of either an almost continuous layer of tightly packed collenchyma or, more usually, alternating strips of collenchyma and parenchyma. The inner cortex consists of loosely packed parenchyma with prominent intercellular spaces that communicate with the air spaces found beneath the stomata. The cells of the cortex, particularly the parenchyma, have abundant chloroplasts which impart a green colour to this region in the living state. The innermost layer of the cortex does not form a visibly differentiated endodermis as in the root, but frequently contains abundant starch grains even when the neighbouring cells may have few or none. This layer may thus be recognizable as a continuous *starch sheath*. The radial and transverse walls of the starch sheath fit tightly together and there are no intercellular spaces passing from the cortex to the more deeply seated tissues. When no starch accumulates in the innermost layer of the cortex the delimitation of the cortex from the interfascicular parenchyma may be difficult or impossible.

The central core of ground tissue, the pith, is usually composed of parenchyma with prominent intercellular spaces, at least in the central part. In some plants the elongation of the young stem causes the central cells of the pith to be torn apart, in which case a hollow pith results.

Vascular system

The vascular system of the stem is made up of a number of vascular bundles which develop behind the shoot apex from their meristematic precursors, the procambial strands. These vascular bundles, as previously pointed out, are continuous with those in the leaves because the shoot develops as a series of leaf–stem units. Although in a transverse section (T. S.) of an internode each vascular bundle appears to be an independent unit, it must be remembered that there is branching and fusing of bundles at the nodes (Fig. 5.5). It is by means of the nodal connections that the continuity of the vascular system is established throughout the shoot.

In each vascular bundle the xylem and phloem are arranged on the same radial plane, with the xylem on the inside and the phloem on the outside and not on alternating radial planes as in the root. This arrangement of the

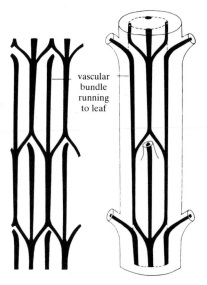

vascular
bundle
running
to leaf

Fig. 5.5 Vascular system of the primary stem of *Anabasis articulata*. (a) Vascular system spread out on one plane. (b) Stem as it would appear if all the tissues except the vascular bundles were made transparent. To simplify the diagram the course of the posterior vascular bundle has been omitted, but it is the mirror image of the front one. The arrangement of the bundles in a T.S. can be seen at the level of the cut. Note the pattern of branching and fusing at the nodes. (From Fahn 1974, p. 217)

xylem and phloem on the same radius is described as *collateral*. Between the xylem and the phloem, in the middle of the bundle, is a thin layer of undifferentiated cells, the fascicular cambium, which retain the merisstematic condition of the procambial strands. The constituent elements of the vascular bundles are similar to those in the root but they are arranged in a different way. The first-formed xylem, the protoxylem, is found on the inner side of each bundle adjacent to the pith and, in longitudinal section, can be seen to consist of tracheids and vessels with annular and helical thickenings (Fig. 5.6). The protoxylem may show clear signs of having been damaged as a result of being stretched while the surrounding tissues were still increasing in length. The later-formed xylem elements, the metaxylem, mature in a centrifugal direction and the xylem is consequently described as endarch. In transverse section the metaxylem vessels are arranged in straight rows which extend up to the fascicular cambium and are separated

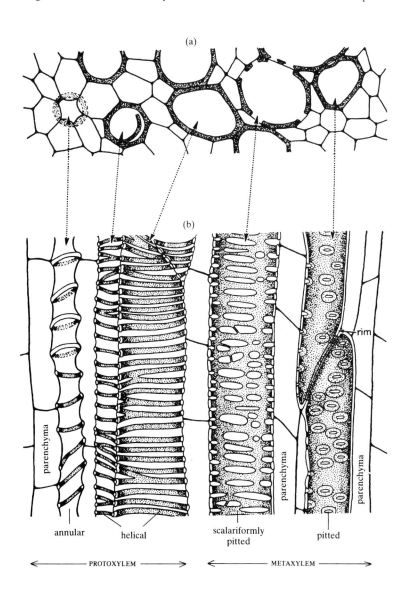

Fig. 5.6 Protoxylem and metaxylem in (a) transverse and (b) longitudinal section of *Aristolochia* stem. The vessel with annular thickening is somewhat stretched, and the adjacent parenchyma cells are slightly bulging into it. The helically thickened vessels have a few interconnections among the coils of the helix. (From Esau 1965, p. 232)

from one another by strips of xylem parenchyma. As seen in longitudinal section the metaxylem consists successively of reticulate, scalariform and pitted vessels which thus form a sequence in which an increasing area of the primary wall is covered by secondary wall material.

In the phloem the order of maturation is reversed. The first phloem elements to mature occur at the outer edge of the vascular bundle, and subsequent maturation takes place in a centripetal direction. In accordance with expectation, the phloem consists of sieve-tube members, companion cells, parenchyma and fibres, but in many dicotyledonous plants the protophloem consists entirely of phloem fibres. If this is the case, then protophloem fibres form a cap on the outside of the phloem strand and serve to indicate the outer limit of the phloem. If no protophloem fibres are present, the outer part of the phloem may not be sharply delimited from the cortex because in stems, in contrast to roots, there is no pericycle separating the vascular tissues from the cortex. In the stems of some plants, as for example members of the Cucurbitaceae (the pumpkin family), Convolvulaceae (the morning glory family) and Solanaceae (the tobacco family), phloem strands develop on the inside as well as on the outside of the xylem (Fig. 5.7a). Such bundles with both external and internal phloem are termed *bicollateral* to distinguish them from the more usual collateral type.

Although the vascular system of all dicotyledonous stems can be interpreted as a ring of vascular bundles separated by interfascicular parenchyma, the bundles of most trees and shrubs are so close together that they form a continuous cylinder (Fig. 5.7b). The only indication that such a cylinder can be regarded as being constructed of separate vascular bundles may be the projection of wedge-shaped masses of protoxylem into the pith. In stems with adjacent vascular bundles the fibres in the outermost part of the phloem may likewise form a continuous or nearly continuous cylinder immediately within the cortex

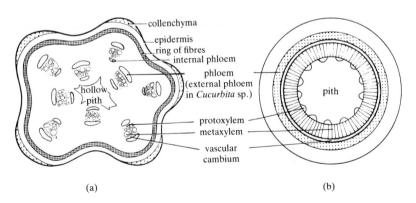

Fig. 5.7 The arrangement of the vascular tissues in two types of primary stem. (a) T.S. stem of *Cucurbita* sp. showing bicollateral bundles with phloem on both the inside and the outside of the xylem. (b) T.S. stem of woody dicotyledon in which the xylem and phloem form a continuous cylinder, with groups of protoxylem projecting into the pith. (a from Coult 1973, p. 37)

Stem structure of monocotyledons

The stems of monocotyledons differ from those of dicotyledons in the arrangement and structure of their vascular bundles. The stem of *Zea mays* (maize) will be taken as a typical example. In a transverse section of an internode (Fig. 5.8) the vascular bundles are not arranged in a ring, but appear to be scattered irregularly throughout the section. The bundles are

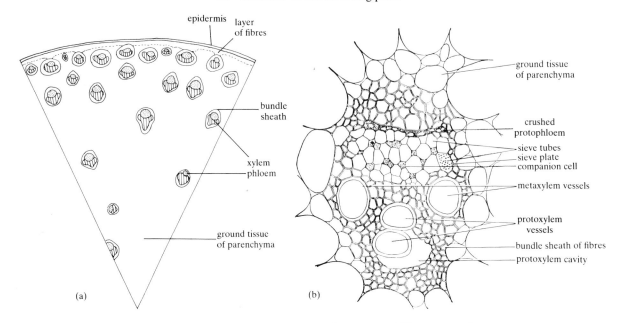

Fig. 5.8 Zea mays, maize. (a) T.S. of a segment of stem, showing the scattered distribution of the vascular bundles. (b) T.S. of a single vascular bundle from central part of stem. (b from Holman, R. M. and Robbins, W. W. (1933), *Elements of botany*, p. 82, (John Wiley & Sons)

very numerous, sometimes up to 200 or more, but are more crowded towards the outside. Below the epidermis is a narrow layer of fibres, one or two cells thick, but there is no obvious division of the ground tissue into cortex and pith.

The individual bundles are collateral as in most dicotyledons, but an important difference is that the metaxylem and metaphloem differentiate right up to each other so that no fascicular cambium is left between them. As a result secondary growth cannot occur in monocotyledonous bundles which are therefore described as closed in contrast to the open type of dicotyledons. In transverse section the xylem appears as a V- or Y-shaped group of vessels with the protoxylem at the base and pointing towards the centre of the stem. Elongation of the young maize stem is rapid, and the earliest protoxylem elements are torn apart so as to form an irregular cavity (the protoxylem cavity) in which the remnants of the spiral and annular thickenings can sometimes be seen. The phloem, which lies between the arms of the V or Y of the xylem, consists entirely of sieve tubes and companion cells. The companion cells are very distinct, being smaller than the sieve-tube members and usually square in cross-section. Each bundle is surrounded by a sheath of fibres which is particularly thick around the peripheral bundles. These peripheral bundles are so extensively thickened and so closely packed that fibrous tissue forms an almost complete cylinder which is continuous with the thin layer of fibres immediately below the epidermis.

The above description of an internode of *Zea mays* is typical of the majority of herbaceous monocotyledons, but in the primary stem of a number of monocotyledons, particularly arborescent species, there is a peripheral band of ground tissue outside the region of scattered vascular bundles. It is in this cortical region that a cambium arises in the few monocotyledons, native to the tropics, which undergo a type of secondary growth.

Chapter 6

Secondary growth

In small dicotyledonous annuals and most monocotyledons all the cells of the adult plant body are produced at the apical meristems, and these plants therefore complete their life cycle entirely by primary growth. In most dicotyledons, however, particularly the woody perennials which continue to grow year after year, the primary plant body is augmented by the formation of additional *secondary tissues* which increase the thickness of the plant axis (Fig. 6.1). As explained in Chapter 3 (p. 27) this secondary growth is brought about mainly by the development of a lateral meristem, the vascular cambium, which increases the amount of vascular tissues in stems and roots. If the plant axis is thickened to any appreciable extent, secondary growth also involves the formation near the surface of the plant axis of another meristematic layer, the phellogen, the activity of which produces a protective tissue, periderm, to replace the epidermis of the primary body. The formation of additional vascular tissues and periderm by secondary growth makes possible the development of large, much branched plant bodies such as trees. The features of secondary growth will be described mainly with reference to stems.

Fig. 6.1 Secondary thickening of stems. Comparison of silk-cotton tree (left), which shows secondary growth, with oil palm (right) which does not. (Redrawn from Irvine 1952, p. 37)

Secondary growth in dicotyledonous stems

Structure of the vascular cambium

Although the fully formed vascular cambium is a continuous layer, more or less circular in cross-section, it nevertheless has a double origin. In stems it arises partly within the bundles from procambial tissue that has not differentiated into primary xylem and phloem, and partly between the bundles from tangential bands of mature parenchyma cells which dedifferentiate and revert to the meristematic condition. Hence it is customary to refer to the two regions of a recently formed vascular cambium as fascicular and interfascicular cambium respectively (Figs. 6.2 and 6.3). In the stems of woody (arborescent) dicotyledons, where the vascular bundles are coalescent, it is impossible to distinguish between fascicular and interfascicular regions.

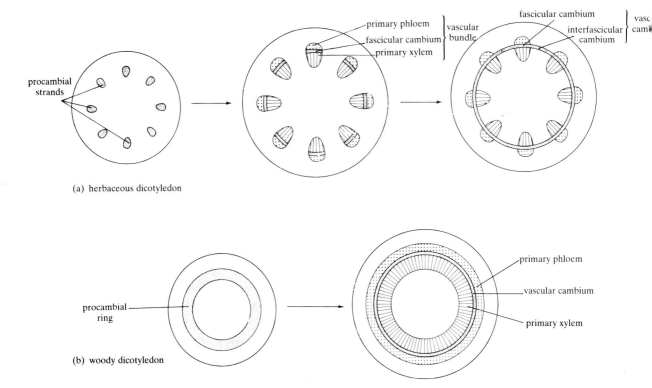

(a) herbaceous dicotyledon

(b) woody dicotyledon

Fig. 6.2 Formation of the vascular cambium in (a) herbaceous dicotyledonous plants, and (b) woody dicotyledonous plants.

The cells of the vascular cambium are highly vacuolate and thus quite different in appearance from those of apical meristems, on which the usual concept of meristems is based. Cambial cells are of two types, fusiform initials and ray initials (Fig. 6.4). *Fusiform initials* are elongated longitudinally, have wedge-shaped ends, and are wider tangentially than radially so that they appear rectangular in cross-section. They form elements such as tracheids, vessel members, fibres and sieve-tube members, that is, vascular elements which are arranged vertically in the stem or root. It will be noted that fusiform initials are similar in shape to the cells to which they give rise. This is associated with the fact that when secondary growth occurs elongation of the stem is already complete. Consequently the cells derived from fusiform initials, unlike those from an apical meristem, cannot grow in harmony with the surrounding tissues, and can elongate only by intruding their growing ends between one another. *Ray initials* are much shorter than fusiform initials, and more or less cubical in shape. They are also much less numerous than fusiform initials, and appear as 'islands' of small cells when the cambium is viewed tangentially (Fig. 6.4c). They give rise to parenchyma cells, all or most of which elongate in the horizontal direction. The mass of cells cut off from an island of ray initials forms a horizontal strand of parenchyma, called a *vascular ray*, which extends both inwards (as a xylem ray) and outwards (as a phloem ray) from the cambium (see Fig. 6.14). Rays serve for the storage and radial transport of food materials.

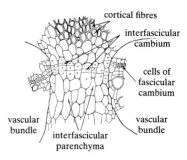

Fig. 6.3 T.S. young stem of *Aristolochia* between two vascular bundles, showing formation of the interfascicular cambium from parenchymatous cells of the ground tissue. (From Holman, R. M. and Robbins, W. W. (1933), *Elements of botany*, p. 66, published by John Wiley & Sons)

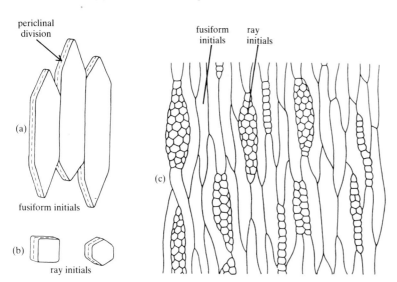

Fig. 6.4 Vascular cambium. (a) and (b) Diagrams of fusiform and ray initials, the broken lines indicating the plane of division involved in the formation of phloem and xylem cells (i.e. periclinal division). (c) T.L.S. vascular cambium showing arrangement of cambial initials as seen in surface view. (c from Fahn 1974, p. 328)

Activity of the vascular cambium

Secondary growth is brought about by the tangential division of cambial cells which cut off new xylem elements inwards and new phloem elements outwards. The cells of the secondary xylem and secondary phloem are thus arranged, at least initially, in regular radial rows. The vascular cambium is only one cell thick, the same cell cutting off daughter cells both inwards and outwards. At the end of each division, one of the daughter cells (the initial) remains meristematic and perpetuates the cambium, while the other (the derivative) becomes differentiated into either a xylem or a phloem cell. If the derivative lies towards the outside of the stem it contributes to the secondary phloem, if towards the inside then it contributes to the secondary xylem. The divisions are not alternate, however, and invariably many more cells of secondary xylem are formed than of secondary phloem. Figure 6.5 illustrates the way in which the division of a cambial initial, followed by differentiation of the derivatives, adds new cells to the secondary xylem and phloem.

When the cambium is dividing actively the formation of new cells is so rapid that there is a zone of meristematic cells on either side of the cambium, differentiating into xylem inwards and phloem outwards. It is thus difficult, if not impossible, to distinguish the true initials from their nearest derivatives, and the term cambial zone (Fig. 6.6) may be conveniently used to

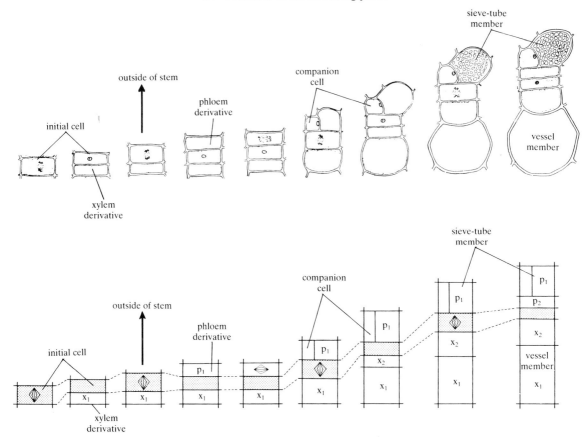

Fig. 6.5 The activity of a single cambial initial, as seen in T.S. of a stem undergoing secondary growth, showing how the derivatives give rise to cells of secondary xylem (x_1 and x_2) and secondary phloem (p_1 and p_2). The lower diagram explains the sequence of divisions of the cambial initial; the broken lines show how the initial cell (dotted) is continually displaced towards the outside of the stem.

include both. The true initial, in so far as it can be distinguished, is the narrowest member of a radial file of cells crossing the cambial zone. Three-dimensionally, therefore, the cambial initials and their nearest derivatives together form a continuous cylinder of thin-walled vacuolate cells between the xylem and phloem. In most plants the fascicular and interfascicular regions of the cambium both form fusiform and ray initials so that the secondary xylem and phloem form continuous cylinders. In many woody vines (e.g. *Aristolochia*), however, the fascicular regions form only fusiform initials and the interfascicular regions only ray initials. In this case the vascular system

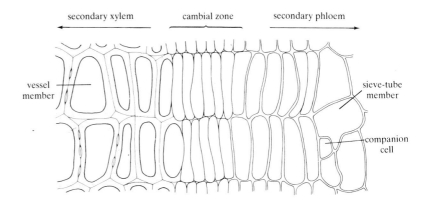

Fig. 6.6 T.S. cambial zone of a young twig

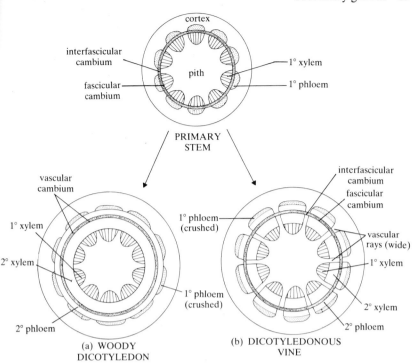

Fig. 6.7 Two types of secondary thickening in dicotyledonous stems. (a) Woody dicotyledon in which both the fascicular and interfascicular regions of the vascular cambium produce essentially the same kinds of tissue. (b) Dicotyledonous vine in which, at least initially, the fascicular regions form only fusiform initials and the interfascicular regions only ray initials. The alternation of thick-walled tracheary tissue with thin-walled ray tissue in the stem of dicotyledonous vines may be associated with the twining habit adopted by these plants.

is divided into wedge-shaped strands (each associated with a primary xylem strand next to the pith), which are separated from one another by wide vascular rays (Fig. 6.7).

Structure of secondary xylem and phloem

Secondary xylem and phloem contain the same basic cell types as their primary counterparts, but the cells of these secondary tissues are organized into two distinct systems, the *axial* (longitudinal or vertical) and the *radial* (transverse or horizontal) systems. In the axial system the cells are orientated with their long axes parallel to the axis of the stem, while in the radial system the cells are arranged at right angles to this axis. The organization of secondary xylem and phloem into axial and radial systems of cells reflects the structure of the vascular cambium, from which both tissues are derived. The fusiform initials of the vascular cambium give rise to the elements of the axial system, and the ray initials to the elements of the radial system (i.e. the vascular rays). In dicotyledons the axial system of the secondary xylem contains vessel members, tracheids, fibres and axial parenchyma; the axial system of the secondary phloem contains sieve-tube members, companion cells, axial parenchyma, and usually fibres as well. The radial systems of both the secondary xylem and secondary phloem consist of radial (ray) parenchyma only. These cells are usually in direct contact with the axial parenchyma so that together they form a continuous network of living cells.

The relationship between the two systems can be seen from the appearance they present in the three kinds of sections that are used for interpreting the structure of any axial organ (Fig. 6.8). In the transverse section (T.S.) the cells of the axial system are cut transversely whereas the ray cells are exposed longitudinally. When an axis like a stem or a root is cut lengthwise, two kinds

T.S.

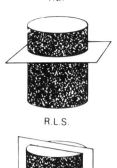

R.L.S.

T.L.S.

(a)

TRANSVERSE SECTION

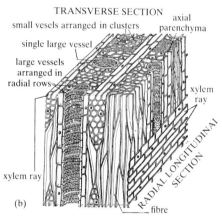

small vesels arranged in clusters

axial parenchyma

single large vessel

large vessels arranged in radial rows

xylem ray

xylem ray

fibre

TANGENTIAL LONGITUDINAL SECTION

Fig. 6.8 The structure of secondary xylem. (a) The three planes of section (transverse, T.S.; radial longitudinal, R.L.S; tangential longitudinal, T.L.S.) which are necessary to construct a three-dimensional picture of the cellular structure of any axial organ. (b) Three-dimensional diagram of a cube of wood from a dicotyledonous tree (*Cercis siliquastrum*) showing the appearance of the axial and radial systems of the secondary xylem in different planes. (b from Fahn 1974, p. 357)

of longitudinal sections can be obtained: the radial longitudinal section (R.L.S.) along a radius, and the tangential longitudinal section (T.L.S.) parallel to a tangent. Both give the same view of the axial system because they expose the cells of this system longitudinally, but they give two very different views of the vascular rays. The R.L.S. shows the rays as horizontal bands lying at right angles to the elements of the axial system, whereas the T.L.S. cuts the rays end-on and so reveals their width and height (Fig. 6.8b).

The composition, arrangement and size of the cells forming the secondary xylem and phloem vary between different plants and, even within the same plant, they may vary considerably according to the time of year when the secondary tissues are formed. Such seasonal variation is common in trees of temperate regions, where there is a marked alternation of hot and cold seasons, but is apparently not a regular feature of trees in the tropics, even where there is a marked alternation of wet and dry seasons. The seasonal variation, if any, is more conspicuous in the secondary xylem, where the elements formed early in the growing season are usually larger and have thinner walls than those formed at the end of the season. When cambial activity is resumed at the beginning of the next growing season, the larger and thinner-walled xylem cells contrast sharply with the smaller and thicker-walled xylem formed at the end of the previous season. This conspicuous boundary between one season's growth and the next is called a *growth ring* (Fig. 6.9). In many plants a concentric zoning of cells is also apparent in the secondary phloem, but these rings do not necessarily correspond with the growth rings seen in the secondary xylem (Fig. 6.13).

Wood anatomy

A knowledge of the structure of secondary xylem is essential to an understanding of the properties of different timbers, since the uses to which a particular wood can be put depend ultimately on its microscopic structure. In order to meet the scarcity of timber relative to the expanding demands of industry, new forest areas are being opened up in many parts of the world with a view to using alternative timbers. On the basis of existing knowledge about the relations between the anatomy and properties of wood, microscopic examination of these timbers is one of the techniques by which their qualities and possible uses can be predicted rather than found by trial and error. In view of this economic application of plant anatomy, it is appropriate to consider secondary xylem in somewhat more detail than is given in the above account of its general structure.

Although secondary xylem consists of only four basic cell types, the relative proportions and arrangement of these vary tremendously between different trees but are nevertheless more or less constant for a particular species of tree. The overall pattern, including any reaction to seasonal variation, is inherited, and this provides the wood anatomist with a means of both identifying woods and recognizing evolutionary relationships between woody plants. There are three main anatomical features, discussed below, which are used to characterize different woods, and their investigation requires sections cut in the three planes already mentioned. The first two of these features reflect the distribution of the elements of the axial system and are therefore determined from the transverse section; the third feature concerns the radial system and is determined from the two longitudinal sections.

1. Arrangement of the vessels   A vessel in cross-section is called a pore because it looks like a hole or pore on the end face of a plank of wood. When the pores in the first-formed xylem of a growth ring are much larger and more numerous than in the rest of the xylem of the same growth ring, the wood is called *ring porous* (see Fig. 7.4b). When, on the other hand, the pores are approximately equal in diameter and more or less evenly distributed throughout the wood, or when the change in pore diameter between the first-formed and later-formed vessels of a growth ring is gradual, the wood is called *diffuse porous* (see Fig. 7.4c). These two patterns of vessel distribution represent extreme types, and many transitional patterns occur.

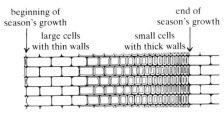

Fig. 6.9 One method by which a growth ring can be formed. (After Galston 1964, p. 71)

2. Arrangement of the axial parenchyma   The distribution of the axial parenchyma in relation to the vessels is another characteristic feature that is studied in transverse sections of wood. Two main patterns are recognized: *paratracheal* where the axial parenchyma is consistently associated with the vessels, and *apotracheal* where the axial parenchyma is independent of the vessels. Both these patterns are subdivided into different types (Fig. 6.10).

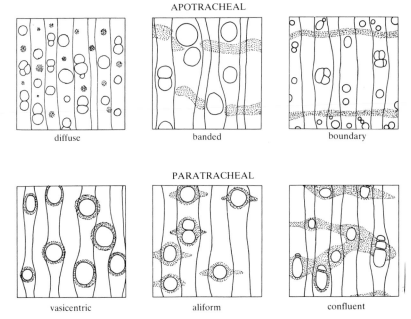

Fig. 6.10 Distribution of axial parenchyma (shown as dotted areas) with respect to the vessels, as seen in transverse sections. Apotracheal parenchyma is said to be *diffuse* when it occurs as small isolated strands or single cells scattered irregularly among the xylary fibres, *banded* when it is in the form of tangential bands alternating with fibres, and *boundary* (a special case of banded) when it forms a more or less continuous layer at the end, or sometimes the beginning, of a growth ring. Paratracheal parenchyma is subdivided into the following types: *vasicentric*, forming a complete sheath around the vessels; *aliform*, vasicentric with wing-like tangential extensions; and *confluent*, coalesced aliform forming irregular tangential or diagonal bands. (Based on Jane 1970, p. 117)

3. Structure of the vascular rays   Whereas the spatial relations of the components of the axial system are determined from the transverse section, the essential features of the radial (ray) system are deduced from the appearance which the cells of this system present in R.L.S. and T.L.S.

The form of the parenchyma cells that compose a ray is determined from the R.L.S. According to the direction in which the longest axis of these cells points in a R.L.S., they can be classified into two main types (Fig. 6.11). A ray parenchyma cell which is elongated radially is *procumbent*, one which is elongated vertically is *upright*. If all the cells in a ray are procumbent the ray is said to be *homogeneous*, whereas if both procumbent and upright cells are present it is *heterogeneous*. The width and height of a ray are determined from the T.L.S., because this section cuts the ray end-on. When a ray is only one cell wide it is said to be a *uniseriate* ray, when two cells wide *biseriate*, and when more than two cells wide *multiseriate*. A multiseriate ray narrows towards its upper and lower edges where it is usually uniseriate. The ray system of a plant may consist of only one type of ray (either all homogeneous or all heterogeneous) or of a combination of the two types of rays.

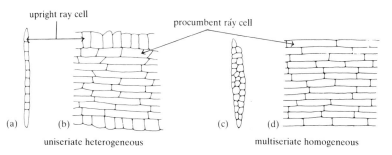

Fig. 6.11 Two types of vascular rays as seen in tangential longitudinal (a and c) and radial longitudinal (b and d) sections. (From Esau 1977, p. 111)

Secondary thickening of the stem

In a primary dicotyledonous stem the xylem and phloem lie adjacent to each other within the vascular bundles but, as a result of the formation and subsequent activity of the vascular cambium, the primary xylem becomes enclosed within a complete cylinder of secondary vascular tissues, and the primary phloem is consequently pushed outwards (see Fig. 6.14). As the cylinder of secondary xylem expands in thickness the vascular cambium moves outwards and thus increases its circumference. This increase in circumference is accomplished mainly by division of the fusiform initials to produce more fusiform initials, but it also involves the formation of new ray initials (Fig. 6.12). The cambial divisions that add to the secondary xylem and phloem are tangential, but those that increase the circumference of the cambium itself are either radial longitudinal or radial oblique. In the latter case the daughter cells elongate by apical intrusive growth until they eventually lie side by side, at least for part of their length.

Fig. 6.12 Increase in circumference of vascular cambium by the formation of new fusiform and ray initials. (a) Radial longitudinal division of a fusiform initial to give two daughter initials lying side by side. (b) Radial oblique division of fusiform initial followed by apical intrusive growth (growing apices are stippled). (c) Formation of a uniseriate column of ray initials by repeated transverse division of a fusiform initial.

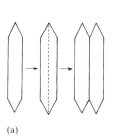

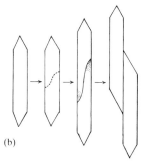

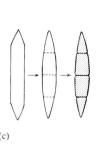

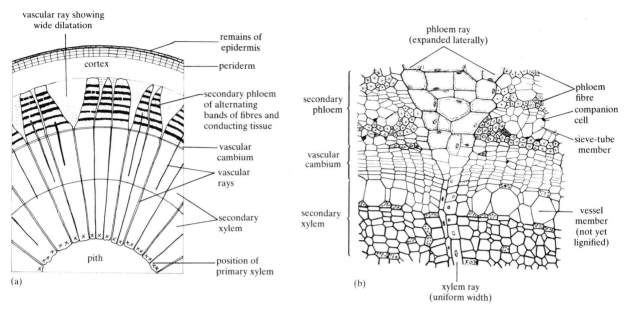

(a)

(b)

Fig. 6.13 Dilatation of vascular rays. (a) L.P. diagram of sector of 2–year-old stem of *Hibiscus* showing dilatation of phloem rays and concentric zonation of tissues in the secondary phloem. (b) H.P. drawing (of *Tilia europaea* which is similar in its stem anatomy to *Hibiscus*) showing radial division and tangential expansion of ray cells in the region of the secondary phloem.

If the growth layers in the xylem near the pith are compared with those further out, the ratio between the number of cells in the rays and of those in the axial system is more or less constant. Since there has been an increase in the total number of cambial cells, this constant ratio can only result from the formation of new ray initials in the vascular cambium. The new ray initials are derived from fusiform initials by a process which, although complicated, involves transverse subdivision of a fusiform initial into a vertical file of cells, which then divide radially if the ray is biseriate or multiseriate. Since new rays are continually being produced as the core of secondary xylem increases in girth, it follows that rays extend into the secondary xylem and phloem for distances which vary according to their age (see Fig. 6.14d, e). In the xylem a ray remains more or less uniform in thickness but in the older parts of the phloem where the stem is increasing in circumference, it usually widens out by further radial division of its cells and their expansion tangentially. This response of the phloem rays to the increasing girth of the stem is known as *dilatation* (Fig. 6.13).

The essential feature of cambial activity is the insertion of a continually enlarging cylinder of secondary tissues (chiefly secondary xylem) between two regions of primary tissue that were originally adjacent to one another, and this has the inevitable result of increasing the thickness of the stem (Fig. 6.14). The effect of such an increase is different on the two sides of the vascular cambium. Inside the vascular cambium the secondary xylem and all the tissues internal to it remain more or less intact because the vascular cambium continues to displace itself further outwards. Outside the vascular cambium, however, the secondary phloem and all the tissues external to it are subjected to progressively greater stresses. The thin-walled cells of the primary phloem are crushed and become non-functional, but the cortical cells may keep pace for a time with the expansion by cell enlargement and division in the radial direction. The epidermal cells, usually being unable to divide, are soon stretched to their limit, rupture and die. Before this happens, however, the epidermis is replaced by a protective tissue of secondary origin

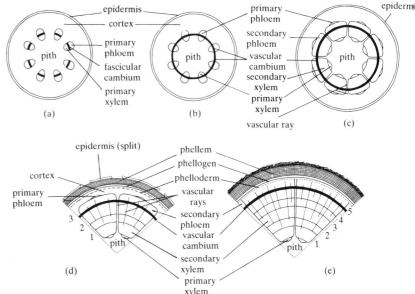

Fig. 6.14 Secondary thickening in a dicotyledonous stem. (a) Primary condition. (b) Transitional stage in which the vascular cambium has assumed the form of a complete cylinder but has not yet become active. (c) Stem after 1 year of secondary growth. The first-formed vascular rays extend from the pith to the cortex. (d) Part of a 3-year-old stem. The vascular rays are of various lengths according to how recently they have been formed; the epidermis is here shown as still present, but in fact may have been shed long before this stage is reached. (e) Part of a 5-year-old stem. The current phellogen has developed within the secondary phloem, and all primary tissues external to the original vascular cambium, i.e. primary phloem, cortex and epidermis, have been shed.

which is produced by another lateral meristem, the *phellogen* or cork cambium.

Phellogen and the formation of periderm

The phellogen cuts off new cells in much the same way as the vascular cambium, forming *phellem* or cork towards the outside of the stem and *phelloderm* or secondary cortex towards the inside. Unlike the vascular cambium, however, the phellogen gives rise to many more derivatives on its outer than on its inner side. These three zones of secondary tissue (phellem, phellogen and phelloderm) are collectively called the *periderm*.

The phellogen always originates near the surface of the stem, but the position varies in different plants. In most plants it forms in the outermost layer of the cortex, but it may arise anywhere between the epidermis and the outer region of the phloem. Because the phellogen develops in a region of mature primary tissues, the layer of parenchymatous or collenchymatous cells in which it is initiated must first dedifferentiate and revert to the meristematic condition. This is accomplished by two tangential divisions resulting in the formation of a radial file of three cells, of which the middle cell is the phellogen initial that continues to divide (Fig. 6.15). The cells of the phellogen, like those of the vascular cambium, are highly vacuolate but, unlike those of the vascular cambium, are all of one kind. The phellogen is therefore relatively simple in structure. In cross-section it appears as a continuous ring of rectangular, tangentially flattened cells which can be seen in T.L.S. to be polygonal in outline. The phellogen is often called the cork cambium because the phellem from certain trees, notably *Quercus suber* (the cork oak), is the source of commercial cork.

The cells of the phellem or cork are characterized by having suberized secondary walls. Suberin is a fatty substance which renders the walls almost completely impermeable to water and gases, with the result that phellem cells are dead at maturity. Brown pigments commonly occur in both the walls and the cell cavities. Phellem is a very compact tissue with no intercellular spaces;

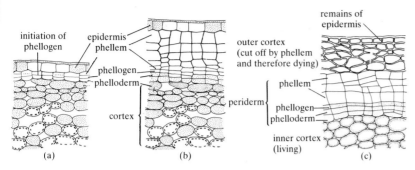

Fig. 6.15 Formation of phellogen and periderm in dicotyledonous stems. (a) Origin of phellogen in the outermost layer of the cortex. (b) Same segment of stem as in (a) but phellogen has now cut off about six phellem cells on its outer side and one or two phelloderm cells on its inner side. (c) Phellogen formed in cortex several layers below epidermis (e.g. in *Triplochiton* and *Ixora*). Note the clearly defined boundary between the primary tissues of the inner and outer cortex and the secondary tissues of the periderm.

it forms an insulating layer near the plant surface which replaces the epidermis when the latter is ruptured. With the formation of even a thin layer of phellem, the tissues external to it become cut off from their supplies of food and water and consequently die. The inner product of the phellogen, the phelloderm, consists of living cells with non-suberized walls. These cells resemble the adjacent cortical parenchyma but can usually be distinguished from the latter because, being formed from the phellogen, they are arranged in regular rows with the same radial alignment as the phellogen and phellem. In some plants the phelloderm may develop collenchymatous thickenings or become sclereids.

Lenticels

With the insertion of an impermeable layer of phellem near the surface of the stem, gaseous exchange between the living internal tissues and the air would become impossible if it were not for the development of perforations, called *lenticels* (Fig. 6.16). A lenticel is a localized area of periderm where

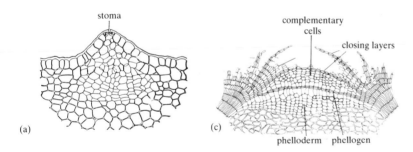

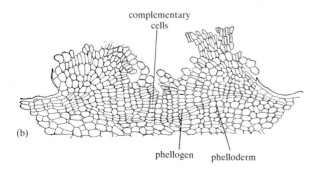

Fig. 6.16 Lenticels. (a) Early stage of lenticel formation in mulberry (*Morus alba*); the phellogen has just arisen under a stoma and the development of the first complementary cells has caused the epidermis to bulge outwards. (b) Mature lenticel of mango (*Mangifera indica*) with complementary cells only. (c) Mature lenticel of *Prunus avium* with closing layers. (a and b from Brown 1935, pp. 169–170; c from Fahn 1974, p. 405)

the phellogen is more active than elsewhere and, instead of forming typical phellem cells, produces a loose, powdery tissue (the *complementary* or *filling* tissue) consisting of non-suberized, rounded cells with numerous air spaces between them. In the regions where lenticels develop the cells of the phellogen have slightly rounded corners, so that there is gaseous continuity right through the periderm; elsewhere the phellogen cells fit tightly together with no intercellular spaces between them. In stems the first lenticels to form normally develop directly beneath the stomata of the epidermis, and active production of complementary cells in these positions forces the overlying cells outwards and ruptures the epidermis. Eventually the epidermis is thrust back, and the mass of complementary cells becomes freely exposed. Later-formed lenticels often arise in positions which coincide with the outer ends of the vascular rays, thereby facilitating aeration of the living cells of the secondary vascular tissue. In many woody stems lenticels appear as small raised dots on the surface, but in others they may form conspicuous, transversely elongated slits up to 3 cm long.

Two types of lenticel are distinguished. In one type the complementary tissue, although permeated by large intercellular spaces, remains more or less intact, and cells of the same type are produced throughout the growing season (e.g. *Persea americana*, avocado pear). In the other, several layers of loose, powdery, complementary cells alternate with thin layers (closing layers) of compact suberized cells with very small intercellular spaces (e.g. *Prunus avium*, the wild cherry of Europe and western Asia). The closing layers prevent the loose complementary cells from dropping out of the lenticel. The phellogen forms several bands of each type of tissue during a single growing season. The closing layers are continually being broken by the outward pressure resulting from the production of new complementary cells, but at least one closing layer always remains intact.

Bark

As the stem increases in thickness, the periderm must either keep pace with the increase in diameter by radial longitudinal division of the living cells of the phellogen and phelloderm, or rupture. If the periderm is not ruptured, the original phellogen may persist for the entire life of the tree, and when this happens the periderm is usually very thin and the surface of the trunk remains smooth. If, as in most trees, the periderm ruptures, a new phellogen is formed further inside the stem, usually in a deeper layer of the cortex. This process is repeated as the tree increases in diameter and successive phellogens arise deeper and deeper in the cortex (Fig. 6.17). Since the cortex is a primary tissue and therefore not replaced, phellogen layers are eventually formed in the secondary phloem (including the phloem rays) and in most mature trees it is in this layer that the functional phellogen occurs.

The tissues on the outside of older stems are loosely referred to as bark. This is a non-technical term but nevertheless a useful one if properly defined. Bark may be applied most appropriately to all tissues outside the vascular cambium. Owing to the death of cells isolated outside the most recently formed periderm, a distinction can be made between an outer dead bark (or *rhytidome*) and an inner living bark. In older trees the outer bark is a mass of dead tissue in which layers of crushed secondary phloem alternate with the layers of periderm, any remains of the original primary tissues having been shed. The outer bark, being composed of dead tissue, cannot expand tangentially as the trunk increases in girth, and its outermost layers

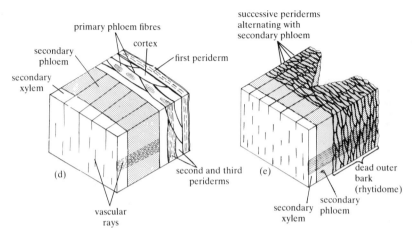

Fig. 6.17 Stages in the development of bark.
(a) The first periderm forms as a complete cylinder beneath the epidermis. (b) Additional periderms are formed deeper in the cortex. (c) Periderms are being laid down within the secondary phloem; some of the earliest formed periderms have already been shed. (d) and (e) Three-dimensional diagrams showing earlier and later stages in the development of bark.

are usually shed by being torn along the planes of weakness between successive periderms. The inner bark of older trees consists of any phelloderm produced towards the inside of the innermost phellogen, and of the currently functional secondary phloem produced towards the outside of the vascular cambium. The inner bark remains more or less constant in thickness because, although it is eroded on the outside by the formation of successive phellogens, it is renewed on the inside by the activity of the vascular cambium.

In a few trees each new phellogen is laid down as a continuous or almost continuous ring around the outside of the trunk, and trees (e.g. *Bursera simaruba*, West Indian birch) which shed their bark as large papery sheets clearly come into this category; this type of bark is called *ring bark*. In most trees, however, the successive phellogens arise in the form of curved, overlapping patches, with the result that the outer bark of many of them is shed as scale-like flakes of tissue; a common tree producing this type of bark, appropriately called *scaly bark*, is the guava (*Psidium guajava*). The bark of some trees (notably those species of *Eucalyptus* which are called 'stringy barks') is intermediate between these two types, the bark peeling off in long narrow strips.

Although the bark type characteristic of different trees is partly due to the way in which successive layers of phellogen are laid down, it also depends on how rapidly the outer bark is produced and how rapidly it is destroyed on its outer surface. In some trees the bark is worn away by a process of gradual disintegration through weathering. In other trees, where layers of thin-walled secondary phloem provide planes of weakness between the periderms, the bark is shed in sheets (ring barks) or scales (scaly barks). When there is a considerable accumulation of outer bark before it is destroyed, the increasing girth of the trunk causes the development of furrows which are such a characteristic feature of the bark of most trees.

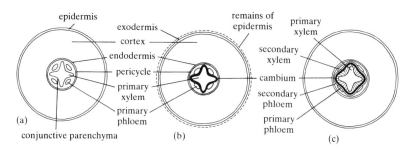

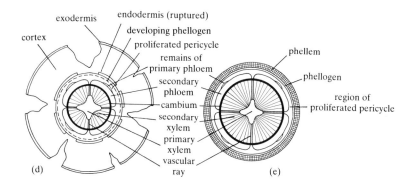

Fig. 6.18 Secondary growth in a tetrarch root. (a) Primary condition. (b) Initiation of vascular cambium. (c) Vascular cambium is active, especially between the groups of primary xylem. (d) Vascular cambium (now circular in cross-section) is uniformly active around its circumference; the pericycle is proliferating, and the endodermis and all tissues external to it have split and shortly will be shed. (e) Secondarily thickened root with an outer layer of periderm.

Secondary growth in roots

The roots of most woody dicotyledons increase in thickness by secondary growth and, as in stems, this results from the development and subsequent activity of two lateral meristems, the vascular cambium and phellogen. When fully formed these meristems are structurally and functionally similar to their counterparts in the stem but, because the arrangement of primary tissues differs in the two organs, the manner in which they arise in the root is different from that in the stem. In roots the vascular cambium is initiated between the phloem strands and the central core of xylem (Fig. 6.18), where bands of conjunctive parenchyma resume meristematic activity and cut off files of cells (i.e. secondary xylem cells towards the inside and secondary phloem cells towards the outside). Resumption of meristematic activity then spreads sideways from these first-formed cambial strips towards the protoxylem poles, until the two edges of each strip meet the pericycle. The isolated strips of cambium are then joined together into one continuous layer when the pericycle cells outside the protoxylem poles also become meristematic. As a result of being formed in this way the vascular cambium is at first convoluted (with the same shape as the central core of primary xylem) but, because secondary xylem forms and becomes active first opposite the primary phloem strands, it soon assumes a circular outline. The fully formed vascular cambium does not differ fundamentally from that of the stem; it divides tangentially (periclinally) to produce secondary xylem inwards and secondary phloem outwards, and radially (anticlinally) to increase its circumference. It is worth noting, however, that in many roots the sectors of the vascular cambium which originate in the pericycle consist entirely of ray initials, with the result that wide vascular rays radiate outwards from the protoxylem poles. Rays also develop at other points on the vascular cambium but they are usually less prominent than the rays opposite the protoxylem poles. As secondary growth proceeds and the cylinder of secondary vascular tissues increases in thickness, the primary phloem is soon crushed. For a limited time, however, the crushed strands of primary phloem can usually be identified by their position on the radial planes midway between the wedges of primary xylem. The primary xylem, being on the inside of the enlarging cylinder of secondary vascular tissues, remains intact and can usually be identified even in old roots.

The formation of the phellogen follows that of the vascular cambium. After the vascular cambium has become a complete cylinder by division of the pericycle cells opposite the primary xylem wedges, a wave of cell division spreads around the circumference of the pericycle so that a cylinder of meristematic tissue is formed. In the outer layers of this proliferated pericycle a phellogen develops; this phellogen, like that of the stem, divides periclinally to produce a three-layered periderm. The formation of the periderm results in the death of the endodermis, cortex and epidermis which are sloughed off. The overall effect of secondary growth in the root is that older roots come to resemble older stems in their anatomy. Apart from the arrangement of the primary xylem (which is exarch in roots, but endarch in stems) it is often difficult to distinguish an older root from an older stem.

Secondary growth in monocotyledons

In most monocotyledons no secondary growth occurs and the entire plant body consists of primary tissues. However, the stems of a few woody monocotyledons (notably species of *Agave, Aloe, Dracaena, Pandanus* and *Yucca*) undergo secondary thickening but, because the vascular bundles of monocotyledonous stems lack a vascular cambium, this is brought about by the formation of a special type of cambium which behaves differently from the vascular cambium characteristic of dicotyledonous stems. The monocotyledonous cambium arises in the ground tissue external to the region of primary vascular bundles, and then divides periclinally to give radially arranged files of cells (Fig. 6.19). Nearly all the secondary tissue so formed lies towards the inside of the cambium, the small amount of tissue produced on the outside developing into parenchyma. The internal product of the cambium consists of a parenchymatous ground tissue (*conjunctive tissue*) in which vascular bundles, similar in structure to the primary vascular bundles, differentiate. The junction between the primary and secondary tissues is, however, clearly visible because, whereas the parenchyma of the primary ground tissue has no definite arrangement, the parenchyma of the secondary conjunctive tissue (unlike the cells of the secondary vascular bundles) is arranged in regular radial rows. The essential feature of secondary thickening in monocotyledons is therefore the formation of a cylinder of new secondary vascular bundles (each composed of xylem and phloem) embedded in a mass of parenchymatous ground tissue. The essential feature of secondary thickening in dicotyledons, by contrast, is the insertion of a cylinder of secondary vascular tissues (secondary xylem on the inside and secondary phloem on the outside) between the xylem and phloem of the primary vascular bundles.

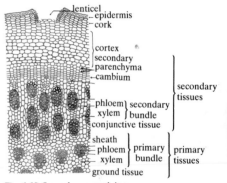

Fig. 6.19 Secondary growth in a monocotyledonous stem (*Dracaena*).(From Dutta 1964, p. 225)

Chapter 7

The gross structure of timber

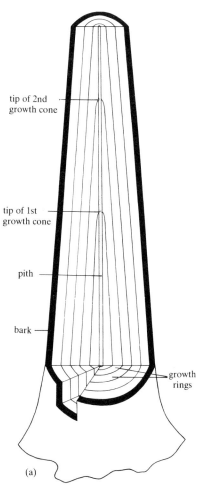

tip of 2nd growth cone

tip of 1st growth cone

pith

bark

growth rings

(a)

Fig. 7.1 Growth cones. (a) Trunk of a young tree showing five growth cones in longitudinal section and cross-section. (b) (*Facing page*.) Wedge-shaped block, taken from position indicated in (a). showing the general layout of the main structural components of wood. (Redrawn from E. W. J. Phillips (1948). The biology and properties of wood. *New Biology* **4**, 76, Penguin Books, England)

The macroscopic features of a piece of wood are best understood with reference to the growing tree. A convenient starting point is the seedling tree towards the end of its first year of growth. At this stage it consists, above ground, of a main stem which ends in a terminal bud and bears along its length a number of lateral buds in the axils of the leaves. During the course of its first year's growth the stem will have increased in length due to the activity of the apical meristem and in girth due to the activity of the vascular cambium; a layer of phellem will probably have been formed just beneath the surface due to the development and activity of a phellogen. The stem is thickest at its base where the vascular cambium developed first and has formed more wood than nearer the apex. It will thus be apparent that the wood of the stem is really in the form of a very elongated cone, open at its apex where it joins the apical meristem, and hollow in the centre because of the presence of the pith.

In most trees the activity of the apical meristem and vascular cambium is periodic, depending on the alternation of hot and cold seasons (in temperate regions) or of wet and dry seasons (in tropical regions). The cessation of growth after a period of active growth is usually indicated by the development in the wood of a distinct growth line. In the second growing season increase in length and girth is continued and a new cone of wood is laid down on top of the first year's cone. This is repeated each successive season so that, if we assume there is one growing season per year, at the end of 5 years there will be five cones, one inside the other, with the pith running up through the middle of them. At this stage the tree will have the structure shown in Fig. 7.1. This diagram is oversimplified because no branches are shown, but the formation of branches will be considered later in the chapter. Although this cone-in-cone arrangement is always present and some trees, particularly when young, are tapered, the trunk in most older trees is practically columnar because the cones are so long and narrow (Fig. 7.2).

If the trunk of a young tree is cut across, the cones will appear as a series of concentric circles provided the growth lines are sufficiently distinct to indicate the limits of each cone. Usually the cones are distinct because the first-formed wood of each cone differs in some way from that which is formed later in the season. These rings are best called *growth rings*, especially in tropical trees where the alternative but less accurate term annual ring may not apply. Where the growth rings are clearly defined the first-formed wood in a ring is called *early wood*, the wood formed later in the ring being known as *late wood*. These terms are preferable to the more common ones, spring wood and summer wood, because they do not imply that the wood is formed at a particular season.

Growth rings can also be seen on the longitudinal surfaces of wood (Fig. 7.3). If the longitudinal cut is a radial one passing through the pith in the centre, the growth rings will appear as a series of more or less parallel lines. If the longitudinal cut is tangential, however, the rings will show as a series of /\s one inside the other.

Formation of growth rings

The reason why the growth rings of some trees show up as distinct lines varies according to the species of tree. In the conifers, which produce the so-called *softwoods* composed largely of tracheids, the tracheids of the early wood are wider and have much thinner walls than those of the late wood, and this marked difference in cell size and wall thickness results in a clearly defined

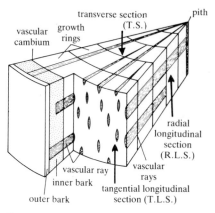

Figure 7.1(b)

Fig. 7.2 Trees with different shapes of trunk. (a) The conical shape of this young baobab tree (*Adansonia digitata*) reflects very clearly the basic cone-in-cone structure of all tree trunks. (b) Mature trees of *Eucalyptus* sp. with columnar trunks, which contrast strikingly with the baobab in (a). (Both photographs original)

growth ring. Among angiosperms, which produce the *hardwoods* containing xylem vessels, there are two different types of growth rings. In ring-porous woods the rings are distinct because the vessels of the early wood are much larger and more numerous than those of the late wood which consists mainly of fibres. The ring-porous condition appears to be highly specialized and is more or less confined to a few genera of north-temperate trees (e.g. *Castanea*, *Quercus* and *Ulmus*). In diffuse-porous woods, i.e. those in which the vessels are more or less evenly scattered throughout the season's growth, growth rings are not always clearly defined but, when they are present, they are due to a thickening and perhaps flattening of the fibres or parenchyma at the end of the growing season. Between these two extreme types of growth ring no sharp line can be drawn and many woods are of an intermediate type. Figure 7.4 shows photomicrographs of the main types of growth rings.

It is a common idea that the age of a tree can be determined by counting the number of growth rings. This is true for north-temperate trees, which grow only during the favourable season and remain dormant for the rest of the year, but information about the growth of tropical trees is scanty and in only a few species is it known whether or not growth rings, if present, are

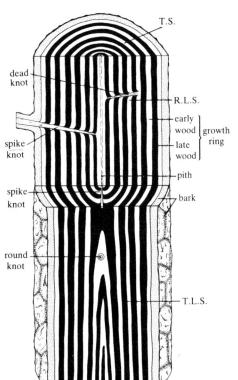

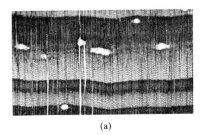

(a)

Fig. 7.3 Diagram of a tree trunk cut in different planes. T.S., transverse surface, or end grain. R.L.S., radial longitudinal surface, or quarter sawn. T.L.S., tangential longitudinal surface, or plain sawn. (Redrawn from Jane 1970, p. 71)

Fig. 7.4 Transverse sections showing the structure of growth rings in (a) slash pine (*Pinus caribaea*), a softwood, (b) teak (*Tectona grandis*), a ring-porous hardwood, and (c) West Indian mahogany (*Swietenia mahagoni*), a diffuse-porous hardwood. (Crown copyright, and reproduced by permission of the Controller of Her Majesty's Stationery Office and the Director of the Royal Botanic Gardens, Kew, England)

seasonal. Even if the age of a tree can be assessed by counting the growth rings, certain precautions must be observed. The first precaution, so obvious that it is easily overlooked, is that the ring count must be made sufficiently low down the trunk to include the first year's cone, i.e. below the height which the seedling tree reached in its first year of growth. Secondly, it must be borne in mind that two growth rings may be formed in one season. It sometimes happens that drought or premature defoliation (which might be caused, for example, by a plague of caterpillars) causes the cambium to become inactive early in the growing season. When cambial growth is resumed, the interruption in growth is registered by a change in the structure of the wood which may be manifest as a growth ring, called a *false ring* (Fig. 7.5). These false rings are less distinct than true growth rings, but might easily be included in a ring count by an inexperienced observer. The third precaution to note is that, especially in old trees, an area of the cambium may cease to function for one or more years, and until it becomes active again no new wood is produced in this area. The growth rings laid down by the neighbouring cambium are discontinuous but merge into an older ring in the inactive region. Should a ring count be made through such a region of discontinuous rings it will give an estimated age which is less than the real age.

The width of growth rings varies enormously. At one extreme, young fast-growing conifers sometimes have growth rings as much as 1 cm wide, while at the other extreme there is on record a case of a naturally dwarfed *Picea sitchensis* (Sitka spruce) which at 98 years old had a trunk only 2 cm in diameter, i.e. having about 100 growth rings to the centimetre. Ring width is determined largely by the rate of growth, the widest rings being formed under conditions of optimum growth. Growth rate is partly an inherited character and partly influenced by factors in the immediate environment of the tree such as shading, water, temperature and soil. Some trees like *Guaiacum officinale* (lignum vitae) grow slowly, while others such as *Ochroma lagopus* (balsa) grow extremely rapidly. For a particular species of tree a sudden increase in the width of the growth rings often indicates that neighbouring trees have fallen or been removed so that, with the disappearance of competition, the tree has been able to grow much faster.

Branch formation and knots

Some of the lateral buds on the seedling shoot may develop into branches, which increase in length and girth in the same way as the main trunk. Where branches develop, new wood is added as a continuous layer over trunk and branches, and consequently each growth cone of the trunk becomes rather like a multi-armed overcoat or, more accurately since it lies beneath the bark, like a multi-armed undergarment (Fig. 7.6). As the trunk increases in thickness, the bases of any branches growing from it gradually become embedded in the trunk and form the *knots* seen when the tree is sawn into timber. If a branch dies, as it commonly does when shaded, its more distal parts will probably become brittle and eventually break off, but the proximal stub will gradually be buried in the ever-widening trunk. In a dead branch there is, of course, no cambial activity and thus the wood of a knot formed from a dead branch is not continuous with the wood of the trunk. Such a knot is loose and may even fall out of a plank; it is termed a *dead* or *loose* knot. A knot formed from a living branch is tightly fixed in the wood of a plank; it forms a *live* or *tight* knot. After a dead knot has become embedded, the trunk subsequently produces clean timber free from knots. Thus clean timber is formed in the outer parts of a large trunk, and knotty timber is found in the centre.

Since knots are formed from lateral branches, they cross the trunk in a radial direction, either horizontally or somewhat obliquely downwards or upwards according to the slope of the branches from which they were formed. Thus a radial longitudinal cut will cut through the knots along their length; such knots are known as spike knots. In a tangential longitudinal cut the knots are cut either transversely or somewhat obliquely and form round knots.

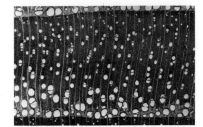

Fig. 7.4b

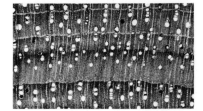

Fig. 7.4c

Heartwood and sapwood

The older wood of most trees, forming the central portion of the trunk and large branches, gradually loses its ability to conduct sap and frequently undergoes a marked change in its properties. The xylem parenchyma and xylem rays die, and the whole central region may become hardened due to the deposition of such substances as tannins and resins. The result is that usually the central wood becomes darker and clearly marked off from that nearer the cambium. The two zones are then distinguished as *heartwood* and *sapwood* respectively (Fig. 7.7). It is heartwood which, because of its dryness and hardness, provides the timber of commerce. In some trees, called sapwood trees, the entire wood is of a uniform light colour. The more central wood of sapwood trees is nevertheless physiologically, although not visibly, heartwood. The extent of the sapwood is fairly constant for any particular species, being narrow in some and wide in others. In all trees, however, the diameter of the heartwood increases as new layers of sapwood are progressively formed on the outside.

Although heartwood and sapwood are so different in their properties, the only histological difference between them is that in the transformation from sapwood to heartwood the xylem vessels may become blocked by numerous bladder-like ingrowths or *tyloses* (Fig. 7.8). During heartwood formation differences in pressure may arise between a vessel and the adjacent parenchyma cells, and the pit membrane may be forced into the cavity of the vessel to form a small bladder. As a result of this occurring from many pits, the vessel becomes completely blocked with numerous tyloses.

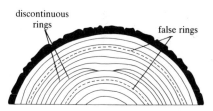

Fig. 7.5 False and discontinuous growth rings as seen on the transverse surface of a log.

Figure in wood

The term figure is applied to any feature which relieves the uniform appearance of a piece of timber. Two structural features, growth rings and vascular rays, play an important part in the production of figure. Since the figure caused by these features depends on the way in which the log is sawn, it will be convenient to mention the methods of converting logs into planks or boards. The least wasteful method is by a 'through-and-through' cut, which means that a series of parallel cuts is made through the entire log (Fig. 7.9). Apart from the few boards cut from the central region of the log, this method gives boards which are cut tangentially or nearly so. Such boards are technically known as *plain-sawn*. For some purposes, however, it is desirable that the log should be cut radially, or *quarter-sawn*. Methods for obtaining quarter-sawn boards are also indicated in Fig. 7.9.

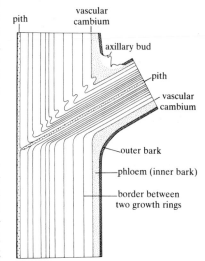

Fig. 7.6 The relationship between the secondary tissues of a branch and those of the main stem. The base of the branch is buried progressively deeper as the main stem increases in thickness. (Adapted from Eames and MacDaniels 1947, p. 198)

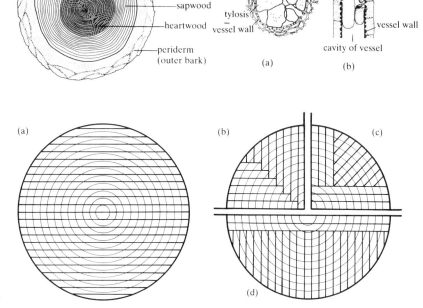

Fig. 7.7 (*left*) Diagrammatic T.S. woody stem showing the major regions in the wood and bark. (From Morey 1973, p. 11)

Fig. 7.8 (*right*) Tyloses. (a) T.S. of xylem vessel filled with tyloses. (b) L.S. of vessel showing continuity between lumina of tyloses and neighbouring parenchyma cells. Note stages in the development of tyloses. (From Fahn 1974, p. 348)

Fig. 7.9 Methods of sawing logs into timber. (a) Through-and-through cut giving mostly plain-sawn boards. (b), (c) and (d). Three methods for obtaining quarter-sawn boards; method d, although giving a smaller proportion of quarter-sawn boards than methods b and c, is the most practical method.

Growth rings form a prominent feature on the longitudinal surfaces of wood wherever the early-wood and late-wood are visibly distinct, as in many softwoods. Figure due to distinct growth rings is most attractive on plain-sawn boards where the rings appear as a series of $\wedge$s or $\cap$s; on quarter-sawn faces the rings appear as more or less parallel lines (Fig. 7.10a, b). Another type of figure is associated with irregularities in the growth rings. In some trees the growth cones, because of localized areas of greater or lesser cambial activity, have bulges and depressions on their surface, so that on a plain-sawn surface the bulges appear as more or less regular ellipses or circles. This *blister figure* (Fig. 7.10c) is sometimes well shown on plywood made from certain north-temperate softwoods, especially *Pseudotsuga taxifolia* (Douglas fir) and some *Pinus* spp. (e.g. pitch pine).

Some woods have large rays (up to 3 cm high) which are responsible for a very characteristic figure. The woods of *Grevillea robusta* (silky oak) and of most species of *Casuarina* provide examples. Since the rays run in a radial direction, figure caused by rays is conspicuous only on radially cut surfaces where the rays are cut lengthwise and appear as silvery streaks and flecks (Fig. 7.10d). For rays to produce a distinctive figure, however, they must not only be large but also differ markedly in colour from the rest of the wood.

In the above account it has been possible to mention only the more obvious features of wood. Like most common materials wood is taken very much for granted. Occasionally the 'grain' in a piece of furniture may be admired but it is unusual to consider that there could be anything interesting about the boards in a packing case. Nevertheless, a great deal can often be learnt about the growth of trees from observing almost any piece of wood.

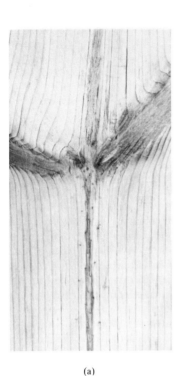

(a)

(b)

(c)

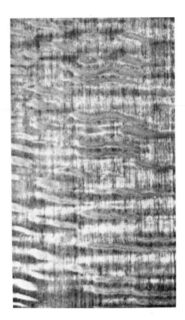

(d)

Fig. 7.10 Figure in wood. (a) and (b) Figure in a softwood (*Pinus sylvestris*) due to a sudden difference in colour and texture between the early and late woods of a growth ring. (a) Quarter-sawn plank (R.L.S.) showing growth rings (parallel lines), spike knots, and pith. (b) Plain-sawn plank (T.L.S.) showing growth rings (Λ-shaped in the centre line) and round knot. (c) Plain-sawn (tangential) face of a pitch pine (*Pinus* sp.) showing blister figure caused by irregularities in the surface of the growth cones. (d) Quarter-sawn plank (R.L.S.) of silky oak (?*Grevillea robusta*) showing figure caused by large rays; the fine vertical lines are vessels. (a and b original ; c and d from Jane 1970, pp. 243 and 245 respectively)

Chapter 8

The foliage leaf

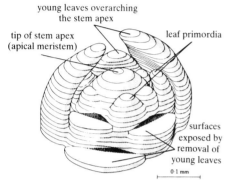

young leaves overarching the stem apex

tip of stem apex (apical meristem)

leaf primordia

surfaces exposed by removal of young leaves

0·1 mm

Fig. 8.1 Terminal bud of flax (*Linum*) to show formation of leaves as bulges of tissue on the flanks of the apical meristem at the stem tip. Three young leaves have been cut away to expose the shoot apex. The diagram has been constructed from a series of scale drawings of serial sections cut across a terminal bud; the transverse lines indicate the thickness of each section. (From R. F. Williams (1970) *Australian Journal of Botany*, **18**, 167–73)

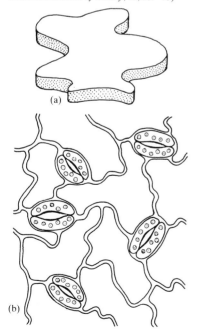

(a)

(b)

The term *leaf*, when applied to a plant, is commonly given to any green, flattened structure which is freely exposed to air and light, but the true distinguishing feature of a leaf is that it originates as a lateral bulge of tissue on the apical meristem of a bud, and thus forms an integral part of the shoot system (Fig. 8.1). Certain organs that are very different in appearance from ordinary green leaves are formed in this manner and are in fact modified leaves. Examples of these are the sharp spines of cacti, the membranous scales on the underground rhizome of *Canna*, and the fleshy scales of an onion bulb. The sepals, petals, stamens and carpels of a flower are also considered by some to be modified leaves on a compact modified stem. The leaf, in the botanical sense of the term, is thus a very variable organ, but only the normal foliage leaf is considered here.

Structure of the foliage leaf

Like the root and stem, the leaf consists of the usual three tissue systems, but these are modified to permit the gaseous exchanges of photosynthesis, which is the primary function of the leaf. Basically the leaf blade or lamina consists of a protective epidermis and an intervening region of ground tissue known as the mesophyll (literally, the middle leaf), which is permeated by the vascular tissue. The mesophyll is also permeated by an elaborate system of intercellular air spaces. Unlike both root and stem, however, the leaf is an organ of limited growth (i.e. it grows to a certain size and then stops).

Epidermis

The epidermis covers the upper and lower surfaces of the leaf and is continuous with the epidermis of the stem. It usually consists of a single layer of cells, although in some leaves, particularly evergreen leaves, it may become two to several cells thick by periclinal divisions in the original surface layer. In such a multiple epidermis only the outermost layer assumes epidermal characteristics, the underlying cells being thought to function as a water-storage tissue. The epidermis consists basically of flattened cells which fit tightly together along their edges so as to form a continuous skin over the leaf. In many leaves these cells have a wavy outline, because the anticlinal walls (i.e. those at right angles to the surface) go on extending after cell enlargement has stopped (Fig. 8.2). When seen in surface view epidermal cells vary considerably in shape, although in most monocotyledonous leaves they are elongated in the direction of the long axis of the leaf.

Epidermal cells are directly exposed to the air, and a fatty substance, cutin, impregnates their outer walls and forms a separate non-cellular layer, the cuticle, over the exposed surface. The presence of cutin plays an important role in restricting evaporation of water (transpiration) from the leaf. Because the cuticle is inelastic and the epidermal cells fit tightly together, the epidermis helps to maintain the rigidity of the leaf by offering resistance to the swelling of the underlying mesophyll cells.

Stomata and stomatal movement

At frequent intervals the epidermis is pierced by small openings, called stomatal pores, which are concerned with gaseous exchange between the atmosphere and the system of intercellular spaces in the underlying mesophyll tissue. Below the stomatal pores there are large intercellular spaces in the mesophyll which are called *substomatal cavities*. Each stomatal pore is surrounded by two specialized epidermal cells called guard cells which, unlike the other cells of the epidermis, contain chloroplasts. The aperture of the pore can be opened or closed by changes in shape of the guard cells. It is common practice to apply the term stoma to the entire unit, the pore and the two guard cells. In some plants the guard cells are surrounded by ordinary epidermal cells, in others the guard cells are flanked or surrounded by so-called *subsidiary cells*, which differ in shape from the other epidermal cells (Fig. 8.3). The number of stomata may range from a few thousand per square centimetre of leaf surface in some species to over a hundred thousand per square centimetre in others. Stomata are commonly much more numerous in the lower epidermis than in the upper, and in many species of plant do not even occur in the upper epidermis. In the more or less broad leaves of dicotyledons the stomata are distributed apparently at random, whereas in the narrow elongated leaves characteristic of monocotyledons they are arranged in regular rows parallel with the long axis of the leaf.

Stomatal movement

The shape of the guard cells is determined by their state of turgor. When these cells are fully turgid the stomatal pore is fully open, but as they become less turgid the pore closes until, when the wall pressure is zero, the two cells come together and the pore is shut. It is clear that such a mechanism has two basic requirements: (1) a means whereby changes in turgor alter not only the volume of the guard cells but also their shape and hence the size of the stomatal pore; and (2) a means of changing the cell turgor.

The first of these requirements is met by an uneven thickening of the walls of the guard cells. In the commonest type of stoma each guard cell has two thick bands of secondary-wall material deposited on the 'inner' wall (i.e. in surface view the one facing the stomatal pore) while the other walls remain thin (Fig. 8.4). Since the thick inner wall is less extensible than the thin 'outer' wall (i.e. the one away from the pore), an increase in turgor of the guard cells relative to that of the surrounding cells causes the outer wall to bulge outwards and the inner wall to become concave. Such a change of shape in two guard cells joined at their ends will cause the pore between them to open. The behaviour of each cell can be compared to the inflation of a balloon to which a wide strip of adhesive tape has been attached along one side. When the guard cells lose turgor the curvature of the inner wall decreases until the pore is eventually closed. Other types of stomata differ

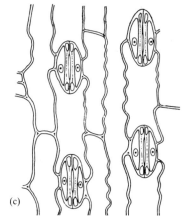

(c)

Fig. 8.2 Epidermis of leaves. (a) Three-dimensional diagram of a single epidermal cell. (b) Surface view of leaf epidermis with ordinary epidermal cells and paired guard cells. The 'jigsaw' pattern of the epidermal cells is typical of most dicotyledonous leaves. (c) Surface view of leaf epidermis of sugar cane (*Saccharum*). The elongation of the epidermal cells in the direction of the long axis of the leaf is typical of many monocotyledonous leaves.

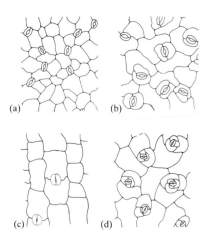

(a)   (b)

(c)   (d)

Fig. 8.3 Surface view of different types of epidermis showing the four main patterns formed by the arrangement of guard cells and subsidiary cells. (a) Irregular-celled pattern (*anomocytic*) having no subsidiary cells, e.g. *Cucurbita*. (b) Parallel-celled pattern (*paracytic*) having two subsidiary cells with their long axes parallel to the guard cells, e.g. *Portulaca*. (c) Cross-celled pattern (*diacytic*) having two subsidiary cells with their common wall at right angles to the guard cells, e.g. many Acanthaceae. (d) Unequal-celled pattern (*anisocytic*) having three subsidiary cells of which one is considerably larger or smaller than the other two, e.g. *Nicotiana*. (a, b and d from Esau 1977, p. 95)

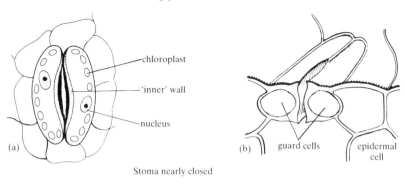

Stoma nearly closed

Stoma fully open

Fig. 8.4 Stomatal movement. (a) and (b) Surface view and three-dimensional diagram of a stoma which is nearly closed. Note that the cell wall bordering the stomatal pore is thicker than that next to the surrounding cells. (c) and (d) Surface view and three-dimensional diagram of a stoma which is fully open. (a and c from Devlin 1975, p. 69)

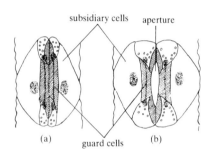

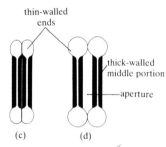

Fig. 8.5 The type of stoma characteristic of grasses. (a) and (b) Surface view of a stoma of sugar cane (*Saccharum*) in closed (a) and open (b) positions. (c) and (d) Diagram of a model illustrating the mechanism whereby the guard cells of a grass stoma are forced apart when their turgor increases. (a and b from Esau 1965, p. 161)

from the 'normal' type just described by the way in which the thickening material is laid down, but the principle on which they work remains essentially the same. In grasses, for example, the guard cells are bone-shaped in outline, with bulbous thin-walled ends and an elongate thick-walled middle portion (Fig. 8.5). When such guard cells take up water, the thin-walled ends increase in volume and so push the inextensible middle portions apart. When they become less turgid the ends decrease in volume and so close the stomatal pore.

The second requirement of the stomatal mechanism, namely a basis for the turgor change, is much harder to account for than the first requirement. In view of (a) the observation that stomata are usually open during the day but closed at night, and (b) the fact that the guard cells are the only cells in the epidermis to contain chloroplasts, it was suggested at the turn of the century that photosynthesis in the guard cells might produce sugar in sufficient concentration to cause osmotic withdrawal of water from the neighbouring cells and consequent opening of the stomatal pore. However, stomatal movement cannot be explained in this way for several reasons. Firstly, the starch/sugar changes which occur in guard cells are exactly the opposite of those which are known to occur in actively photosynthesizing mesophyll cells. In mesophyll cells starch accumulates by day and is converted to sugar by night, whereas in guard cells sugar accumulates by day and is converted to starch at night. Secondly, the opening of stomata in light is not inhibited if plants are kept in carbon-dioxide-free air, i.e. in conditions where net photosynthesis is impossible. Thirdly, and most convincingly, it has been confirmed by the use of radioactively labelled carbon dioxide that guard cells are capable of photosynthesis, but that the rate is much too slow to account for the changes in osmotic potential necessary to cause opening.

If an increase in osmotic potential as a direct consequence of the accumulation of photosynthetic products can be dismissed, what then is responsible for the reversible changes in turgor of the guard cells? The fact that stomata are normally open during the day and closed at night means that they are open when there is a demand by the leaf for carbon dioxide for photosynthesis, and closed when there is no such demand. It was discovered in the 1940s that guard cells are extremely sensitive to small changes in the carbon dioxide concentration of the substomatal air spaces. If the carbon dioxide concentration inside a leaf is experimentally lowered below that of the outside air the stomata will open and, conversely, if it is increased above that of the outside air they will close. Such induced movements occur in both light and darkness, although the magnitude of the response is modified by light intensity. This response of stomata to internal carbon dioxide concentration parallels their response in nature to the alternation of night and day. At night there is a build-up of respiratory carbon dioxide in the substomatal spaces and this would cause the stomata to close, and during the day the reduction in the internal concentration of carbon dioxide by photosynthesis would cause them to open. Although stomata respond to experimentally induced changes in the carbon dioxide concentration in the substomatal spaces, it is presumably the carbon dioxide concentration inside the guard cells which is critical. If this is so, then the presence of chloroplasts in guard cells may be significant because they will fix any intracellular carbon dioxide as soon as the leaf is illuminated and photosynthesis becomes possible.

If stomatal opening depended entirely on the photosynthetic removal of carbon dioxide within the leaf, the action spectrum for stomatal opening should coincide with that for photosynthesis. In fact the discrepancy between the two is considerable, and stomata open much wider in blue light than in red when equal incident light energies are supplied. There is evidence, therefore, that light in the blue region of the spectrum has an effect on stomatal movement which is superimposed on that due to carbon dioxide concentration.

Since carbon dioxide and light are both environmental factors they may be regarded as two external stimuli that trigger off stomatal movement, but how they are involved in changing the turgor of the guard cells is not yet understood. One of the obvious peculiarities of guard cells is that, although they contain chloroplasts, their starch metabolism is out of phase with that of the mesophyll cells. Another peculiarity is the tenacity with which they retain their starch content during periods of prolonged darkness; guard cells can contain abundant starch grains even when the rest of the leaf is suffering from severe carbohydrate starvation. There thus seems little doubt that starch is important in the functioning of guard cells. When it was shown during the first decade of this century that stomatal movement could not be explained in terms of the photosynthetic capacity of guard-cell chloroplasts to produce sugars, the so-called *starch $\rightleftharpoons$ sugar hypothesis* was put forward to fill the gap. According to this hypothesis the aperture of the stomatal pore is controlled by the starch/sugar balance inside the guard cells, and this in turn is regulated by their pH. The condensation of sugar into starch is known to be favoured by pH values of the order of 4–5, whereas the hydrolysis of starch to sugar is favoured by higher pH values. Hence it is suggested that in the dark, when photosynthesis is impossible, respiratory carbon dioxide accumulates in the guard cells and so lowers their pH, the sugar in the guard cells is converted into starch (i.e. the osmotic potential is lowered), water

Table 8.1 Suggested sequence of events involved in stomatal movement, on the basis of the starch $\rightleftharpoons$ sugar hypothesis

	Darkened leaf	Illuminated leaf
Respiratory CO_2 in the intercellular spaces	Accumulates	Used up by photosynthesis
pH of the guard cells	Falls	Rises
Starch $\rightleftharpoons$ sugar balance in the guard cells	Towards starch	Towards sugar
Osmotic potential of the sap in the guard cells	Decreases	Increases
Water content and volume of the guard cells	Decreases	Increases
Width of stomatal aperture	Narrows	Widens

is lost by osmosis to the surrounding epidermal cells, and the stomatal pore closes. In the light the reverse sequence of events is postulated (Table 8.1). Photosynthesis removes any carbon dioxide present, the pH of the guard cells rises, hydrolysis of starch to sugar occurs, water is absorbed by osmosis from the surrounding cells, and the stomatal pore opens. The essential feature of the starch $\rightleftharpoons$ sugar hypothesis is therefore that the pH of the guard cells regulates the starch/sugar balance of an internal store of carbohydrate material. This hypothesis has never been proved wrong, but, equally, direct evidence to support it has never been obtained. Its main failing is that, although in natural conditions there is usually a good positive correlation between the starch content and aperture of the guard cells, in experimental conditions it is possible to induce stomatal movement without any obvious changes in the amount of starch. Such discrepancies led some workers to doubt the validity of the starch $\rightleftharpoons$ sugar hypothesis but it has never been ruled out, if for no other reason than that an accurate method for determining the starch content of guard cells has not yet been devised.

In recent years there has been a renewal of interest in the physiology of stomata and in 1967 a Japanese botanist, Fujino, suggested a new hypothesis called the *active transport hypothesis*. It has been known for 50 years that, when epidermal strips are floated on solutions of known ionic composition, stomatal opening is stimulated by the presence of certain monovalent cations. Relatively little attention was paid to this observation until Fujino produced evidence that active accumulation of potassium ions might cause a turgor increase sufficiently big to account for stomatal opening. Recent work both in Britain and America has confirmed that there is a build-up of potassium ions within guard cells during stomatal opening. Unfortunately current methods for determining the intracellular concentration of cations are not sufficiently accurate to decide whether the uptake could cause a turgor increase large enough to induce stomatal opening. This new hypothesis postulates a possible mechanism of turgor increase, but does not specify how a supply of energy might be made available to the guard cells. Because the starch $\rightleftharpoons$ sugar hypothesis does not satisfactorily explain the correlation between stomatal aperture and starch content of guard cells, the possibility exists that the role of the starch stored within guard cells might be to power the active transport of potassium ions into these cells. Experimental support for this suggestion has been provided by showing that the starch content of guard cells decreases as their potassium content increases, but it is not yet known whether the starch is being used to power the ion uptake or whether it is merely being hydrolysed to sugar. It is possible that *both* active transport of potassium ions *and* hydrolysis of starch contribute to the turgor increase.

The general conclusion to be drawn from the above evidence is that many components are involved in stomatal movement, and no single hypothesis (or combination of hypotheses) has yet succeeded in bringing them together. Indeed it seems certain that no one hypothesis will explain the stomatal behaviour of all plants, because it has recently been shown that the stomata of some plants show an endogenous rhythm of opening and closure, even in constant environmental conditions.

Mesophyll

The mesophyll is primarily a photosynthetic region. It consists of thin-walled parenchyma cells which contain chloroplasts and are separated from one another by intercellular spaces. The mesophyll may be relatively homogeneous but is commonly differentiated into two regions, the palisade and spongy layers. In the leaves of most woody dicotyledons the palisade is located on the upper (adaxial) side of the blade, the spongy layer on the lower (abaxial) side. Such leaves are said to be *dorsiventral* (Fig. 8.6). If the palisade tissue occurs on both sides of the leaf, a condition often found in leaves of plants growing in dry habitats, the leaf is called *isobilateral* (Fig. 8.7). An isobilateral leaf may also result when the palisade is absent or weakly developed, as in many herbaceous dicotyledons and most monocotyledons.

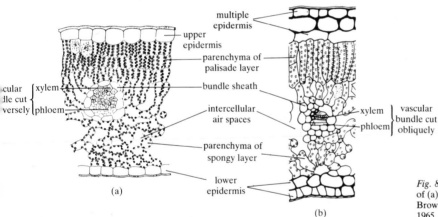

Fig. 8.6 Vertical sections of dorsiventral leaf of (a) *Ixora*, and (b) *Nerium oleander*. (a from Brown 1935, p. 40; b modified from Esau 1965, p. 428)

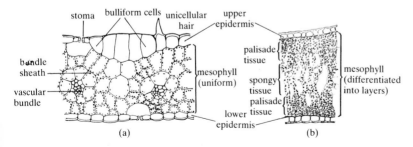

Fig. 8.7 Vertical sections of isobilateral leaves. (a) *Zea mays*, where the mesophyll is a more or less homogeneous tissue. (b) *Eucalyptus*, with a well-defined palisade layer on each surface of the leaf.
The large bulliform cells in the upper epidermis of *Zea mays* leaf occur in parallel longitudinal rows, and changes in their turgor are responsible for the rolling and unrolling of the leaves. (a from Eames & MacDaniels 1947, p. 328; b from Brown 1935, p. 41)

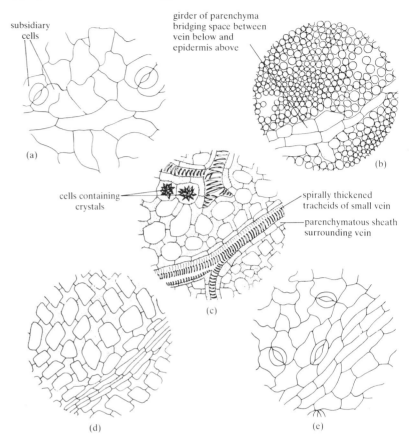

Fig. 8.8 Paradermal sections of leaf of *Ricinus*. (a) Upper epidermis. The cells are somewhat elongated where they overlie the larger veins. (b) Palisade tissue. The palisade cells hardly touch one another, and are absent below the stomata and above the larger veins. (c) Level of the veins, which are of different sizes. The cells between the veins resemble those of the spongy tissue beneath but are organized rather less regularly. (d) Spongy tissue. In *Ricinus* the spongy tissue is unusual in having such a regular structure, due to the dominance of star-shaped cells with six arms, four of which lie in the horizontal plane while the remaining two link with the corresponding arms of the adjacent cells, below and above. (The vertical arms are omitted from the drawing.) (e) Lower epidermis. The stomata are more frequent in this epidermis than in the upper epidermis. (From J. W. Lorch (1962), *School Science Review*, **43**, 434–37)

The palisade layer may be from one to three cells thick, and consists of elongated parenchyma densely packed with chloroplasts. In no other part of the plant are chloroplasts so numerous as in the palisade tissue, which is the main photosynthetic region of the whole plant. The palisade cells are arranged at right angles to the leaf surface and, when viewed in a vertical section of a leaf, resemble the boards of a fence (hence the name palisade). Although in vertical sections the palisade cells appear to be tightly packed, in sections cut parallel to the leaf surface (i.e. paradermal sections) they are seen to be circular and separated by numerous intercellular spaces, rather like sticks in a bundle (Fig. 8.8). These narrow intercellular spaces are in direct communication with those of the spongy tissue.

The spongy layer consists of loosely arranged cells, many of which are irregularly branched and connect with one another at the ends of their branches. As a result of this arrangement the layer has the form of a sponge (hence the name spongy) with an extensive intercellular space system. As already mentioned a particularly large air space is always present beneath each stoma. The volume of air spaces in the spongy mesophyll is often much larger than the volume taken up by the cells. The spongy layer is the main transpiring region of the leaf and its cells usually have relatively fewer chloroplasts than the palisade cells.

The permeation of the mesophyll by an extensive system of intercellular spaces creates an internal surface area many times greater than that between the epidermis and the atmosphere. This large internal surface is one of the

(a)

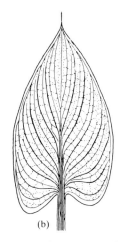
(b)

(c)

Fig. 8.9 Venation patterns of leaves. (a) Reticulate venation of tobacco (*Nicotiana tabacum*) leaf; only the main veins of the network are shown on the left half. (b) Parallel venation of *Zantedeschia* leaf; the cross-connections between the longitudinal veins are shown by dotted lines. (c) Parallel venation of a bamboo (*Arundinaria* sp.) leaf. (a and b from Esau 1965, pp. 433 and 444 respectively)

factors which increases the efficiency of gaseous exchange between the leaf and the atmosphere.

Vascular system

The vascular tissues of the leaf form a complex branching system of strands which ramify in the mesophyll where the palisade and spongy tissues meet. The insertion of the larger strands causes the mesophyll to bulge upwards and especially downwards to form veins. The branching of the main veins gives rise to characteristic patterns (Fig. 8.9). In most dicotyledons the veins form a dense network with the thicker veins branching and becoming progressively thinner as they diverge (*reticulate venation*). The finest veins of all end blindly in vein islets, which are the smallest separate areas into which the leaf is cut up by the veins. In most monocotyledons the veins subdivide at the base of the blade, or along the midrib, and then run more or less parallel towards the apex where they fuse (*parallel venation*). The vein system, whether reticulate or parallel, converges on the petiole rather like the way in which the tributaries of a river are channelled into the main stream. Veins may be regarded as extensions into the leaf of the vascular bundles in the stem, and it is therefore not surprising to find that in leaf bundles the phloem faces the under side of the leaf and the xylem the upper side.

In the larger veins the upper and lower sides of the vascular bundles are covered with a strand of either fibres or collenchyma, which may be so extensive as to reach the upper and lower epidermal layers. When the course of a large vascular bundle is followed towards a blind ending in a vein islet, it is found that the amount of strengthening material diminishes progressively until, when none is left, there is direct contact between the vascular bundles and the mesophyll cells (Fig. 8.10). The mesophyll cells surrounding each vascular bundle usually form a continuous sheath of elongated parenchyma cells which, in dicotyledonous leaves, typically have fewer chloroplasts than the neighbouring mesophyll cells. The bundle sheath is commonly assumed to be important in controlling the transfer of carbohydrates from the photosynthetic cells of the mesophyll to the conducting cells of the vascular bundle. As the blind ending of a vascular bundle is approached, the amount of conducting tissue diminishes until at the extreme end the bundle may consist only of one or a few spiral or reticulate tracheids completely surrounded by the bundle sheath.

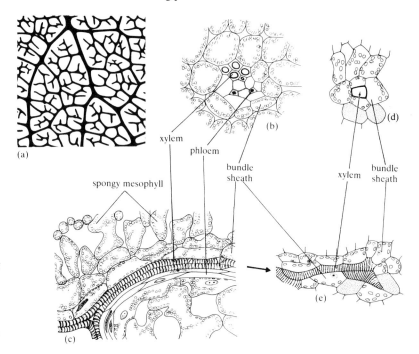

Fig. 8.10 Vein endings. (a) Surface view, much enlarged, of a small portion of a leaf of lime (*Citrus aurantifolia*) to show the vascular bundles ending blindly in the vein islets. (b) and (c) Transverse and longitudinal sections of small bundle with xylem and phloem enclosed in sheath of parenchymatous cells containing chloroplasts. (d) and (e) Transverse and longitudinal sections of bundle ending, consisting of tracheids enclosed by bundle-sheath cells. (b and c from Coult 1977, p. 66)

Development of the foliage leaf

All leaves originate as small bulges of tissue, the leaf primordia, on the flanks of the apical meristem of a bud. As the shoot apex grows, new leaf primordia are initiated in a pattern characteristic for each plant species. In plants with typical stalked leaves the primordium elongates into a finger-like structure that will become the midrib and the petiole of the mature leaf. The cells along opposite edges of this axial structure, which is itself meristematic, then begin to divide extra rapidly to initiate the formation of the lateral flattened lamina. These zones of very rapid cell division are called the marginal meristems (Fig. 8.11); if they are continuous along the full length of the future midrib they give rise to a simple leaf, whereas if they are discontinuous a pinnate leaf results. By the time a leaf is only a few millimetres long its general shape has usually been established.

The main increase in surface area of the lamina is due to the activity of another meristematic tissue, the plate meristem, which is left behind by the marginal meristem as it grows sideways. The activity of the plate meristem consists essentially in anticlinal divisions of the layers of derivative cells cut off by the marginal meristem. Because these divisions are almost entirely anticlinal, the lamina remains more or less uniformly thick except in the regions where procambial strands develop. The plate meristem continues to be active long after the marginal meristem has ceased to grow. In the early stages of development all the cells of a leaf are meristematic, but this initial phase is soon followed by extensive cell enlargement and differentiation. Differences in the rate and duration of cell division and cell enlargement between the various developing tissues create stresses, and these cause cells to split apart along their middle lamellae and air spaces to develop between them. The presence of a more extensive system of air spaces in the spongy

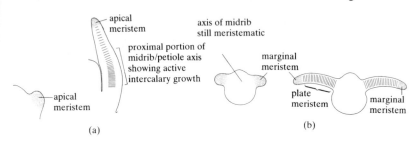

Fig. 8.11 The meristems of a developing leaf, as seen (a) in L.S. through the axis of the primordium, and (b) in T.S. The right-hand diagrams of both a and b represent a later stage of development than the left-hand diagrams. (From Cutter 1971, p. 120)

layer than the palisade layer can be partly accounted for by the fact that the palisade cells continue to divide after the spongy cells have stopped dividing.

During the development of a leaf the vascular system begins to differentiate extremely early, even when the leaf is still only a finger-like primordium. As a result of this early maturation the vascular bundles of a leaf are soon connected with those in the stem, and food material can thus be readily supplied to the rapidly developing tissues of the primordium. By the time the leaf is fully expanded every cell has passed through the three phases of growth (division, enlargement, and differentiation) so that all form part of the leaf's permanent tissues, i.e. there is no meristem in a mature leaf. The great variety of leaf form to be found in different species, or in the same species at different times or under different environmental conditions, is attributable to the relative level and duration of activity of the various localized meristems which, functioning either simultaneously or sequentially, contribute to the growth of a leaf.

Abscission

It is well known that many trees and shrubs shed their leaves at the beginning of the dry season, and by so doing they reduce the exposed transpiring surface of the plant. The process by which a leaf is shed from a branch, without injuring the branch, is called leaf abscission. The leaves of deciduous trees and shrubs are shed, or abscised, as a result of cellular modifications in a narrow region at the base of the petiole where separation of the leaf will subsequently occur. This region is known as the abscission zone, and is often externally recognizable, even in young leaves, by the presence of a shallow groove or by a difference in the colour of the epidermis. Internally, the abscission zone consists of small, thin-walled cells with few or no intercellular spaces (Fig. 8.12), forming a tissue sharply delimited from those above and below it. Within the abscission zone the vascular bundles are sometimes narrower, and sclerenchyma and collenchyma less well developed than in the adjoining tissues.

The shedding of a leaf involves the formation of two layers, firstly a *separation layer* where the actual separation of the leaf takes place, and secondly a *protective layer* which develops beneath the separation layer and protects the surface that becomes exposed when the leaf falls. The separation layer develops in the distal region of the abscission zone days, or even weeks, before abscission takes place. In this layer the cells separate from one another by the dissolution of part (invariably the middle lamellae) or all of their cell walls. Generally, divisions occur in the cells destined to give rise to the abscission tissue, and it is the newly formed cells which are affected

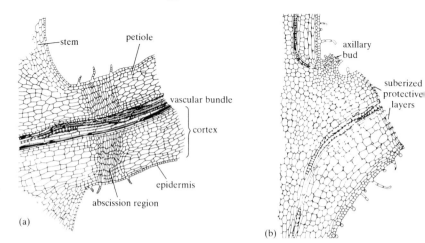

Fig. 8.12 Leaf abscission in a typical dicotyledonous plant such as *Coleus*. (a) L.S. base of petiole cutting through the abscission zone. (b) L.S. leaf base after the abscission of the leaf. (a from J. G. Torrey (1967), *Development in flowering plants*, p. 141, Macmillan Co., New York; b from Fahn 1974, p. 277)

by the dissolution processes. All the living cells of the separation layer, including the parenchyma of the vascular bundles, are subject to dissolution, so that eventually the leaf is attached to the stem only by the dead elements of the vascular bundles. These are finally broken by the weight of the leaf or by the action of wind, and the leaf falls.

The surface which becomes exposed when the leaf falls is protected from desiccation and parasitic invasion by the formation of the protective layer. This layer is formed, usually before abscission occurs, by the deposition in the cells and intercellular spaces of suberin and wound gum, substances which seal the scar by rendering it relatively impermeable to water and gases. The protective layer is sooner or later, often in the year following its formation, replaced by periderm which develops underneath the leaf scar in continuity with the periderm of the rest of the stem.

The structural changes associated with abscission are known to be controlled by the balance between a number of growth regulators, among which the auxin indole-acetic acid (IAA) and the growth inhibitor abscisic acid (ABA) appear to be important. When the modes of action of these growth regulators are more fully understood, it may be possible to interpret the anatomical aspects of abscission in terms of the metabolic changes which are responsible for initiating them.

Suggestions for further reading

Elementary books useful for revision

English, L. R. (1964) *Botanical notes and drawings for West Africa*. Oxford University Press.

A book of fully annotated drawings, supported by a short clear text about plants of the forest and savanna regions. The book is a useful aid for both O-level and A-level biology.

Ewer, D. W. and Hall, J. B. (Eds) (1972/1978) *Ecological biology*. Vol. 1. *Organisms and their environments* (1972). Vol. 2. *The inter-relations of organisms* (1978). Longman.

This attractive, two-volume text was written specifically for A-level students in Africa, and gives an up-to-date account of the principles of biology from an ecological point of view. The book was written in the conviction that ecological problems will play an important role in determining the future of many African and other countries. The subject matter goes beyond A-level requirements, and so the material used should be selected to suit particular syllabuses. Very highly recommended.

Hall, J. R. (1970) *Senior tropical biology*. Longman.

An attractive book intended for the School Certificate examination in biology. Adopts an ecological approach, and is very well illustrated.

Irvine, F. R. (1952) *West African botany*, 2nd edn. Oxford University Press.

A book which dates from the time when botany was an O-level examination subject. The author spent the major part of his working life in West Africa, and had a flair for the macroscopic aspects of the structure of flowering plants. The text is copiously illustrated with line drawings.

Mackean, D. G. (1977) *Introduction to biology*, West African edn. Murray.

Clearly written and well illustrated, this book is generally accepted as being one of the best textbooks designed to cover traditional biology syllabuses for O-level and similar examinations.

Mitchelmore, J. A. (1967) *Tropical biological drawings*. Allen & Unwin.

A slim volume of very useful, annotated, original drawings covering most of the 'tropical' topics listed in syllabuses for O-level biology. This book provides an excellent supplement to both class notes and practical work.

Robertson, E. T. and Gooding, E. G. B. (1963) *Botany for the Caribbean*. Collins.

A short, well illustrated, elementary book which gives a very good account of the external morphology of angiosperms, but is much less successful in the physiology section.

Stone, R. H. and Cozens, A. B. (1969) *New biology for tropical schools*. Longman.

An elementary text which emphasizes function, ecological relationships and social biology rather than structure and systematics. The drawings of the morphology of specified plants or parts of plants are very good.

Textbooks of general botany

These books are listed as a separate category because they cover the whole field of botany and would otherwise have to be repeated at the end of each section. The following list includes only those books which the author has found to be particularly valuable.

Adams, P., Baker, J. J. W. and Allen, G. E. (1970) *The study of botany*. Addison-Wesley.

An interesting general textbook in which much of the information is presented through an analysis of the reasoning by which the information was obtained.

Arnett, R. H. and Braungart, D. C. (1970) *An introduction to plant biology*, 3rd edn. C. V. Mosby.

An undergraduate textbook, noteworthy for the way in which the illustrations are used to back up the text. Written specifically for North American students.

Bell, P. R. and Coombe, D. E. (Translators) (1976) *Strasburger's textbook of botany*. Translated from the 30th German edn. Longman.

'Strasburger' has continued to maintain its time-honoured position as a comprehensive, balanced and modern textbook by numerous revisions, frequently so extensive as to result in virtually a new book. This latest revision, like its predecessors, is both a textbook and a reference work. Because of its wide coverage, in the tropics it is more valuable as a reference work than as a standard text.

Brown, W. H. (1935) *The plant kingdom*. Ginn & Co., USA.

This textbook of general botany has been out of print for many years, but it is extremely useful to anybody studying botany in the tropics because many of the illustrations are drawn from tropical material. The author, who was at the University of the Philippines, took great care to ensure that the illustrations in his text should be of the best quality.

Dutta, A. C. (1979) *Botany for degree students*, 5th edn. Oxford University Press (Bombay, Calcutta, Delhi and Madras).

A comprehensive textbook which has been found useful by many Indian students, for whom the book is written. By some modern standards, too much emphasis is given to the morphological aspects of the subject.

McClean, R. C. and Ivimey-Cook, W. R. (1951–73) *Textbook of theoretical botany*. Vol. 1 *Algae to gymnosperms; morphology and anatomy of angiosperms* (1951). Vol 2 *Floral biology and the families of angiosperms* (1956). Vo. 3 *Palaeobotany, genetics, physiology* (1967). Vol. 4 *Plant ecology, plant geography* (1973). Longman.

A monumental reference book, which is very readable and well worth consulting on most topics.

Muller, W. H. (1979) *Botany: a functional approach*, 4th edn. Collier Macmillan.

A comprehensive, modern textbook which combines the presentation of biological concepts of general educational value with a discussion of such vitally important, and fortunately also fashionable, topics as conservation, food chains and population problems.

Priestley, J. H., Scott, L. I. and Harrison, E. (1964) *An introduction to botany*, 4th edn. Longmans.
Although firmly based in the temperate zone and becoming somewhat out of date in places, this book is still stimulating to read because of the emphasis it places on the significance of development in the interpretation of form and structure.
Raven, P. H., Evert, R. F. and Curtis, H. (1976) *Biology of plants*, 2nd edn. Worth Publishers, Inc.
Probably the best general textbook of plant biology currently available. It is lavishly illustrated, and the text is extremely well written. Although there is inevitably a North American bias, the book is written in very broad terms.
Simon, E. W., Dormer, K. J. and Hartshorne, J. T. (1973) *Lowson's textbook of botany*, 15th edn. University Tutorial Press.
The standard British textbook designed to meet the needs of A-level students. In its entirely revised format, 'Lowson' is a stimulating, easily readable and modern introduction to botany, with only the essential minimum of technical terminology.

Books on the structure of the flowering plant

Cutler, D. F. (1978) *Applied plant anatomy*. Longman.
A short book which enlivens the study of plant anatomy by emphasizing the numerous ways in which the subject can be applied to solve many important everyday problems. The lists of plants in which particular anatomical features are to be found will prove very useful to teachers of plant anatomy.
Cutter, E. G. (1971–78) *Plant anatomy: experiment and interpretation.* Part 1 *Cells and tissues.* 2nd edn. (1978). Part 2 *Organs* (1971). Edward Arnold.
One of the first, and still one of the few, books which seeks to relate the anatomical structure of plants to the processes of growth and metabolism occurring within them.
Dodge, J. D. (1971) *An atlas of biological ultrastructure*. Edward Arnold.
A small book of well chosen illustrations which give a good idea of the extent of structural detail revealed by the electron microscope.
Eames, A. J. and MacDaniels, L. H. (1947) *An introduction to plant anatomy*, 2nd edn. McGraw-Hill.
This standard textbook has, after nearly half a century, been replaced by more modern books. It is still worth consulting as a model of the more descriptive aspects of plant anatomy.
Esau, K. (1965) *Plant anatomy*, 2nd edn. John Wiley & Sons, Inc.
Comprehensive in coverage, and beautifully illustrated with line drawings and photomicrographs, this advanced book is the standard reference work on the subject.
Esau, K. (1977) *Anatomy of seed plants*, 2nd edn. John Wiley & Sons, Inc.
A shorter and more up-to-date text than the same author's standard reference work, this is the best introductory book in the field.
Fahn, A. (1974) *Plant anatomy*, 2nd edn. Pergamon Press.
A well-illustrated, modern textbook covering all aspects of plant anatomy.

Because this book was originally written for students studying in Israel, many of the examples cited refer to plants whose distribution extends into the tropics.

Gemmell, A. R. (1969) *Developmental plant anatomy*. Edward Arnold.

As indicated by its title, this booklet emphasizes that plant anatomy is no longer merely a description of tissues and their distribution in an organ, but an attempt to interpret structure in terms of the developmental processes underlying growth and differentiation.

Hurry, S. W. (1965) *The microstructure of cells*. John Murray.

A small collection of excellent electron micrographs which was one of the first books to make available to schools and colleges the revolution in knowledge of cellular organization that took place in the 1950s.

Jane, F. W. (1970) *The structure of wood*, 2nd edn. revised by K. Wilson and D. J. B. White. Adam & Charles Black.

A comprehensive and well illustrated book, which has deservedly survived the death of its author. The best book on wood structure.

Morey, P. R. (1973) *How trees grow*. Edward Arnold.

A very useful summary of modern research on the growth of trees, and on the structure of wood.

O'Brien, T. P. and McCully, M. E. (1969) *Plant structure and development*. Macmillan.

One of the best pictorial treatments of the cellular structure and development of flowering plants. The photographic illustrations are outstanding and the text is very lucid.

Robards, A. W. (Ed.) (1974) *Dynamic aspects of plant ultrastructure*. McGraw-Hill.

The definitive work on plant ultrastructure. A unique collection of sixteen review papers on organelles and tissues by the world's leading specialists in each field.

Roland, J-C. and Roland, F. (1980) *Atlas of flowering plant structure*. Longman.

An impressive collection of high quality line-drawings, photomicrographs, and transmission and scanning electron micrographs which demonstrate in a convincing manner the relation between plant structure and function. This book is a useful supplement to the approach followed in most textbooks.

Part II

Physiology of the flowering plant

Chapter 9

Diffusion and osmosis

The physical process of diffusion (of which osmosis is a special case) plays such an important part in the physiology of plants that it is essential to have a clear understanding of it, but in order to do this some general properties of matter must first be considered. It is well known that all substances, elements and compounds alike, consist ultimately of small particles. These particles have two important general properties, namely: (1) the power of independent movement; and (2) the tendency of like particles to attract one another. These two properties oppose each other. The power of independent movement tends to separate the component particles of a substance, whereas the forces of attraction tend to bring them together. The outcome of the interplay between these opposing tendencies (whether, for instance, the tendency for independent movement is greater than that for attraction, or vice versa) determines the physical state of a substance. As an approximation it may be said that if the tendency for free movement is predominant the substance will exist as a gas, if the tendency for attraction is predominant the substance will exist as a solid, while if both tendencies are about equally matched the substance will exist as a liquid. There are two important factors which determine whether a given substance behaves as a solid, a liquid or a gas: (1) the inherent mobility of the particles (e.g. those of oxygen are very mobile whereas those of iron are held tightly together); and (2) the temperature of the substance (e.g. the application of heat may turn a liquid into a gas by increasing the power of independent movement of the particles).

Diffusion

If the particles of a substance are able to move freely without being hindered by the forces of attraction, then in the course of time they will distribute themselves evenly throughout the space available. Until such an even distribution exists there will be more particles moving from regions where they are more concentrated to regions where they are less concentrated, than vice versa: this net movement of particles in a given direction is called diffusion. The greater the difference in concentration between two regions, i.e. the steeper the concentration gradient, the faster will be the rate of diffusion. When equilibrium has been attained the particles continue to move just as freely as before, but there is no diffusion because there are as many particles entering a given region as there are leaving it, i.e. the equilibrium is a dynamic one. Since the particles of a gas are in constant motion it follows that the ability to diffuse is a property of all gases.

Gaseous diffusion is demonstrated whenever a gas tap is turned on at one end of a room and the smell of gas is detected shortly afterwards at the other end of the room.

An important feature of the diffusion process is that particles of different substances diffuse independently of one another. This fact may be illustrated in the following way. Suppose that there are two adjoining rooms of equally small size connected by a closed door, and that in one room there are 30 people dressed in red and in the other 10 people dressed in green. Imagine now that the door connecting the two rooms is opened. The obvious thing to do to relieve the congestion in the room with the 30 people dressed in red would be for 10 of them to enter the other room which contains the 10 people dressed in green, so that there would be 20 people in each room. If, however, both groups of people were to 'diffuse' like gaseous molecules, they would both simultaneously invade the room not occupied by themselves, behaving as if the other group were not there. At equilibrium there would be 15 people dressed in red and 5 people dressed in green in each of the two rooms. It is important to grasp this idea of independent diffusion because we shall have occasion to refer to it again later in this chapter.

Diffusion in solutions

As has been said above, the free movement of the particles of liquids and solids is offset to a lesser or greater extent by forces of attraction, so that liquids and solids cannot diffuse in the way that gases do. If, however, the forces of attraction can be overcome, then free movement will assert itself and diffusion becomes possible. This is what happens when a solid is dissolved in a liquid.

The simplest way of demonstrating the diffusion of a solute is to add water to a few crystals of a coloured soluble salt (such as copper sulphate or potassium permanganate) placed in the bottom of a glass cylinder. The diffusion of the solute particles which pass into solution is made apparent by the water becoming coloured. The colour change is first evident around the crystals but, as the solute particles diffuse into the surrounding water, the coloration spreads outwards until eventually all the liquid is uniform in colour. In order to appreciate what is happening in this demonstration, let us assume that the coloured salt, the solute, occupies the left-hand half of a horizontal column (top of Fig. 9.1) and is in contact with water, the solvent, which occupies the right-hand half of the column. If, at zero time, the concentrations of the solute and solvent are plotted against distance along the column, the graph (a) shown in Fig. 9.1 will be obtained. There will be no solute on the right-hand side and no solvent on the left-hand side. If, however, the concentrations are plotted when the advancing coloration has just reached the right-hand end of the column, the curves will have the shape indicated in graph (b). Solute particles have migrated progressively into the right-hand side so that there will be a decreasing concentration of them from left to right along the column. At the same time as this solute migration has been taking place, solvent particles have been migrating into the left-hand side of the column so that there will be a decreasing concentration of solvent from right to left. Both solute and solvent will continue to diffuse independently down their concentration gradients until the whole system is of a uniform concentration throughout. This equilibrium position

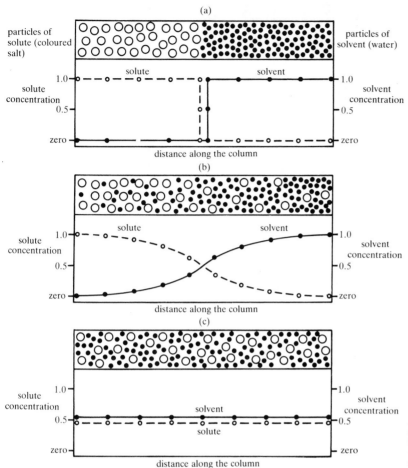

Fig. 9.1 Relation between concentration and distance along diffusion column (a) at zero time, (b) when the coloration has just reached the right-hand end of the column, and (c) at equilibrium.

is represented by graph (c). The important point illustrated by these graphs is that, in solutions, the diffusion of the solvent must be considered as well as that of the solute.

Osmosis

Having learnt the two facts, namely: (1) that both the solute and the solvent particles of a solution are capable of diffusion; and (2) that substances diffuse independently of one another; it is now possible to consider what happens when two solutions with different concentrations of the same solute in the same solvent are separated by various types of membrane. For the purpose of this discussion it will be convenient to think of the membrane as stretched across a U-tube as shown in Fig. 9.2. One hundred ml of 3 per cent sucrose solution are put into the left-hand arm and 100 ml of 1 per cent sucrose solution into the right-hand arm, and the two solutions are separated by a membrane through which both water and dissolved particles of sucrose can pass freely. The concentration of sucrose particles is greater in the left-hand arm, where there are 3 g of sucrose in 100 ml of solution, than in the right-hand arm, where there is only 1 g of sucrose in the same volume

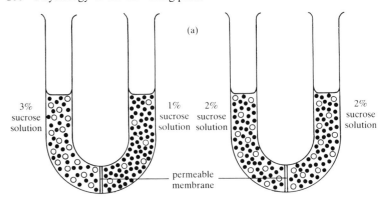

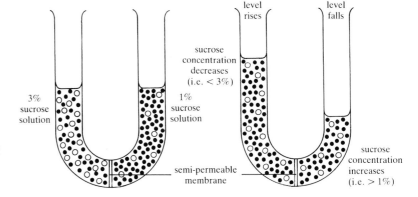

Fig. 9.2 Diagrams to illustrate (a) diffusion of both solute and solvent across a permeable membrane, and (b) diffusion of solvent across a semi-permeable membrane, i.e. osmosis.

of solution. The concentration of water particles, on the other hand, is greater in the right-hand arm because there are fewer particles of sucrose and hence more particles of water per unit volume of solution in this arm than in the left-hand arm. Since the membrane allows the free passage of both water and dissolved sucrose particles, water will diffuse from the right into the left arm and sucrose from the left into the right arm. At equilibrium the concentration of water particles will be equal on both sides of the membrane, and this will also be true of the sucrose particles, i.e. there will be a 2 per cent sucrose solution on both sides of the membrane. Such a membrane, which permits both the solvent and solute particles to pass through it, is called a *permeable* membrane.

It is possible to prepare membranes which permit water to pass freely but restrict the passage of sucrose. A membrane which behaves in this way is described as *semi-permeable*, the prefix semi- implying that the membrane is permeable to only one of the two components of a solution. Let us now see what happens when a semi-permeable membrane separates 100 ml of 3 per cent sucrose solution in the left arm from 100 ml of 1 per cent sucrose solution in the right arm. As before, water diffuses through the membrane from the less concentrated solution on the right-hand side to the more concentrated solution on the left, but the sucrose cannot pass in the reverse direction because the membrane is impermeable to it. As a result of the diffusion of water through the membrane from right to left, the volume of the solution in the left arm increases and the level rises, while the volume in the right arm decreases and the level falls. The solution in the left arm therefore becomes more dilute while that in the right arm becomes more concentrated. Such a movement of water by diffusion across a semi-permeable membrane is an example of *osmosis*. This process may be defined as the net movement of solvent from a weaker solution (or the pure solvent) to a stronger solution when two such solutions are separated by a semi-permeable membrane, i.e. one which is permeable to the solvent but not to the solute. From what has been said it should be clear that osmosis is not a process distinct from diffusion, but merely a convenient term to describe the diffusion of a solvent across a semi-permeable membrane.

Osmotic pressure and osmotic potential

Although a semi-permeable membrane stretched across a U-tube is a useful way of explaining what osmosis is, the situation does not remain as simple as this demonstration implies because, immediately osmosis occurs, another factor begins to operate and this affects the movement of the solvent particles. A more elaborate but more realistic way of demonstrating osmosis is to use a piece of apparatus called an *osmometer*. One type of osmometer (also known as a *Pfeffer cell* after the German botanist Pfeffer who first devised such an apparatus) consists of a cylindrical porous pot in the walls of which a gelatinous precipitate of copper ferrocyanide is laid down. Copper ferrocyanide is semi-permeable to solutions of many crystalline substances and, when supported within the wall of a porous pot, forms a rigid semi-permeable membrane. The membrane is prepared by filling the pot with potassium ferrocyanide solution and standing it in a solution of copper sulphate. The two solutions enter the pores from each side of the wall and, where they meet in the middle, undergo double decomposition to form a gelatinous precipitate of copper ferrocyanide which is the membrane. After washing in distilled water, the porous pot with a rigid semi-permeable membrane now present in its wall is ready for use.

The osmometer is used as follows. The pot is filled to the brim with a solution of sucrose (say a 1 per cent solution), and the top is plugged with a rubber bung pierced by a long vertical glass tube as shown in Fig. 9.3. The sucrose solution is thereby forced a short way up the tube and the level is marked. The osmometer is then immersed in a beaker of pure water. After a short time the level of the solution in the tube begins to rise and, if nothing else were to happen, would continue to do so until the solution reached infinite dilution. In fact the level remains stationary after a certain height is reached, at which point osmotic equilibrium has been attained. From the previous discussion it will be readily understood that the osmotic movement of water into the osmometer depends on the difference in water concentration on the two sides of the copper ferrocyanide membrane. However, the rise of the solution in the vertical tube to accommodate the incoming water results in the establishment of a hydrostatic pressure on the solution in the osmometer. This rise of solution in the tube against the force of gravity therefore implies the imposition of some pressure on the solution inside the porous pot by the pure water outside it.

In order to understand the pressure responsible for this rise it is necessary to supplement our previous ideas on diffusion. This is a convenient place, therefore, to introduce the concept of *water potential* which expresses the energy status of water in a system. Every component of a biological system

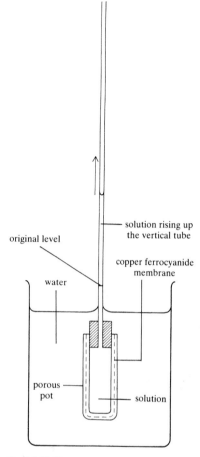

solution rising up
the vertical tube

original level

copper ferrocyanide
membrane

water

porous
pot

solution

Fig. 9.3 Pfeffer osmometer.

possesses free energy, or molecular activity which is capable of doing work. The chemical potential of a substance is the free energy per mole of that substance (a mole is the amount of a substance in grams that is numerically equal to the molecular weight), and thus measures the energy with which that substance will react or move. Substances always move down a free-energy gradient, losing energy as they do so, and equilibrium is only reached when further movement does not result in any further loss of energy. Since the chemical potential of a substance increases with an increase in concentration of its particles (e.g. the pressure exerted by a gas is directly proportional to the number of particles per unit volume), it follows that diffusion can be interpreted in terms of differences in chemical potential between two regions rather than of differences in concentration. On this basis diffusion is the net movement of the particles of a substance from a region of higher to a region of lower chemical potential. The chemical potential of the water in a system is called the water potential of that system, and differences in water potential are therefore the 'driving forces' responsible for the movement of water. The symbol for water potential is the Greek letter psi ψ and, because it can be measured in pressure units as well as energy units, it has traditionally been expressed in atmospheres (atm) although bars (1 atm = 1.01 bars) are now the standard units. By definition, pure water has the highest free energy status of any *free* aqueous system (i.e. not subjected to pressure or other influences), and therefore the highest ψ for such a system. By convention the value of ψ at atmospheric pressure is defined as zero.

The only factors which influence the water potential of pure water are pressure and temperature. If pressure is imposed on water in a closed vessel by means of a piston, the value of ψ will be increased by the amount of the imposed pressure. The reason is that the imposed pressure increases the free energy of the water particles so that they move faster and their water potential is thereby increased. Correspondingly if water is subjected to a 'negative pressure' or tension, ψ will be decreased by the amount of the negative pressure. Although the value of ψ is also influenced by temperature (e.g. diffusion will occur from a warmer to a cooler region of a liquid, even though the concentrations in the two regions are initially the same), it is not necessary to consider the effect of this factor if all parts of the system are at the same temperature. In this discussion, therefore, the influence of temperature will be omitted.

In an aqueous solution, however, ψ is influenced by a third factor, namely the presence of dissolved solute particles. One effect of the addition of a solute is to reduce the number of water particles per unit volume so that, because the potential of water movement depends on the number of moving particles per unit volume, ψ is also reduced. Hence the greater the concentration of solute particles the lower the value of ψ.

Molecular activity capable of exerting pressures is less obvious in liquid systems than in gases, because in liquid systems such pressures only become apparent when a solution is separated by a semi-permeable membrane from the pure solvent or a more dilute solution. Nevertheless the concept of water potential is extremely helpful in the interpretation of diffusion phenomena in liquids.

Returning to the osmometer it is now possible to interpret what is happening in terms of differences in water potential. The value of ψ for the solution in the osmometer, as a result of the presence of dissolved sugar molecules, is less than that of the pure water outside. Water will therefore diffuse inwards towards the region of lower ψ. It will also be apparent that the pressure causing the rise of the solution in the vertical tube results from the greater value of ψ for the pure water outside the porous pot than that for the solution inside it. As water enters the osmometer and the solution rises in the tube, the value of ψ of the solution increases for two reasons, namely: (1) dilution by the incoming water particles; and (2) the imposition of a hydrostatic pressure on the solution by the weight of the ascending column of liquid. The solution will continue to rise until the value of ψ of the solution becomes equal to that of the pure water on the other side of the membrane. This interpretation of osmosis in terms of water potential brings out the fact that the occurrence of osmosis depends not simply on the existence of a gradient in water concentration, but also on the pressure to which the water is subjected. For this reason the attainment of osmotic equilibrium does not imply an equalization of solvent concentrations (i.e. the same ratio of solvent to solute particles) on the two sides of a semi-permeable membrane, but what it does imply is an equalization in the values of ψ.

Since the solution rises in the tube of the osmometer as a result of osmosis, the hydrostatic pressure that is imposed on a given solution at osmotic equilibrium is called the *osmotic pressure*. For instance, a 0.1 molal solution of sucrose (N.B. a molal solution is the molecular weight in grams of solute dissolved in 1 litre of water) inside the osmometer produces an osmotic pressure of 2.6 bar at 20 °C. Osmotic pressure may be defined quantitatively either as 'the maximum hydrostatic pressure that develops when a given solution is separated from the pure solvent by a semi-permeable membrane', or, because at equilibrium the head of water exactly balances the tendency of water to enter the solution, as 'the external pressure which must be applied to a solution to prevent osmosis occurring when it is separated from the pure solvent by a semi-permeable membrane'.

The magnitude of an observed osmotic pressure obviously depends on the concentration of the solution inside the osmometer, because the value of ψ of a solution decreases as the concentration of solute increases. It is misleading, however, to say that a solution has an osmotic pressure because the forces causing osmosis are exerted not by the solution, but by the solvent upon the solution when the two are separated by a semi-permeable membrane. The development of an osmotic pressure is thus not a property of the solution itself but of the system, solution/semi-permeable membrane/ solvent. The property of the solution which is measured by the osmotic pressure may conveniently be called the *osmotic potential*; this term conveys the idea that the solution is only potentially capable of inducing a pressure when it is placed in an osmometer. The osmotic potential, which is numerically equal but opposite in sign to the osmotic pressure, represents the tendency of pure water to enter a solution across a semi-permeable membrane, and therefore measures the amount by which the value of ψ of a solution is less than that of pure water at the same temperature and atmospheric pressure.

The closed osmotic system

The osmometer is an example of what is called an open osmotic system because there is a tube up which the solution may rise as more water enters. Let us now consider an example of a closed osmotic system in which there is

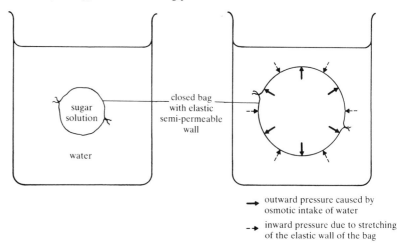

outward pressure caused by
osmotic intake of water

inward pressure due to stretching
of the elastic wall of the bag

Fig. 9.4 The closed osmotic system.

no such tube to accommodate the incoming water. If a sugar solution completely fills a closed bag with an elastic, semi-permeable wall, and the bag is immersed in pure water (Fig. 9.4), water at first enters readily but soon the bag begins to swell and its wall becomes stretched. By offering resistance to further stretching the distended wall of the bag exerts an inward pressure on the solution, which has the effect of increasing the value of ψ of the solution. As more water enters the bag, the wall is progressively stretched until a point is reached at which, provided the bag does not burst, the inward pressure of the wall equals the osmotic potential of the sugar solution contained within the bag. At this point the value of ψ of the solution will be the same as that of the pure water outside, and so there will be no net movement of water into the bag.

The closed type of osmotic system differs from the open type mainly in the way in which the pressure that develops in the solution as a result of osmosis is applied. In the open type of system pressure is applied by the production of a hydrostatic head of water, in the closed type by the development of an inward pressure of the wall. The closed type of osmotic system finds a close parallel in the living plant cell, the water relations of which will be considered in the next chapter.

Chapter 10

The water relations of the plant cell

The walls of living plant cells are always permeated, and sometimes also surrounded, by an aqueous solution, which is continuous from cell to cell and thus forms a network throughout the plant. From the point of view of its relation to this solution, a plant cell can usefully be compared to the closed type of osmotic system described in the previous chapter. The two cytoplasmic membranes, the plasmalemma on the outside and the tonoplast on the inside, are both freely permeable to water but relatively impermeable to solutes, so that, for the sake of simplicity, the entire cytoplasmic layer may be regarded as a single, continuous, semi-permeable membrane. The cell wall, on the other hand, is an almost completely permeable membrane because its cellulose framework is pierced by numerous pores which, although too small to be visible under the microscope, are nevertheless well above the molecular range of size. The significance of the cell wall lies in the fact that it is more or less rigid and therefore tends to resist any increase in the size of the cell. The vacuolar sap of the central vacuole is a solution of various substances dissolved in water. In essence, therefore, we have a solution inside the cell separated from an external solution by two membranes, an inner semi-permeable one and an outer permeable one. Given such a system, it is clear that, if there is any difference in water potential (ψ) between the two solutions, water will diffuse from the region of higher to the region of lower water potential.

Plasmolysis and turgor

Let us consider the behaviour of a single cell when placed in pure water and in different concentrations of a solution of some solute which is non-toxic to protoplasm and which does not readily pass through living membranes. An example of such a solute is sucrose. Suppose first the cell is immersed in a sucrose solution which is more concentrated than the vacuolar sap. Assuming the cytoplasm to be completely impermeable to the solutes both inside and outside the cell, ψ of the vacuolar solution will be greater (i.e. less negative) than ψ of the external solution, and water will therefore diffuse outwards. (Remember that, as explained in the last chapter, the value of ψ decreases as the concentration of dissolved solute increases). As a result of this outward passage of water the central vacuole will shrink, and the protoplast and adhering cell wall will also contract with it. If there is a considerable decrease in the volume of the vacuole, the protoplast will contract away from the cell wall. While it is contracting, the protoplast passes through a series of irregular shapes towards a final rounded form which it

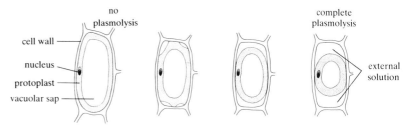

Fig. 10.1 Successive stages in the plasmolysis of a plant cell.

assumes under the influence of surface forces (Fig. 10.1). The cell wall, once it has ceased to be under tension, does not continue to contract with the protoplast because it is more rigid. The space that forms between the wall and the contracting protoplast becomes filled with external solution, which passes freely through the permeable wall. This phenomenon of contraction is known as *plasmolysis*, and a cell whose protoplast shows any degree of contraction from the cell wall is said to be *plasmolysed*. It should be realized that there are degrees of plasmolysis from incipient plasmolysis, where contraction is just detectable at only one or a few points around the cell, to complete plasmolysis, where the protoplast has broken away completely from the cell wall. From what has been said it will be clear that the immediate cause of plasmolysis is the presence of an external solution that is more concentrated than the vacuolar sap. Such a solution is said to be *hypertonic* to the vacuolar sap.

If a cell which has been plasmolysed is transferred into pure water, the gradient in ψ is reversed and water will diffuse into the cell. The protoplast will usually regain its original shape, being apparently none the worse for temporary plasmolysis. Water, as indeed any solution less concentrated than the vacuolar sap, is therefore *hypotonic* to the vacuolar sap. It will be evident that by immersing similar plant cells in a suitable range of solutions of slightly increasing concentration, there will be one concentration which itself fails to cause plasmolysis but which is only slightly weaker than the next one up the series which just causes plasmolysis. The former solution is said to be *isotonic* with the vacuolar sap because its osmotic potential is theoretically equal to the osmotic potential of the vacuolar sap.

If a living plant cell had an elastic wall, it would continue to expand when placed in water until it eventually burst. However, because the cell wall is in fact only slightly extensible, the wall exerts an increasing pressure inwards (appropriately called the *wall pressure*) against the cytoplasm and cell vacuole as the protoplast is distended outwards. A point is eventually reached at which the inwardly directed wall pressure (tending to force water out of the cell) becomes equal to the osmotic potential of the vacuolar sap (tending to cause water to enter the cell), so that the cell will be unable to absorb any more water. A cell whose wall shows any degree of stretching is said to be *turgid* or in a state of *turgor*, and a cell which has reached the point of water saturation is therefore fully turgid or in the state of full turgor. Just as there are degrees of plasmolysis, so there are degrees of turgor ranging from zero turgor to full turgor.

The water-absorbing capacity of plant cells

Although plasmolysis and turgor have been explained in terms of differences in ψ inside and outside the cell, the absolute values of ψ cannot readily be determined. However, it is relatively simple to measure the amount by which the water potential of a solution differs from that of pure water at atmospheric pressure and at the same temperature as that of the system under consideration. It is for this reason that the value of ψ for pure water at atmospheric pressure is defined as zero, and all other values of ψ are measured from this standard of reference. A reduction from this zero value of ψ for pure water is therefore a negative number, and an increase from it is a positive number.

We have seen that the value of ψ of a solution is determined primarily by two components: (1) the presence of dissolved solute, i.e. the osmotic potential; and (2) the pressure exerted on the solution. The symbol for osmotic potential is ψ_s, the subscript s denoting solute. Since the presence of solute reduces the water potential of pure water, ψ_s always has a negative value. The symbol ψ_p refers to the pressure exerted on a solution (the subscript p standing for pressure) and, by analogy with the term osmotic potential, this may be called the pressure potential. ψ_p usually has a positive value because the effect of applying pressure is normally to increase the water potential of a solution, but it will be negative if the solution is subjected to a tension.

The ability of a plant cell to absorb water from its surroundings is determined by the value of its water potential. If, for example, a partially turgid cell has a vacuolar sap with an osmotic potential (ψ_s) of -10 bar and a wall pressure or pressure potential (ψ_p) of $+6$ bar, then the water potential of the cell would be -4 bar ($-10 + 6 = -4$). Such a cell would therefore have a water absorbing capacity of 4 bar. If this cell is placed in pure water, full turgor would be established when the water potential of the vacuolar sap reaches zero, because at this point the value of ψ inside the cell becomes equal to that of the pure water outside. At full turgor the vacuolar sap still has a ψ_s (now less negative than -10 bar due to dilution by the water absorbed) but this would be offset by a ψ_p, or wall pressure, which is equal in magnitude but opposite in sign, i.e. ψ is zero. From this discussion it will be seen that the water-absorbing capacity of a cell is always equal to the osmotic potential of the cell sap (ψ_s) plus the pressure potential of the wall (ψ_p). This relationship can be expressed by the following equation:

$$\psi = \psi_s + \psi_p$$

where

ψ = the water potential, or water-absorbing capacity, of the cell
ψ_s = the osmotic potential of the vacuolar sap
ψ_p = the pressure potential, or wall pressure
and all three terms may be expressed in bars.

The components of the above equation vary as a plant cell gains or loses water. When, for example, a cell in its normal slightly turgid state is placed in an external solution that is not isotonic with the vacuolar sap, the value of ψ will change because the values of both ψ_s and ψ_p alter as a result of the entry or exit of water. The manner in which these quantities vary in relation to one another is shown graphically in Fig. 10.2, which may be interpreted as follows. A cell immersed in a solution isotonic with the vacuolar sap will assume the condition (represented by the point A on the graph) where the wall pressure is zero but yet the cell shows no signs of plasmolysis. In this 'neutral' condition the cell is said to be at the point of zero turgor. At this point the vacuolar sap has an unspecified osmotic potential which is represented by the point B.

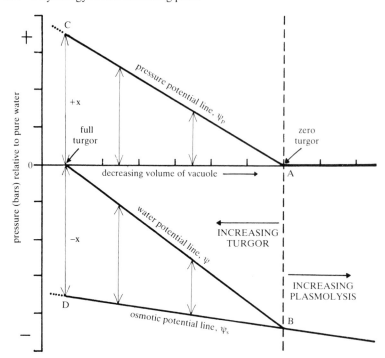

Fig. 10.2 Changes in osmotic potential, pressure potential (wall pressure) and water potential which accompany changes in volume of the vacuole of a plant cell.

If such a cell is now placed in pure water, water will enter the cell by osmosis with two consequences. Firstly, the volume of the vacuole will increase with the result that the osmotic potential of the vacuolar solution will become less negative along the ψ_s line (towards the left from point B) as a result of dilution. Secondly, the wall pressure will increase along the ψ_p line (towards the left from point A) until at point C the positive value of ψ_p becomes equal to the negative value of ψ_s. When point C is reached the vacuole will have swollen to its maximal volume, and the cell will be fully turgid. At full turgor the cell cannot absorb any more water, even though the vacuolar sap will still have a definite ψ_s represented by the point D.

Throughout the course of water uptake by the cell, the value of ψ is equal to the sum of the corresponding points on the ψ_s and ψ_p lines. It steadily becomes less negative as the cell increases in turgor, until it reaches zero at full turgor.

Since	$\psi = \psi_s + \psi_p$
and at full turgor	$-\psi_s = +\psi_p$
Therefore	$\psi = 0$

Now consider the effect of transferring a cell at the point of zero turgor (i.e. point A) into a solution hypertonic to the vacuolar sap. In such a solution water will leave the cell, the volume of the vacuole will decrease (i.e. the cell will become plasmolysed) and the osmotic potential of the vacuolar solution will correspondingly become more negative along the ψ_s line (towards the right from point B). At the point of zero turgor the wall pressure is zero (i.e. $\psi_p = 0$), so that at this point the water potential of the cell is equal to the osmotic potential of the cell sap.

| At zero turgor | $\psi_p = 0$ |
| Therefore | $\psi = \psi_s$ |

In hypertonic solutions the wall pressure remains zero so that for plasmolysed cells immersed in a solution ψ equals ψ_s. The water-absorbing capacity therefore increases (ψ becomes more negative) as the cell becomes progressively more plasmolysed.

The equation $\psi = \psi_s + \psi_p$ measures the tendency of a cell to absorb water from an external environment of pure water. In practice we are usually concerned with knowing whether a cell will gain or lose water when transferred to a solution and not pure water. When a cell is immersed in a solution, the osmotic potential of the solution opposes the water potential of the vacuolar sap. For cells in solutions, therefore, the equation expressing the water-absorbing capacity of a cell has to be modified as follows:

$$\psi = \psi_s + \psi_p - \psi_s \text{ (outside)}$$

where ψ_s (outside) is the osmotic potential of the external solution. For example, if a cell with a water potential of -10 bar is placed in a solution with a ψ_s (outside) also of -10 bar, no water will be absorbed. Similarly, if the same cell were placed in an external solution with a ψ_s (outside) of say -12 bar, water will leave the cell because its water-absorbing capacity is less (i.e. ψ is less negative) than the osmotic potential of the external solution, ψ_s (outside).

The water relations of plant cells within a tissue

The general principles underlying the water relations of plant cells have been presented above by reference to the behaviour of a single cell under different conditions. This method of treatment is an oversimplification, because most cells (and certainly those of flowering plants) do not exist as isolated units. A cell within a tissue has, in addition to the pressure due to the stretching of its own wall, further constraints imposed on it by the pressures exerted by neighbouring cells. The term ψ_p in the equation defining the water-absorbing capacity of a cell has, therefore, a different implication when applied to an isolated cell from that which it has when applied to a cell within a tissue. In an isolated cell the wall pressure is due entirely to the elasticity of the cell wall, but for a cell within a tissue the wall pressure is a general inward pressure on the protoplast resulting from the combined effect of its own stretched wall and the pressures exerted by the neighbouring cells. Provided this difference in meaning is realized, however, the equation defining the water potential of a cell is just as applicable to a cell within a tissue as to an isolated cell.

Whenever the water potential of a cell in a tissue is less than (more negative than) that of an adjacent cell, water will pass into the former cell causing an increase in its value of ψ. The water potential of the adjacent cell will decrease as a result of losing water, and this will continue until the values of ψ of the two cells become equal. The equilibrium value will seldom be the exact average of the two initial values, and all that can be said is that it will lie somewhere between the two. It follows, therefore, that in a tissue which is not suffering loss of water to the external environment, the values of ψ for all the cells in the tissue will be the same. If, on the other hand, there is a continual loss of water from one region of the tissue (i.e. there exists what is appropriately called a 'sink factor'), equilibration of the ψ values of the separate cells will not be attained and a gradient in ψ will automatically be set up. Thus, although movement of water across a tissue can take place

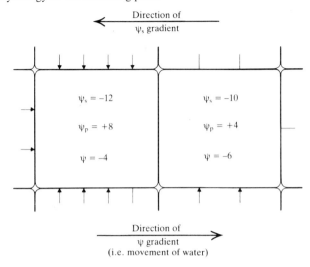

Fig. 10.3 Diagram to demonstrate the theoretical possibility that water can move from cell to cell in the opposite direction from that of a gradient in osmotic potential.

only when such a gradient exists, the demonstration of a ψ gradient in a tissue should be regarded not as a permanent condition but as evidence for the operation of a sink factor at the low end of the gradient.

It cannot be too strongly emphasized that in all questions of water movement from cell to cell it is not a difference in osmotic potentials (ψ_s) but a difference in water potentials (ψ) that is the determining factor. Water always diffuses from a region of higher to a region of lower ψ (i.e. from a region of less negative to a region of more negative ψ) but not necessarily from a region of less negative to a region of more negative ψ_s. It is theoretically possible, for instance, for a cell A to have ψ_s more negative than that of a cell B and yet, because ψ_p is more positive, to lose water to cell B (Fig. 10.3). It goes without saying, therefore, that a ψ gradient does not necessarily (although in fact it usually does) involve a ψ_s gradient, and hence it is *incorrect* to say that water crosses a tissue as a result of a gradient in osmotic potential.

Measurement of the osmotic quantities of plant cells

Of the three components in the fundamental equation $\psi = \psi_s + \psi_p$ only the first two are capable of experimental determination. The wall pressure (ψ_p) cannot be measured directly and has to be estimated by difference.

Measurement of the water potential (ψ) of plant cells

The determination of the water potential of living cells depends on the fact that a tissue will neither gain nor lose water to an external solution which has a ψ_s equal to the mean of the ψ values of its component cells. It is important in determining values of ψ to use as the solute of the external solution a substance such as sucrose to which living cells are not readily permeable.

To determine the mean value of ψ for the cells of a tissue, therefore, comparable samples of the tissue (discs of a storage organ such as sweet potato (*Ipomoea batatas*), cut about 1 mm thick, are very suitable) are weighed and placed in a dilution series of sucrose solutions, maintained at a known constant temperature. A dilution series is a series of solutions

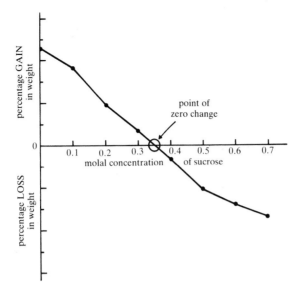

Fig. 10.4 Typical curve obtained from the results of an experiment to determine the mean water potential of the cells of a plant tissue.

graded according to concentration, usually molal concentration. The samples are then left for about an hour for equilibrium to become established, dried rapidly between filter papers, and finally re-weighed. From the weights of the different samples before and after immersion, the percentage change in weight is plotted against molal concentration of sucrose (Fig. 10.4). The reason for expressing the change in weight as a percentage is that the initial weights of the samples may be different. The concentration corresponding to zero change (i.e. 0.35 M in Fig. 10.4) has, at the temperature of the experiment, an osmotic potential ψ_s equal to the mean ψ of the cells of the tissue. The ψ_s values (in bars) of different molal concentrations of sucrose at standard temperatures are given in published tables.

Measurement of the osmotic potential (ψ_s) of the vacuolar sap

The value of ψ_s of the vacuolar sap of living cells can be determined either directly by measuring the freezing point of extracted sap (i.e. the cryoscopic method), or indirectly by the so-called *plasmolytic method*. For purely practical reasons the latter is the more usual method.

In the plasmolytic method thin sections or comparable strips of a tissue (epidermal strips from the coloured lower epidermis of the leaves of *Rhoeo spathacea* or *Zebrina pendula* work very well) are immersed for about half an hour in each member of a dilution series of sucrose solutions, maintained at a known constant temperature, and then observed under the microscope. Provided a suitable range of solutions has been chosen, it will be found that in the stronger solutions all the cells are plasmolysed, while in the weaker solutions none of the cells is plasmolysed. If the percentage of cells showing plasmolysis is plotted against molal concentration of sucrose, an S-shaped curve is obtained (Fig. 10.5). By dropping a perpendicular from the point on the curve corresponding to 50 per cent plasmolysis, the molal concentration of the sucrose solution in which this condition occurs can be read off. The value of ψ_s of this solution, at the temperature of the experiment, is considered to be equal to ψ_s of the vacuolar sap of an 'average' cell in the tissue under investigation.

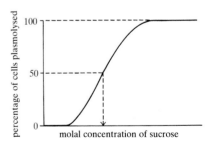

Fig. 10.5 Determination of the osmotic potential (at incipient plasmolysis) of plant cells by the plasmolytic method.

It should be noted that the value obtained by the above method measures ψ_s of the vacuolar sap *at the point of incipient plasmolysis*. This value is usually more negative than the true value of ψ_s of the vacuolar sap, since most tissues are slightly turgid in the natural state and plasmolysis of turgid cells must be preceded by a shrinkage in their total volume. It is only when the cell wall is no longer in a stretched condition (i.e. when $\psi_p = 0$) that the external solution can begin to cause plasmolysis. Thus a cell which has a measured ψ_s of -15 bar at incipient plasmolysis might have had an actual ψ_s of, say, -14 bar in the naturally turgid state. In order to determine the true value of ψ_s of a partially or fully turgid cell it is necessary to make a correction for the difference in volume of the vacuole in the two conditions. The correction factor is usually small, and the value obtained by the plasmolytic method is sufficiently accurate for most purposes.

Chapter 11

The water relations of the plant

A typical plant growing in soil can be thought of as two branching systems, one below and one above ground-level. These two systems are connected by a main axis, most of which is above ground. The subterranean system consists of the repeatedly branching roots which occupy a large hemisphere of soil, the smallest roots being located mainly around the outside of the hemisphere. The aerial system comprises a similar hemisphere, at the surface of which the smaller branches bear a dense canopy of leaves. The smallest roots collectively present a very large surface in contact with the soil, and in a similar way the numerous leaves present a very large surface in contact with the air. Under normal conditions the cells of the roots are surrounded by a soil solution having an osmotic potential usually below –2 bar, and often close to zero, whereas the cells of the leaves and other aerial parts are surrounded by unsaturated air with a water-absorbing capacity of many bars. Since the axis connecting the roots with the leaves is capable of permitting the flow of water through it without undue resistance, it follows that water will inevitably flow down the gradient in water potential extending through the plant from the soil to the air. The whole plant may thus be compared to a wick, by which water is absorbed from the soil by the roots, conducted through the stem, and evaporated from the leaves into the air. This flow of water is known as the *transpiration stream* (Fig. 11.1), and is an inevitable consequence of the structure of the plant in relation to its surroundings. It is common knowledge that plants absorb water, but it is not so generally realized that the bulk of the water absorbed evaporates into the air mainly from the leaves. The reason is obvious because, whereas the water absorbed is supplied to the plant in the visible form of a liquid, the water lost is in the invisible form of water vapour.

It will be convenient to discuss the water relations of plants in terms of output, conduction and intake, that is, under the three headings (1) transpiration, (2) the ascent of sap, and (3) the absorption of water. The reason for dealing with output before input is that, as will be explained later, water is pulled up the stem from above rather than pushed up from below.

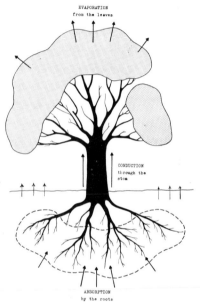

Fig. 11.1 The transpiration stream, resulting from the relationship between a plant and soil water.

Transpiration

Despite the necessity of water for growth and metabolism most land plants are extremely 'wasteful' in their use of water. Most of the water absorbed by the roots is not stored within the plant or used in metabolic processes, but is lost by evaporation into the air. For example, a maize plant absorbs about 200 litres of water during its growth from the seedling stage to maturity, but of this only about 2 litres are present in a mature plant. The process of evaporation from plants is given the special name *transpiration*, but this must not be taken to imply that transpiration is fundamentally different from evaporation from a non-living surface. Although transpiration may take place, at least to a slight extent, from any part of the plant exposed to the atmosphere, in general the greatest loss of water occurs through the leaves.

Two types of leaf transpiration are recognized: (1) cuticular transpiration, in which evaporation of water takes place directly through the cuticle of the epidermis; and (2) stomatal transpiration, in which loss of water vapour occurs through open stomata. The leaf cuticle is relatively impervious to water, and in most species cuticular transpiration accounts for only 10 per cent or less of the total loss of water from leaves. Most of the loss of water must therefore occur through the stomata. As mentioned in Chapter 8, the cells of the mesophyll of a leaf do not fit tightly together, so that there is a system of intercellular air spaces bounded by the water-saturated walls of the mesophyll cells. Water evaporates from these wet walls into the intercellular air spaces, and water vapour then diffuses from the intercellular spaces through the stomata into the outside atmosphere. The area of contact between the mesophyll cells and the intercellular air system is so great that under normal conditions evaporation keeps the air spaces more or less saturated with water vapour. Provided the stomata are open, diffusion of water vapour into the atmosphere must inevitably occur except when the atmosphere itself is equally humid.

The occurrence of transpiration can be easily demonstrated by placing a plant growing in a pot of damp soil under a dry bell jar. The pot and soil surface must be covered with a waterproof material (such as plastic sheeting) so that no water can escape except from the plant. The inside of the bell jar soon turns misty as water condenses on its inner walls. A control bell jar without a plant must be placed alongside to show that the condensation is not due to a fall in temperature of the atmosphere.

Measurement of transpiration

Whereas measurement of the rate of evaporation from a free water surface is relatively straightforward because the process is controlled entirely by environmental factors, measurement of the rate of transpiration is not so easy. The main difficulty is that all methods of measuring transpiration require placing a plant in conditions which affect the transpiration rate. There are four laboratory methods for estimating the rate of transpiration, all of them having inherent errors.

1. Cobalt chloride papers   Essentially this method consists in substituting the loss of water vapour to the air by loss of water vapour to dry cobalt chloride paper, which is bright blue when dry but becomes paler blue and eventually turns pink as it absorbs water. A small piece of bright blue paper is placed on the leaf surface and covered with a microscope slide. The latter is held in place by clipping it to another slide placed in a corresponding position on the other side of the leaf. The time taken for the bright blue paper to change to a standard shade of lighter blue is a measure of the rate at which water is lost by that portion of the leaf covered by the paper. This tech-

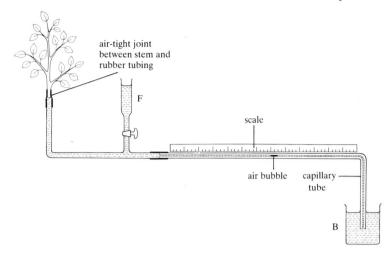

air-tight joint
between stem and
rubber tubing

F

scale

air bubble   capillary
tube

B

Fig. 11.2 A common type of potometer. The
leafy shoot is cut under water and inserted
into the apparatus when this has been
completely filled with water. As the shoot
transpires, water moves through the
apparatus from the supply in the beaker **B**.
A small air bubble is introduced into the
capillary tube by lifting the apparatus
temporarily out of the water in the beaker.
The rate at which the bubble moves along the
capillary tube enables the volume of water
absorbed by the shoot in a given time to be
determined. When the bubble reaches the
left-hand end of the scale it must be moved
back to the right-hand end by admitting more
water into the apparatus through the tap of
the funnel **F**.

nique suffers from the serious disadvantage that the stomata under the paper,
as a result of shading, begin to shut within minutes of being covered. Furthermore, even if the experiment can be completed before the stomata begin
to close, the leaf surface beneath the paper is transpiring into almost completely dry air, a condition seldom encountered in nature. The use of cobalt
chloride papers therefore gives a poor estimate of the actual transpiration
rate of a leaf.

2. Potometer   This apparatus (Fig. 11.2) measures the water intake of a
cut shoot, the assumption being that, when water is freely available to a
plant, the amount taken in is equal to the amount given out by transpiration. Unfortunately the behaviour of a cut shoot may be very different from
that of an intact plant, so that measurements based on this method may not
reflect transpiration under natural conditions. The potometer is useful,
however, for demonstrating the influence of external conditions on transpiration.

3. Collection of water vapour transpired   This method necessitates enclosing the plant, or part of it, in a transparent container so that the water
vapour transpired can be confined. If a potted plant is used, precautions
must be taken to prevent evaporation from the pot and soil surface. A continuous stream of air is drawn through the container and then passed
through previously weighed U-tubes containing a water absorbent (such as
phosphorus pentoxide or calcium chloride). After a given time the U-tubes
are re-weighed. A control experiment, in which the same volume of air is
drawn through a similar apparatus without the plant, must also be set up to
determine the water content of the air stream. From the changes in weight
of the two sets of U-tubes the quantity of water vapour given off by the
plant during the experiment can be determined. The main objection to this
method is that the transpiration rate is affected by the rate of the moving air
stream, i.e. a 'wind factor' (see later) is introduced.

4. Direct weighing   The most satisfactory measurements of transpiration
are obtained from plants growing in pots where precautions have been
taken to prevent evaporation from the pot and soil surface. Water loss from

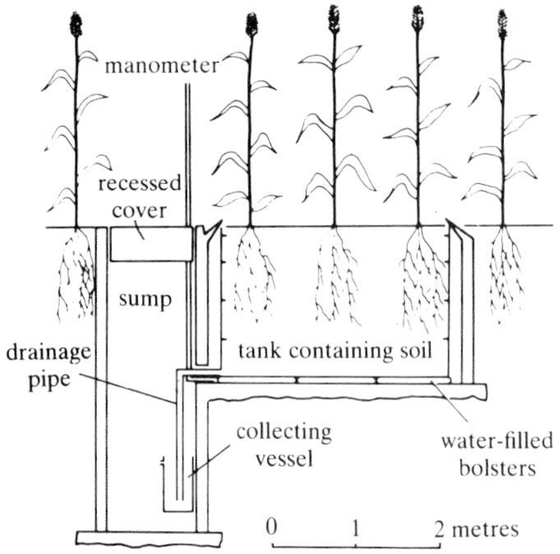

Fig. 11.3 Vertical section of a lysimeter. Any changes in mass of the tank, due to addition of water, drainage or evapotranspiration, result in changes of pressure in the water-filled bolsters and hence in changes of the height of water in the open-ended manometer. (From Rutter 1972, p. 4)

potted plants can be estimated over a convenient period of time by direct weighing. Alternatively, to prevent the soil drying out too much during a prolonged experiment, the transpiration rate can be assessed by finding how much water must be added to the pot to bring the total weight back to the original. Because the plants are growing during an experiment the changes in their weight represent not only losses due to transpiration but also gains due to photosynthesis. The losses in weight due to transpiration are, however, usually several hundred times larger than the gains due to photosynthesis. This error, inherent in the method, is therefore small and can often be neglected.

Agriculturists are interested in transpiration because they want to know how much water different crops need for optimum growth. Water loss from an area of vegetation, however, includes not only that lost by transpiration but also that lost from the surface of the soil by evaporation. The combined loss is called *evapotranspiration* and gives a more meaningful description of the water lost by a crop than does transpiration. Measurement of evapotranspiration on a field scale involves very large pieces of apparatus called *lysimeters*. In one type of lysimeter (Fig. 11.3) the crop is grown in a large water-tight tank which has an open top as much as 10–20 m^2 in area, and is sufficiently deep to allow the normal development of the roots of the crop. The tank is sunk, so that its rim is level with the surrounding soil, in a pit where it rests on a series of water-filled bolsters. Changes in pressure in the bolsters, resulting from increases or decreases in the mass of the tank, are registered by an open-ended manometer which provides hydrostatic balance for the system. Any water percolating to the bottom of the tank drains into a collecting vessel, which is housed in a sump adjacent to the pit. Knowing, over a given period of time, the amount of water supplied to the lysimeter either by irrigation or as rainfall, the amount lost by drainage, and the change in weight of the lysimeter, it is possible to estimate the amount of evapotranspiration that has occurred during the same period. The relative proportions of the total evapotranspiration contributed by evaporation from the soil and by transpiration from the crop can be found by measuring the former in a control lysimeter filled with the same soil-type but kept free of vegetation. It is found that evaporation from the soil usually contributes less than transpiration from the plants to the total evapotranspiration. In dense plant communities evaporation from the soil may be less than 10 per cent of the total evapotranspiration, but in open communities it is a much higher percentage.

Transpiration rates, determined by one or other of the above methods, are expressed in terms of the amount of water lost per unit of plant per unit of time, but the actual units chosen will depend on the purpose for which the measurements are made. Thus the unit of plant may be leaf area, leaf surface (i.e. leaf area multiplied by two because there are two surfaces to a leaf), an entire plant or a unit area of field or forest. Similarly the unit of time may be an hour, a day, a month, or even a year.

Conditions affecting the rate of transpiration

Since transpiration involves the diffusion of water vapour from the intercellular air spaces through the stomata into the atmosphere, it follows that the rate of transpiration will depend upon: (1) the resistance of the pathway to the diffusing water vapour molecules; and (2) the difference in concentration between the water vapour inside and outside the leaf, that is, on the steepness of the diffusion gradient. Any environmental factor which influences either the resistance to diffusion or the diffusion gradient will inevitably affect the rate of transpiration.

In any given leaf the resistance to free gaseous diffusion offered by the intercellular spaces is constant because it is determined by the anatomy of the leaf, but it is generally assumed to be low. Provided the stomata are open, the total resistance along the pathway of the water vapour is therefore also low. When the stomata are closed, however, they create such an extremely high resistance that, even when the outside air is dry, transpiration practically stops. Because of its controlling influence on the opening and closure of most stomata, light probably influences the rate of transpiration more than any other environmental factor. As a result of the response of stomata to light, transpiration tends to be much faster during the day than at night, when it is largely or entirely cuticular.

When the stomata are open and therefore their resistance is minimal, the rate of transpiration is affected by any factors that influence the steepness of the diffusion gradient between the intercellular spaces and the atmosphere. Since under normal conditions the air inside the intercellular space system is kept constantly moist, the steepness of the diffusion gradient depends on the dryness of the atmosphere. The moisture content of air is often expressed in terms of 'relative humidity', which is the amount of water vapour in a given volume of air expressed as a percentage of the amount of water vapour which that volume of air could hold if it were saturated at the same temperature. Since the amount of moisture the air can hold at saturation varies with the temperature (warm air can hold more moisture than cold air), the same relative humidity may refer to widely different moisture contents. Relative humidity is therefore not the best measure of the moisture content of air when we are concerned with the diffusion of water vapour, as in transpiration. For this purpose the actual vapour pressure (mmHg) of the air is the best unit for expressing humidity because it is a measure of the chemical potential, or ability to diffuse, of water vapour molecules. When the stomata are open the rate of transpiration depends on the difference between the vapour pressure of the saturated air inside the leaf and the vapour pressure of the air outside the leaf. Other things being equal, the lower the vapour pressure of the outside air, the faster will transpiration occur.

Temperature influences the rate of transpiration because it has different effects upon the vapour pressure inside and outside the leaf. Leaves tend to acquire the temperature of their surroundings and therefore, because the air in the intercellular spaces is normally maintained in a saturated condition at the prevailing leaf temperature, a rise in atmospheric temperature soon results in an increase in the vapour pressure inside the leaf. The effect of the same rise in atmospheric temperature on the vapour pressure of the outside air is negligible (except in the immediate vicinity of ponds, lakes, etc.), and the vapour pressure gradient through the stomata is therefore steepened and the rate of transpiration increases. Suppose, for example, we have a leaf with open stomata in an atmosphere at 20 °C with a vapour pressure of 10.52 mmHg. At this temperature the vapour pressure inside the leaf will be approximately 17.54 mmHg (this being the saturation vapour pressure at 20 °C), and the difference between the vapour pressure inside and outside the leaf will therefore be 7.02 mmHg (Fig. 11.4). Now suppose that the

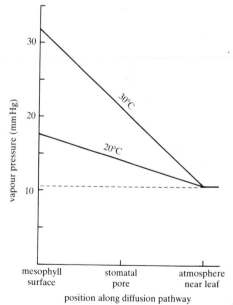

Fig. 11.4 Diagram showing how a rise in temperature increases the vapour-pressure gradient between the atmosphere inside and outside a leaf.

temperature of both the leaf and the atmosphere increases from 20 °C to 30 °C. At 30 °C the vapour pressure inside will increase to about 31.82 mmHg (i.e. the saturation vapour pressure at 30 °C) but the vapour pressure of the atmosphere will remain virtually the same at 10.52 mmHg. The difference between the vapour pressure inside and outside the leaf is now 21.30 mmHg. From these figures it can be seen that water vapour will diffuse out of the leaf at 30 °C at a rate nearly three times as fast as that at 20 °C. In this example we have postulated that the temperatures of the leaf and the surrounding air are the same. In fact leaves in direct sunlight often have temperatures one to several degrees higher than the surrounding air, and therefore light influences transpiration not only by controlling stomatal opening and closure but also by having a secondary effect on leaf temperatures.

The effect of wind on transpiration rate also depends partly on steepening the diffusion gradient. In very still air a thin layer of saturated air forms in the immediate vicinity of actively transpiring leaf surfaces. If the atmosphere as a whole is unsaturated there will be a concentration gradient of water vapour from this saturated layer to the unsaturated atmosphere further away. Under such conditions transpiration is retarded because the layer of saturated air acts as a barrier to the diffusion of water vapour away from the leaf surface. In still air, therefore, there are two resistances to be overcome by water vapour diffusing from the intercellular spaces to the atmosphere outside. These are, firstly, the resistance through the stomatal pores and, secondly, the resistance through the layer of saturated air adjacent to the leaf surface. In still air this second resistance is many times greater than the stomatal resistance and largely controls the rate of transpiration. Under these conditions a change in the stomatal aperture has little effect on transpiration rate except when the stomata are nearly shut and their resistance becomes as large as that of the layer of saturated air. One effect of wind is to prevent the accumulation of water vapour in the vicinity of the leaf and thus, by reducing or removing the second resistance, to steepen the vapour pressure gradient through the stomata and so increase transpiration. In moving air, therefore, the size of the stomatal apertures has a much greater effect on transpiration than in still air (Fig. 11.5). The effect of wind is, however, more complicated than indicated above because its tendency to promote transpiration is offset to some extent by its tendency to cool leaves and so reduce transpiration. Nevertheless the overall effect of wind is always to increase transpiration.

The availability of the water in the soil is yet another environmental factor which influences the rate of transpiration. Whenever soil conditions are such that the supply of water to the mesophyll cells is retarded a fall in transpiration rate is soon evident.

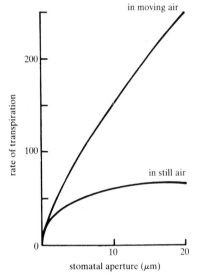

Fig. 11.5 Effect of stomatal aperture on transpiration rate of a *Zebrina* leaf in still and moving air.

Significance of transpiration

Previously there used to be much futile discussion as to whether transpiration is a harmful or a useful process to the plant. The confusion arose from the assumption that transpiration must have a definite function of its own, rather than being a side-effect of the plant's structure, which is the modern view. The structure of the leaf makes it very efficient for the exchange of gases in photosynthesis and respiration but, given the existence of this structure, transpiration is an inevitable consequence whether useful or not.

While it is true that transpiration is sometimes detrimental to plants which may die from excessive water loss during hot dry periods, it is also true that some of the consequences of transpiration may be biologically advantageous to a plant. For instance, the cooling of plant tissues brought about by evaporation may have survival value under certain conditions, and certainly the transpiration stream assists the movement of mineral salts up the plant by transporting anything that happens to be dissolved in it. It should be emphasized, however, that mineral uptake by the roots is an active process distinct from water absorption, and there appears to be no consistent relation between transpiration rate and salt uptake. Whatever interpretation is placed on transpiration, its occurrence in a land plant is inseparable from the necessity for having moist mesophyll cells exposed for the exchange of gases in photosynthesis and respiration.

The ascent of sap

The large amount of water lost by transpiration is normally replaced, as fast as it is lost, by a stream of water which is absorbed from the soil by the roots.

Path of water movement

The actual path of this stream of water is determined by the relative resistances offered by the various tissues of the root, stem or leaf through which the water passes. The structure of the xylem elements suggests that they are the channel of least resistance, and this can easily be verified experimentally in the following ways. (1) If a cut shoot is dipped in a solution of a dye such as eosin (which stains all cell walls with which it comes into contact) the solution will be absorbed up the stem, and subsequent examination of cross-sections reveals that, except near the cut surface where all the cells are stained, only the cell walls of the xylem elements are coloured by the dye. (2) If a short proximal length of the central core of xylem of a cut shoot is exposed by cutting away the surrounding cortical tissues, and the cut end is then dipped into melted gelatine or low-melting-point paraffin wax so that the vessels and tracheids are blocked when it solidifies, the shoot will wilt when placed in water because water absorption is prevented, even though the cortical tissues above the exposed core of xylem are in contact with water. (3) The most conclusive way of showing that the xylem is the path of the transpiration stream, which has the merit of leaving the xylem undamaged, is to supply the root system of an intact plant with water labelled with the radioactive isotope of hydrogen. The bulk of the labelled water moves upwards in the xylem, although some moves outwards from the xylem into the phloem and other adjoining tissues.

Mechanism of the ascent of sap

The problem of how water moves from roots to leaves is presented in its most acute form in the case of tall trees. Atmospheric pressure will not raise water more than about 10 m and yet individual trees of redwood (*Sequoia sempervirens*) and several other species are over 90 m high. In such tall trees water must be raised to heights of approximately 105 m since the roots extend for a considerable distance below soil level. An explanation which can account for water movement in tall trees should also be adequate for plants of smaller stature. Unfortunately the plant physiologist still does not have a complete answer, although theoretically there are two possibilities or a combination of them both. Firstly, water might be pushed up from below, or secondly it might be pulled up from above, or it might be both pushed and pulled.

Root pressure

When vigorously growing stems of many plants are cut off just above ground level, large quantities of sap may be seen to exude from the root stumps. This exudation, or 'bleeding' as it is sometimes called, may last for days or even weeks. The force causing the exudation of sap from the cut surface is called *root pressure* because it must result from the operation of a mechanism located in the roots. If a mercury manometer is attached to a root stump which is bleeding (Fig. 11.6), or alternatively to excised roots of a plant grown in a sterile agar medium, a column of sap will rise in the manometer and its pressure can be measured. Recorded root pressures seldom exceed 2 bar, although values as high as 7 bar have been reported for the roots of tomato (*Lycopersicon esculentum*).

Root pressure is thought to be an osmotic phenomenon which depends on the sap in the xylem having an osmotic potential more negative than that of the soil solution, with the root tissues in between functioning as a semipermeable membrane. This interpretation is supported by the fact that root pressures do not develop if the roots are watered with a solution having an osmotic potential equal to or more negative than that of the xylem sap. However, continuous metabolic activity by the living cells of the root is necessary for the maintenance of root pressures because these are markedly influenced by factors such as oxygen supply and respiratory inhibitors, which have a more immediate effect on the respiration rather than the membrane permeability of living cells. It seems that the living cells of the root, by means of energy released in respiration, actively secrete solutes into the xylem and so maintain the osmotic potential of its sap despite the continued passage of water into it.

Although root pressure is undoubtedly capable of developing positive pressures in the xylem there are several reasons why it cannot be the main mechanism for the ascent of sap in plants, especially tall trees. (1) Many plants, notably most conifers among which are some of the tallest trees, do not exhibit the phenomenon. (2) Known rates of water movement caused by root pressure are too slow to compensate for known rates of transpiration. Indeed negative pressures have often been registered in the stems of rapidly transpiring plants; such negative pressures would not be found if the sap were being pushed up from below. (3) Root pressures are strongly seasonal in occurrence, but the season at which they are most evident usually coincides with the season when the rates of sap movement are lowest. For example, in many trees which 'flush' once a year, root pressures are at a maximum just before, and not at the same time as, the rapid movement of sap up the trunk which accompanies the production of the new crop of leaves.

While root pressure may play some part in water transport in some species at certain seasons, most physiologists tend to regard the phenomenon as an indirect consequence of metabolic activity by the living cells in the

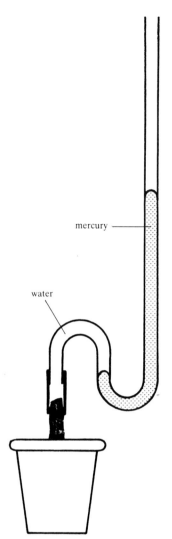

Fig. 11.6 Demonstration of root pressure in the cut stump of a plant. Osmotic uptake of water by the roots of the potted plant causes the mercury to rise in the column.

mercury

water

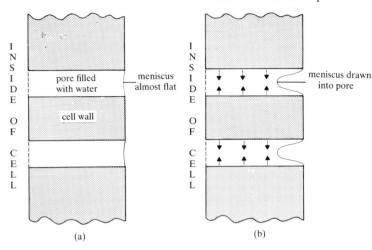

Fig. 11.7 Effect of evaporation from the outer surface of a plant cell on the water potential of the water in the submicroscopic pores of the cell wall. The dotted line drawn at the left-hand end of each pore indicates the continuity of the water in the pores with that inside the cell. (a) Cell wall saturated with water. (b) Evaporation has caused the menisci to be pulled into the pores, causing the water in them to become subject to a negative pressure (shown by the arrows).

root. The explanation for the ascent of sap therefore seems to demand the operation of a mechanism by which water is pulled upwards from above.

Cohesion theory of the ascent of sap

The cohesion theory, first propounded by Dixon and Joly in 1894, supposes that the ascent of sap does not involve any expenditure of energy on the part of the plant because the motive force comes from solar heat which, by supplying the latent heat of evaporation, causes water to evaporate from the leaves. During foliar transpiration water evaporates into the intercellular air spaces of the mesophyll from the water-saturated cell walls, in which water is present in minute pores too small to be seen under the light microscope. Evaporation of water from the surface of a transpiring cell causes the menisci in these submicroscopic pores to be pulled inwards (Fig. 11.7), and this has the effect of imposing a negative pressure on the water so that the water potential (ψ) of the cell decreases. As a result water diffuses into this cell from the adjacent cells, whose water potentials therefore also decrease. A gradient in water potential is thus set up, and this is transmitted from cell to cell until the cells abutting directly on the xylem elements in the vascular bundles of the leaf veins are reached. These last cells then withdraw water from the vessels or tracheids, which contain water with a low solute concentration and therefore a high water potential (i.e. a ψ of nearly zero). The faster water is being transpired, the steeper will be the gradient in water potential and hence the faster will water diffuse from the xylem elements towards the evaporating surfaces. When a mesophyll cell absorbs water from an adjacent tracheid or vessel it does not empty the water-conducting unit. Since this is entirely filled with water, the only way to empty it would be to break the column of water. Such a break could only occur by the creation of a vacuum as a result of *either* the water pulling away from the cell wall *or* the water molecules pulling apart from one another.

Molecules of water, although ceaselessly in motion, are strongly attracted to one another and to any perfectly clean solid surface with which they are in contact. These properties of water are known as cohesion and adhesion respectively. They form the basis of the so-called cohesion theory of the ascent of sap. According to this theory withdrawal of water from the xylem causes a tension or negative pressure to develop in the water and this has the effect of decreasing its water potential. It is supposed that as transpira-

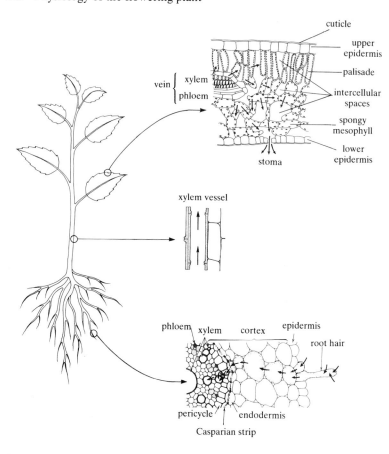

Fig. 11.8 The cohesion theory of the ascent of sap.

tion goes on and water is withdrawn from the xylem, the water columns are stretched into taut threads but do not break because of the mutual attraction (cohesion) between the water molecules. Each column of water behaves like a tiny steel wire which is in a state of tension. Because they are saturated with water, cross-walls in the xylem conduits do not interfere with the continuity of the water columns. The tension which develops at the upper end of the water columns in the transpiring leaves is thus transmitted all the way back to their lower ends in the roots (Fig. 11.8). The water columns are not pulled away from the surrounding walls because of the attraction (adhesion) between the water molecules in the columns and the structural molecules forming the fabric of the walls.

The state of tension in the xylem of transpiring plants can be demonstrated in at least two ways. (1) If the stem of a young wilted plant is placed in a solution of the dye eosin and then pierced with the tip of a scalpel beneath the solution, the dye is instantaneously sucked both upwards and downwards in the xylem from the point of the incision. No such injection of eosin solution occurs if the experiment is repeated with an unwilted plant which has had access to an unlimited supply of water. (2) By fixing a very sensitive calliper (called a *dendrograph*) to the trunk of a tree it can be shown that the diameter of the trunk contracts slightly when transpiration occurs, an observation which agrees with the development of longitudinal tensions in the water-conducting units.

Calculations of the force required to raise water to the tops of the tallest

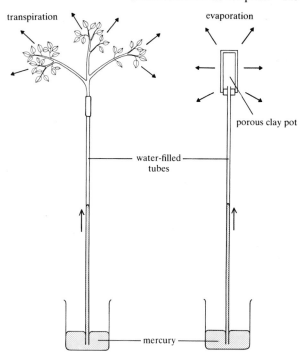

transpiration

evaporation

porous clay pot

water-filled tubes

mercury

Fig. 11.9 Demonstration of the cohesion theory of the ascent of sap. The column of mercury rises up the glass tube as a result of either transpiration from a leafy twig (left) or evaporation from a porous clay pot (right).

trees, taking into account the overall resistance of the pathway for water movement, indicate that the tensions developed must be of the order of 20 bar. Experimental estimates of the tensile strength of pure water and expressed xylem sap, taking into account the average diameter of the xylem conduits and the nature of their walls, vary from as high as 300 bar for water to as low as 30 bar for sap. Although these figures do not measure the tensile strength of sap under the conditions existing in the plant, they do indicate that the cohesive and adhesive properties of water are capable of withstanding the tensions required by the cohesion theory for maintaining the continuity of the water columns, even in the tallest trees.

The fundamental concept of the cohesion theory, namely that the ascent of sap can be explained in purely physical terms, is supported by the fact that the behaviour of the living plant can be simulated by a non-living model (Fig. 11.9). If a cut shoot is attached to a glass tube full of water with its lower end dipping in mercury, then under conditions of rapid transpiration the mercury will be drawn up the tube to heights significantly greater than 77 cm (the height equivalent to 1 bar). If a porous pot completely filled with water is substituted for the transpiring shoot in the above apparatus, the mercury is similarly drawn up the tube as water evaporates from the porous pot into the surrounding air. The physical conditions affecting the transpiration rate of the cut shoot also have a comparable effect on the rate of evaporation from the porous pot. The non-living model thus behaves in the way postulated by the cohesion theory.

Although the available evidence thus leaves little doubt that evaporation at the leaf surfaces is the motive force for the ascent of sap, there are several serious objections to the cohesion theory, notably that nobody has yet devised a method of measuring the tensions in the xylem of an intact plant. Until the existence of substantial tensions has been confirmed by direct measurements the cohesion theory must remain a theory.

The absorption of water

Although plants normally obtain their water supply from the soil, the water in the soil exists in various states which are not equally available to the plant. The problem of absorption therefore involves a preliminary consideration of soil–water relations.

Soil–Water Relations

In all soils the framework of mineral particles is permeated by a complex system of spaces, or *soil pores*, of different sizes. These pores occupy from about 30 per cent of the soil volume in sandy soil to about 50 per cent in clay soils, and may be considered as a water reservoir which is alternately filled and emptied. Until about 50 years ago it was thought that as water was depleted from the upper layers of a soil it was replaced by capillarity from the moist soil below. It is now known, however, that the capillary rise of water is limited to about 1 m and its rate of movement is exceedingly slow. This means that soil water only becomes available to plants when the pore spaces are colonized by roots.

If rain falls on a dry, freely draining soil, a front of water percolates downwards through it to a depth which will depend on the amount of water which falls per unit area and the type of soil. After several days, downward movement of water will cease and there will be a relatively sharp boundary between an upper layer of uniformly moist soil and the dry soil beneath. In uniformly moist soil the water fills the smaller pores in which it is held by capillary and surface tension forces, and forms a thin film on the surface of the particles surrounding the larger pores. This retained water is called *capillary water* (Fig. 11.10). The larger pores, apart from the surface films on the surrounding soil particles, are filled mainly with air because water drains rapidly out of them under the influence of gravity. The capillary water, like the *gravitational water* (i.e. that which drains away under gravity), is freely available to plants but, when it is depleted, there is nevertheless some water still left which is adsorbed on the colloidal particles of the clay and humus. This *hygroscopic* or *bound water* is held by strong physical forces and is not available to plants.

On the basis of the states in which water may be present in soil, it is possible to define two indices which express the percentage of water in a given soil available for plant growth. These are the field capacity and the wilting percentage. A soil is said to be at *field capacity* when it contains the

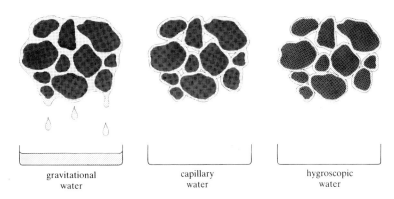

Fig. 11.10 The different states in which water may be present in soil. Gravitational water will usually occur only for short periods during a rain storm or when soil is irrigated. It is primarily the capillary water which is available to, and used by, plants.

gravitational water   capillary water   hygroscopic water

maximum amount of capillary water, i.e. the maximum amount which it can retain against gravity. At this point the soil reservoir is full even though the larger pore spaces will be occupied mainly by air, from which plant roots obtain the oxygen necessary for their respiration. Thus a soil at field capacity, although uniformly moist, is also well aerated; hence soils at or near to field capacity are the most favourable for the growth of most kinds of plants.

The *wilting percentage* is the percentage of water in any soil which cannot be removed by plant roots. It is measured by growing a plant in a sample of soil enclosed within a waterproof pot until permanent wilting of the plant sets in (i.e. the plant does not recover overnight from the wilting of the previous day). The water content in the soil at the wilting percentage varies widely from soil to soil, from less than 5 per cent in coarse sandy soils to about 20 per cent in fine clay soils. The wilting percentage of a given soil can be determined fairly accurately and the value obtained is about the same regardless of the kind of plant used in making the test.

From what has been said about the forces tending to retain water in soil it will be clear that as a soil dries out, the water potential of the soil solution will decrease and so render absorption more difficult. The value of ψ in a waterlogged soil is virtually zero, at field capacity less than -1 bar, and at the wilting percentage, on the average, about -15 bar (Fig. 11.11). As the water content of a soil drops below its wilting percentage, the water potential decreases rapidly and soon reaches values of minus several hundred bars. Irrespective of the type of soil or the kind of plant, however, plants cannot remove water from soil particles if a pressure of more than about 15 bar is required for its release. This critical stage, which coincides with the wilting percentage, is reached at a higher water content in clay soils than in sandy soils, because clay soils retain more water in the bound state unavailable to plants. A plant may thus be able to absorb water from a sandy soil containing 10 per cent of water while quite unable to obtain any from a clay soil with the same percentage of water.

Relation of root growth to water absorption

Since under normal drainage conditions any gravitational water rapidly drains out of the surface layers of the soil, it is the capillary water which forms the bulk of a plant's water supply. There is little movement of capillary water in soil and therefore, once most of this fraction has been absorbed by the root hairs, the root tips must continue to grow if they are to maintain contact with untapped supplies of soil water. We must thus visualize a plant as expanding further and further into a soil mass, with its growing roots leaving behind them soil sucked dry of its available water. This is not a hit-and-miss process because some roots are attracted by the presence of water, a response (called hydrotropism) about which virtually nothing is known.

It will be clear from the above that the large amount of water taken in by a plant depends on an enormous area of contact between the roots and the sites of capillary water (i.e. the smaller pore-spaces and the water films around the soil particles). Unless one goes to the trouble to try and excavate an entire root system in an undamaged condition, it is difficult to appreciate its extent. Measurements by several workers show that it is not uncommon for a root system to increase in length by many kilometres per week during the growing season. In the light of such measurements it is

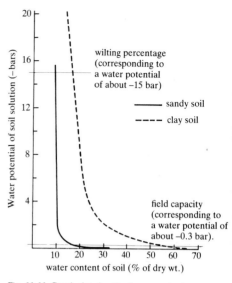

Fig. 11.11 Graph showing the force required to remove water from two contrasting types of soil as the water content is progressively reduced below saturation. Soil water is available to plants growing in a sandy soil over an approximate range of water content from 8 to 18 per cent of the dry weight of soil, whereas the corresponding range for plants growing in a clay soil is from about 20 to 55 per cent of the dry weight of soil.

possible to understand how roots advancing through the soil absorb sufficient water to compensate for that lost by transpiration.

Mechanism of water absorption by the root

According to the cohesion theory of the ascent of sap the tension generated at the upper end of the water columns in the transpiring leaves is transmitted down the xylem into the roots. The lower ends of the xylem conduits, however, do not open directly into the soil but are surrounded on all sides by the living tissues of the outer part of the root. The absorption of water from the soil and its transfer into the xylem occur in these living tissues.

How water absorption by these living cells is linked with conduction through the xylem may be explained by a reversal of the process occurring in transpiring leaves. As soon as the tension in the water columns causes the water potential in them to drop below that of the adjacent pericycle cells in the root tip, water passes from these cells into the xylem. This causes further cell to cell movement in a lateral direction so that a gradient in water potential is set up from the pericycle on the inside to the root hairs and other epidermal cells on the outside. As soon as the water potential of these peripheral cells falls below that of the soil solution, water will be absorbed by osmosis from the soil. The net result is that water is taken in from the soil by the upward pull of the water columns in the xylem (Fig. 11.8, bottom).

From the point of view of water absorption, the root can thus be regarded as a simple osmotic system, with its outer tissues collectively acting as a semi-permeable membrane through which water diffuses from the soil solution into the xylem. It is particularly important to realize that the osmotic potential of the contents of the individual cells is irrelevant in this picture. It is the gradient in water potential set up by the withdrawal of water from the pericycle cells into the xylem which causes the movement (see water relations of plant cells within a tissue, p. 109).

In conclusion it is interesting to consider the anatomical pattern of the root in relation to water absorption. Whereas in the stem the phloem always lies outside the xylem, in the root the xylem and phloem occur as alternating strands. This distribution makes it possible for water moving inwards from the epidermis to enter the xylem without having to cross the phloem, where another type of conduction is occurring (see Chapter 20).

Relative rates of water absorption and transpiration

Simultaneous measurements of the rate of transpiration from the shoot of a plant and the rate of water absorption by the roots show that an increase in rate of transpiration is accompanied by a similiar increase in rate of absorption. However, on warm bright days, when transpiration rates are high at noon (i.e. when the evaporating power of the air is reaching its maximum), the rate of absorption typically lags behind the rate of transpiration presumably because of the resistance of the root cells to the passage of water (Fig. 11.12). As a result there is a gradual increase in the tension of the water columns and a gradual diminution in the water content of the leaves. This water deficit may lead to *temporary wilting* of the leaves, especially in 'soft' leaves which depend upon turgor for their support. If the supply of soil water is adequate, absorption of water during the late afternoon and

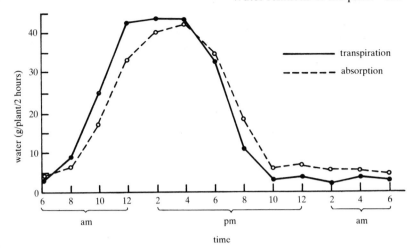

Fig. 11.12 Rates of transpiration and water absorption by a sunflower plant during a hot summer day and the following night. Note the lag between transpiration and absorption. (From P. J. Kramer (1937), *American Journal of Botany*, **24**, 10–15).

night exceeds transpirational loss and the leaf cells gradually regain their turgor. Temporary wilting, caused by a temporary excess of output over intake of water, is to be distinguished from *permanent wilting*, caused by a permanent deficiency of soil water. Plants recover from temporary wilting during the night, but will not recover from permanent wilting unless the supply of soil water is replenished.

At one time it was believed that stomata behaved in such a way as to minimize the risk of wilting, closing when water was scarce in the leaf and opening when it was plentiful. This is not always so. When a leaf wilts rapidly the epidermal cells lose water faster than the guard cells, with the result that the latter bulge into the epidermal cells and the stomata open even more widely. This condition is, however, short-lived, for soon the guard cells also lose water and then the stomata close. Once a leaf has wilted, therefore, still further losses of water are prevented, so that wilting may itself retard severe drying up and permanent injury. Since closure of the stomata stops carbon dioxide getting into the leaf as effectively as it stops water vapour getting out, it follows that conditions which promote wilting simultaneously limit the rate of photosynthesis. To the layman it might appear that, under conditions when water loss is too rapid to be made up by the roots, plants put water economy before photosynthesis. In scientific terms it would be more appropriate to say that the check in growth which accompanies water shortage is an inevitable consequence of the structure and function of the plant body.

Chapter 12

Chemical reactions and enzymes

A living cell does not consist of a static collection of molecules which, once synthesized at an appropriate site within the cell, continue to function indefinitely until 'wear-and-tear' requires their replacement. By the use of radioactive isotopes it has been shown that the individual molecules of a tissue are continually being broken down and new ones synthesized to take their place. In other words, every living cell is constantly re-creating itself from materials which it obtains from its surroundings. How does the living cell manage to carry out the large number of complex chemical reactions necessary for the synthesis of the highly specific macromolecules of which it is composed? Before attempting to answer this question, it is first necessary to understand why any chemical reaction takes place.

Chemical reaction

An atom consists of a central positively charged region, the *nucleus*, which contains positively charged particles, *protons*, and electrically neutral particles, *neutrons*. The nucleus is surrounded by a 'cloud' of *electrons*, which are negatively charged particles moving around it. The number of electrons is equal to the number of protons in the nucleus, so that the atom as a whole is electrically neutral. The electrons move in a way which can be described only in mathematical terms, but for many chemical purposes it is sufficient to think of the electrons as being grouped in *shells* of increasing size, rather like the orbits of planets around the sun. The energy level of an electron shell increases with its distance from the nucleus. To raise an electron to a higher energy level (i.e. to an electron shell further away from the nucleus) requires an input of energy, and the electron thereby gains potential energy. When it returns to a lower level, its potential energy is released. An atom in which one or more electrons have been raised to a higher energy level is said to be in an *excited* state.

With certain exceptions (the so-called noble gases) atoms do not normally exist singly but are chemically combined with one or more other atoms, of the same or of different elements, to form *molecules*. A molecule is the smallest particle of a substance capable of existing independently and retaining the properties of that substance. The atoms of a molecule are held together by electrical forces of attraction, called *chemical bonds*, which involve the sharing or transfer of electrons between atoms. The strength of a bond between two atoms depends mainly on the particular pair of atoms concerned but also, when the molecule consists of more than two atoms, on the surrounding atoms. All molecules are in constant random motion, which is greatest in gases and least in solids. The rate of this movement is, however, directly related to temperature, the motion always becoming faster as the temperature is increased.

Molecules are not completely stable structures, and if two or more of them collide in the course of their random motion they may react with one another. Assuming that a reaction is possible as a result of collision, it will be readily appreciated that the rate of the reaction depends on the concentration of the reacting molecules and on the temperature. Molecules are obviously more likely to meet the more there are of them in a given space (i.e. their concentration) and the faster they are moving (i.e. their temperature). Whether or not they react is also affected by their orientation when they collide. Many collisions will result not in reaction but in mere rebounding of the colliding molecules.

When a chemical reaction occurs, the reacting molecules change into different kinds of molecules. It is convenient to distinguish four main types of reactions.

1. *Synthesis*, in which two or more molecules combine to form a single larger molecule.

   $A + B \rightarrow AB$

2. *Decomposition*, in which one molecule breaks up into two or more smaller molecules. Decomposition is the reverse of synthesis.

   $AB \rightarrow A + B$

3. *Exchange*, in which one or more of the atoms of one molecule change places with one or more of the atoms of another molecule.

   $AB + CD \rightarrow AD + BC$

4. *Rearrangement*, in which the number and types of atoms in a molecule remain the same but the arrangement of the atoms changes.

   $$\begin{array}{ccc}
   CHO & & CH_2OH \\
   | & & | \\
   H - C - OH & \rightleftharpoons & C = O \\
   | & & | \\
   CH_2OH & & CH_2OH
   \end{array}$$

   (glyceraldehyde)   (dihydroxyacetone)

Since a reaction between colliding molecules makes, breaks, or rearranges the bonds (which represent chemical energy) between some of the atoms, it follows that reactions are accompanied by energy changes. In a chemical synthesis, for example, the formation of the bond which unites the reacting molecules into a larger molecule requires energy to be supplied from the outside. Such a reaction, which may be represented in general terms as

$A + B + energy \rightarrow AB$

is called an endergonic (energy-requiring) reaction. Conversely, in a chemical decomposition at least one bond is broken and the energy of that bond is released

$AB \rightarrow A + B + energy$

This is an example of an exergonic (energy-yielding) reaction.

The energy changes occurring in the formation and breaking of bonds can be compared to pushing a stone up a hill and then allowing it to roll down. Pushing a stone to the top of a hill requires the expenditure of kinetic energy, and the stone thereby gains a certain amount of potential energy. If the stone is then allowed to roll down the hill, this potential energy is released as an equal amount of kinetic energy. The amount of kinetic energy converted into potential energy in pushing a stone to the top of a hill is determined by the height of the hill and not its slope, which merely affects the speed with which the stone rolls down or is pushed up the hill. Exergonic reactions may be compared to the rolling of a stone downhill, and endergonic reactions to the pushing of a stone uphill.

Most chemical reactions will not occur spontaneously and require to be 'triggered off' before they will start. The reason is that every reaction has an energy barrier which must be overcome if reaction is to take place. This means that the energy level of the reacting molecules must be raised by a certain amount, called the *activation energy*, before they will react. The activation energy can be regarded as the energy required to loosen the original bonds before new ones can be formed. The burning of wood is an exergonic reaction but, as anybody who has tried to start a wood fire knows, the wood will not react with atmospheric oxygen (i.e. burn) until its temperature is raised above a certain point. When that point has been reached, the reaction continues of its own accord. The amount of heat that has to be applied to the wood before it will catch fire represents the activation energy of the reaction. Once lit, the wood releases far more energy than had to be put into it to start it burning, but it would never have started to burn without the initial 'push'.

The course of a chemical reaction can be represented graphically as a path over an 'energy hill' (Fig. 12.1a) from the reactants to the products, i.e. from left to right along the horizontal axis. Consider an exergonic or 'downhill' reaction in which a molecule AB decomposes into molecules A and B. The molecule AB contains a certain amount of potential energy by virtue of its raised position on the energy hill. Before a molecule of AB can roll down the hill to yield the products A and B, it must first acquire an appropriate amount of activation energy (represented by E_A) to gain the added push necessary to start it rolling down the hill. Once a molecule of AB has been activated it rolls down the energy hill and, in the process of

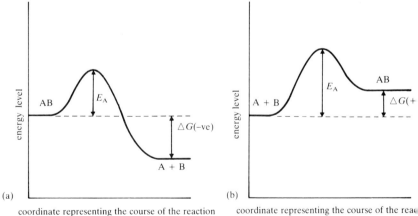

Fig. 12.1 The 'energy hill' for (a) an exergonic and (b) an endergonic reaction.

(a) coordinate representing the course of the reaction

(b) coordinate representing the course of the rea[...]

doing so, decomposes to produce molecules of A and B at the bottom. The products A and B are at a lower potential energy level than the reactant AB, and the energy difference between the two levels is called the change in free energy because it represents the net energy change of the reaction. The free-energy change is denoted by the symbol $\triangle G$, which in this reaction has a negative value because energy is given out by the reaction.

For an endergonic reaction the energy hill is drawn the other way round (Fig. 12.1b) because energy must be fed into the reaction to make it proceed, and the value of $\triangle G$ is positive. In an endergonic reaction the reacting molecules also require to be activated before they will react, but here the activation energy is always greater than the net energy gained in the reaction (i.e. $+\triangle G$) because, as in exergonic reactions, it represents the difference between the energy level of the reactants and the maximum energy level reached during the reaction. (N.B. the activation energy is always measured from the energy level of the reactants.)

The values of E_A and $\triangle G$ are constant for any given reaction under the same conditions of temperature and pressure, but they vary greatly from one reaction to another. A reaction which takes place at room temperature has a low activation energy because the heat energy at this temperature provides the reactant molecules with enough energy of motion for them to react on colliding with one another. Many reactions, however, have a higher activation energy and require the application of heat energy before they will start because, at room temperature, no molecules in the reaction mixture have sufficient energy of motion to become activated. When the reaction mixture is heated, the energy of motion of the molecules becomes equal to or greater than the activation energy, and reaction occurs. Once a reaction has started, the activation energy must be maintained if the reaction is to continue. For many reactions this means that continued heating is necessary, but for some exergonic reactions (of which the burning of wood is an obvious example) the heat released by the reaction is sufficient to sustain the reaction without further heating being necessary.

Reversible reactions

In the above discussion of energy in relation to chemical reactions, it has been assumed that once the requisite conditions are satisfied the reaction will go to completion. This is not always true because the molecules of a substance do not all have the same kinetic energy. They have a range of energies which are distributed about the most frequent energy (the actual value of which will increase with temperature because molecular motion is always proportional to the environmental temperature) according to a pattern which is called the *Boltzmann distribution* (Fig. 12.2). The curve of this distribution is not symmetrical. There are more molecules to the right of the peak than to the left, so the average energy will not be the energy of the greatest number of molecules (shown at the peak) but will be somewhat larger than this. The long 'tail' of the Boltzmann distribution also means that a small fraction of the molecules have energies much larger than the average.

Consider the course of the hypothetical reaction A + B → AB, shown graphically in Fig. 12.3. Here an activation energy of 10 units is required for the reaction to go from left to right, and a net release of 10 units occurs. The activation energy for the reverse reaction will be 20 units, and a net input of 10 units will be required to drive the reaction from right to left. The

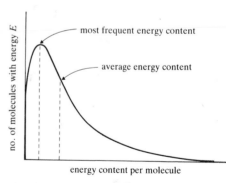

Fig. 12.2 The *Boltzmann distribution* of the molecular energies in a population of molecules. The above curve applies only to a given temperature. If the temperature is raised, the spread of the curve increases, the peak (and hence the average energy) shifts to the right and, because the total number of molecules remains unchanged, also diminishes in height.

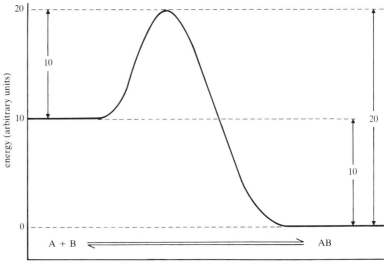

coordinates representing the course of the reaction

Fig. 12.3 Energy diagram for a reversible reaction.

energy relations of the system as a whole therefore favour the formation of AB. In spite of this, it is nevertheless possible that some AB molecules at the high end of the Boltzmann distribution may have enough energy to overcome the energy barrier of the reverse reaction (i.e. E_A of 20 units) and thus to decompose into A and B. If this is so, the reaction will occur in both directions at the same time. The reverse reaction can, of course, start as soon as any AB is formed but, because its rate depends on the concentration of AB, it will be slow at first but will become faster as progressively more AB is formed by the forward reaction. Eventually a state of *dynamic equilibrium* will be reached, at which the rates of the forward and reverse reactions are the same. A reaction in which a state of equilibrium may be reached between the reactants and the products is called a *reversible reaction*. Such a reaction is written as A + B $\rightleftharpoons$ AB, where the double arrow indicates reaction occurring in both directions at the same time.

The fact that reversible reactions do not go to completion raises the question of how big is the yield of a reversible reaction. The actual, as a proportion of the potential , yield will of course vary from one reaction to another, being very much greater for some reactions than for others. To express how far a reversible reaction can go it is necessary to describe the equilibrium condition quantitatively in mathematical terms. The most convenient expression is the *equilibrium constant* (K) which is derived as follows.

For the general reaction A + B $\rightleftharpoons$ C + D the rate of the forward reaction is proportional to the product of the concentrations of A and B. Therefore, representing the concentration of a chemical by the shorthand convention of enclosing it in a square bracket, the rate of the forward reaction at any given temperature is k_1 [A] [B], where k_1 is a constant called the *rate constant* for the reaction. Similarly the rate of the reverse reaction at the same temperature is k_2 [C] [D], where k_2 is another rate constant.

At equilibrium $k_1 [A][B] = k_2 [C][D]$

Therefore $\dfrac{k_1}{k_2} = \dfrac{[C][D]}{[A][B]}$

or $\qquad K = \dfrac{[C][D]}{[A][B]}$

where K is the equilibrium constant. Expressed in words the equilibrium constant K consists of the product concentrations multiplied together, divided by the reactant concentrations multiplied together.

Every reversible reaction has a characteristic value of K, which is the same whether one starts with A and B only, with C and D only, or with a mixture of all four. Because K expresses the balance between the products and the reactants when equilibrium is reached, its numerical value indicates how far the equilibrium position lies towards one side or the other. If the reaction goes almost to completion K is high, whereas if the reaction goes only a little way before equilibrium is reached K is small. In other words, K is a measure of the yield of a reversible reaction.

Catalysis

One way of speeding up a chemical reaction is to raise the temperature, but this is not the only possible way. A reaction will go faster if the same products can be obtained by an alternative pathway involving a lower activation energy. This can sometimes be achieved by performing the reaction in the presence of a substance which assists the course of the reaction but remains unchanged at the end. Consider, for instance, an exergonic reaction A → B + C which has a fairly high activation energy and requires the application of external heat to take place. The same result could be achieved if another mechanism, which takes place in two stages and involves a third substance, X, can be found.

$$A + X \ \rightarrow \ AX \qquad\qquad (1)$$
$$AX \ \rightarrow \ B + C + X \qquad\qquad (2)$$

Added together $\qquad\qquad A \ \rightarrow \ B + C \qquad\qquad (3)$

In this example X takes part in the reaction by combining with A to form an unstable intermediate AX but, since this immediately decomposes to regenerate X, the amount of X remains unchanged at the end of the reaction. For this reason only a small quantity of X is required in the reaction mixture to convert many molecules of the reactant A into the products B and C.

If the activation energies of reactions (1) and (2) are both sufficiently low that the heat energy at room temperature enables them to occur spontaneously, then reaction (3), which is the sum of reactions (1) and (2), will also occur spontaneously. When this happens, the overall reaction A → B + C takes place spontaneously at room temperature in the presence of X but fails to do so in its absence. The way in which the energy hill of this reaction is modified by the presence of X is shown in Fig. 12.4. Notice that, although X lowers the activation energy, it does not alter the net energy

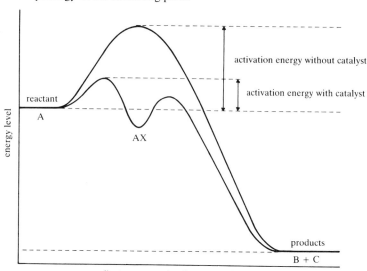

Fig. 12.4 Curves to illustrate how the energy of activation is lowered by the participation of a catalyst (X) in the mechanism of the reaction.

change of the reaction (i.e. its free-energy change, $\triangle G$). Any substance like X which speeds up a chemical reaction by lowering its activation energy and which emerges unchanged from the reaction is called a *catalyst*, and the phenomenon whereby a chemical reaction is assisted in this way is known as *catalysis*. There are many different types of catalysts, and some of them are widely used both in the laboratory and in chemical industry.

Since a catalyst does not alter the final outcome of a reaction but merely the rate at which it occurs, it follows that a catalyst cannot change the position at which equilibrium is reached in a reversible reaction. If a catalyst is present at equilibrium, it increases the rates of both the forward and reverse reaction equally, and so the value of the equilibrium constant remains the same.

Having dealt with the general properties of catalysts, we are now in a position to examine a particular group of catalysts, collectively known as enzymes, which are the 'tools' used by living cells to perform the complex reactions necessary for the maintenance of life.

Enzymes

Following the discovery in the early nineteenth century of the phenomenon of catalysis, a Swedish chemist, Berzelius, suggested in 1836 that the numerous chemical reactions occurring in living organisms might depend upon the presence of catalysts within the tissues. However, no convincing evidence to support his suggestion was brought forward until 1897 when two Germans, the Buchner brothers, carried out *in vitro* the fermentation of sugar to produce alcohol, a reaction which had previously been regarded as inseparably linked to the activity of a living organism. The Buchner brothers, while attempting to prepare an extract of yeast (which is a unicellular fungus) for medicinal purposes, ground actively growing yeast in a mortar to break up the cells and then squeezed the liquid out of the mixture in a press. To prevent the growth of micro-organisms in the extract they added sugar as a 'preservative' (a procedure regularly followed in making jam) and, to their surprise, the resulting solution started to ferment almost as readily as if intact living yeast cells had been present. Since the yeast juice was undoubtedly non-living they concluded that it must have fermented the sugar catalytically. The name 'enzyme' (literally, in yeast) was coined for the postulated catalyst in the juice, because it was assumed that the active agent responsible for the *in vitro* fermentation must be the same as that inside living yeast cells. It is now known that the fermentation of sugar to alcohol is a very complicated process which involves about twelve consecutive reactions, each promoted by a distinct catalyst. Hence yeast juice contains not one catalyst, as originally thought, but a mixture of at least twelve different catalysts. As more and more catalysts were discovered and extracted from other kinds of cells, the word enzyme became accepted as a general term for the entire class of biological catalysts.

It is now recognized that Berzelius was even more correct in his suggestion than he realized, because one of the most important biochemical discoveries of this century is that practically every chemical reaction occurring in living cells is catalysed by a specific enzyme. Most biochemical reactions would, on their own, take place extremely slowly within the ranges of temperature (about 20–40 °C) and pH (usually between 5.0 and 7.5) in which living cells must exist. In addition, biochemical reactions are often highly specific, and yield only one of a range of possible products. The secret of the cell's capacity to speed up the highly specific reactions essential for its maintenance lies in the existence inside the cell of a very large number of specific enzymes. It is thus clear that enzymes are essential to all living systems.

Characteristics of enzymes

Although many different solutions containing enzymic capability were prepared during the 20 odd years after the Buchner brothers demonstrated that an extract of yeast could ferment sugar catalytically, the chemical nature of enzymes continued to be unknown. It was not until 1926 that Sumner succeeded in isolating the first enzyme (urease) in crystalline form and showed that it was a protein. Since then hundreds of enzymes have been isolated in a pure or semi-pure state, and all have proved to be proteinaceous. Since no exception has been found, it is reasonably safe to assume that all enzymes are proteins, although many also have a non-protein component, or a cofactor, which is essential to their catalytic activity. Enzymes may therefore be defined as proteins with catalytic properties.

In discussing reactions catalysed by enzymes (as opposed to those catalysed by non-biological catalysts) it is customary to refer to the substance on which the enzyme acts as the *substrate* rather than as the reactant; this convention will be followed from now on. In common with all catalysts, enzymes show the following *general characteristics*:

1. They increase the rate of a chemical reaction by lowering its activation energy and thus increasing the proportion of molecules which have enough energy to react at any one time (Fig. 12.5). It should be stressed, however, that enzymes can only catalyse reactions which are energetically possible, because they do not drive reactions one way or the other but merely accelerate the attainment of equilibrium. The value of the equilibrium constant K for a particular reaction is always the same, irrespective of whether it takes place in the presence or the absence of an enzyme.

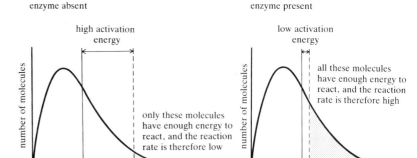

Fig. 12.5 Graph showing how, by lowering the activation energy of a reaction, the presence of an enzyme increases the proportion of molecules capable of reacting at any particular time and consequently the rate at which the reaction proceeds.

2. Although enzymes take part in the reactions which they catalyse, they do not appear to undergo any net change because they are regenerated at the end of the reaction process and so can be used over and over again. The activity of a purified enzyme is sometimes expressed as its *turnover number*, which is the number of substrate molecules acted upon by one molecule of enzyme in one minute in the presence of an excess of the substrate compound. For most enzymes this number is over 1000 and for some over 1 000 000.

3. As a consequence of (2) only a small amount of an enzyme is needed to have a far-reaching effect on the reaction it catalyses. It is not surprising, therefore, that enzymes are often present in cells in amounts which are infinitesimal compared with the amounts of substrate participating in the enzyme-catalysed reactions.

In addition to the above general characteristics, enzymes have their own *special characteristics* which serve to distinguish them from all other catalysts. The most remarkable of these additional characteristics is that enzymes are highly 'selective' in the kind of reaction they will catalyse, although different enzymes vary in this respect. Some enzymes will act on a group of closely related substrates, but others will act on one substrate only. An example of the less specific type of enzyme is provided by the carboxyesterases which catalyse the hydrolysis of many aliphatic esters.

$$R^1.COOR^2 + H_2O \rightleftharpoons R^1.COOH + R^2.OH$$

Almost any carboxyesterase will hydrolyse virtually any aliphatic ester (e.g. ethyl acetate where $R^1=CH_3$ and $R^2=C_2H_5$, or methyl butyrate where $R^1=C_3H_7$ and $R^2=CH_3$) because the specificity of carboxyesterases does not extend much beyond the presence of an ester linkage within an aliphatic substrate molecule. The nature of the rest of the molecule, i.e. the chemical identities of the aliphatic R^1 and R^2 radicals, is relatively unimportant. By contrast, most enzymes are much more sensitive in the type of substrate with which they will react. For example, maltose and cellobiose are two very similar disaccharides, both composed of two D-glucose units joined by 1,4 glycosidic linkages. The enzyme maltase will catalyse the hydrolysis of maltose but not of cellobiose because it acts only upon an α1,4 glycosidic linkage. Similarly the enzyme cellobiase catalyses the hydrolysis of cellobiose but not of maltose because it acts only upon a β1,4 glycosidic linkage.

HOCH₂ ... HOCH₂ ... HOCH₂ ... HOCH₂

α-glucose unit   α-glucose unit   β-glucose unit   β-glucose unit
(rotated through 180°)

maltose   cellobiose

An extreme case of specificity is shown by lactic dehydrogenase which oxidizes lactic acid to pyruvic acid:

CH₃ — H—C—OH — COOH $\xrightarrow[+2H]{-2H}$ CH₃ — C=O — COOH

L-lactic acid   pyruvic acid

Not only does this enzyme react with lactic and no other acid, but it also discriminates between the two optical isomers (see p. 149) because it is absolutely specific for the L-isomer.

Other special properties of enzymes include characteristic responses to heat and changes of pH, which depend primarily on the fact that enzymes are proteins. It will be more appropriate to consider these properties in the context of the next section.

The rate of enzyme reactions

An enzyme-catalysed reaction can be represented in the simplest possible terms as $S \rightleftharpoons P$, where S is the substrate and P the products. The course of such a reaction can be followed by measuring either the disappearance of S or the formation of P. Sometimes this is relatively easy. For example, in the hydrolysis of ethyl acetate by a carboxyesterase enzyme

$$CH_3COO \overbrace{C_2H_5 + HO} H \rightleftharpoons CH_3COOH + C_2H_5OH$$

ethyl acetate   water   acetic acid   ethyl alcohol

the rate at which the reaction takes place may be measured simply by titrating the acetic acid in samples of the reaction mixture taken at known times after the start of the reaction. Although the experimental procedure for estimating the rate of most enzyme reactions is considerably more complicated than this, the principle remains the same.

When the amount of product formed is plotted against time a *progress curve* of the reaction is obtained. In most cases this curve has the form shown in Fig. 12.6. At first the curve is more or less linear, indicating a constant rate of reaction, but later it flattens out. The main reason why the rate falls off is that, as the reaction proceeds, the substrate is progressively used up, although other changes in the reaction mixture (such as a change in its pH) may also be involved. To obtain a reliable measure of the rate of an enzyme reaction it is therefore necessary to determine it over a short time interval immediately after the enzyme is added to the substrate. Ideally the rate should be measured at the exact moment the enzyme is added, but this

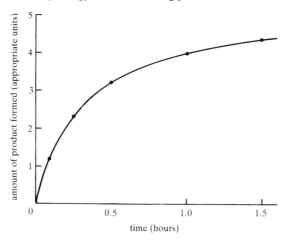

Fig. 12.6 A typical progress curve for an enzyme-catalysed reaction.

is not a practicable proposition. However, since the initial part of the prog-ress curve is linear, extrapolation of the line through zero time gives an esti-mate of the rate of reaction at zero time. This rate is referred to as the *ini-tial reaction velocity*, and approximates very closely to the rate of an en-zyme-catalysed reaction before any change in the concentration of the subs-trate has occurred. In the presence of excess substrate and very low enzyme concentration, the rate of formation of minute but measurable quantities of product is linear for a relatively long time. Under these conditions the mi-cro-analytical methods now available enable the initial reaction velocity to be readily estimated with great accuracy.

Investigation of the rate of enzyme action under different experimental conditions provides valuable clues about enzymes and how they work. In designing an experiment to discover the effect of any one factor on the velocity of an enzyme reaction, it is of course essential that the other con-trolling factors are kept constant. Thus, if we are investigating the effect of substrate concentration, then the enzyme concentration, temperature and pH must all be constant. The type of information that can be gained from such investigations is well illustrated by the effect which substrate concen-tration has on the rate of enzyme reactions. With a fixed concentration of enzyme, an increase of substrate will at first cause a very rapid rise in the rate of reaction. As the substrate concentration is increased further, howev-er, the increase in the reaction rate begins to slow down until eventually an almost constant rate is attained. A plot of initial reaction velocity against substrate concentration thus gives a curve like that shown in Fig. 12.7, which is mathematically called a rectangular hyperbola because the constant rate is theoretically reached only at infinite substrate concentration. Such a curve is defined (along the *y* axis, or ordinate, of the graph) by the max-imum or limiting velocity, V_{max}, which is the constant rate reached when the substrate concentration is very high, and (along the *x* axis, or abscissa) by the concentration of substrate at which the rate is half the maximum velocity. The value of this substrate concentration is called the Michaelis constant, K_m. The values of these two quantities, V_{max} and K_m, specify the shape of the curve which is obtained for any given enzyme reaction at any one particular enzyme concentration. The existence of a limiting substrate concentration, above which the rate of reaction remains constant, can be explained by assuming that the substrate (S) and the enzyme (E) are bound for a finite time as an enzyme–substrate complex (ES), which then decom-

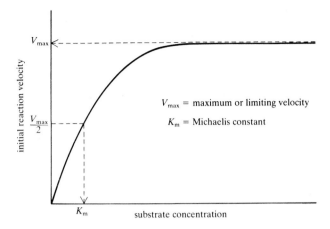

V_{max} = maximum or limiting velocity

K_m = Michaelis constant

Fig. 12.7 Influence of substrate concentration on the rate of an enzyme-catalysed reaction.

poses into the products (P) and liberates the enzyme; the latter is then available to repeat the cycle. The overall reaction can be formulated as

$$E + S \rightleftharpoons ES \rightarrow E + P$$

from which it can be seen that the rate of the entire process depends on the rate of the breakdown of ES. At low substrate concentrations, where there is an excess of enzyme, the enzyme is not working at full capacity and hence increasing the substrate concentration will also increase the rate by increasing the concentration of ES. At high substrate concentrations the enzyme is working at full capacity, and increasing the substrate concentration will not result in an increase in rate because further increase in the concentration of S does not greatly increase the concentration of ES. A marked increase in rate can therefore only be achieved by increasing the concentration of E to react with the excess substrate. The two situations are represented diagrammatically in Fig. 12.8. This theoretical interpretation of the dependence of reaction velocity on substrate concentration is now known to be correct, because in certain instances the formation and breakdown of an enzyme – substrate complex has been proved experimentally.

Fig. 12.8 Diagrammatic representation of the effect of substrate concentration on enzyme activity, assuming that enzyme concentration is constant. Note that both 'saturating' and high-substrate concentrations give the same yield of product (P) per unit of time.

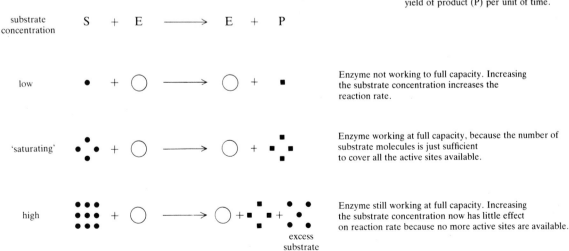

substrate concentration   S + E ⟶ E + P

low — Enzyme not working to full capacity. Increasing the substrate concentration increases the reaction rate.

'saturating' — Enzyme working at full capacity, because the number of substrate molecules is just sufficient to cover all the active sites available.

high — excess substrate — Enzyme still working at full capacity. Increasing the substrate concentration now has little effect on reaction rate because no more active sites are available.

Of the other factors influencing the rate of enzyme reactions, only temperature and pH will be considered. The responses to these two factors are characteristic properties of enzymes, resulting mainly from the fact that enzymes are proteins.

Effect of temperature   In common with all chemical reactions, the rate of an enzyme-catalysed reaction is affected by temperature. As a general rule, chemical reactions go about twice as fast for every 10 °C increase in temperature (i.e. they have a Q_{10} of about 2), and this is true for enzyme reactions up to about 35 °C. However, because enzymes are proteins, temperatures above this value cause progressive denaturation of the enzyme and hence destroy its catalytic function. As the temperature is raised beyond about 35 °C, therefore, the increase in rate of an enzyme reaction begins to fall off. Since denaturation may often have a Q_{10} of several hundred, the loss of catalytic properties is very fast and, in typical cases, begins at about 35 °C and is complete as 60 °C is approached. Consequently the reaction rate reaches a maximum somewhere around 45 °C but, because thermal denaturation becomes increasingly important beyond this temperature, it then shows a sharp decline until the reaction stops as 60 °C is approached. The twofold effect of temperature on the rate of a typical enzyme reaction is shown in Fig. 12.9.

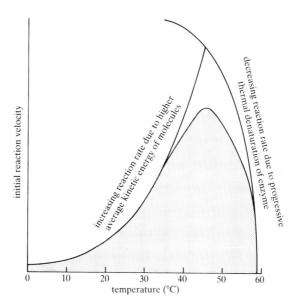

Fig. 12.9 Effect of temperature on the rate of an enzyme-catalysed reaction. A rise in temperature always tends to increase the rate of the reaction but beyond about 35 °C it also increases the rate of thermal denaturation of the enzyme. The shaded area represents the outcome of this two-fold effect of temperature.

In considering the effect of temperature on enzyme reactions it is essential to take the time factor into account. If the reaction is allowed to continue for any length of time at any temperature where denaturation of the enzyme is occurring, then the rate will gradually fall off. Hence, for any specified period of time there will be an *optimum* temperature, i.e. one particular temperature at which the greatest amount of chemical change is brought about in a given time by a given amount of enzyme under a given set of experimental conditions (Fig. 12.10).

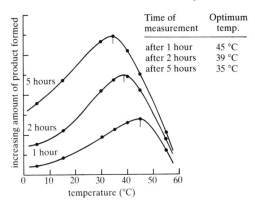

Time of measurement	Optimum temp.
after 1 hour	45 °C
after 2 hours	39 °C
after 5 hours	35 °C

Fig. 12.10 Relationship between time and the optimum temperature of an enzyme-catalysed reaction. The longer an enzyme is left to react at a particular temperature, the larger will be the amount of product formed. However, since a rise in temperature beyond about 35 °C increases the rate of thermal inactivation of an enzyme, the higher the temperature at which the reaction is carried out the sooner will the reaction rate fall off. Hence the optimum temperature will be higher if the yield of the reaction is measured after a short time (i.e. before the enzyme has had time to become markedly inactivated by high temperatures) than if this yield is measured after a longer time.

Effect of pH As a general rule, an enzyme is active only within a relatively narrow range of pH. As shown in Fig. 12.11 there is a definite peak of activity over a particular range of pH values, the optimum pH, on either side of which the activity falls off sharply with changes in pH. Some enzymes exhibit an optimal range extending over several pH units, whereas others have a much narrower pH optimum. The optimal range of nearly all enzymes is near to neutrality, between about pH 5 and 7. Extremes of acidity and alkalinity, which lead to denaturation of proteins, lead also to the inactivation of enzymes. These changes in activity are irreversible, whereas those observed in the immediate vicinity of the optimum pH are usually reversible.

The activity/pH relationships of enzymes can be largely attributed to changes in their ionic condition. An enzyme, like all proteins, may exist in several different ionic forms, and the particular form which predominates depends on the pH of the reaction medium. The form predominating at the optimum pH is either the only one to have catalytic properties or the one which is most active. Generally speaking, an enzyme is most stable at or near its optimum pH, the observed value of which does not vary with the time factor in the way that its optimum temperature does. The optimum pH is therefore a more characteristic feature of a particular enzyme than its optimum temperature.

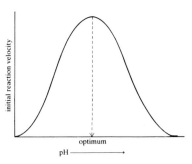

Fig. 12.11 Effect of pH on enzyme activity.

Mechanism of enzyme action

While it is known that enzyme action involves the formation and subsequent breakdown of an enzyme–substrate complex, it is probably true to say that for no single enzyme reaction is the precise mechanism fully understood. Nevertheless certain general features seem clear. Enzymes are proteins and, as such, have very large molecules whereas the substrates on which they act have, by comparison, relatively small molecules. For example, the enzyme invertase has a molecular weight of 270 000 whereas its substrate, sucrose, has a molecular weight of 342. The tremendous difference in molecular size suggests that only a small part of an enzyme molecule is directly involved in the formation of the enzyme–substrate complex. This critical part is called the *active site* because it is presumably here that the substrate becomes attached to the enzyme and chemical reaction occurs. Since enzymes are highly selective with regard to the substrates on which they act, it has further been suggested that the active site consists of a very specific molecular pattern which is complementary in some way to the molecular pattern of its substrate.

Enzymes are globular proteins, i.e. their molecules are coiled and folded into complex, roughly spherical structures. It is postulated that, at a given pH value, the configuration of the polypeptide chain and of the electrical charges carried by the amino acids on it is such that, at one particular place on the surface of the enzyme molecule, there exists a space which is complementary to the shape of a particular substrate molecule. This space is the active site. Since direct contact between enzyme and substrate is necessary for the formation of the enzyme – substrate complex, the substrate molecule is thought to be attracted by chemical forces (such as electrostatic charges) into this space, where it is activated prior to forming the products. The products, whose molecular patterns do not conform with that of the active site, are therefore repelled into the surrounding solution, leaving the active site free to 'trap' another substrate molecule (Fig. 12.12).

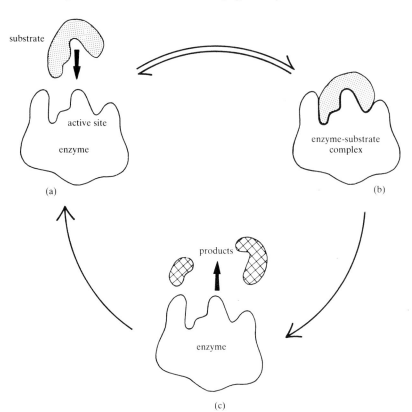

Fig. 12.12 The active-site model of enzyme action. (a) Substrate molecule is attracted to active site. (b) Enzyme–substrate is activated and reaction takes place. (c) Products are repelled from active site which is now free to repeat the cycle.

This interpretation of enzyme action in terms of an active site agrees with the behaviour of enzymes as proteins. The ability of a protein to function as an enzyme is determined by the particular shape of its molecule, and hence by the bonds which are holding the molecule in its three-dimensional structure. The highly complex but precise shape of the tertiary structure of a protein molecule thus offers an explanation of why enzymes should be so selective in the type of reaction they catalyse. They are selective because their active sites can combine with only one kind of substrate molecule (absolute specificity) or, at most, only a group of chemically related substrates (group specificity). The continued existence of an active site implies the maintenance of the three-dimensional structure of the molecule in which it is present. Any factor causing denaturation of proteins (such as high temperature, or extremes of pH) should therefore inactivate enzymes, and this is precisely what happens. The active-site model also suggests that the remainder of the molecule, not directly involved in combining with the substrate, may play an essential part in enzyme action by holding the active site in the correct shape. The fact that enzymes are always large molecules, even though their active sites occupy only a very short length (perhaps a sequence of up to twelve amino-acid units) of the entire polypeptide chain, is therefore not as paradoxical as might at first sight appear.

Since other properties of enzyme activity (e.g. inhibition) can also be interpreted in terms of the formation of an enzyme – substrate complex at an active site, there seems little doubt that the model is a valid one. Rapid progress is being made towards an understanding of the chemical nature, and likely mechanism of action, of the active sites of several enzymes, and it will probably not be long before an exact chemical account of many types of enzyme action can be given. For example, the activity of the enzyme chymotrypsin, which catalyses the hydrolysis of peptide bonds, is known to depend on histidine and serine residues at positions 57 and 195 respectively in the polypeptide chain. The three-dimensional structure of the enzyme is also known in some detail, and biochemists are now in a position to postulate a reaction mechanism.

Inhibition

The activity of an enzyme depends on a close fit between the substrate molecule and the active site. Therefore anything which interferes with the molecular architecture of the active site may prevent the formation or breakdown of the enzyme-substrate complex and thus reduce the rate of an enzyme-catalysed reaction. Compounds which reduce the rates of enzymic reactions by interfering with the structure of the active site are called *inhibitors*. Inhibitors may be classified into two main groups, *competitive* and *non-competitive*.

In competitive inhibition the structure of the inhibitor molecule is sufficiently like that of the normal substrate to enable it to compete with the normal substrate for possession of the same active site on the enzyme. Once bound at the active site, however, the inhibitor does not form free enzyme and the reaction product, and thus blocks the active site for occupation by the normal substrate. An example of competitive inhibition is the inhibition of succinic dehydrogenase by malonic or glutaric acid, the homologues below and above succinic acid in the dibasic-acid series. Succinic dehydrogenase catalyses the removal of hydrogen (dehydrogenation) from succinic acid to give fumaric acid (see top of p. 144).

The inhibitory effect of either malonic or glutaric acid can be lessened by increasing the concentration of succinic acid, because now a greater proportion of the enzyme molecules will react with the normal substrate rather than with the inhibitor. Numerous cases of competitive inhibitors are known, and their study has contributed greatly to an understanding of the mechanism of enzyme action.

Although substrates combine with enzymes at the active site, the func-

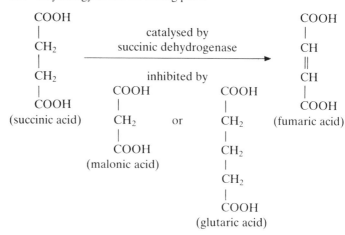

tioning of that site depends on the integrity of the rest of the enzyme molecule. In the non-competitive type of inhibition, the inhibitor may work by attaching itself somewhere on the enzyme surface, away from the the active site, in such a way that the structure of the active site is changed and therefore prevented from performing its normal catalytic role. Enzymes whose integral structure and therefore catalytic activity depend on sulphydryl (–SH) groups will be non-competitively inhibited by such heavy-metal ions as those of mercury (Hg^{2+}) and silver (Ag^{2+}), because these ions combine with the –SH groups of the enzyme. Similarly, enzymes which require metal ions (e.g. Mg^{2+}) for activity cease to function when reagents capable of binding the essential ion are added. Since non-competitive inhibitors interact not only with the free enzyme but also with the enzyme–substrate complex, their effects cannot be altered by increasing the concentration of the substrate. In fact, a study of the effects of a standard concentration of inhibitor on the rate of an enzymic reaction at different substrate concentrations provides a means of distinguishing between non-competitive and competitive inhibitors.

Inhibitors have been extensively used to block one reaction in a metabolic pathway, so as to cause the accumulation of an identifiable intermediate, thereby assisting in the elucidation of the steps involved in the overall process. The results of such 'blocking' experiments must be interpreted with extreme care because there are very few inhibitors which are specific for only one enzyme. Nevertheless, there is very strong experimental support for the claim that many drugs function because they exert a specific inhibitory effect on some critical enzyme in a tissue. Penicillin acts in this way by inhibiting one of the enzymes responsible for the synthesis of the bacterial cell wall.

Enzyme cofactors

Some enzymes consist entirely of amino acids and are therefore pure protein. Many enzymes, however, are active only if they contain one or more non-protein components, called cofactors. A cofactor may be either an organic molecule, often a vitamin derivative, or a metal ion. Some enzymes have cofactors which are firmly attached to the protein and therefore difficult to remove, whereas others have cofactors which may be easily separated from the protein. Cofactors that are firmly attached to the protein are called *prosthetic groups*, whereas those that are easily separated from it are termed coenzymes. When the protein component (the apoenzyme) and the coenzyme of an enzyme are separated, neither portion possesses the catalytic activity of the enzyme–cofactor complex (the holoenzyme). By mixing the apoenzyme and coenzyme together, the fully active holoenzyme can often be reconstituted. The same cofactor may be associated with different apoenzymes. Since the resultant holoenzymes are specific for different reactions, it must be the nature of the apoenzyme rather than that of the cofactor which determines the specificity of the catalysed reaction.

Cofactors play a direct part in the functioning of enzymes as catalysts, usually by serving as an intermediate carrier of functional groups, of specific atoms or of electrons. For example, the coenzyme NAD (nicotinamide adenine dinucleotide), which contains the amide of the vitamin nicotinic acid as an essential component, functions as the cofactor for a large number of dehydrogenases, i.e. enzymes which catalyse the transfer of hydrogen atoms $(e^- + H^+)$ from a donor molecule to an acceptor molecule other than oxygen. The coenzyme NAD is bound to the dehydrogenase protein relatively loosely during the catalytic reaction, and therefore functions more as a substrate than as a prosthetic group. In order for such a NAD-linked dehydrogenation reaction to continue, the reduced NAD has to be re-oxidized by another enzyme. Thus for the coenzyme to serve as a carrier of hydrogen atoms, two separate enzymes must be involved. The action of NAD as a coenzyme can be expressed as follows.

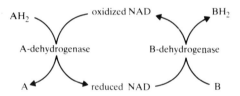

In enzymes requiring metal ions as prosthetic groups the metal ion may be the primary catalytic centre. For example, certain coenzymes, which are associated with oxidation in cells, contain iron. They are called cytochromes, and their ability to undergo reversible oxidation–reduction reactions depends on the interconversion of ferrous and ferric ions attached to the enzyme molecules. The interconversion of ferrous and ferric ions in cytochromes underlies one of the most fundamental processes in respiration.

$$Fe^{2+} \underset{\text{reduction}}{\overset{\text{oxidation}}{\rightleftharpoons}} Fe^{3+} \quad + \quad e^-$$

(ferrous ion)   (ferric ion)   (electron)

Naming and classification of enzymes

In a concerted effort to provide a systematic basis for naming the numerous enzymes which were being discovered, the Commission on Enzymes of the International Union of Biochemistry proposed in 1961 that such well established but chemically uninformative names as *papain*, which is a protease enzyme obtained from the sap of papaw (*Carica papaya*), should be abolished unless they can be equated with some specific enzyme, and that thenceforth all enzymes should be named and classified on the basis of the reaction they catalyse. On this international system an enzyme may be given two names, a *trivial* name which is usually short and convenient for everyday use, and a *systematic name* which identifies the reaction it catalyses. For example, the enzyme which splits fructose-1,6-diphosphate into two molecules of triose phosphates during the anaerobic breakdown of carbohydrates in living cells is normally called aldolase. This is the recommended trivial name, but the systematic name is ketose-1-phosphate aldehyde lyase, which is used only when a completely unambiguous definition is required as in international research journals. To classify the hundreds of enzymes now known, the following six classes are used.

1. Oxidoreductases   These enzymes catalyse oxidoreduction reactions. Most of them require prosthetic groups or coenzymes, which participate in the catalytic reaction by serving as intermediate electron carriers. The coenzymes include NAD^+, $NADP^+$, FAD, and various metal ions such as iron, copper and molybdenum. Some oxidoreductases, the oxidases, use molecular oxygen as the electron acceptor, whereas others, the dehydrogenases, remove hydrogen atoms from their substrates and transfer them to some acceptor other than oxygen.

2. Transferases   These catalyse the transfer of a particular chemical group from one substance to another. The transaminases, which transfer amino groups, are typical examples.

$$
\begin{array}{cccc}
R & R^1 & R & R^1 \\
| & | & | & | \\
H-C-NH_2 \;\;+ & C=O & \rightleftharpoons \quad C=O & + \;\; H-C-NH_2 \\
| & | & | & | \\
COOH & COOH & COOH & COOH
\end{array}
$$

(α-amino acid)   (α-keto acid)   (α-keto acid)   (α-amino acid)

An important subclass of this group are the kinases which catalyse the phosphorylation of their substrates by transferring a phosphate group, usually from ATP, thereby activating an otherwise metabolically inert compound for further transformations.

3. Hydrolases   The hydrolases catalyse reactions involving the cleavage of a molecule by the addition of water.

$$AB + H_2O \rightleftharpoons AH + BOH$$

There are a large number of hydrolases, which are subdivided into a number of subclasses, including the following.
(a) Carbohydrases which hydrolyse polysaccharides into oligosaccharides, and oligosaccharides into monosaccharides, e.g.

$$\text{sucrose} + H_2O \xrightarrow{\text{invertase}} \text{glucose} + \text{fructose}$$

(b) Esterases which promote the hydrolysis of esters, e.g.

$$\text{triglyceride} + H_2O \xrightarrow{\text{lipase}} \text{fatty acids} + \text{glycerol}$$

(c) proteases which catalyse the hydrolytic breakdown of proteins into smaller units, eventually to amino acids, e.g.

$$\text{dipeptide} + H_2O \xrightarrow{\text{peptidase}} \text{amino acids}$$

4. Lyases   These catalyse the removal of groups from their substrates leaving double bonds, or the addition of groups to double bonds. Examples are decarboxylases which remove CO_2 from carboxyl groups, and dehydrases which remove a molecule of water.

$$
\begin{array}{ccc}
\begin{matrix} CH_3 \\ | \\ C=O \\ | \\ \overline{(COO)}H \end{matrix}
&
\underset{\text{decarboxylase}}{\overset{\text{pyruvic}}{\rightleftharpoons}}
&
\begin{matrix} CH_3 \\ | \\ C=O \\ | \\ H \end{matrix}
\quad + \quad CO_2
\end{array}
$$

(pyruvic acid)   (acetaldehyde)

$$
\begin{array}{ccc}
\begin{matrix} CH(OH)COOH \\ | \\ CH_2 COOH \end{matrix}
&
\overset{\text{fumarase}}{\rightleftharpoons}
&
\begin{matrix} CHCOOH \\ \| \\ CHCOOH \end{matrix}
\quad + \quad H_2O
\end{array}
$$

(malic acid)   (fumaric acid)

5. Isomerases   These are enzymes which catalyse the interconversion of isomers.

$$
\begin{array}{ccc}
\begin{matrix} CH_2OH \\ | \\ C=O \\ | \\ CH_2O\,\text{P} \end{matrix}
&
\underset{\text{isomerase}}{\overset{\text{triose phosphate}}{\rightleftharpoons}}
&
\begin{matrix} CHO \\ | \\ HCOH \\ | \\ CH_2O\,\text{P} \end{matrix}
\end{array}
$$

(dihydroxyacetone   (glyceraldehyde-
phosphate)   3-phosphate)

6. Ligases (or synthetases)   These catalyse the joining together of two molecules (usually by condensation), coupled with the exergonic hydrolysis of ATP. They enable the chemical energy in ATP to be used for driving endergonic reactions.

$$
\begin{array}{ccc}
\begin{matrix} CH_2COOH \\ | \\ CH_2 \\ | \\ CHNH_2 \\ | \\ COOH \end{matrix}
\; + NH_3 + ATP
&
\underset{\text{synthetase}}{\overset{\text{glutamine}}{\longrightarrow}}
&
\begin{matrix} CH_2CONH_2 \\ | \\ CH_2 \\ | \\ CHNH_2 \\ | \\ COOH \end{matrix}
\; + H_2O + ADP + P_i
\end{array}
$$

(glutamic acid)   (glutamine)

Chapter 13

Carbohydrates

This group of substances gets its name from the fact that the first members to be chemically analysed contained only carbon, hydrogen and oxygen, with the last two elements in the ratio of 2 : 1 as in water. They were therefore regarded as hydrates of carbon which could be represented by the general formula $C_x(H_2O)_y$ where x and y are the same or consecutive whole numbers, e.g. glucose $C_6H_{12}O_6$ or sucrose $C_{12}H_{22}O_{11}$. This definition was subsequently shown to be inadequate because some compounds whose molecular composition does not agree with this general formula have the characteristic chemical and physical properties of carbohydrates (e.g. deoxyribose $C_5H_{10}O_4$), and other substances which conform to the formula clearly belong to other groups of compounds (e.g. lactic acid $C_3H_6O_3$). Although the name carbohydrate is retained, it is now used as a collective term for polyhydroxy aldehydes or polyhydroxy ketones and derivatives of these compounds.

Carbohydrates are classified according to their complexity into the three groups, monosaccharides, oligosaccharides and polysaccharides. *Monosaccharides* are simple sugars which cannot be hydrolysed into smaller units; the prefix mono- indicates one basic unit. *Oligosaccharides* are compound sugars which consist of two to about six monosaccharides joined together (oligo- means a few); they can be hydrolysed into the monosaccharide subunits of which they are composed. Sugars, whether monosaccharides or oligosaccharides, are colourless crystalline substances which are readily soluble in water and usually have a sweet taste.

Polysaccharides do not have the characteristic properties of sugars, being usually amorphous, often insoluble in water, and lacking a sweet taste. They are formed by joining together several hundreds to thousands (poly- means many) of monosaccharide subunits, often all of the same type, to form large complex molecules of high molecular weight. Like oligosaccharides, they can be broken down into their constituent monosaccharide subunits by hydrolysis.

Monosaccharides

The monosaccharides are divided into two groups: (1) the *aldoses*, which always contain a terminal aldehyde (–CHO) group and can be represented by the molecular formula $CH_2OH–(CHOH)_n–CHO$ where n can be any whole number from 1 to 7; and (2) the *ketoses*, which always contain a sub-terminal ketone (>CO) group and can be represented by the molecular formula $CH_2OH–(CHOH)_n–CO–CH_2OH$ where n can be zero (i.e. the >

CHOH group in the formula is missing) or any whole number from 1 to 4, exceptionally 5 or 6. The possession of alcohol groups ($-CH_2OH$ and $>$ CHOH) and either an aldehyde or a ketone group means that monosaccharides exhibit many of the properties characteristic of these functional groups. One of the important properties of monosaccharides is their ability to function as reducing agents. This property is usually demonstrated by the formation of a red precipitate on warming with Fehling's solution. The Fehling's reaction is also an important test in determining the structure of oligosaccharides, some of which (like all monosaccharides) are reducing sugars whereas others are non-reducing sugars.

Monosaccharides may also be classified according to the number of carbon atoms in the molecule. The simplest monosaccharides have three carbon atoms and are called trioses. Tetroses, pentoses, hexoses and heptoses contain 4, 5, 6 and 7 carbon atoms respectively, and monosaccharides with 8 and 9 carbon atoms also exist. As will be described shortly, still further variation results from differences in the spatial arrangement of the hydrogen ($-H$) and hydroxyl ($-OH$) groups attached to the carbon atom of the $>$CHOH groups that occur along the length of the molecule.

Optical Isomerism

The two simplest monosaccharides are the aldotriose, glyceraldehyde, and the ketotriose, dihydroxyacetone.

$$\begin{array}{ll}
_1CHO & _{*1}CH_2OH \\
| & | \\
H - _2C^* - OH & _2C = O \\
| & | \\
_3CH_2OH & _3CH_2OH \\
(\text{glyceraldehyde}) & (\text{dihydroxyacetone})
\end{array}$$

By convention, the aldehyde group is written at the top of the formula and the ketone group next to the top. For reference purposes the carbon atoms along the length of a monosaccharide chain are numbered sequentially as 1, 2, 3, etc., and referred to as C-1, C-2, C-3, etc. The numbering starts with the carbon atom of the aldehyde group for an aldose and with the terminal $-CH_2OH$ group adjacent to the ketone group for a ketose. In glyceraldehyde the C-2, which is indicated by an asterisk, has four different groups ($-CHO, -H, -OH$ and $-CH_2OH$) attached to it and is consequently asymmetric. This means that, when a molecule of glyceraldehyde is considered as a three-dimensional structure, there are two possible arrangements in space of these four groups. The two arrangements are related to each other as the left hand is to the right, and represent two distinct molecules which are mirror images of each other. Whichever way either one of them is orientated it is impossible to superimpose it on the other in such a way that all four groups coincide (Fig. 13.1). The substance glyceraldehyde therefore exists in two possible forms, which are different compounds because the configuration of their molecules, i.e. the arrangement of their atoms in space, is different. These two compounds have similar chemical and physical properties (e.g. the same boiling point, melting point, solubility in various solvents, etc.) but can be distinguished from each other by the way in which they rotate the plane of plane-polarized light. For this reason they are refer-

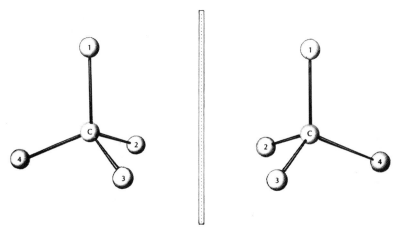

plane of an imaginary mirror

Fig. 13.1 Optical isomerism. When a carbon atom in an organic compound is attached to four different groups (1, 2, 3 and 4), the compound can exist in two forms which are mirror images of each other.

red to as *optical isomers*. If polarized light is passed through a solution of one of the isomers its plane of polarization is rotated to the left, while a solution of the other isomer will rotate the plane of polarization to the right. These two optical isomers can be distinguished on paper by a convention in which the – OH group on C-2 is written on the right or the left of the formula. The isomer whose configuration is represented by writing the – OH group on the right of C-2 is said to have the D configuration, whereas the isomer with the – OH written on the left of C-2 has the L configuration. The designation of the two compounds as D-glyceraldehyde and L-glyceraldehyde refers to the spatial arrangement of the groups around the asymmetric C-2, and does not indicate the direction in which the plane of polarized light is rotated by the compounds.

$$_1CHO \qquad\qquad _1CHO$$
$$| \qquad\qquad\qquad |$$
$$H - _2C - OH \qquad HO - _2C - H$$
$$| \qquad\qquad\qquad |$$
$$_3CH_2OH \qquad\qquad _3CH_2OH$$
(D-glyceraldehyde)   (L-glyceraldehyde)

Glyceraldehyde and dihydroxyacetone can be regarded, in theory, as the parent molecules from which aldoses and ketoses with longer carbon chains are built up by inserting one or more >CHOH groups between C-1 and C-2 for aldoses and between C-2 and C-3 for ketoses. Thus the addition of one >CHOH group to glyceraldehyde and to dihydroxyacetone gives an aldotetrose and a ketotetrose respectively.

$$_1CHO$$
$$|$$
$$H - _2C - OH \longleftarrow ------ \text{ >CHOH group added}$$
$$| \qquad\qquad\qquad\qquad\quad \text{to glyceraldehyde}$$
$$H - _3C - OH$$
$$|$$
$$_4CH_2OH$$
(an aldotetrose)

$$_1CH_2OH$$
$$|$$
$$_2C = O$$
$$|$$
$$H - _3C - OH \longleftarrow ------ \text{ >CHOH group added}$$
$$| \qquad\qquad\qquad\qquad\quad \text{to dihydroxyacetone}$$
$$_4CH_2OH$$
(a ketotetrose)

Further lengthening of the chain results in pentoses with five carbon atoms, and then hexoses with six carbon atoms. Each time the length of the carbon chain is increased by the addition of a >CHOH group, a new asymmetric carbon atom is introduced into the molecule so that the number of possible optical isomers is doubled. In the tetroses derived from glyceraldehyde there are two asymmetric carbon atoms and thus there are four possible isomers, two derived from D-glyceraldehyde and two from L-glyceraldehyde.

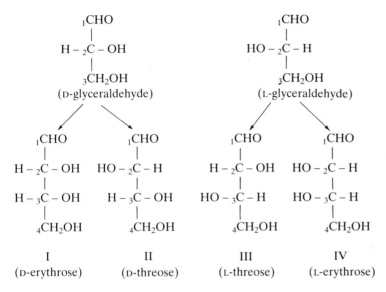

As can be seen from their structural formulae, compounds I and IV are mirror images and thus are related to each other in the same way as D-glyceraldehyde is to L-glyceraldehyde; they have similar chemical and physical properties but differ in their behaviour to polarized light. The same is true for compounds II and III. A compound and its mirror image are known as enantiomers (or enantiomorphs). Because compounds I and IV have different chemical and physical properties from compounds II and III, the two pairs are given separate names, erythrose and threose respectively. Optical isomers which are not related to each other as enantiomers are called diastereoisomers. Thus D-erythrose is an enantiomer of L-erythrose but a diastereoisomer of D-threose and L-threose. For any particular monosaccharide the number of possible isomers is given by the expression 2^n where n is the number of asymmetric carbon atoms within the molecule; hence in the aldohexoses, with four asymmetric carbon atoms, there are 2^4 or 16 possible isomers.

Dihydroxyacetone, as can be seen from its structural formula, has no asymmetric carbon atom, and so it is not until a >CHOH group has been introduced into the molecule as C-3, to give a ketotetrose, that optical isomers are possible. The two isomers can be represented as shown. As in glyceraldehyde, the D- and L-isomers are distinguished on paper by writing the −OH group on the right and left respectively of the asymmetric carbon atom, which in ketotetrose is C-3.

With the existence of a large number of optical isomers in carbohydrates, it is necessary to have a standard method of labelling their structural formulae. By international agreement it has been decided that all monosaccharides should be classified as D or L according to whether they are chemically derived from D- or L-glyceraldehyde, which are the agreed 'parental' configurations. This leads to a general rule that, when the carbon atoms of a monosaccharide molecule are numbered in the conventional manner from the carbonyl end downwards, the $-OH$ group on the highest numbered asymmetric carbon atom is to the right in the D-compound, and to the left in the L-compound. For example, in an aldohexose like glucose, the $-OH$ group is written on the right of C-5 in a D-hexose and on the left of it in an L-hexose, because in aldohexoses C-5 is the asymmetric carbon atom furthest away from the carbonyl end of the molecule. The formulae for some common hexoses are given.

$$
\begin{array}{cccc}
_1CHO & _1CHO & _1CHO & _1CH_2OH \\
| & | & | & | \\
H-_2C-OH & HO-_2C-H & HO-_2C-H & _2C=O \\
| & | & | & | \\
HO-_3C-H & H-_3C-OH & HO-_3C-H & HO-_3C-H \\
| & | & | & | \\
H-_4C-OH & HO-_4C-H & H-_4C-OH & H-_4C-OH \\
| & | & | & | \\
H-_5C-OH & HO-_5C-H & H-_5C-OH & H-_5C-OH \\
| & | & | & | \\
_6CH_2OH & _6CH_2OH & _6CH_2OH & _6CH_2OH \\
\text{(D-glucose)} & \text{(L-glucose)} & \text{(D-mannose, a} & \text{(D-fructose,} \\
 & & \text{diastereoisomer of} & \text{a ketohexose)} \\
 & & \text{glucose)} &
\end{array}
$$

The aldohexoses which are diastereoisomers of glucose differ from it by having the –H and –OH groups on carbon atoms 2 to 4 arranged differently from the positions they occupy in glucose. These diastereoisomers also exist as D- and L-enantiomers in which the $-OH$ group on C-5 is on the same side of the formula as in the D- and L-isomers of glucose. The assignment of a particular monosaccharide to either the D- or the L-series is thus merely a convenient means of relating its configuration to that of the two isomers of the standard reference substance, glyceraldehyde.

Ring structure of monosaccharide molecules

Up to this point monosaccharide molecules have been represented by straight-chain formulae, but such representations are inconsistent with some of the observed properties of monosaccharides. In aqueous solution, for example, many monosaccharides react as if they had one more asymmetric carbon atom than is indicated by the structure of their straight-chain formulae. When a carbon atom is considered in three dimensions, its four valency bonds point to the four corners of a regular tetrahedron, and this means that the carbon atoms along the length of a monosaccharide molecule are joined together through an angle of about 109 ° and not through 180 ° as represented in a straight-chain formula. As a consequence of these bond angles in a monosaccharide molecule the two ends of the carbon chain may be brought close together in space. A characteristic feature of the carbonyl group is that it is capable of reacting with the hydroxyl group of an alcohol to produce a type of compound called a hemiacetal, a reaction which can be shown as follows.

$$
\begin{array}{ccccc}
\text{R} & & \text{H} & & \text{R} \qquad\qquad \text{OH} \\
\diagdown & & | & & \diagdown \qquad\qquad \diagup \\
\quad \text{C} = \text{O} & + & | & \rightarrow & \text{C} \\
\diagup & & | & & \diagup \qquad\qquad \diagdown \\
\text{H} & & \text{O}\!-\!\text{R}^1 & & \text{H} \qquad\qquad \text{O}\!-\!\text{R}^1 \\
\text{(an aldehyde)} & & \text{(a primary alcohol)} & & \text{(a hemiacetal)}
\end{array}
$$

A hemiacetal may be formed in a monosaccharide molecule when the carbonyl group at one end of the chain comes into a position where it can react with one of the hydroxyl groups at the other end. The reaction produces a cyclic molecule in which an oxygen bridge ($-\text{O}-$) joins the two carbon atoms involved in forming the hemiacetal. The cyclic molecules are usually either five-membered rings (with four carbon atoms and an oxygen bridge) or six-membered rings (with five carbon atoms and an oxygen bridge). The structures are called furanose and pyranose rings respectively. In D-glucose, for example, ring completion between C-1 and the oxygen atom of the $-\text{OH}$ group on C-5 gives a six-membered (pyranose) ring.

$$
\begin{array}{ll}
\begin{array}{l}
\text{H}\quad\text{O} \\
\diagdown\!\!/\!/ \\
\;_1\text{C} \\
| \\
\text{H}-\,_2\text{C}-\text{OH} \\
| \\
\text{HO}-\,_3\text{C}-\text{H} \\
| \\
\text{H}-\,_4\text{C}-\text{OH} \\
| \\
\text{H}-\,_5\text{C}-\text{OH} \\
| \\
\;_6\text{CH}_2\text{OH}
\end{array}
&
\begin{array}{l}
\text{H}\quad\text{OH} \\
\diagdown\!\!/ \\
\;_1\text{C} \\
| \\
\text{H}-\,_2\text{C}-\text{OH} \\
| \\
\text{HO}-\,_3\text{C}-\text{H}\qquad\text{O} \\
| \\
\text{H}-\,_4\text{C}-\text{OH} \\
| \\
\text{H}-\,_5\text{C} \\
| \\
\;_6\text{CH}_2\text{OH}
\end{array}
\end{array}
$$

(open chain form of D-glucose)   (pyranose ring of D-glucose)

The true spatial relations of the cyclic molecules of monosaccharides are better represented in a manner proposed by the English chemist Haworth. In the Haworth formulae a six-membered ring is represented as a hexagon in a plane projecting at right angles to the plane of the paper on which the formula is written, the front half of the hexagon nearest the reader being indicated by thicker lines. Haworth formulae for both the six- and five-membered rings of D-glucose are given in Fig. 13.2, but the following discussion of the Haworth arrangement will be restricted to the six-membered ring. The carbon atoms are numbered clockwise from the right-hand corner of the hexagon so that C-2 and C-3 are in front of the paper, and C-5 and the oxygen bridge are behind it. The $-\text{H}$ and $-\text{OH}$ groups that are written on the left-hand side of the asymmetric carbon atoms in the straight-chain formula are represented as projecting above the plane of the ring, and the corresponding groups on the right-hand side as projecting below. However, there is a discrepancy in the relative position of the $-\text{H}$ group at C-5 which appears in the Haworth formula below instead of above the plane of the ring. This apparent anomaly is due to the twisting which takes place when C-5 forms an internal hemiacetal. In forming a ring structure from the straight-chain molecule, C-5 must be rotated so that the oxygen atom in the $-\text{OH}$ group on this carbon atom is brought upwards into the plane of the

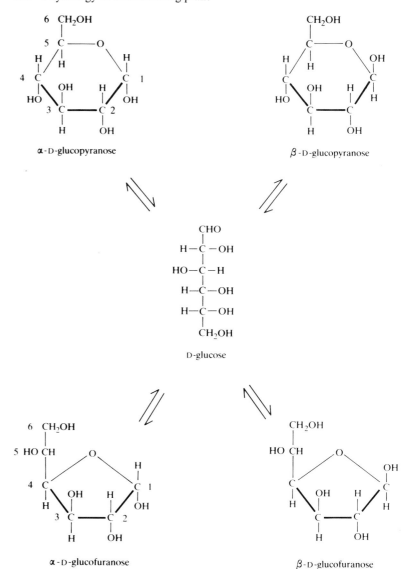

first five carbon atoms. Consequently the −H group attached to C-5 is shifted downwards to the other side of the chain because this carbon atom has been rotated through more than a 90 ° angle. Because the −OH group on C-5, which determines whether a molecule belongs to the D- or L-series, is involved in forming the oxygen bridge of the pyranose ring, the configuration of D-glucopyranose (the name given to the pyranose form of D-glucose) may be recognized by the position of the −CH$_2$OH group at C-6 which is not involved in the ring. When the −CH$_2$OH group is written above the plane of the ring the compound is D-glucose, and when written below it the compound is L-glucose. It is valid to use this terminal group to recognize the D or L configurations of the pyranose form of aldohexoses, because it is joined to the same carbon atom (i.e. C-5) which distinguishes between the two configurations in the straight-chain formulae.

From the Haworth formulae for D-glucose shown in Fig. 13.2, it will be noted that the formation of a ring structure creates a new asymmetric centre at C-1 because this now has four different groups attached to it. This means that there are two possible forms of D-glucopyranose, one in which the −OH group at C-1 is below the ring and another in which it is above the ring. These two isomers are called the α and β forms respectively, so that the full description of the two forms of D-glucopyranose is α- D-glucopyranose and β- D-glucopyranose. The distinction between α and β forms of monosaccharides is extremely important in the biological world, because the two forms frequently give rise to derivatives with very different properties. For example, starch is a storage material derived from the α form of D-glucopyranose, whereas cellulose is a structural material derived from the β form of the same compound.

Oligosaccharides

Oligosaccharides are formed by joining together a definite but small number, usually less than six and never more than ten, of monosaccharide molecules. Linkage, which occurs only when the monosaccharides are in the ring form, involves the removal of the elements of water, i.e. condensation, between the reducing group of one monosaccharide and a hydroxyl group, often at C-4, of another monosaccharide. Such a linkage, which forms an 'oxygen bridge' between the monosaccharide residues, is known as a *glycosidic bond*. (Note: whenever any building-block compound is assembled into a larger molecule, it is thereafter called a residue, e.g. a glucose residue in a polysaccharide, an amino-acid residue in a protein, etc.) The formation of a glycosidic bond can be shown as follows.

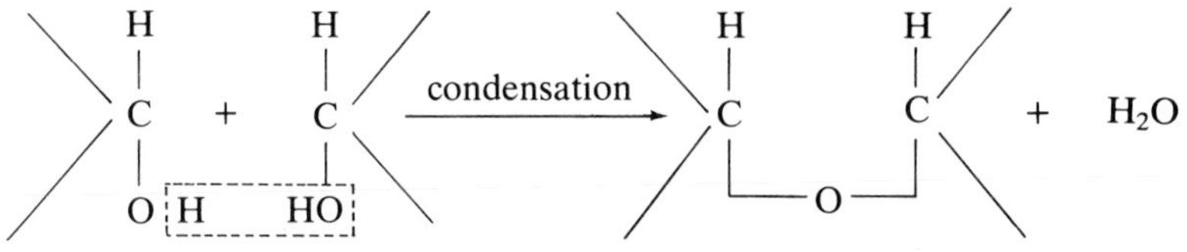

(glycosidic bond)

Oligosaccharides are classified into disaccharides, trisaccharides, and so on, according to the number of monosaccharide molecules linked together. Only disaccharides will be considered here.

The disaccharides of biological importance are composed of hexose units so that their formation may be represented by

$$C_6H_{12}O_6 + C_6H_{12}O_6 \rightarrow C_{12}H_{22}O_{11} + H_2O$$

The reversal of this reaction is effected by either a specific enzyme or dilute acid, both of which catalyse the hydrolysis of the glycosidic linkage.

The two monosaccharide units of a disaccharide may be alike, as in maltose which on hydrolysis gives two molecules of D-glucose, or different, as in sucrose which hydrolyses into one molecule of D-glucose and one of D-fructose.

There are only a few naturally occurring disaccharides, of which sucrose is the most important, but other disaccharides, notably maltose and cello-

maltose

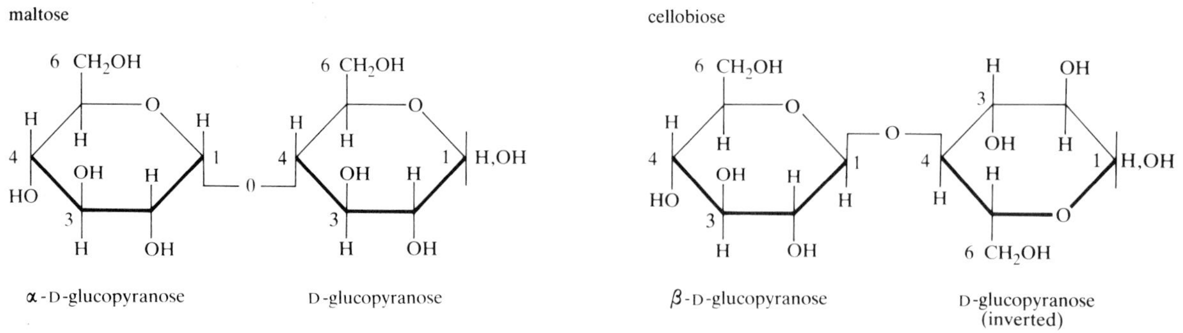

cellobiose

6 CH$_2$OH   6 CH$_2$OH   6 CH$_2$OH   H   OH

H   H   O   H   H   H   O   H   H   O   OH   H

4   OH   H   1   4   OH   H   1 H,OH   4   OH   H   1   4   H   1 H,OH

HO   3   O   3   HO   3   H   O

H   OH   H   OH   H   OH   6 CH$_2$OH

α-D-glucopyranose   D-glucopyranose   β-D-glucopyranose   D-glucopyranose (inverted)

Fig. 13.3 Comparison of the structure of maltose with that of cellobiose. (Note: where both α and β forms may be present at the C-1 or C-2 of a monosaccharide residue, neither configuration is specified and this fact is indicated in a Haworth formula by writing H.OH at the appropriate carbon atom.)

biose, are encountered as hydrolysis products of polysaccharides. This second type of disaccharide illustrates a general principle that linear polysaccharides, and the backbone of branched polysaccharides, can be represented as repeating disaccharide units.

If the glycosidic bond of a disaccharide is formed in such a way that the reducing group of one of the monosaccharide components is left free (or, more accurately, potentially free because of the equilibrium existing between the free carbonyl group and the hemiacetal linkage), the disaccharide which is formed will give many of the reactions characteristic of the carbonyl group, including the reduction of Fehling's solution. If, on the other hand, the glycosidic bond involves the reducing groups of both the monosaccharide components, the disaccharide is non-reducing and will not form a red precipitate on warming with Fehling's solution. The reducing disaccharides constitute by far the larger group, the only well-known non-reducing disaccharide being sucrose.

The elucidation of the structure of a disaccharide requires the solution of three problems. Firstly, the identity of the monosaccharide components and whether they are present in the pyranose or furanose ring form; secondly, which carbons are involved in the glycosidic bond between the two monosaccharide components; and thirdly, the configuration of the glycosidic bond as either α- or β-glycosidic. On the basis of the answers to these problems, the structures assigned to some common disaccharides will be discussed.

Maltose is formed from starch by the action of amylase enzymes, and is itself hydrolysed by the enzyme maltase or by acids to give D-glucose as the sole product. It has been shown that the two glucose residues are both in the pyranose form, and the glycosidic bond between them involves the C-1 in the α configuration of one residue and the C-4 of a second residue; this second residue may be in either the α or the β form. For this reason maltose is described as an α(1→4) glycoside, where the bracketed numbers indicate which carbons are linked. Maltose is a reducing sugar because the hemiacetal hydroxyl (i.e. the potential carbonyl group) of the second residue is not involved in forming the glycosidic bond, and therefore retains its reducing character. The structure of maltose is shown in Fig. 13.3.

The structure of maltose should be compared with that of cellobiose, a disaccharide which does not occur free in nature but can be obtained under certain conditions of acid hydrolysis from cellulose. Although maltose and cellobiose are two entirely different disaccharides, they differ structurally from each other only in the configuration of the link between the two glucose residues which is α in maltose and β in cellobiose (Fig. 13.3). The technique of X-ray diffraction analysis shows that in cellobiose the plane of the ring of the second glucose residue is inverted, as shown by the fact that its

C-6 is below the plane of the ring whereas the C-6 of the first residue is above it.

Sucrose (cane sugar) is the most abundant and widespread sugar in the plant world. It consists of a molecule of D-glucose linked to one of D-fructose by a glycosidic bond between C-1 of α-D-glucopyranose and C-2 of β-D-fructofuranose.

α-D-glucopyranose   β-D-fructofuranose

Since the glycosidic bond involves the potential reducing groups of both the glucose and fructose residues, sucrose is a non-reducing sugar. Sucrose is readily hydrolysed into its monosaccharide components by the catalytic action of dilute acid or of the enzyme invertase. Although fructose exists in sucrose in the furanose form, fructopyranose is the form which is obtained on hydrolysis. D-fructose is an exceptional monosaccharide in that it always assumes the furanose form when serving as a glycosidic component in an oligosaccharide.

Polysaccharides

Polysaccharides, like oligosaccharides, consist of monosaccharide units joined together by glycosidic bonds. The difference is mainly one of molecular size, several hundreds (or even thousands) of monosaccharide residues often contributing to the macromolecule of a polysaccharide. The molecules of some polysaccharides are linear chains, those of others are branched chains.

Chemically, polysaccharides are commonly classified into homoglycans or heteroglycans according to whether they are built from units of one type of monosaccharide or from a mixture of monosaccharide units. Homoglycans can be further subdivided according to the particular monosaccharide unit obtained on complete hydrolysis. Thus a polysaccharide which yields only the hexose D-glucose on hydrolysis is called a glucan, whereas one that yields the pentose D-xylose is called a xylan. From a plant-biological point of view, however, it is probably more useful to classify polysaccharides on a functional basis into structural polysaccharides and food-reserve polysaccharides. Cellulose and starch are the obvious examples of these two types.

Cellulose

Cellulose is the major component of plant cell walls and, as such, it is the most abundant naturally occurring organic substance. Chemically, cellulose is a glucan, being built up of β-D-glucose units linked by (1→4) glycosidic bonds to form linear chain-like molecules. The chain length, i.e. the number of component D-glucose units, often described as the degree of polymerization (DP), may be determined by various physical and chemical methods. The results vary according to the source of the cellulose, and the method used to extract it. It is therefore impossible to give an exact figure for the molecular weight of cellulose, because its molecules are of various lengths. The best that can be given is an average figure, but estimated molecular weights of between 350 000 and 550 000 imply that there are between 2000 and 3000 glucose residues in one molecule of naturally occurring cellulose.

Partial hydrolysis of cellulose yields the disaccharide cellobiose (derived from the paired residues shown in parentheses in the formula) and a group of larger oligosaccharides. A cellulose molecule can therefore be pictured as a long thin rod with the primary alcohol radicals at C-6 of the β-D-glucose units projecting alternately above and below the plane of the pyranose rings. Its structure may be represented as shown.

cellobiose unit

The part played by cellulose in the structure of plant cell walls is discussed in Chapter 2 (p. 17).

Starch

Starch is structurally related to cellulose, but nevertheless fulfils a very different function. It is the commonest form in which the carbohydrates produced by photosynthesis are stored for future metabolic use. Glycogen, with a very similar structure, performs a comparable function in the animal body.

Starch, like cellulose, is a glucan (i.e. glucose is the only monosaccharide obtained on complete hydrolysis) but, unlike cellulose, it is a mixture of two polymers. Of these, one is a linear polysaccharide called amylose, whereas the other is a branched polysaccharide called amylopectin. The proportions of the two polymers, and the degree of polymerization of each of them, vary according to the plant from which the starch is obtained. Most starches, as synthesized by plants, contain approximately 20–30 per cent of amylose and the balance of amylopectin.

Amylose has a chain length of between 50 and 1500 D-glucose units, all linked by α(1→4) bonds (Fig. 13.4). Amylose is therefore related structurally to cellulose in the same way as maltose is to cellobiose, the (1→4) glycosidic bonds of amylose and maltose being in the α configuration whereas those of cellulose and cellobiose are in the β configuration. As a consequence of the α configuration of amylose, all the primary alcohol radicals at C-6 of the D-glucose units project in the same direction away from the plane of the pyranose rings. X-ray crystallography has shown that, in the crystalline state, the molecular chains of amylose are not straight but coiled into a helix. The blue-black coloration produced by the addition of aqueous iodine

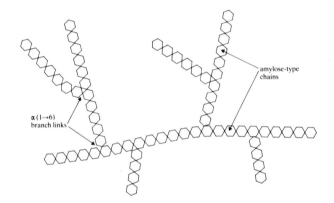

amylose

(a)

maltose unit

amylopectin

$\alpha(1\rightarrow4)$

$\alpha(1\rightarrow6)$

(b)

$\alpha(1\rightarrow4)$

amylose-type chains

$\alpha(1\rightarrow6)$ branch links

(c)

Fig. 13.4 Polymers of starch. (a) Structure of amylose. (b) α (1 → 6) link at a branching point in the molecule of amylopectin. (c) Diagram to show general structure of amylopectin; each hexagon represents a glucose residue.

to starch is due to the ability of the amylose component to hold iodine within the coils of the helix by forming an adsorption complex. The precise colour depends on the length of continuous helix in the core of which iodine can be held.

Amylopectin has a branched-chain structure, which is formed from the linear chain of amylose by detaching short lengths, averaging about twenty-five glucose units, and then re-attaching them as short branch-chains by $\alpha(1\rightarrow6)$ linkages. A branching point is represented in Fig. 13.4b. Amylopectin stains reddish-purple with iodine, presumably because the lengths of unbranched amylose-type helix are much shorter than in amylose itself. Amylopectin molecules vary in size from 2000 to 220 000 glucose units, and form multiple-branched structures (Fig. 13.4c) which are very much larger and denser than the helical molecules of amylose.

In a living plant cell the molecules of amylose and amylopectin are synthesized in special leucoplasts, and deposited in the form of granules. Although starch functions as reserve food material in all plants in which it is formed, the detailed structures of starch from different plants are so specific that an experienced observer can usually identify the plant from which a sample of starch comes by the size and shape of the granules (see Fig. 2.14).

Chapter 14

Proteins

Proteins are exceedingly complex substances which always contain the elements carbon, hydrogen, oxygen and nitrogen, and often sulphur as well. Their molecular weights are very high, ranging from about 5000 to 1 million or more. Proteins play a vital role in the chemistry of life, a fact which was recognized in the early nineteenth century when the Greek word *proteios*, meaning 'primary', was used as a basis for the name of these substances. Although there are literally thousands of different proteins, which vary greatly in their properties and biological functions, all proteins are constructed of subunits called *amino acids*. These are joined end to end to form long thread-like molecules, which in some proteins occur as straight chains, but in others are coiled up into roughly spherical structures. In order to understand the nature of proteins, it is first necessary to consider the structure and properties of amino acids.

Amino acids

As their name implies, amino acids contain an amino ($-NH_2$) group and a carboxyl ($-COOH$) group. Although several hundred substances containing these two groups are known, it is a remarkable fact that only about twenty occur naturally as the building blocks of proteins. These are all α-amino acids, which means that the amino group is attached in the alpha position, i.e. on the carbon atom next to the $-COOH$ group in the molecule. Amino acids are derived from fatty acids by substituting an $-NH_2$ group for one of the $-H$ atoms in the hydrocarbon chain, and thus the general formula for α-amino acids is

$$\begin{array}{c} NH_2 \\ | \\ R - C - COOH \\ | \\ H \end{array}$$

where the group R, representing the rest of the molecule, is different in each amino acid. The chemical nature of R, frequently called the side chain, provides a convenient basis for dividing the naturally occurring amino acids into three main groups, the neutral, basic, and acidic amino acids. To show the range of structure that occurs in the side chain, some of the twenty common amino acids are listed in Table 14.1. Three-letter abbreviations, which are commonly used for these amino acids, are also shown in the table.

The members of the first main group, which contains the majority of the common amino acids, have only one $-NH_2$ and one $-COOH$ group and are classified as *neutral* amino acids because the $-NH_2$ and $-COOH$ groups

Table 14.1 Some of the common amino acids found in proteins

R	Name	Abbreviation	R	Name	Abbreviation
			NEUTRAL		
Aliphatic			**Aromatic**		
H–	glycine	gly		phenylalanine	phe
CH_3–	alanine	ala			
CH_3 $\quad$CHCH$_2$– CH_3	leucine	leu	*R contains a hydroxyl group*		
				tyrosine	tyr
C_2H_5 $\quad$CH– CH_3	isoleucine	ile			
R contains a hydroxyl group			**Heterocyclic**		
$HOCH_2$–	serine	ser		tryptophan	try
R contains sulphur					
$HSCH_2$–	cysteine	cys			
			BASIC		
NH_2–$(CH_2)_4$–	lysine	lys			
NH_2–C–NH–$(CH_2)_3$– $\quad\;$NH	arginine	arg			
			ACIDIC		
HOOC–CH_2–	aspartic acid	asp			
HOOC–CH_2–CH_2–	glutamic acid	glu			

(R of phenylalanine)

(R of tyrosine)

(R of tryptophan)

neutralize each other. This group is subdivided into several series according to whether the R group (1) is aliphatic, aromatic or heterocyclic, and (2) has other functional groups attached to it. The simplest substitution for R is a hydrogen atom, giving glycine which therefore has the formula H–$CH(NH_2)COOH$. The next amino acid in order of complexity is when R is a methyl group, giving alanine with the formula

$$CH_3 - \underset{\underset{\displaystyle H}{|}}{\overset{\overset{\displaystyle NH_2}{|}}{C}} - COOH$$

In phenylalanine (aromatic) one of the H atoms of the methyl group in alanine is replaced by a benzene ring. Leucine and isoleucine have aliphatic hydrocarbon side chains which are branched. Serine (aliphatic) and tyrosine (aromatic) each have a hydroxyl group in the side chain. Cysteine is an example of an amino acid containing sulphur in addition to the four elements common to all amino acids. This acid deserves special comment because its sulphydryl group, –SH, is highly reactive and, on oxidation, combines with the corresponding sulphydryl group of another cysteine molecule to give a double amino acid, cystine.

H S – CH_2 – $CH(NH_2)COOH$

oxidizing agent

$\xrightarrow{\text{oxidation}}$

S–CH_2–$CH(NH_2)COOH$

H S – CH_2 – $CH(NH_2)COOH$
(two molecules of cysteine)

S–CH_2–$CH(NH_2)COOH$
(one molecule of cystine)

In this reaction the two –H atoms of the –SH groups are split off and the two –S atoms become joined by what is called a *disulphide bridge*, –S — S –. This linkage is very important in maintaining the structure of certain protein molecules.

The other two main groups of amino acids can be discussed together. Amino acids with an additional –NH_2 group in the side chain are known as the *basic* amino acids. By contrast, amino acids with an additional –COOH group are the *acidic* amino acids; they are even called acids (e.g. aspartic acid) because they undergo the usual reactions of carboxylic acids, including salt and ester formation. Biochemically, however, the most important reaction of the extra –COOH group on the side chain is with ammonia to give the corresponding amide.

$$
\underset{\text{(ammonia)}}{\begin{array}{c} H \\ \diagdown \\ N\text{–}H \\ \diagup \\ H \end{array}}
+
\underset{\text{(aspartic acid)}}{\overset{\overset{\displaystyle NH_2}{|}}{HOOC\text{–}CH_2\text{–}C\text{–}COOH}}
\;\rightarrow\;
\underset{\text{(asparagine)}}{\overset{\overset{\displaystyle NH_2}{|}}{\begin{array}{c} H \\ \diagdown \\ N\text{–}OC\text{–}CH_2\text{–}C\text{–}COOH \\ \diagup \\ H \end{array}}}
+ H_2O
$$

There are only two naturally occurring acidic amino acids, aspartic acid and glutamic acid, and both of them are frequently found in the form of their amides, asparagine and glutamine. These two amides are still amino acids because it is the extra –COOH group on the side chain, and not the diagnostic one attached to the α-carbon atom, which is involved in their formation.

Isomerism

From the general formula for α-amino acids it can be seen that a –COOH group, a –NH$_2$ group, a –H atom, and an R group are all attached to the α-carbon atom. In other words, with the exception of glycine where the R group itself is a –H atom, all α-amino acids possess an asymmetric carbon atom at the α position because four different groups are attached to it. Hence they show optical isomerism and can exist as both D– and L– forms (Fig. 14.1). Although two isomeric forms are possible, it is a remarkable fact that all the amino acids present in proteins occur exclusively in the L configuration.

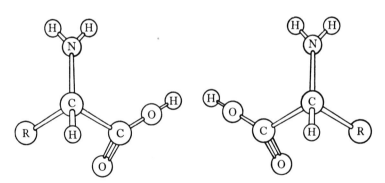

Fig. 14.1 Optical isomerism in α-amino acids. The L configuration is shown on the left, and the D configuration on the right.

A few amino acids, for example isoleucine, have two asymmetric carbon atoms in their molecules, giving four possible optical isomers. Nevertheless only one of them occurs naturally in proteins.

Properties of amino acids

The properties of amino acids result from both the nature of the side chain, and the presence of an amino and a carboxyl group. The properties due to the side chain depend on the functional groups present. For instance, a side chain which is an alkyl group will be relatively inert, whereas one possessing a sulphydryl group, –SH, or an additional –NH$_2$ or –COOH group will be more reactive. Properties resulting from the possession of reactive side chains become important when amino acids containing them are linked together to form proteins. Of the properties due to the –NH$_2$ and –COOH groups possessed by every amino acid on the α-carbon atom, only the ability of amino acids to act as electrolytes will be considered here.

Amino acids in aqueous solution react with both acids and bases because they have at least one ionizable basic –NH$_2$ group ($\rightarrow$ –NH$_3^+$) and one ionizable acidic –COOH group ($\rightarrow$ –COO$^-$). Since they can behave both as acids and bases, amino acids are described as *amphoteric electrolytes*. Although glycine has so far been written as H–CH(NH$_2$)COOH, this formula is not a correct representation of its structure either in the crystalline form or in solution. This is because glycine exists as an ionic compound in which both the –NH$_2$ and –COOH groups are ionized simultaneously, so that the correct formulation is

H–CH(NH$_3^+$)COO$^-$

Ions which have two opposite charges like this are called *zwitterions*, a name derived from the German for 'ions of both kinds'.

When an acid is added to a neutral solution of glycine, i.e. the pH is reduced, H$^+$ ions from the acid react with the –COO$^-$ ion to form the uncharged –COOH group. Hence the only charge left on the glycine molecule is the positive charge of the –NH$_3^+$ ion, so that the molecule becomes H–CH(NH$_3^+$)COOH and carries a net positive charge. Conversely, if a base is added to a neutral solution, i.e. the pH is increased, H$^+$ ions are liberated from the charged –NH$_3^+$ group and react with the excess of OH$^-$ ions in the solution to form water. The glycine molecule now becomes H–CH(NH$_2$)COO$^-$ and carries a net negative charge. Thus, depending on the pH of the solution that surrounds it, glycine can exist in positively charged, negatively charged, or neutral forms. The pH at which the net charge is zero, i.e. where the glycine exists in the zwitterion form, is called the *isoelectric point*. These changes of glycine from one form to another are a general property of all amino acids and can be summarized as follows:

$$
\begin{array}{ccc}
\text{NH}_3^+ & \text{NH}_3^+ & \text{NH}_2 \\
| \quad \xrightarrow{\text{add base}} & | \quad \xrightarrow{\text{add base}} & | \\
\text{R}-\text{C}-\text{COOH} \xrightleftharpoons[\text{add acid}]{} & \text{R}-\text{C}-\text{COO}^- \xrightleftharpoons[\text{add acid}]{} & \text{R}-\text{C}-\text{COO}^- \\
| & | & | \\
\text{H} & \text{H} & \text{H}
\end{array}
$$

zwitterion

low pH (acid) isoelectric point high pH (alkaline)

Proteins

The peptide bond

Proteins are built up from amino acid molecules linked end to end to give long chains. The linkages are formed by joining the carboxyl group of one amino acid to the amino group of another, with the elimination of a molecule of water between them (i.e. it is a condensation reaction).

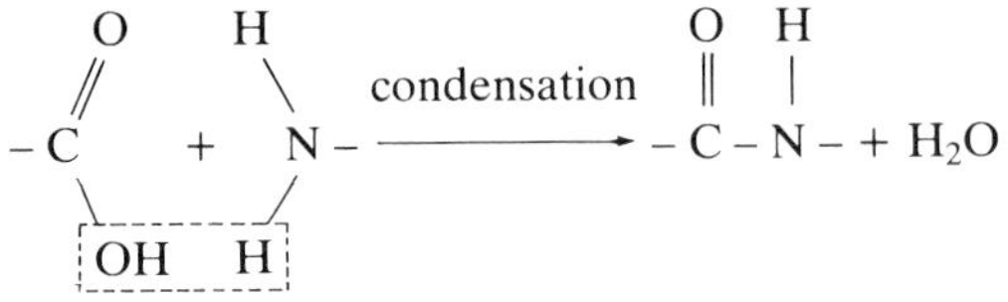

$$-\overset{\displaystyle O}{\underset{\displaystyle \text{OH}}{C}} + \overset{\displaystyle H}{\underset{\displaystyle H}{N}-} \xrightarrow{\text{condensation}} -C-N- + H_2O$$

The linkage, $-CO-NH-$, by which the two amino acids are joined together is called the *peptide bond*. For instance, the equation of peptide bond formation between the two simplest amino acids, glycine and alanine, is

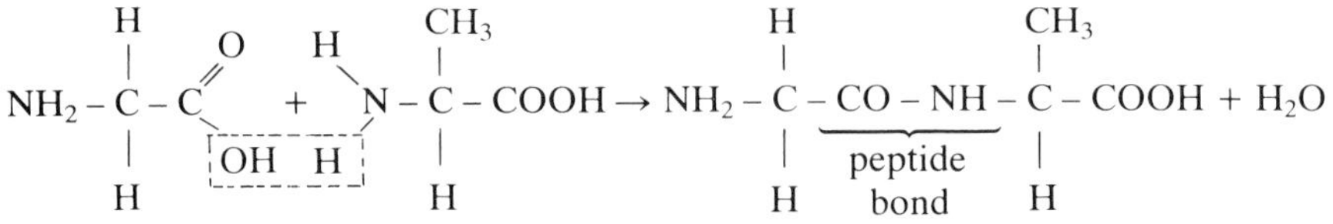

$$NH_2-\overset{\displaystyle H}{\underset{\displaystyle H}{C}}-\overset{\displaystyle O}{\underset{\displaystyle \text{OH}}{C}} + \overset{\displaystyle CH_3}{\underset{\displaystyle H}{N-C-COOH}} \rightarrow NH_2-\overset{\displaystyle H}{\underset{\displaystyle H}{C}}-\underset{\displaystyle \text{peptide bond}}{CO-NH}-\overset{\displaystyle CH_3}{\underset{\displaystyle H}{C}}-COOH + H_2O$$

(glycine)   (alanine)   (glycyl-alanine)   (water)

Because glycyl-alanine is formed by the linkage of two amino acids through one peptide bond it is called a dipeptide. From the formula for glycyl-alanine it will be seen that the $-NH_2$ group of the glycine residue and the $-COOH$ group of the alanine residue are left intact, and therefore both of them are free to form a further peptide bond with another amino acid. For example, with another molecule of alanine, either of the two following reactions is possible:

glycyl-alanine + alanine $\rightarrow$ glycyl-alanyl-alanine + H_2O
alanine + glycyl-alanine $\rightarrow$ alanyl-glycyl-alanine + H_2O

and 2 alanine), are different compounds with different physical and chemical properties, because the sequence of the amino acids is different. Such linking can in theory be continued indefinitely to give peptide chains of various lengths. When many amino acids are linked into a chain the molecule is called a polypeptide, and when a polypeptide has reached a length of about fifty amino acids it is generally regarded as a protein. There is no sharp dividing line between polypeptides and proteins but, as a general rule, one can say that polypeptides have a molecular weight of 1500 or less whereas proteins have a molecular weight of 5000 or more.

A polypeptide chain may be regarded as consisting of a 'backbone' of repeating — CH — CO — NH — units linked together like a linear 'jigsaw'. The R groups, specifying the nature and order of the amino acids along the chain, are each attached to a CH group, and they are arranged on alternate sides of the backbone of the chain (Fig. 14.2).

A polypeptide chain always contains a free $-NH_2$ group at one end (the N-terminal end) and a free $-COOH$ group at the other (the C-terminal end), and the convention is to write the formula starting with the N-terminal end.

A polypeptide chain can be split into its component amino acids by hydrolysis (which is the opposite of its synthesis by condensation), i.e. a molecule of water is added to each peptide bond so that one $-COOH$ group and one $-NH_2$ group are re-formed. Hydrolysis can be carried out either chemi-

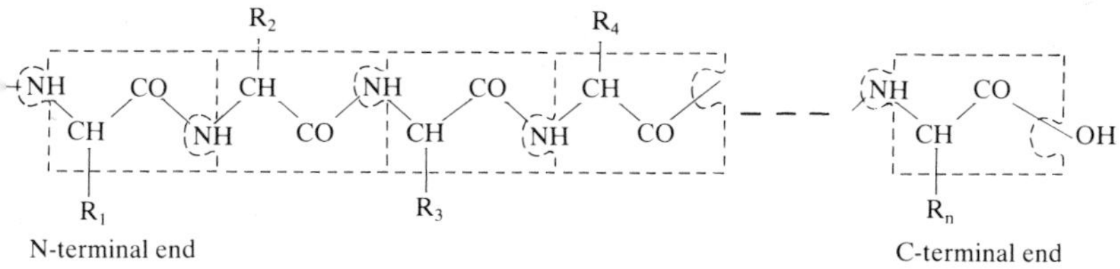

Fig. 14.2 The backbone of a polypeptide chain. The backbone is represented as a linear 'jigsaw' composed of interlocking amino-acid 'pieces'.

cally (by boiling with HCl) or biologically (with peptidase enzymes). If hydrolysis is not complete, the products may consist of di-, tri-, and higher polypeptides in addition to free amino acids.

Structure of proteins

Proteins are elaborate polypeptides, each particular protein consisting of a specific sequence of amino acids. Some proteins consist of a single polypeptide chain, but other consist of a number of chains which are linked together. Cross-linking also occurs between various points along the length of an individual chain. For purposes of description it is convenient to recognize four separate levels of protein organization.

Primary structure

Although there are only about twenty different amino acids to be found in proteins, there may be several hundred amino-acid residues in any one protein molecule, some of them recurring many times along the length of the chain and others occurring only once or a few times. The sequence of amino acids along the chain is characteristic for each protein, and all the properties of a protein, at whatever level of structure, are ultimately determined by this sequence.

The amino-acid composition of a protein can be determined by completely hydrolysing the molecule into its constituent amino acids, and then analysing the relative amounts of each. Provided the molecular weight of the protein is also known, such analyses enable the number of individual amino acids in the molecule to be calculated. However, this gives no information about the order in which the amino acids occur along the peptide chain, and the determination of this sequence presents a much more difficult problem.

The standard approach to determine the amino-acid sequence of a protein is as follows. Firstly, the amino acids at the N- and the C-terminal ends of the chain are identified by chemical or enzymic methods or both. Next the peptide chain is broken into smaller fragments of various lengths by using the enzyme trypsin, which hydrolyses only those peptide bonds which involve the C-terminal end of arginine or lysine residues. Once the peptide chain is broken into smaller fragments, the task of analysing these is of course much simpler. The fragments are separated, and their amino-acid sequences established by enzymic or chemical methods which split off the amino acids, one at a time, from the ends of the fragments. Another sample of the protein is then broken in a different way by using a second method of partial hydrolysis, which splits the peptide chain at other specific points and thus yields a different set of fragments, which are separated and analysed in the same way as was adopted following the hydrolysis with trypsin. The final stage in determining the amino-acid sequence is to deduce, from the overlaps between the two sets of fragments, the sequence of the fragments in the original protein. If, for example, four fragments have been obtained from a protein chain by different methods of splitting and have been found to have the sequences A–B–D–C, B–D–C–F–E, C–F–E, and F–E–G–C, then it is reasonably certain that somewhere in the complete chain there is a sequence A–B–D–C–F–E–G–C. This jigsaw method of identifying and matching fragmented peptides was the method by which the primary structure of a protein was first successfully determined by Frederick Sanger and his associates in the 1950s. Sanger worked with the protein hormone, bovine insulin, which is a very small protein (molecular weight 6000) containing only fifty-one amino acids in its molecule. Nevertheless it took Sanger 10 years to determine the structure of bovine insulin, and he was awarded a Nobel prize in 1959 for his pioneer work. Progress using Sanger's method was inevitably slow, but the primary structures of seven proteins, including the protein subunit of the tobacco mosaic virus with 157 residues, were determined along similar lines. More recently, by an ingenious combination of chemical and highly sensitive fluorescent techniques, a method has been devised whereby amino acids are chopped off a peptide chain one at a time and readily identified. As a result of the application of this much quicker method, the primary structures of over 200 proteins are now known.

Secondary structure

One of the most significant features of amino acids is the number of reactive points in their structure which make possible a variety of chemical and physical linkages. As a result, certain atomic groupings along a peptide chain form linkages with one another so that the chain, instead of being extended as a straight rod, is held in a definite coiled or folded conformation (three-dimensional shape). It is this ability to form organized structures of great complexity and variety, inherent in the chemical and physical properties of amino acids, which is the basis for the functional diversity of proteins in living matter. The secondary structure of a protein describes the way in which the axis of a polypeptide chain is folded over a distance of several amino-acid units. Because the bond angles of the peptide bonds between consecutive amino-acid residues are fixed, polypeptide chains are not free to take up random conformations, but must exist in one of a few preferred stable arrangements. The most widespread of these arrangements is like an extended spiral spring and is called the α-helix (Fig. 14.3). The most important type of linkage responsible for the formation of the α-helix is called *hydrogen bonding*. This type of bond arises from the tendency of hydrogen, when attached to oxygen or nitrogen, to share the electrons of a neighbouring oxygen. The attached oxygen or nitrogen draws electrons away from the hydrogen, leaving it with a positive bias. Hydrogen bonding may thus be regarded as a weak form of electrostatic linkage, in which there is electron sharing instead of electron transfer. Hydrogen bonding is usually indicated by a dotted line in the following way:

$$>C=O \ldots H–N<$$

In the α-helix the turns of the spiral are kept in place by hydrogen bonds between the >CO and >NH groups on adjacent turns of the helix. Although hydrogen bonds are individually weak, they are so numerous that their total binding effect is highly significant in maintaining the stability of the α-helix.

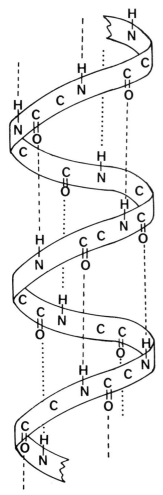

Fig. 14.3 The α-helix conformation of a protein, resulting from the formation of hydrogen bonds between >NH and >CO groups along the polypeptide chain. Only the C atom of the repeating >CHR group of the backbone of the chain is shown.

Tertiary structure

In addition to the helical arrangement of a polypeptide chain along its length, further folding, called tertiary structure, can occur and convert the chain as a whole into a compact, more or less spherical structure. This three-dimensional folding is brought about mainly by linkages between the R-group side chains projecting from the backbone of the polypeptide chain. The most important cross-linkage is the covalent disulphide bridge of cystine, formed between two cysteine residues (see p. 161). It is possible that cystine is built into the peptide chain during synthesis, but it is more likely to be formed in position by oxidation of neighbouring cysteine units. The polypeptide chains of proteins are so long that they may bend back on themselves in such a way that a disulphide bridge can be formed between two cysteine residues which are widely separated along the length of the same chain (Fig. 14.4). When this happens, the polypeptide chain assumes a stable folded arrangement. Disulphide bridges may also be formed between cysteine residues in different polypeptide chains, thus holding the chains together in a stable structure.

Lys·Glu·Thr·Ala·Ala·Ala·Lys·Phe·Glu·Arg· 10
Gln·His·Met·Asp·Ser·Ser·Thr·Ser·Ala·Ala· 20
Ser·Ser·Ser·Asn·Tyr·Cys·Asn·Gln·Met·Met· 30
Lys·Ser·Arg·Asn·Leu·Thr·Lys·Asp·Arg·Cys· 40
Lys·Pro·Val·Asn·Thr·Phe·Val·His·Glu·Ser· 50
Leu·Ala·Asp·Val·Gln·Ala·Val·Cys·Ser·Gln· 60
Lys·Asn·Val·Ala·Cys·Lys·Asn·Gly·Gln·Thr· 70
Asn·Cys·Tyr·Gln·Ser·Tyr·Ser·Thr·Met·Ser· 80
Ile·Thr·Asp·Cys·Arg·Glu·Thr·Gly·Ser·Ser· 90
Lys·Tyr·Pro·Asn·Cys·Ala·Tyr·Lys·Thr·Thr· 100
Gln·Ala·Asn·Lys·His·Ile·Ile·Val·Ala·Cys· 110
Glu·Gly·Asn·Pro·Tyr·Val·Pro·Val·His·Phe· 120
Asp·Ala·Ser·Val 124

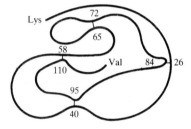

Fig. 14.4 The primary structure of bovine ribonuclease, showing the disulphide bridges between different parts of the chain. The folding of the chain is arbitrarily arranged so as to fit on to the page.

Another type of linkage involved in tertiary structure is electrostatic bonding (alternatively called salt bridges) which occurs, if the pH conditions are suitable, between an ionized amino group on the side chain of one amino acid (e.g. lysine) and an ionized carboxylate group on the side chain of another amino acid (e.g. glutamic acid). This type of linkage is not as strong as a covalent bond and can be broken, for example, by a change in pH of the surrounding medium.

On the basis of their tertiary conformation proteins may be classified into two major groups, fibrous and globular. Fibrous proteins, as their name implies, consist of straight polypeptide chains which are regularly arranged to give long fibres or flat sheets. They are insoluble in water, and may be regarded as the basic structural material (e.g. muscle, tendons, etc.) of the animal body. By contrast, globular proteins have molecules which are folded into complex spherical shapes. Most of them are relatively soluble in water and, generally speaking, they take an active part in physiological processes. They include the enzymes, many proteins with a transport function (e.g. myoglobin), and some hormones.

Denaturation   When proteins are 'harshly' treated (see below) they lose the ability to carry out their function and, if they are of the globular type, they become far less soluble. They are said to be *denatured*. Curdled milk and the white of a boiled egg are examples of denatured proteins.

The explanation for denaturation is that harsh treatments disrupt the tertiary structure of a protein. Any agent leading to the weakening or breakage of the linkages which maintain the correct tertiary structure will be a denaturant. Against a background knowledge of the forces maintaining tertiary structure, it is now possible to explain the action of individual denaturants. Thus, reducing or oxidizing agents denature many proteins because they break disulphide bridges, and extremes of pH denature by charging or discharging ionizable groups. The influence of temperature is less specific because an increase in temperature, by increasing the random movement of the atoms of a protein molecule, leads to an increased likelihood of all the bonds being broken.

Quaternary structure

Some proteins consist of two or more protein molecules loosely associated into a supramolecular complex. Each component protein has its own primary, secondary and tertiary structure, but when assembled in a specific arrangement to form a 'super-protein', the composite unit has structural and chemical properties which the constituent subunits do not possess. The term quaternary structure refers to the way in which the individual polypeptide chains are arranged in relation to one another. Quaternary structures are usually held together by Van der Waals' forces (i.e. the weak forces of atomic attraction between neighbouring molecules or, in macromolecules like proteins, between different regions of the same molecule) and/or electrostatic bonds between oppositely charged sites on separate chains, rather than by the much stronger covalent type of bonds. An example of a protein with a quaternary structure is the blood pigment haemoglobin, in which four protein molecules are held together by electrostatic bonds to form a globular structure which is moderately stable (Fig. 14.5).

To summarize the four levels of protein organization, the analogy of a long flexible spring tied into a knot may be helpful. The primary structure of a protein would then correspond to a linear strand of steel wire, its secondary structure to the regular coil of the spring, and its tertiary structure to the three-dimensional shape of the knot. If a second spring were intertwined with the first, the spatial relationship of the two springs to each other would constitute the quaternary structure.

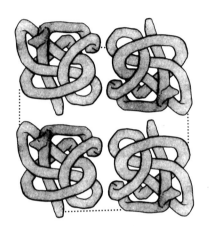

Fig. 14.5 The quaternary structure of a globular protein containing four polypeptide chains. (Modified from Conn and Stumpf 1966, 2nd edn, p. 89)

Chapter 15

Nucleic acids

Nucleic acids occur in every living cell, where they play an essential part in the production of proteins. They are also the genetic material which is passed from one generation to the next, thereby enabling every daughter cell to make proteins similar to those of its parent. The name nucleic acid was given to these substances because they were originally isolated from cell nuclei. When they were also found in the surrounding cytoplasm this name was nevertheless retained as a general one for the whole group of these substances, irrespective of where they are located. Even though nucleic acids were first discovered just over a hundred years ago, it was not until the 1930's, when their possible biological importance became apparent, that they were studied intensively. About this time it became clear that there were two distinct types of nucleic acids. One of the types, originally thought to be confined to plant cells and hence called plant nucleic acid, was shown to contain the pentose sugar, ribose. The other type, originally isolated only from animal cells and hence called animal nucleic acid, was shown to contain not ribose but a related deoxygenated sugar, deoxyribose (deoxy-means 'without oxygen'). About 1940, however, as a result of the development of histochemistry and differential centrifugation as techniques for investigating the chemistry of cell structure, the distinction between plant and animal nucleic acids was shown to be untenable because every cell, plant and animal, was found to contain both types. The names were therefore changed to refer to the identity of the sugar present. Plant nucleic acid was renamed ribonucleic acid, now referred to by the abbreviation RNA. Similarly animal nucleic acid was renamed deoxyribonucleic acid, now known as DNA. The results obtained by the new techniques showed also that DNA is found in all cell nuclei and, except in a few special organelles (e.g. chloroplasts and mitochondria), is not found outside the nucleus. By contrast, RNA is found mainly in the cytoplasm although a small amount is also present in the nucleus, especially in the nucleolus. There are many different DNAs and RNAs, but it is customary to refer to all types of DNA molecules collectively as DNA and all types of RNA molecules as RNA.

DNA and frequently RNA have extremely large molecules (with molecular weights up to several million) but, like all biological macromolecules, they are composed of repeating subunits joined together to form long chains. The subunits of nucleic acids are called *nucleotides* but, unlike amino acids and monosaccharides which are the building blocks of proteins and carbohydrates, each nucleotide is itself composed of three still smaller subunits, a sugar, a phosphate radical, and a nitrogenous base.

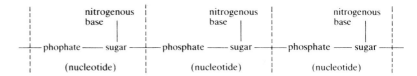

Nucleotides and nucleotide chains

Just as proteins are formed by the linking together of hundreds, or some-times thousands, of amino acids to form a polypeptide chain, so DNA and RNA consist of an even larger number (tens or hundreds of thousands) of nucleotides joined together to form a polynucleotide chain. Further, just as a polypeptide chain consists of a 'backbone' of repeating peptide units with projecting side chains, so a polynucleotide chain consists of a 'back-bone' of repeating sugar-phosphate units with the nitrogenous bases pro-jecting as side chains from the sugar residues.

In RNA the sugar component of the chain is ribose, and in DNA it is 2-deoxyribose (i.e. ribose with the O atom removed from the C-2 position). Both sugars are present in the furanose (five-membered ring) form:

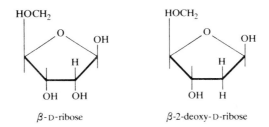

In both RNA and DNA a nucleotide is formed by attaching a phosphate group to the C-3 and a nitrogenous base to the C-1 of the sugar. Hence a ribonucleotide (i.e. a subunit of RNA) can be represented thus:

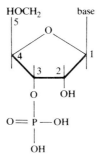

When many nucleotides are linked together to form a polynucleotide chain, the phosphate at C-3 of one nucleotide condenses with the −OH group at C-5 of the next nucleotide, so that the backbone of the chain con-sists of alternating sugar and phosphate groups. Adopting the common practice of abbreviating the phosphate group as ℗, a polynucleotide chain of RNA can be drawn like this:

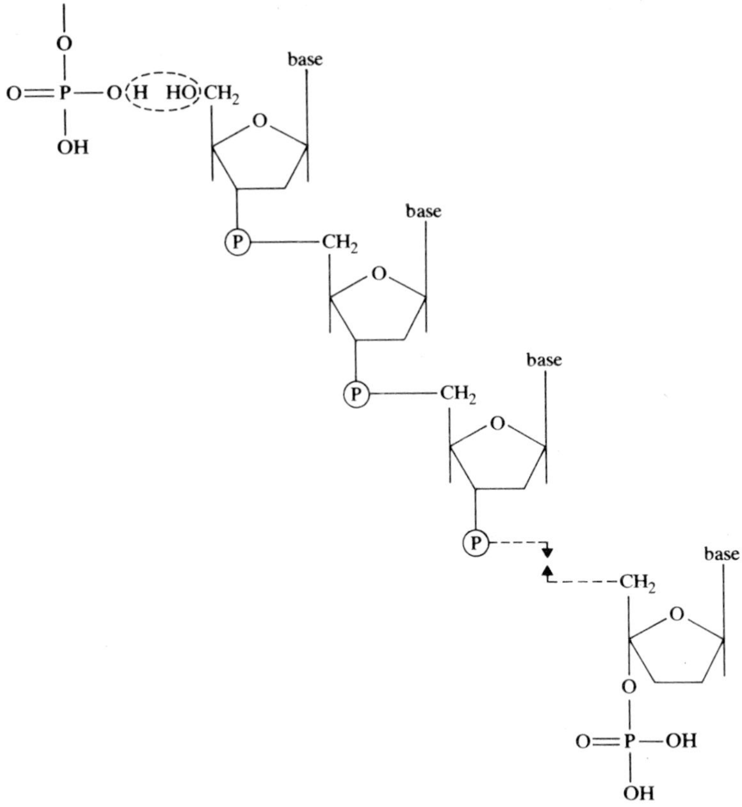

The essential pattern of a polydeoxyribonucleotide chain (i.e. DNA) is similar to that of the above formula except that at C-2 in the sugar ring there is a H atom instead of a –OH group. There is, however, another major chemical difference between RNA and DNA and this concerns the types of bases which form the side chains. These bases all contain nitrogen and are derivatives of two heterocyclic compounds, pyrimidine and purine (Fig. 15.1). Pyrimidine has a six-membered ring which contains nitrogen in the 1 and 3 positions. Purine is a larger molecule with two rings, one a pyri-

Fig. 15.1 The five nitrogenous bases which occur in DNA and RNA.

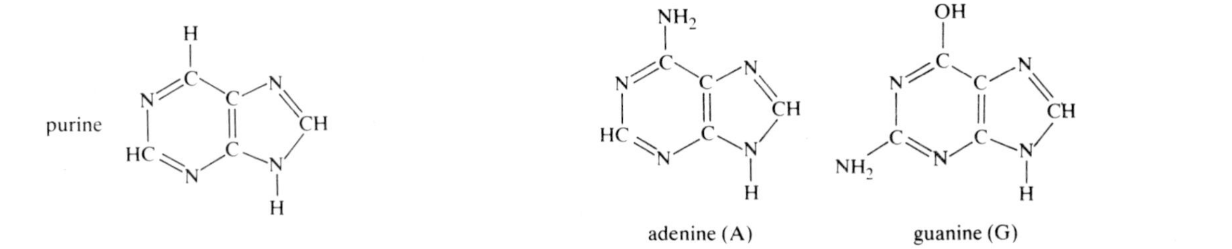

midine and the other a five-membered ring with nitrogen atoms in the 7 and 9 positions. There are five bases (Fig. 15.1) which occur in RNA and DNA, three of them, cytosine (C), uracil (U) and thymine (T), are substituted pyrimidines (i.e. pyrimidine in which other atoms or groups are attached to the ring) and two, adenine (A) and guanine (G), are substituted purines. When forming part of a nucleotide, these bases are always linked to the C-1 of the sugar through a nitrogen atom. In the pyrimidine derivatives this linkage is through the N atom at position 3 whereas in the purine derivatives it is through the N atom at position 9.

If molecules of RNA and DNA are dismantled by hydrolysis into their component building blocks, it is found that each yields a mixture of only four out of the possible five bases. Both yield the two substituted purines, adenine and guanine, and the substituted pyrimidine, cytosine, but the fourth base in RNA is uracil whereas in DNA it is thymine. Thus the four bases occurring in RNA and DNA are:

	RNA	*DNA*
Purines	Adenine	Adenine
	Guanine	Guanine
Pyrimidines	Cytosine	Cytosine
	Uracil	Thymine

As already mentioned, a polynucleotide chain of RNA or DNA consists of tens or hundreds of thousands of nucleotides. Although there is only one base in each nucleotide and there are only four different types of bases in any one chain, the number of possible ways of arranging these bases along the length of such long chains is astronomically large. It is this variation in the sequence of bases along the backbone of a polynucleotide chain which is responsible for the differences between individual RNA molecules and individual DNA molecules. Because the structure of these molecules is directly related to their function in protein synthesis and heredity, we shall postpone further discussion of nucleic acids until Chapter 39 where the bare biochemical facts discussed here will fit together to make biological sense.

Chapter 16

Metabolism and respiration

General principles of metabolism

Metabolism is the word used to describe all the chemical reactions taking place in living cells. The metabolism of any cell, even the smallest bacterium, is a highly integrated activity which involves a fantastically large number of reactions, each catalysed by a specific enzyme. Fortunately this complexity is not as bewildering as might at first appear, because the machinery of metabolism can be understood in terms of a few general principles. From a biochemical standpoint a living cell can be regarded as a chemical machine which maintains itself, grows and reproduces by tapping sources of matter and energy in its environment. These exchanges of matter and energy are brought about by a relatively few types of chemical reaction (e.g. hydrolysis, decarboxylation, dehydrogenation and esterification), and these in turn fall into a relatively few types of reaction sequences, or pathways, with clearly defined functions (e.g. the conversion of small soluble molecules into large insoluble ones suitable for storage, the synthesis of new structural components or of enzymes, or the breakdown of fuel molecules to liberate energy that can be harnessed for the performance of useful work). The complex metabolic machinery of cells will only make sense if the underlying general principles are clearly understood.

Autotrophs and heterotrophs

The energy required by all living organisms is ultimately derived from sunlight, but only green plants and photosynthetic bacteria can trap and utilize light energy to manufacture carbohydrates, fats and proteins from such simple inorganic compounds as carbon dioxide (CO_2), water (H_2O), and ammonia (NH_3) in the environment. The energy of the manufactured carbohydrates and fats can be released again when it is required. Green plants and photosynthetic bacteria are known collectively as autotrophs because they are energetically self-supporting. There is a small group of chemosynthetic bacteria which are also autotrophic because, independent of light, they make use of energy produced by inorganic oxidations taking place in the environment. All other organisms, which include all animals and fungi and most bacteria, are heterotrophs because they have no primary source of energy, and depend on organic substances (food) produced by the autotrophs. Some heterotrophs, e.g. herbivores, depend directly on autotrophs, but for others, e.g. carnivores, the dependence is indirect. The chemical reactions which occur during the breakdown of foodstuffs serve two main

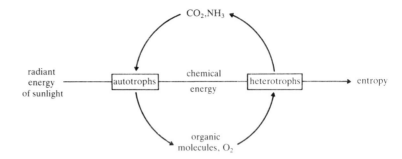

Fig. 16.1 The flow of energy and matter through the biological world.

purposes: (l) they provide, from complex starting materials, the simpler fragments from which cell constituents are synthesized; and (2) they yield the energy needed for such chemical syntheses and for the many other energy-requiring life processes.

Although the energy for all cells comes ultimately from the sun, autotrophs and heterotrophs are nevertheless mutually dependent. While it is true that all animals depend on materials that have been synthesized by plants, it is also true that plants use some of the waste-products (e.g. CO_2, NH_3) released by animals, and so matter, unlike energy, circulates between them (Fig. 16.1).

Energy, movement and work

Newton's First Law of Motion states that any material body, irrespective of whether it is large (a boulder) or small (an electron), has no power of itself to change its state of rest or uniform motion. To do this a force must act, and work is said to have been done. Any agent capable of doing work is said to possess energy. Since energy is a capacity it need not, at any given moment, be accomplishing work. Energy that is inactive but still capable of doing work is called *potential energy*, whereas energy that is doing work is called *kinetic energy*. Energy can appear in several forms such as heat and light, or chemical, mechanical and electrical energy. These forms of energy are interconvertible so that, for example, light can be converted to heat, or heat to electricity. Energy flow and the transformation of energy from one form to another are governed by the laws of the branch of physics called thermodynamics. The first two laws of thermodynamics state, in essence, that energy can neither be created nor destroyed, and that the effect of physical and chemical changes is to increase the disorder, or randomness (i.e. *entropy*) of the universe. Do the chemical workings of cells obey these laws? A plant grows by gathering atoms from the air and soil and arranging them in an orderly fashion to form shoots and roots. This appears to contravene the second law of thermodynamics, but it only appears to do so because an important part of the system has been ignored. If the increase in entropy of the sun, which ultimately provides the energy for the growth of the plant, is also taken into account, then there is a total increase of entropy. Furthermore this apparent reversal of the second law of thermodynamics is only temporary because the plant eventually dies and its atoms are dispersed at random, i.e. entropy increases. In short, living organisms neither consume nor create energy, they can only transform it from one form to another. They absorb energy from the environment in a form useful to them, and they return an equivalent amount of energy to the environment in a biologically less useful form.

Of the various forms in which energy can appear, heat energy can perform useful work only as long as it flows from a hotter to a colder body. The greater the temperature difference through which heat flows, the greater is the amount of work that can be performed. This temperature differential is denoted by the symbol $\triangle T$ where the capital Greek letter delta ($\triangle$) signifies 'the change in a given quantity', which in this case is temperature. Conversely, the less the $\triangle T$ the smaller is the amount of work that can be performed, and where $\triangle T=0$, i.e. where the system is *isothermal*, heat energy is incapable of doing work. Cells are isothermal systems, and therefore heat cannot perform work in them.

The concept of free energy

The useful energy capable of doing work under isothermal conditions is called free energy, and is denoted by the symbol G. The letter G is used in honour of J. W. Gibbs who put forward the concept a century ago. The units of G are calories or kilocalories. The concept of free energy was developed to express the *total* energy possessed by a chemical substance, i.e. the sum of its so-called heat content (a term which implies not its temperature but its potential to release heat when it undergoes chemical reaction) and the energy it contains by virtue of its organized state. The free-energy content of a given substance A cannot be measured experimentally. If, however, A is converted into B in the reaction $A \rightleftharpoons B$ it is possible to measure the difference in free-energy content, $\triangle G$, between A and B. If the free-energy content of the product (G_B) is less than the free-energy content of the reactant (G_A), then $\triangle G$ will have a negative value.

$$\triangle G = G_B - G_A$$
$$= \text{negative quantity when } G_A > G_B$$

Similarly, if B is converted back to A the reaction will involve an increase in free energy, and $\triangle G$ will be positive. Chemical reactions which occur spontaneously have a negative $\triangle G$, whereas those with a positive $\triangle G$ proceed only if energy is supplied to the system to drive the reaction.

The value of $\triangle G$ for a reaction changes if the temperature or concentration of the reactants is varied. To use the concept of free energy for predicting the tendency of a reaction to proceed, it is therefore necessary to measure $\triangle G$ under standard conditions. These are defined as a temperature of 25°C and a concentration of 1 mol for each of the reactants and products. Since biological reactions nearly always occur around neutrality, biologists usually specify a pH of 7 as a third standard condition. Defined in this way, the *standard free energy* of a reaction is indicated by adding the superscript zero, to give the symbol $\triangle G°$. Thus $\triangle G°$ for the reaction $A \rightleftharpoons B$ is the change in free energy when A is converted to B at 25°C and pH 7 under conditions such that the concentrations of A and B are both 1 molar throughout the reaction. Even though such conditions as unit molarity are unnatural, $\triangle G°$ is a useful concept because it expresses the driving force of a reaction, i.e. the *tendency* of A to be converted to B in a mixture containing 1 mol of each. If, for example, we know that $\triangle G°$ is large and negative, we can predict that, in a mixture of A and B, most of A will be converted to B, because there is a stronger tendency for A to be converted into B than vice versa. The numerical value of $\triangle G°$ is, in fact, proportional to the logarithm of the equilibrium constant K (see p. 132). Since equilibrium constants can be readily determined from analytical measurements, the $\triangle G°$ value of most biochemical reactions is calculated from the equilibrium constant. Table 16.1 compares some values of K with the corresponding values of $\triangle G°$. Note that, because the relationship is a logarithmic one, a large change in K corresponds to a relatively small change in $\triangle G°$.

ATP, the carrier of chemical energy

Energy is made available to cells by various means such as the breakdown of foodstuffs or the conversion of solar into chemical energy, but it can only be used by cells in more or less exclusively one form. The situation can be compared to lighting an electric bulb. The energy used by a lighted bulb may be derived from the burning of coal, from the disintegration of radioactive nuclei, or from the kinetic energy of a waterfall, but the bulb cannot be lit unless the various forms of energy – thermal, nuclear, or kinetic – are transformed into an electric current, the only form of energy usable by the bulb. In a similar sort of way the carrier for nearly all the chemical energy transferred in cells is one particular chemical compound, adenosine triphosphate (or ATP for short). In order to understand how ATP transfers energy from one molecule to another, it is necessary to consider briefly its molecular structure (Fig. 16.2).

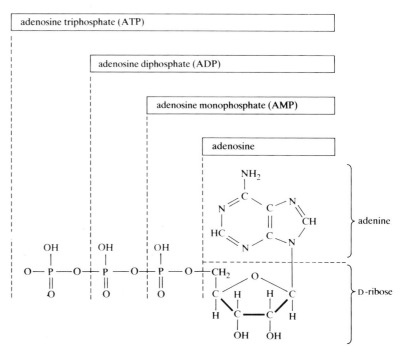

Fig. 16.2 Structural formula of adenosine triphosphate.

Adenosine itself consists of two chemical subunits, namely adenine (a N-containing ring compound) linked to a 5-carbon sugar, D-ribose. It can combine, by the removal of the elements of water, with orthophosphoric acid to form a phosphate ester called adenosine monophosphate (AMP). This compound can be further phosphorylated to yield firstly adenosine diphosphate (ADP) and then adenosine triphosphate (ATP). The main point to notice here is that the three phosphate groups are attached in a row to the adenosine residue. Using $\textcircled{P}$ as shorthand for a phosphate group, ATP can be written as $A - \textcircled{P} - \textcircled{P} - \textcircled{P}$.

If AMP is hydrolysed into adenosine and phosphoric acid, often represented by the letters P_i, the amount of energy released by 1 mol is about 3.4 kcal (i.e. $\triangle G° = -3.4$ kcal/mol). If, however, ADP is hydrolysed to AMP and phosphoric acid, the energy released by 1 mol is more than twice as much, namely about 7.3 kcal/mol. If ATP is hydrolysed to ADP and phosphoric acid the energy released by 1 mol is also about 7.3 kcal/mol.

$$AMP + H_2O \rightleftharpoons adenosine + P_i \qquad \triangle G° = -3.4 \text{ kcal/mol}$$
$$ADP + H_2O \rightleftharpoons AMP + P_i \qquad \triangle G° = -7.3 \text{ kcal/mol}$$
$$ATP + H_2O \rightleftharpoons ADP + P_i \qquad \triangle G° = -7.3 \text{ kcal/mol}$$

Conversely, the formation of ATP from ADP and phosphoric acid is an energy-requiring (endergonic) reaction.

$$ADP + P_i \rightleftharpoons ATP + H_2O \qquad \triangle G° = +7.3 \text{ kcal/mol}$$

Thus there is an essential difference in the linking of the three phosphate groups in ATP, the two terminal bonds being called 'energy-rich' to distinguish them from the 'energy-poor' bond of AMP. A curly line is often used to denote an energy-rich bond, so that the formula of ATP can be written as $A - \textcircled{P} \sim \textcircled{P} \sim \textcircled{P}$. It should be pointed out, however, that the symbol $\sim$ ('squiggle' bond) implies, not that the energy of the molecule is somehow concentrated in the bond shown, but that an unusually large amount of energy is liberated when the bond energies of the whole molecule are redistributed as a result of the splitting of that particular bond by hydrolysis.

The significance of ATP lies in the fact that the terminal phosphate group can, under the influence of appropriate enzymes, be transferred to other compounds together with a larger or smaller proportion of the energy associated with this 'energy-rich' bond. Thus, to take a specific example, ATP can react with D-glucose to yield glucose-6-phosphate and ADP.

This reaction can be envisaged as taking place in two steps, an exergonic step and an endergonic step. The exergonic step, which yields free energy, is the hydrolytic removal of the terminal energy-rich phosphate group from ATP, and the endergonic step, which requires an input of free energy, is the phosphorylation of glucose. The sum of the two steps gives the above equation for the overall reaction.

$$ATP + H_2O \rightleftharpoons ADP + P_i \qquad \triangle G° = -7.3 \text{ kcal/mol}$$
$$glucose + P_i \rightleftharpoons glucose\text{-}6\text{-}phosphate + H_2O \qquad \triangle G° = +3.3 \text{ kcal/mol}$$

$$ATP + glucose \rightleftharpoons ADP + glucose\text{-}6\text{-}phosphate \qquad \triangle G° = -7.3 + 3.3$$
$$= -4.0 \text{ kcal/mol}$$

Although ATP is the best known energy-rich compound, there are many others including some which are not organic phosphate esters. A feature which most of them share with ATP, and which is related to their high free energy of hydrolysis, is the possession of an ester bond linking two atoms, each of which forms a double bond with another atom. Two examples of this arrangement are:

$$
\begin{array}{cc}
\underset{\underset{\displaystyle \text{OH} \quad \text{OH}}{|\qquad |}}{-\,\overset{\overset{\displaystyle O}{\|}}{P}-O-\overset{\overset{\displaystyle O}{\|}}{P}-OH} &
\underset{\underset{\displaystyle \qquad \text{OH}}{\qquad |}}{-\,\overset{\overset{\displaystyle O}{\|}}{C}-O-\overset{\overset{\displaystyle O}{\|}}{P}-OH}
\end{array}
$$

(present in ATP)   (present in the respiratory intermediate 1,3-diphosphoglycerate)

The breaking of such ester linkages by hydrolysis results in the separation of this double bonded arrangement with the consequent release of considerable energy.

Table 16.1 The relationship between the equilibrium constant K and the standard free-energy change $\triangle G°$

K	$\triangle G°$ (kcal per mol)
1000	−4.09
100	−2.73
10	−1.36
1	0
0.1	+1.36
0.01	+2.73
0.001	+4.09

Table 16.2 Values for $\triangle G°$ of hydrolysis of some phosphorylated compounds

Compound	$\triangle G°$ (kcal mol^{-1})	Phosphate-transfer potential*
Phosphoenolpyruvate	−14.8	14.8
1, 3-Diphosphoglycerate	−11.8	11.8
Acetyl phosphate	−10.1	10.1
ATP (to ADP)	− 7.3	7.3
Fructose-6-phosphate	− 3.8	3.8
Glucose-6-phosphate	− 3.3	3.3
Glycerol-1-phosphate	− 2.2	2.2

* Defined as $-\triangle G°$

If all the naturally occurring phosphates are listed in order of decreasingly negative values for $\triangle G°$ of hydrolysis, the values range from about −15 to about −2 kcal/mol (Table 16.2). Compounds at the higher end of the scale, often called high-energy phosphates, tend to be more completely hydrolysed at equilibrium than those at the lower end, often classed as low-energy phosphates. ATP, with a $\triangle G°$ of −7.3 kcal/mol, is near the middle of the range. Compounds such as 1,3-diphosphoglycerate, which are above ATP in the scale, are said to have a high phosphate-transfer potential because they have a strong tendency to lose their phosphate group. Compounds such as glucose-6-phosphate, which are below ATP, have a stronger tendency to hold on to their phosphate group because the acceptor molecules (e.g. glucose) have a higher affinity for phosphate; such phosphates have a low phosphate-transfer potential.

The fact that ATP has an intermediate $\triangle G°$ value is very significant in relation to its function as a carrier of chemical energy. Any compound above ATP on the scale will tend to lose its phosphate group to ADP, provided an appropriate enzyme is available to catalyse the transfer reaction. Similarly

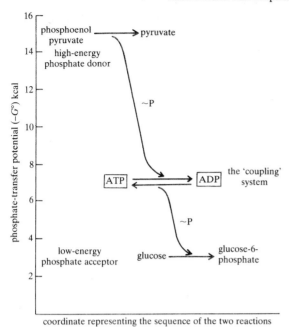

Fig. 16.3 ATP as a coupling agent in the phosphorylation of glucose.

ATP will tend to donate its terminal phosphate to an acceptor molecule lower on the scale than itself.

Many reactions of metabolism may be linked together by the product of one reaction forming the reactant of another. Such reactions are said to be *coupled* through a common intermediate, and the coupling mechanism may provide a means by which chemical energy can be transferred from one reaction to another. Endergonic reactions will not occur spontaneously under the isothermal conditions prevailing in cells, but they can be driven by coupling them with reactions which are exergonic. The significance of ATP is that it frequently acts as the common intermediate between coupled reactions.

The action of ATP as a carrier of chemical energy can be illustrated by two reactions which occur during carbohydrate respiration (Fig. 16.3). In the first of the reactions a phosphate group is transferred, by the action of a specific phosphotransferase enzyme, from phosphoenolpyruvate (PEP) to ADP. The ATP so formed then transfers, in a second enzymic reaction, a phosphate group to glucose forming glucose-6-phosphate.

$$\text{phosphoenolpyruvate} + \text{ADP} \rightleftharpoons \text{pyruvate} + \text{ATP}$$
$$\text{ATP} + \text{glucose} \rightleftharpoons \text{ADP} + \text{glucose-6-phosphate}$$

$$\text{phosphoenolpyruvate} + \text{glucose} \rightleftharpoons \text{pyruvate} + \text{glucose-6-phosphate}$$

The net result of coupling these two reactions through ATP is that glucose is converted into an energized form, glucose-6-phosphate, which can take part in reactions (e.g. the biosynthesis of starch) in which free glucose, with its lower energy content, is unable to participate.

Because ATP functions as a coupling agent in far more cellular reactions than any other substance with high-energy phosphate bonds, it serves as a link between sources of energy available to a living cell and the chemical, osmotic and other work which is associated with the maintenance, growth and reproduction of cells. For this reason the ADP–ATP system may be compared to a 'current account' in an energy bank, into which the cell can deposit energy until it needs to withdraw 'ready cash' to do chemical or other work.

In transferring its energy to other molecules, ATP loses its terminal phosphate group as inorganic phosphate (P_i) and becomes ADP. The resynthesis of ATP, i.e. the phosphorylation of ADP, is achieved by using energy from food or sunlight in three distinct ways:

1. Substrate-level phosphorylation, in which the hydrolysis of an energy-rich compound, such as phosphoenolpyruvate or succinyl coenzyme A, is coupled to the phosphorylation of ADP. No oxygen is required for this type of phosphorylation. It is the only method of energy conservation available to micro-organisms which respire anaerobically, but it is also involved in the breakdown of carbohydrates in both plants and animals.
2. Oxidative phosphorylation, in which the formation of ATP from ADP is associated with a chain of oxidation reactions (the electron transport chain) that are an essential part of aerobic respiration. Oxygen is required for this type of phosphorylation and water is the final product. Oxidative phosphorylation takes place in the mitochondria of both plant and animal cells, and is the main method of energy conservation for non-photosynthetic cells living under aerobic conditions.
3. Photosynthetic phosphorylation, in which the energy of light is converted into chemical energy in the form of ATP. This type of ATP synthesis takes place in the illuminated chloroplasts of green plants.

Biological oxidation

A large number of the reactions occurring in living cells, including those which yield biologically useful energy, are oxidative reactions, and some knowledge of the concept of oxidation is thus essential to an understanding of the process of respiration. Oxidation, and therefore reduction which is the opposite of it, can be defined in several ways. The oldest definition is that oxidation is a reaction in which a substance combines with oxygen. The formation of a fatty acid from an aldehyde is an example.

R.CHO + O → R.COOH
(aldehyde)   (fatty acid)

Another definition is that oxidation is the removal of hydrogen from a substance (dehydrogenation). As an example, catechol is oxidized to *ortho*-quinone.

(catechol)   (*o*-quinone)

In fact, the addition of oxygen and the removal of hydrogen cannot be regarded as alternative methods of oxidation, because ultimately they are often the same; thus the mechanism for the oxidation of an aldehyde to a fatty acid may involve the removal of hydrogen rather than the addition of oxygen, taking place as follows.

$$R.CHO \quad + \quad H_2O \longrightarrow R.CH \begin{cases} OH \\ OH \end{cases}$$

$$R.CH \begin{cases} O\!\!\mid\!\!H \\ O\!\!\mid\!\!H \end{cases} + O \longrightarrow R.COOH \quad + \quad H_2O$$

The situation is further complicated by the fact that oxidation is not re-stricted to the addition of oxygen or the removal of hydrogen. For example, if iron is heated with sulphur, it forms ferrous sulphide, and this reaction is also regarded as an oxidation. To arrive at a complete definition of oxida-tion it is necessary to take into account the transfer of electrons. According to this approach, oxidation is defined as the loss of electrons from an ele-ment or compound, and thus an oxidizing agent (or oxidant) is an acceptor of electrons. On this definition the conversion of a ferrous ion to a ferric one is an oxidation reaction.

$$Fe^{2+} \longrightarrow Fe^{3+} \quad + \quad e^-$$

(ferrous ion)   (ferric ion)   (electron)

In fact, the above equation underlies one of the most fundamental oxidative processes in cells. Dehydrogenation also fits the definition quite readily when it is realized that a hydrogen atom is merely an electron combined with a hydrogen ion (or proton).

$$AH_2 \rightarrow A + 2H^+ + 2e^-$$
$$ \text{proton} \quad \text{electron}$$

To understand oxidation in terms of the transfer of electrons, consider the formation of water from oxygen and hydrogen (Fig. 16.4). An atom of oxygen consists of 2 inner and 6 outer negatively charged electrons revolv-

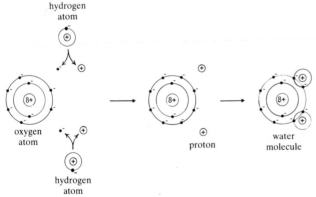

oxygen atom   proton   water molecule

hydrogen atom   hydrogen atom

Fig. 16.4 The transfer of electrons involved in the formation of water from oxygen and hydrogen.

ing round a nucleus of 8 positively charged protons. The full complement of electrons for the outer shell of an oxygen atom is 8, and so an oxygen atom needs 2 more electrons to fill its outer shell. The hydrogen atom, which is the simplest of all atoms, consists of 1 proton and 1 electron. The removal of this electron (i.e. oxidation) from each of 2 hydrogen atoms, and the acceptance of these 2 electrons by an oxygen atom, would result in the formation of an O^{2-} ion and two H^+ ions, i.e. protons. These 3 ions unite to give a molecule of water with no net charge. The stages of this sequence may be written as:

$$2H \rightarrow 2H^+ + 2e^-$$
$$2e^- + O \rightarrow O^{2-}$$
$$2H^+ + O^{2-} \rightarrow H_2O$$

Reduction is the opposite of oxidation, and can therefore be defined as the gain of electrons by an element or compound. By the same token, a reducing agent (or reductant) is a donor of electrons. Since reduction is the reverse process of oxidation, it follows that every oxidation must be accompanied by a simultaneous reduction. In the formation of water from oxygen and hydrogen, for example, the oxygen is reduced by the gain of electrons from the hydrogen, and the hydrogen is oxidized by the loss of electrons. (The electrons of the hydrogen atoms are lost in the sense that they are shared by the oxygen atom). For this reason, oxidation–reduction reactions are often referred to by the abbreviated term *redox* reactions.

Oxidation–reduction (redox) reactions are biologically important because molecules must be oxidized before they will yield useful energy. Using two dots to indicate a pair of electrons that are transferred, a redox reaction can be represented as follows.

A: + B ⇌ A + B:
(reduced) (oxidized)   (oxidized) (reduced)

Since electrons are transferred from A to B, the reduced form of A is an electron donor (reducing agent) and the oxidized form of B an electron acceptor (oxidizing agent). If the reaction proceeds spontaneously from left to right, it must involve a decrease in free energy. This means that the electrons must have passed from a higher energy state in the reduced form of A to a lower energy state in the reduced form of B.

The relative ease with which a substance donates electrons is a measure of its ability to function as a reducing agent and be oxidized. Reactions in which electrons are indicated as being donated but in which no electron acceptor is specified are called half-reactions, because the complete reaction cannot take place unless an appropriate oxidizing agent, i.e. electron acceptor, is present to provide the complementary half of the reaction. The potential of a half-reaction to act as an electron donor can be determined experimentally under clearly defined, standard conditions, and the value (expressed in volts) thus obtained is called the standard reduction potential (E_o). Different half-reactions have different values of E_o, and it is therefore possible to arrange them in order of magnitude on a scale. It is necessary to have some standard for comparison, and the zero point on the scale has been fixed by arbitrarily assigning a potential of 0.0 volts (V), at pH 0, for the half-reaction of hydrogen, $H^+ + e^- \rightleftharpoons \frac{1}{2}H_2$. With this as a standard it is possible to measure the reduction potential of any compound relative to that of hydrogen.

A list of the reduction potentials of the half-reactions of some biologically important compounds, including several coenzymes and substrates to be discussed later in this chapter, is given in Table 16.3. The scale is arranged in order of increasing E_o, which (because of the conventions employed) means that the substances at the top with high negative values of E_o have electrons with high potential energy. In other words, the direction of electron transfer in any redox reaction is always *down* the scale from a substance nearer the top to a substance nearer the bottom. When any two half-reactions are coupled, the one with the more positive E_o and therefore nearer the bottom of

Table 16.3 The standard reduction potentials (E_0) of some half-reactions of biological importance

Half-reaction (written as a reduction)	E_0 at pH 7 (volts)
$2H^+ + 2e^- \rightarrow H_2$	$-0.421*$
$NAD^+ + 2H^+ + 2e^- \rightarrow NADH + H^+$	-0.320
Acetaldehyde $+ 2H^+ + 2e^- \rightarrow$ ethanol	-0.197
$FAD + 2H^+ + 2e^- \rightarrow FADH_2$	-0.06
Fumarate $+ 2H^+ + 2e^- \rightarrow$ succinate	-0.031
Cytochrome b $Fe^{3+} + e^- \rightarrow$ cytochrome b Fe^{2+}	$+0.030$
Cytochrome c $Fe^{3+} + e^- \rightarrow$ cytochrome c Fe^{2+}	$+0.235$
Cytochrome a $Fe^{3+} + e^- \rightarrow$ cytochrome a Fe^{2+}	$+0.385$
$\frac{1}{2}O_2 + 2H^+ + 2e^- \rightarrow H_2O$	$+0.816$

* The E_0 values in this table apply to standard biological conditions, i.e. at pH 7.0. The E_0 of this half-reaction at pH 0 is 0.000 volts, which is the standard with which other reduction potentials are compared.

the scale will go in the direction shown (i.e. as a reduction) driving the half-reaction with the less positive E_0 backwards (i.e. as an oxidation). For example, consider the overall reaction resulting from the coupling of the two half-reactions involving acetaldehyde and the coenzyme NAD^+. The acetaldehyde half-reaction will go as a reduction

acetaldehyde $+ 2H^+ + 2e^- \rightarrow$ ethanol

because it has the higher E_0. The NAD^+ half-reaction will be driven as an oxidation

$NADH + H^+ \rightarrow NAD^+ + 2H^+ + 2e^-$

which is the opposite direction from that given on the scale. The overall reaction, therefore, is

acetaldehyde $+ NADH + H^+ \rightarrow$ ethanol $+ NAD^+$

It will be noticed that the substances at the bottom of the scale, i.e. with those at the top are reducing agents. The position of the oxygen half-reaction at the bottom of the scale indicates that molecular oxygen has the strongest electron affinity of any acceptor on the scale or, alternatively, that the electrons in water have less potential energy than those in any other reduced substance listed on the scale. For any pair of half-reactions, the difference between their E_0 values ($\triangle E_0$) measures, under standard conditions, the tendency of electrons to be transferred from the reduced form of the substance higher up the scale to the oxidized form of the substance lower down the scale. This difference measures the tendency of a redox reaction to take place, and therefore will be closely related to the decrease in free energy ($\triangle G°$) of the reaction (see p. 175 above) which also measures the tendency of a reaction to proceed. There is a mathematical formula for calculating the $\triangle G°$ of a reaction from the E_0 values of the reactants, but for the present purpose it is sufficient to note merely that the further apart two substances are on the scale the greater is the potential yield of free energy from the reaction between them.

Respiration

One of the requirements for the maintenance of life is a continuous supply of energy. This energy is obtained by 'tapping' the chemical energy that is built into the organic molecules synthesized by photosynthesis. The process by which such energy is released and made to serve the purposes of the cell is known as *respiration*. It would be logical to consider first the way in which energy is obtained and stored (i.e. photosynthesis) and then to examine the ways in which it is released (i.e. respiration). However, an appreciation of the energetics of cell chemistry requires an insight into the interrelationships of many different chemical events, and it is probably simpler to obtain this insight by dealing first with the ways in which the energy contained in organic molecules is converted into a form usable by cells.

Normally the respiration of plant cells consists of the oxidation of organic molecules by atmospheric oxygen to form carbon dioxide and water. For this reason the usual method of respiration is sometimes qualified by the adjective *aerobic*. The respiration of glucose, for example, can be represented by the following equation.

$$C_6H_{12}O_6 + 6O_2 \rightarrow 6CO_2 + 6H_2O \qquad \triangle G° = -686 \text{ kcal}$$

It will be seen from this equation that 686 kcal of energy are released in the process. This same amount of energy is also liberated when the gram-molecular weight (mole) of glucose is burned to carbon dioxide and water, but combustion and aerobic respiration are two fundamentally different processes. In combustion, the glucose has to be heated to a high temperature before its energy is liberated in one 'burst' as heat and light. In aerobic respiration, however, the oxidation is carried out at a much slower rate and at ordinary temperatures. In addition, the energy is released in a succession of small amounts and, although some of this energy is lost as heat, a significant part of it is conserved as ATP and subsequently used by cells for chemical synthesis and other cell processes. The summary equation for respiration is therefore misleading because respiration, unlike combustion, is not a single reaction but takes place in a large number of stages, each catalysed by its own appropriate enzyme. The glucose molecule is dismantled in an orderly sequence, one part of the molecule being changed at a time. When respiration is seen as the controlled release of energy in small amounts, the necessity for a long chain of carefully integrated reactions becomes apparent. Some of the energy-yielding reactions of respiration are coupled to the synthesis of ATP and it is this coupling which is responsible for part of the energy released during respiration being conserved instead of being lost as heat. The essential function of respiration is thus seen to be the production of ATP molecules.

Many different organic compounds can be used as the starting material for respiration, but discussion will be restricted to the respiration of glucose because this illustrates clearly the processes involved. The respiration of glucose can be divided into three stages (Fig. 16.5). In the first stage, glucose is incompletely oxidized to yield the 3-carbon compound pyruvic acid. This stage is called *glycolysis* (literally the lysis or splitting of glucose) because its essential feature is the breakdown of glucose into two 3-carbon fragments. In the second stage, the carbon atoms of pyruvic acid are completely oxidized to CO_2 in a cyclic series of reactions known as the tricarboxylic acid (TCA) cycle. During this stage hydrogen atoms (i.e. electrons plus H^+ ions) are transferred from intermediate compounds formed in the TCA cycle to a specific carrier molecule, called nicotinamide adenine dinucleotide (NAD). This compound is used by all cells as an electron acceptor in the oxidative reactions of respiration.

$$\text{reduced substrate} \quad \rightarrow \quad \text{oxidized substrate} + 2H^+ + 2e^-$$
$$NAD^+ + 2H^+ + 2e^- \quad \rightarrow \quad NADH + H^+$$

$$\text{reduced substrate} + NAD^+ \quad \rightarrow \quad \text{oxidized substrate} + NADH + H^+$$

In the third stage, electrons are removed from the reduced NAD produced in Stage II (and also in one reaction of Stage I), and transferred via a system of carriers, collectively called the electron transport chain, to oxygen to form water.

In the absence of oxygen, the end product of glycolysis, i.e. pyruvic acid, does not enter the TCA cycle but follows an alternative sequence of reactions. In yeast (a fungus), for example, pyruvic acid is converted in two stages into ethanol. The overall pathway from glucose to ethanol is an example of *fermentation*, the name given to any metabolic pathway by which

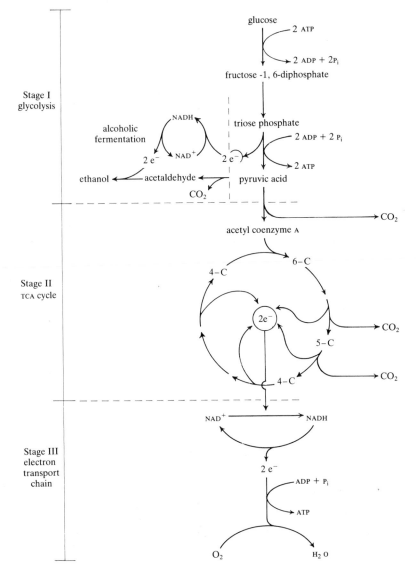

Fig. 16.5 The three stages of respiration. (The pathway followed in alcoholic fermentation is shown to the left of stage I).

organisms extract chemical energy from organic food molecules in the absence of oxygen. Pyruvic acid is thus a key compound in both respiration and alcoholic fermentation, and many plants can switch from one to the other according to the availability of oxygen.

Stage I of respiration: glycolysis

Although glycolysis occurs in all organisms except primitive bacteria, it has been studied in greatest detail in yeast and animal muscle cells. The same sequence of reactions is found in both types of cells.

Glycolysis takes place in ten consecutive steps, each catalysed by a different enzyme or enzyme complex. In order to obtain energy in the form of ATP from the breakdown of glucose, it is first necessary to 'spark off' the reaction sequence by putting ATP into it. During step 1 the primary alcohol group ($-CH_2OH$) at position C$-$6 of glucose reacts with the terminal phosphate group of ATP, forming glucose-6-phosphate and ADP. For convenience, the phosphate (or, more precisely, phosphoryl) group is represented by (P). The glucose-6-phosphate is then rearranged (isomerized) to form fructose-6-phosphate (step 2). In this reaction, the aldehyde group ($-CHO$) at C$-$1 is reduced to a primary alcohol group ($-CH_2OH$) as a result of the simultaneous oxidation of the secondary alcohol group ($>CHOH$) at C$-$2 to a keto group ($>CO$). The formation of a primary alcohol group at C$-$1 makes possible step 3, which is a repetition of the phosphorylation effected in step 1.

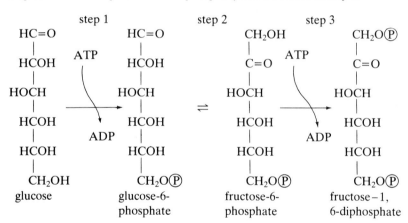

The first three steps of glycolysis have thus converted, at the expense of two molecules of ATP, a molecule of free glucose into one of fructose$-$1,6-diphosphate, with a phosphate group at each end. Fructose$-$1,6-diphosphate is now split into two smaller fragments.

Splitting the hexose molecule   In step 4 the 6-carbon sugar fructose$-$1,6-diphosphate is split between C$-$3 and C$-$4 to form two 3-carbon fragments, one an aldehyde (glyceraldehyde-3-phosphate) and the other a ketone (dihydroxyacetone-phosphate). These two fragments, which may collectively be called triose phosphates, are readily interconvertible by an isomerization (step 5) comparable to that in step 2. This interconvertibility has the effect that energy can be obtained from both halves of the original glucose molecule by the same sequence of reactions.

$$\text{CH}_2\text{O(P)}$$
$$|$$
$$\text{C}=\text{O} \qquad\qquad \text{dihydroxyacetone-phosphate}$$
$$|$$
$$\text{CH}_2\text{OH}$$

CH₂O(P)
|
C=O
|
HOCH   step 4   step 5
------+---¬
HCO¦H
| L----
HCOH   CHO
| |
CH₂O(P)   HCOH   glyceraldehyde-3-phosphate
                    |
                  CH₂O(P)

fructose-1,6-diphosphate

The formation of ATP   In the second half of glycolysis, comprising steps 6 to 10, there is a net gain of ATP, achieved by the oxidation of glyceralde-hyde-3-phosphate, one of the triose phosphates formed in step 4. Because the two triose phosphates are interconvertible, both follow the same pathway and steps 6 to 10 are therefore repeated twice to complete the breakdown of one molecule of glucose.

From the point of view of energy production step 6 is an important reac-tion. In this step glyceraldehyde-3-phosphate is simultaneously oxidized and phosphorylated, and the energy liberated by the oxidation of the aldehyde group is conserved to form the high-energy compound 1,3-diphos-phoglycerate, i.e. it has a high free energy of hydrolysis. The elec-trons removed from the aldehyde group during its oxidation are accepted by the coenzyme NAD^+, which is thereby reduced to give $NADH + H^+$. The NAD^+ is bound to the enzyme (glyceraldehyde-3-phosphate dehyd-rogenase) which catalyses the overall reaction. In step 7 the high-energy phosphate on C–1 of 1,3-diphosphoglycerate is transferred to ADP to form ATP and 3-phosphoglycerate. The overall result of steps 6 and 7, therefore, is that the energy liberated during the oxidation of an aldehyde group (–CHO in glyceraldehyde-3-phosphate) to a carboxylic acid group (–COOH in 3-phosphoglycerate) is conserved as the phosphate bond energy of an ATP molecule.

           step 6   step 7
    CHO   COO (P)   COOH
     | NADH | ATP |
Pᵢ + HCOH   ⟶   HCOH   ⟶   HCOH
     | NAD⁺ | ADP |
  CH₂O(P)   CH₂O(P)   CH₂O(P)
glyceraldehyde-3-phosphate 1,3-diphosphoglycerate 3-phosphoglycerate

The 3-phosphoglycerate from step 7 now undergoes an intramolecular rearrangement (step 8), followed by the removal of the elements of water (step 9), to form another high-energy phosphate compound, phosphoenol-pyruvate (often abbreviated as PEP), which acts as a second source of ATP in glycolysis. During step 10 the phosphate group of PEP is readily transferred to ADP to yield pyruvic acid and ATP. Since steps 6 to 10 occur twice for every molecule of glucose entering the glycolytic pathway, there is a net gain of two ATP molecules (four produced minus two initially invested) from the breakdown of each glucose molecule into two molecules of pyruvic acid.

step 8		step 9		step 10	
COOH		COOH		COOH	COOH
$\|$		$\|$	$-H_2O$	$\|$ ATP	$\|$
HCOH	$\rightleftharpoons$	HCO $\circled{P}$	$\rightleftharpoons$	CO$\circled{P}$	C=O
$\|$		$\|$	$+H_2O$	$\|$	$\|$
CH_2O $\circled{P}$		CH_2OH		CH_2 ADP	CH_3
3-phospho-glycerate		2-phospho-glycerate		phosphoenol-pyruvate	pyruvic acid

Alcoholic fermentation

Glycolysis enables ATP to be produced as the overall result of steps 6 and 7 but this is only made possible by the reduction of NAD^+. This coenzyme is present in cells in only very minute quantities, and therefore, once it has been used up in step 6, no further molecules of glucose can enter the glycolytic pathway until it is regenerated. In anaerobic systems this means that electrons must be transferred from reduced NAD to some organic acceptor molecule, which is consequently reduced. In yeast, pyruvic acid is first decarboxylated to form acetaldehyde and CO_2 (step 11), and it is the acetaldehyde so formed which accepts electrons from reduced NAD to yield ethanol and oxidized NAD (step 12). NAD^+ thus continually cycles between step 6 and 12, being reduced in step 6 and oxidized in step 12.

	step 11		step 12	
COOH			NAD^+	CH_2OH
$\|$		CHO		$\|$
C=O	$-CO_2$	$\|$		CH_3
$\|$	$\longrightarrow$	CH_3	NADH	ethanol
CH_3				
pyruvic acid		acetaldehyde		

The alcoholic fermentation of glucose by yeast is only one of many fermentations which occur in nature. In the so-called lactic acid bacteria, and in muscle cells which are working hard in the absence of adequate supplies of oxygen, the NAD is replenished by pyruvic acid itself acting as the acceptor molecule.

$$CH_3.CO.COOH + NADH + H^+ \rightarrow CH_3.CH(OH).COOH + NAD^+$$
pyruvic acid   lactic acid

End products besides ethanol and lactic acid are formed in fermentations carried out by other organisms.

Most organisms break down their food either by fermentation or by aerobic respiration according to the availability of oxygen. However, some organisms (obligate anaerobes) respire exclusively by fermentation process-es, and are killed by molecular oxygen. Such organisms are relatively rare, the majority depending in normal circumstances on aerobic respiration.

Stage II of respiration: the tricarboxylic acid cycle

Alcoholic fermentation, like all fermentations, releases only a fraction of the energy contained in the glucose molecule. The reason is that acetalde-hyde, the last intermediate in the reaction sequence, acts as the terminal acceptor for the electrons which are removed during the energy-yielding reaction (step 6) of glycolysis. The release of ATP is thus small, because the carbon in the products of fermentation is still in a highly reduced state. Ox-idation is important in the provision of energy only in so far as it is accom-panied by the fission of carbon–carbon bonds. The production of ethanol as the end product of fermentation thus represents a 'sink' for chemical ener-gy, which is lost to the cell when the ethanol leaves it. The situation would be energetically much more favourable if all the carbon–carbon bonds could be split to give carbon dioxide, so that all the usable chemical energy in the glucose molecule could be transformed into ATP. By causing the electrons which are removed during the oxidation of glucose to combine with an environmental acceptor, atmospheric oxygen, respiration achieves precisely this more favourable yield of energy.

	$\triangle G°$	Net gain of ATP per molecule of glucose broken down
Alcoholic fermentation $C_6H_{12}O_6 \rightarrow$ $2C_2H_5OH + 2CO_2$	– 52 kcal	2 molecules
Respiration $C_6H_{12}O_6 + 6O_2 \rightarrow$ $6H_2O + 6 CO_2$	– 686 kcal	38 molecules

In the presence of oxygen, the pyruvic acid produced in glycolysis is directed into the TCA cycle where it is completely oxidized. To enter the TCA cycle, however, it must be converted into acetyl coenzyme A (acetyl CoA). Stage II of respiration will therefore be described under two head-ings: (i) the conversion of pyruvic acid to acetyl CoA; and (ii) the TCA cy-cle.

The conversion of pyruvic acid to acetyl coenzyme A

Although not itself part of the TCA cycle, the oxidative decarboxylation of pyruvic acid to acetyl CoA is the essential step by which the partially oxi-dized glucose molecule enters the cycle. The reaction is very complicated and involves the participation of several enzymes and coenzymes, collective-ly called the pyruvate dehydrogenase complex. The most important coen-zyme in this complex is coenzyme A. From the point of view of its involve-ment in the oxidative decarboxylation of pyruvate, the most significant feature of the coenzyme A molecule is that it is terminated by a –SH group because it is here that an acetyl group becomes attached. The molecule can thus conveniently be written in equations as CoA–SH. The oxidation which accompanies the removal of CO_2 from pyruvic acid requires the presence of NAD as the oxidizing agent. The overall reaction may be written as follows

$$CH_3COCOOH + CoA-SH + NAD^+ \rightarrow CH_3CO\sim S-CoA +$$

pyruvic acid   coenzyme A   acetyl CoA

$$CO_2 + NADH + H^+$$

Acetyl CoA, which is the end product of this reaction, is another high-energy compound, a fact which is indicated in the above equation by the squiggle bond between the acetyl group and the coenzyme A part of the molecule. It is the energy locked in this high-energy bond which is ultimately responsible for driving the TCA cycle.

The tricarboxylic acid cycle

The complete oxidation of acetyl CoA occurs in a complex sequence of reactions which is usually known as the tricarboxylic acid (TCA) cycle, but is also called the Krebs cycle after its discoverer, Sir Hans Krebs. The reaction sequence is called a 'cycle' because, unlike the reactions of the glycolytic pathway, the end product (here, the 4-carbon oxaloacetic acid) combines with more of the starting material (here, acetyl CoA) and re-enters the reaction sequence as soon as it is formed. The sequence is called the TCA cycle because the first intermediate in the cycle is a tricarboxylic acid.

The relatively large number of stages in the TCA cycle reflects the chemical difficulties which arise in 'dismembering' an organic compound with only two carbon atoms in such a way that its energy is released in a number of controlled steps. These problems have been solved by the simple but effective device of combining the 2-carbon compound with a 4-carbon acceptor to give a 6-carbon compound, which is much easier to dismember and oxidize in stages. Each turn of the TCA cycle is initiated by the formation of citrate (with 6 carbon atoms) from oxaloacetate (with 4 carbon atoms) and the acetyl radical (with 2 carbon atoms) of acetyl CoA. Subsequent reactions result in the re-formation of oxaloacetate and two molecules of carbon dioxide. The carbon atoms that go into the formation of CO_2 are lost to the cell, but the accompanying oxidations (whereby electrons in conjunction with H ions are removed from intermediate compounds in the TCA cycle and transferred ultimately to oxygen to form H_2O) are responsible for generating most of the ATP in the cell.

The TCA cycle (Fig. 16.6) can conveniently be divided into nine main steps, although some of them involve more than one chemical reaction. In step 1, acetyl CoA reacts with oxaloacetate to yield citrate and to regenerate coenzyme A.

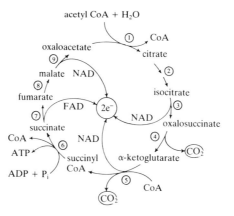

Fig. 16.6 The tricarboxylic acid cycle (Krebs cycle). The numbers refer to the reactions described in the text.

step 1

$$CH_3CO{\sim}SCoA + \underset{\underset{CH_2COOH}{|}}{\overset{\overset{O}{||}}{C}{-}COOH} + H_2O \longrightarrow \underset{\underset{CH_2COOH}{|}}{\overset{\overset{CH_2COOH}{|}}{HO{-}C}{-}COOH} + HSCoA$$

acetyl CoA   oxaloacetate   citrate   coenzyme A

Citrate undergoes isomerization (step 2) to form isocitrate, which is oxidized by the coenzyme NAD (step 3) to form oxalosuccinate. This is very unstable and consequently loses carbon dioxide (step 4) to become the 5-carbon acid α-ketoglutarate.

step 2   step 3   step 4

$$\underset{\underset{\underset{citrate}{CH_2COOH}}{|}}{\overset{\overset{CH_2COOH}{|}}{HO{-}C}{-}COOH} \longrightarrow \underset{\underset{\underset{isocitrate}{HO{-}CH(COOH)}}{|}}{\overset{\overset{CH_2COOH}{|}}{H{-}C}{-}COOH} \overset{NADH}{\underset{NAD^+}{\longrightarrow}} \underset{\underset{\underset{oxalosuccinate}{O{=}C{-}COOH}}{|}}{\overset{\overset{CH_2COOH}{|}}{C{-}COO{-}H}} \longrightarrow \underset{\underset{\underset{\alpha\text{-ketoglutarate}}{O{=}C{-}COOH}}{|}}{\overset{\overset{CH_2COOH}{|}}{CH_2}} + CO_2$$

α-Ketoglutaric acid, which resembles pyruvic acid in having a keto group (>CO) next to its acidic group (–COOH), undergoes an oxidative decarboxylation reaction with coenzyme A (step 5) similar to that undergone by pyruvic acid, the product being the 4-carbon succinyl~S–CoA. This is a high-energy compound analogous to acetyl~S–CoA formed from pyruvic acid. Succinyl~S–CoA, unlike acetyl~S–CoA, then transfers its energy-rich bond to ADP, regenerating coenzyme A and forming succinate and ATP (step 6)

$$\underset{\underset{\underset{\alpha\text{-ketoglutarate}}{O{=}C{-}COOH}}{|}}{\overset{\overset{CH_2COOH}{|}}{CH_2}} + CoA{-}SH + NAD^+ \overset{step\ 5}{\longrightarrow} \underset{\underset{\underset{succinyl\ CoA}{O{=}C{\sim}S{-}CoA}}{|}}{\overset{\overset{CH_2COOH}{|}}{CH_2}} + CO_2 + NADH + H^+$$

$$\underset{\underset{\underset{succinyl\ CoA}{O{=}C{\sim}S{-}CoA}}{|}}{\overset{\overset{CH_2COOH}{|}}{CH_2}} + ADP + P_i \overset{step\ 6}{\longrightarrow} \underset{\underset{succinate}{CH_2COOH}}{\overset{CH_2COOH}{|}} + ATP + CoA{-}SH$$

coenzyme A

The remaining three steps of the TCA cycle serve to regenerate oxaloacetate, the initial 4-carbon acceptor of acetyl CoA, from succinate. This involves the oxidation of the hydrocarbon group (>CH$_2$) in the α position (i.e. next to the –COOH group) of succinate to a carbonyl group (>CO). In the first of the three stages succinate is oxidized to fumarate by the removal of two hydrogen atoms (step 7), which are accepted not by NAD but by another coenzyme called FAD. (The only functional difference between NAD and FAD which is noteworthy in a general description of respiration is that the subsequent oxidation of each molecule of reduced FAD is coupled to the formation of two molecules of ATP, whereas the oxidation of reduced NAD results in the formation of three molecules of ATP.) The fumarate then undergoes molecular rearrangement by the addition of the elements of water across the double bond (step 8) to form malate. Finally, two hydrogen atoms are removed from malate and accepted by NAD^+ (step 9), resulting in the formation of oxaloacetate and reduced NAD. The formation of oxaloacetate completes the TCA cycle, which can now begin again with the formation of citrate.

step 7

step 8

step 9

$$\begin{array}{c} CH_2COOH \\ | \\ CH_2COOH \end{array} \xrightarrow[\substack{FAD}]{\substack{FADH_2}} \begin{array}{c} CHCOOH \\ \| \\ CHCOOH \end{array} \xrightarrow{+H_2O} \begin{array}{c} CH(OH)COOH \\ | \\ CH_2COOH \end{array} \xrightarrow[\substack{NAD^+}]{\substack{NADH}} \begin{array}{c} O \\ \| \\ C-COOH \\ | \\ CH_2COOH \end{array}$$

succinate   fumarate   malate   oxaloacetate

The TCA cycle has been described with reference to glucose, but it must be emphasized that its significance is not confined to the respiration of glucose. All food materials, after they have been broken down into their constituent units (e.g. storage polysaccharides into monosaccharides, proteins into amino acids, etc.), are incompletely oxidized to form, apart from carbon dioxide and water, one of only three possible substances, all of which are intermediates in the TCA cycle. The first, acetic acid in the form of acetyl CoA, is by far the most common product; it is the product of two-thirds of the carbon incorporated into carbohydrates and glycerol, all of the carbon in most fatty acids, and approximately half of the carbon in amino acids. The other two possible substances, α-ketoglutarate and oxaloacetate, are breakdown products of several amino acids. Thus the acetyl CoA which enters the TCA cycle can be derived not only from glucose but from any organic substrate which is degraded to acetyl fragments. Similarly α-ketoglutarate and oxaloacetate resulting from the breakdown of amino acids can feed directly into the TCA cycle at the appropriate points.

Besides serving as a source of energy for living cells, the reactions of the TCA cycle are also used as a source of metabolic intermediates which provide starting materials for the biosynthesis of important cell constituents. Thus α-ketoglutarate becomes aminated to yield the amino acid glutamic acid which, in turn, takes part in the synthesis of other amino acids. Similarly oxaloacetate is converted into aspartic acid, which is also a precursor of a number of amino acids as well as of pyrimidines.

As a consequence of the addition and removal of intermediates in the TCA cycle, the levels of the different intermediates may fluctuate with time. However, a supply of oxaloacetate is essential to maintain the operation of the cycle and the cell possesses several ancillary mechanisms to accomplish this.

Stage III of respiration: the electron transport chain

Two of the reactions leading up to the formation of acetyl CoA and four of the reactions in the TCA cycle involve the dehydrogenation of an intermediate substance. These dehydrogenations are carried out at the expense of NAD^+ which is reduced to NADH (except in the dehydrogenation of succinic acid, where the function of NAD^+ is taken over by the coenzyme FAD). The NADH must subsequently be re-oxidized to replenish the supply of NAD^+.

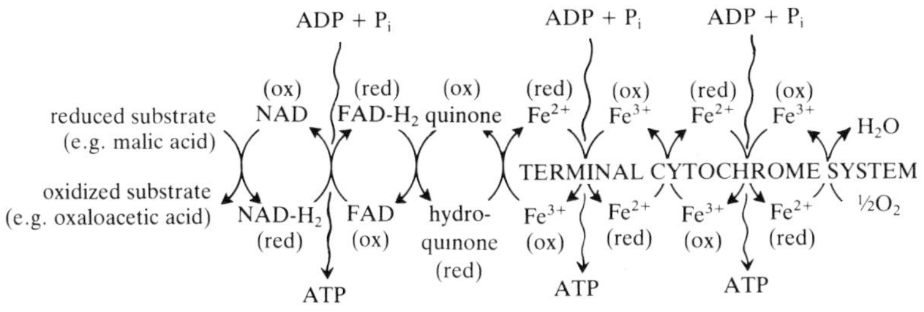

Fig. 16.7 The electron transport chain, showing how it is coupled at three sites with the synthesis of ATP (Fe^{2+} and Fe^{3+} = prosthetic group of a cytochrome molecule; ox = oxidized; red = reduced).

The overall reaction ($NADH + H^+ + \frac{1}{2}O_2 \rightarrow NAD^+ + H_2O$) releases a large amount of energy ($\triangle G^\circ = -52$ kcal), and takes place in a series of enzyme-catalysed steps which are directly coupled to ATP production. The essential point of the sequence is that every time hydrogen atoms or electrons are passed from one intermediate to the next a small amount of energy is released, and at certain points in the sequence this can be coupled to the synthesis of ATP. The sequence of reactions is known as the electron transport chain (Fig. 16.7). At the end of the chain of oxidation–reduction reactions, oxygen itself acts as the electron acceptor and the electrons, now at a low energy level, are disposed of as water. The last reaction in the electron transport chain is therefore

reduced carrier + oxygen → oxidized carrier + water

Between NADH and molecular oxygen at least three types of electron carriers (flavoproteins, a quinone, and cytochromes) are known to participate in the chain of oxidations. Each carrier is alternately oxidized and reduced, i.e. the oxidized form of the carrier accepts electrons from the reduced carrier preceding it in the series, and in turn donates electrons to the next carrier in the series. The carriers taking part in the final reactions of the sequence are called *cytochromes*, and consist of a protein with a non-protein component, or prosthetic group, containing iron in an ionic form. During the sequence of oxidation–reduction reactions the ion of iron in each cytochrome alternately exists in its oxidized Fe^{3+} (ferric) and reduced Fe^{2+} (ferrous) forms. The last cytochrome in the series (cytochrome a, also known as cytochrome oxidase) is able to donate electrons to oxygen, which is thus reduced to water

$$2Fe^{2+} + 2H^+ + \tfrac{1}{2}O_2 \rightleftharpoons 2Fe^{3+} + H_2O$$

Oxidative phosphorylation

Oxidative phosphorylation is the mechanism by which the packets of energy liberated along the electron transport chain are conserved as ATP. As previously mentioned, the tendency of the oxidized form of a substance to act as an electron acceptor, under standard conditions, can be expressed as its standard reduction potential E_o (in units of volts). The E_o values for the carriers of the hydrogen transport chain range from -0.32 V for the half-reaction $NAD^+ + 2H^+ + 2e^- \rightarrow NADH + H^+$ to $+0.82$ V for the half-

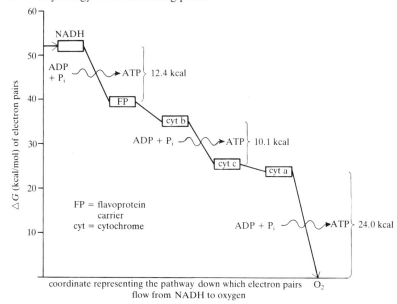

Fig. 16.8 The three steps along the electron transport chain where the decrease in free energy is sufficient to generate a molecule of ATP from ADP and phosphate.

reaction $\frac{1}{2}O_2 + 2H^+ + 2e^- \rightarrow H_2O$. The values for the intermediate carriers lie between these extremes. Reduced NAD^+, the first carrier, is the most electronegative substance involved in the chain, while oxygen, the final hydrogen acceptor, is the most electropositive substance. Since electron transfer in any redox reaction is always in the direction leading to an increase in the positive value of E_o, the electron transport chain consists of a highly ordered series of substances which act as electron carriers between the reduced form of the most electronegative member of the chain (i.e. reduced NAD^+) and the oxidized form of the most electropositive member (i.e. oxygen). The pathway of electron transport is shown in Fig. 16.8. Each step in the chain is accompanied by a decrease in standard free energy ($\triangle G°$) proportional to the difference between the standard reduction potentials ($\triangle E_o$) of the two consecutive carriers involved. The transfer of electrons from NADH at the start of the chain to oxygen at the end results in a $\triangle E_o$ of -1.14 V which, by applying the appropriate equation, can be shown to correspond to a loss of free energy of 52 kcal per mole of NADH oxidized.

For any chemical reaction to occur spontaneously its overall free-energy change must be negative. Since $\triangle G°$ for the reaction $ADP + P_i \rightarrow ATP$ is $+7$ kcal/mol, it will not take place unless it is coupled to a reaction with a $\triangle G°$ more negative than -7 kcal/mol. The 52 kcal/mol of free energy liberated along the electron transport chain is theoretically sufficient to allow the synthesis of seven molecules of ATP ($7 \times 7 = 49$ kcal). However, if a quantitative study is made of the electron transport chain, it is found that only three molecules of ATP are formed per atom of oxygen consumed. The energy that is not conserved as ATP is lost as heat. Although it is not yet clear how some of the energy released is recovered in the form of ATP, the sites of ATP synthesis along the chain are known. The electron pathway shown in Fig. 16.8 is oversimplified because certain carriers whose site of action is not well established have been omitted, but it will be noted that the three sites of ATP synthesis coincide (as indeed they must) with the three steps which have a $\triangle G°$ of more than -7 kcal/mol. With the information that three molecules of ATP are formed per pair of electrons passed down the electron transport chain, i.e. per atom of oxygen consumed, it is possible to calculate the energy yield of aerobic respiration.

Energy yield of respiration

For every molecule of pyruvate fed into the TCA cycle 4 molecules of reduced NAD, 1 molecule of reduced FAD and 1 molecule of ATP are formed. Collectively these account for 15 molecules of ATP. Since each glucose molecule yields 2 molecules of pyruvate, the total output from the aerobic oxidation of pyruvate is 30 ATP molecules per glucose molecule. It will be recalled that step 6 of glycolysis is also responsible for the production of 1 molecule of reduced NAD, and that this step is repeated twice for every glucose molecule entering the glycolytic pathway. There is also a net gain of 2 ATP molecules from the glycolytic breakdown of each glucose molecule. The glycolysis of glucose to 2 molecules of pyruvate therefore yields 8 molecules of ATP. Adding these to the 30 obtained from the oxidation of 2 pyruvate molecules via the TCA cycle, the overall total is 38 molecules of ATP per molecule of glucose oxidized.

With $\triangle G°$ for ATP synthesis at +7 kcal/mol, the total energy trapped during the aerobic respiration of glucose is $38 \times 7 = 266$ kcal/mol. Since $\triangle G°$ for the complete oxidation of glucose to carbon dioxide and water is -686 kcal/mol, the efficiency of ATP synthesis in aerobic respiration is $\frac{266}{686} \times 100 =$ about 40 per cent. This level is efficiency compares very favourably with that of the most up-to-date oil-fired power stations.

Location of respiration within the cell

The complex enzymic sequences of respiration have been described without reference to where they are taking place inside the cell. In order to discover which parts of the cell contain the enzymes necessary for the component reactions, fragmented cells are centrifuged into a series of fractions containing different cellular constituents. These are then tested separately for the presence of enzymes known to be involved in glucose metabolism.

It is found that the enzymes associated with the TCA cycle, the electron transport chain and oxidative phosphorylation are all located within the mitochondrial fraction, and not in any of the fractions containing other parts of the cell. By contrast, the enzymes of glycolysis are located entirely in the cytoplasmic fraction (the supernatant), which remains after all the subcellular particles have been removed by centrifugation. A preparation of cell cytoplasm will catalyse the breakdown of glucose and produce pyruvic acid, whereas a preparation of mitochondria will not oxidize glucose but will oxidize pyruvic acid, liberating CO_2 and, under appropriate conditions, synthesizing ATP from ADP and phosphate.

Localization of function within the mitochondrion has proved very difficult. It is known that the enzymes of the TCA cycle are found free in the fluid that fills its interior. Pyruvic acid, generated by glycolysis in the cell cytoplasm, enters the mitochondria where it reacts with the TCA enzymes which oxidize it to CO_2. The CO_2 diffuses out of the mitochondria and eventually out of the cell (Fig. 16.9). The NADH, which is simultaneously generated during the operation of the TCA cycle, remains inside the mitochondria. The enzymes of the electron transport chain and oxidative

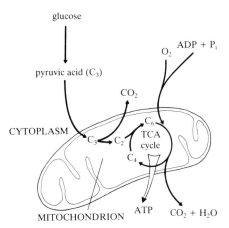

glucose

ADP + P$_i$

O$_2$

pyruvic acid (C$_3$)

CO$_2$

CYTOPLASM

C$_6$

TCA cycle

C$_3$ C$_2$

C$_4$

MITOCHONDRION   ATP   CO$_2$ + H$_2$O

Fig. 16.9 The location of the glycolytic and oxidative stages of respiration within the cell.

phosphorylation are part of the actual structure of the inner membranes (cristae) of the mitochondria, and it is on the surface of these membranes that ATP is produced before passing into the cytoplasm. Chemical analysis of mitochondria shows that the various enzymes linking NADH to oxygen along the electron transport chain are present in fixed proportions, a condition prerequisite for the smooth running of such a production line. Fragmentation of mitochondria breaks the continuity of ATP production, and many of the components of the electron transport chain and oxidative phosphorylation cannot be isolated as functional units separate from intact mitochondria. Hence it is possible to regard a single mitochondrion as one giant macromolecule, of which the individual enzymes and electron carriers are mere components.

It will be apparent that the mitochondrion is highly specialized structurally to carry out particular metabolic sequences. Such structural specialization in relation to function is known as *compartmentation*.

Chapter 17

Photosynthesis

In essence photosynthesis is the process whereby green plants synthesize organic compounds from carbon dioxide and water. The process can take place only in the presence of light (hence the prefix photo-) and through the agency of the green chlorophyll pigments which, as we have already seen, are located in special cytoplasmic organelles called chloroplasts. The overall reaction can be represented by the following equation:

$$CO_2 + H_2O + light\ energy \xrightarrow{\text{chlorophyll}} (CH_2O) + O_2$$

(carbon   (water)   (organic   (oxygen)
dioxide)   matter)

In this equation (CH_2O) is a generalized formula to denote organic matter which is commonly, but not necessarily, starch or some other carbohydrate. The organic matter, which is the important product of photosynthesis, contains more potential chemical energy (i.e. energy that can be made available by chemical change) than the carbon dioxide and water from which it is formed, the difference being obtained by the absorption of light energy. In photosynthesis, therefore, the radiant energy of light is transformed into the chemical energy of a stable organic compound such as a carbohydrate.

The site of photosynthesis In the presence of light, photosynthesis can occur in any green part of a plant, but in a typical land plant it is the leaves with their large surface area and abundant chloroplasts which are the main centres of the process. The carbon dioxide used in photosynthesis is obtained from the atmosphere which usually contains about 0.03 per cent by volume of this gas. Since the cuticle is relatively impervious to gases, it follows that the carbon dioxide must enter a leaf mainly through the stomata. Once inside a leaf, the carbon dioxide diffuses through the system of intercellular air spaces and dissolves in the water that saturates the walls of the mesophyll cells. It then diffuses, or is actively moved by protoplasmic streaming, through the water of the cytoplasm to the chloroplasts where, in the presence of light, photosynthesis takes place (Fig. 17.1).

When a leaf is actively photosynthesizing, the concentration of carbon dioxide at the chloroplast surface will diminish and a carbon dioxide gradient will be set up between the chloroplast surface and the external atmosphere. Provided the stomata are open, carbon dioxide will therefore diffuse into the leaf, the rate depending on the steepness of the concentration gradient which, in its turn, will depend upon the activity of the chloroplasts. At

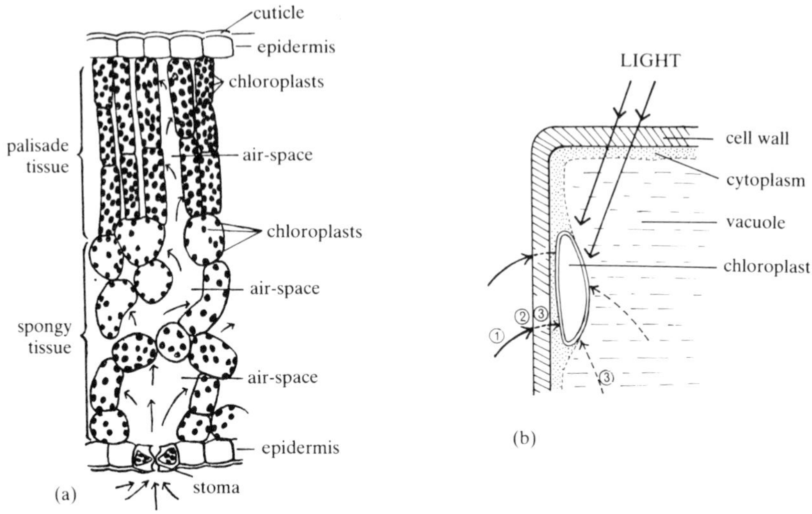

Fig. 17.1 The passage of CO_2 through a leaf during photosynthesis. (a) Small part of a typical dorsiventral leaf. The arrows show the paths of diffusion of CO_2. (b) Diagrammatic section through one chloroplast lying against the wall of a palisade cell. The numbered arrows show the diffusion of CO_2 to the lighted chloroplast. (1) Part of the diffusion path is gaseous diffusion through the system of air spaces; at (2) CO_2 dissolves in the water permeating the cell walls; at (3) there is diffusion through the watery medium of cell wall, cytoplasm, and vacuole.

the same time as the carbon dioxide concentration decreases at the chloroplast surface the oxygen concentration increases, and an oxygen gradient is set up in the opposite direction to that of the carbon dioxide. The inward diffusion of carbon dioxide is therefore accompanied by a simultaneous outward diffusion of oxygen.

The water used in photosynthesis is normally readily available since the water content of the cytoplasm of healthy mesophyll cells is always high. Water is continually supplied to the leaf from the soil through the xylem system, the terminal branches of which permeate the leaves so that all the component cells are in close proximity to their water supply. The vascular tissue of the network of veins in a leaf not only supplies the leaf with water, but also provides the pathway by which the photosynthetic products are removed from the leaf and transported through the phloem to other parts of the plant.

Starch formation

In most dicotyledonous plants, though in only a few monocotyledons, starch begins to accumulate in the leaves shortly after the onset of rapid photosynthesis. For this reason, dicotyledons usually have what may be described as *starch leaves*, whereas monocotyledons usually have *sugar leaves*.

The starch test   The presence of starch within a leaf is easier to detect than the presence of sugars, and, provided a plant with starch leaves is used as experimental material, the demonstration of starch formation can be used as a convenient method to show the necessity for such factors as carbon dioxide, light and chlorophyll in photosynthesis. To test for starch the leaf is dropped for 1 minute in boiling water, boiled in 70 per cent alcohol until completely decolorized, washed with tap water and then treated with aqueous iodine (i.e. a solution of iodine in potassium iodide) which gives a blue-black colour with starch. The depth of the coloration gives a very approximate indication of the concentration of starch present. When using starch formation as a test for the occurrence of photosynthesis it is first necessary to deplete the leaves of any starch already present. Unless this is done it is difficult to distinguish, purely by coloration, between the amounts of starch present before and after the experimental treatment. Removal of starch from leaves is accomplished by keeping them, while still attached to the parent plant, in complete darkness for 24 to 36 hours. During this period any starch originally present is hydrolysed to sugar, which is removed by the combined action of respiration and the translocation of sugar from the leaves to the rest of the plant (see below). The absence of starch in leaves that have been kept in complete darkness should be confirmed by testing one of the leaves.

Although the formation of starch within a leaf can be used as a convenient index of the occurrence of photosynthesis, it should be clearly understood that starch formation is a sequel to photosynthesis and not an integral part of the process. Photosynthesis is a chemical synthesis which involves the absorption of light energy and occurs only in the presence of chlorophyll. Starch formation from sugar, on the other hand, is a condensation reaction which can go on as readily in the dark as in the light. It is also independent of the presence of chlorophyll for it is freely carried out by the leucoplasts (colourless plastids) found in the non-green cells of storage tissues. The function of chloroplasts as starch formers is therefore quite separate from their primary function in the photoreduction of carbon dioxide. That starch formation is a sequel to photosynthesis rather than a stage of it can be easily demonstrated by floating starch-free leaves on a glucose solution in the dark, when it will be found that starch formation readily takes place.

Relation of photosynthesis to other activities in the leaf

So far in this discussion photosynthesis has been considered as an isolated process, but it must be remembered that in the living plant photosynthesis is only one of several major activities going on in the green leaf. In addition to the synthesis of starch and other organic compounds from the immediate products of photosynthesis, photosynthesis is also accompanied by respiration and the continuous translocation of photosynthetic products away from the leaves to other parts of the plant.

Daily cycle in starch content of starch leaves   In a typical starch leaf carbohydrate is formed during the day much more rapidly than it is removed by respiration and translocation, with the result that it accumulates as starch. There will therefore be a progressive deposition of starch in the chloroplasts during the course of the day. At night, when photosynthesis ceases, respiration and translocation of carbohydrate continue, with the result that the content of starch within the leaf falls during the night until little or none remains in the morning (Fig. 17.2). It is the occurrence of this daily cycle in the starch content of a leaf which underlies the experimental procedure for destarching leaves. The removal of starch from a leaf by translocation at night can easily be demonstrated by detaching one leaf from a plant at sunset (when the leaves have a high starch content, as shown by the application of the starch test to a leaf removed at this time) and leaving it all night in a damp vessel beside the parent plant. A starch test of this leaf on the following morning will show it to be almost as full of starch as when it was gathered, whereas a leaf left on the parent plant will be found to contain little or no starch.

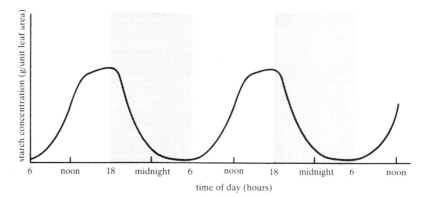

Fig. 17.2 Diagrammatic representation of the daily cycle in starch content of a dicotyledonous leaf attached to the parent plant.

Apparent photosynthesis   The gaseous exchange involved in photosynthesis is the opposite of that in respiration. When any green plant is kept in total darkness, it gives out carbon dioxide as a result of respiration. If the plant is now placed in bright light, it takes in carbon dioxide for photosynthesis and, as under these conditions the rate of photosynthesis is faster than the rate of respiration, there is a net intake of carbon dioxide, i.e. the respiratory output is masked. By diminishing the light intensity it is possible to reduce photosynthesis to a rate at which the photosynthetic intake and respiratory output exactly balance one another. The light intensity at which there is no net exchange of carbon dioxide is called the *compensation point* (Fig. 17.3). By diminishing the light still more there will be a net output of carbon dioxide, which progressively increases in amount until in complete darkness this is entirely due to output by respiration. The fact that respiration is going on all the time alongside photosynthesis means that the net intake of carbon dioxide into the leaf during photosynthesis equals the actual absorption of carbon dioxide by the cells minus the simultaneous but smaller output in respiration. In practice, however, it is often more useful to know the extent of the 'apparent' or net rather than the 'actual' or gross photosynthesis, because it is the former which represents the total gain to the plant in the form of crop yields.

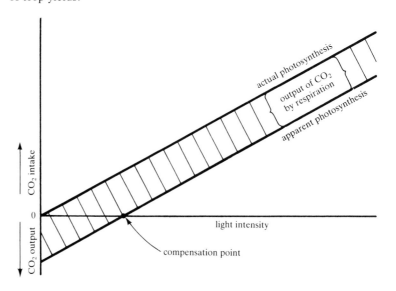

Fig. 17.3 Diagram to illustrate concepts of apparent photosynthesis and of the compensation point.

When a plant is kept at the compensation point, all the organic matter produced by photosynthesis is consumed by respiration and there is no excess available for growth. For growth to occur, a plant must live in light intensities which allow photosynthesis to exceed respiration, i.e. in light intensities above the compensation point. In natural conditions many leaves receive suboptimal quantities of light because they are shaded by other leaves, and this is the reason why the growth of tree seedlings in a forest is often retarded until an opening appears in the canopy. Plants which normally grow on the forest floor beneath the tree canopy are adapted to shade conditions by having low compensation points (say 3 per cent of full sunlight). They can produce an excess of organic nutrients at low levels of light, but are usually poorly adapted to strong light because their photosynthesis reaches its maximum rate at light intensities well below full sunlight (Fig. 17.4). By contrast, plants which grow naturally in well-lit habitats have higher compensation points (say 10 per cent of full sunlight) and can use bright light more efficiently. They are well adapted to sunny conditions but poorly adapted to shade.

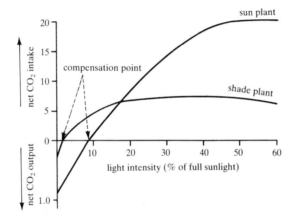

Fig. 17.4 Light relations of sun and shade plants. Note that the shade plant not only has a lower compensation point, but also cannot make such good use of higher light intensities. The axis showing the net CO_2 output has been exaggerated for clarity.

Rate of Photosynthesis

Measurement of the rate of photosynthesis

From the summary equation for photosynthesis it is obvious that the rate of the process could theoretically be measured in terms of either the consumption of carbon dioxide and water, or the production of organic matter and oxygen. In practice, since photosynthetic tissues are constantly subject to considerable fluctuations in water content due to activities other than photosynthesis (e.g. transpiration), it is impossible to detect the relatively small decreases in water content due to photosynthesis. Increases in organic matter (which are usually assessed by measuring the dry weight) can be determined, but only by sacrificing test material for each determination. Therefore methods based on changes in the amount of either carbon dioxide or oxygen, which in any case are more convenient to measure than organic matter, are the ones most frequently used. The principle underlying both methods is the same and is as follows.

The experimental plant material is contained in a transparent chamber through which it is illuminated, and a steady current of air is passed through the chamber. The volume of air passing over the photosynthesizing material per unit of time is measured by a gas flow meter. A small measured sample of the ingoing air is removed and its content of either carbon dioxide or oxygen is measured. A comparable sample is removed from the outgoing air and similarly analysed. From the difference in the two sets of readings the rate of carbon dioxide intake or of oxygen output can be calculated. Since small changes in carbon dioxide concentration are easier to detect than small changes in oxygen concentration, intake of carbon dioxide is usually measured in preference to output of oxygen.

Blackman's principle of limiting factors

In order to understand how the various factors controlling photosynthesis affect its rate, a knowledge of what F. F. Blackman has called the principle of limiting factors is necessary. Early research workers who investigated the rate of photosynthesis studied one factor at a time without reference to the influence of the other factors. It is not surprising, therefore, that separate workers investigating the effect of any one particular factor sometimes obtained vastly different results. This chaotic situation was not clarified until 1905, when F. F. Blackman enunciated his principle of limiting factors as follows: 'When a process is conditioned as to its rapidity by a number of separate factors, the rate of the process is limited by the pace of the slowest factor.'

Blackman's principle may be illustrated by considering a theoretical example of a plant photosynthesizing at various levels of carbon dioxide concentration under conditions of low but constant light intensity and a suitable temperature. As the carbon dioxide concentration is raised from zero, the rate of photosynthesis increases until a certain concentration is reached, above which no increase in rate takes place however much the carbon dioxide concentration is raised. This does not mean that carbon dioxide can have no further effect, but merely that all the available light energy is being used up. At this concentration the process can be said to be carbon dioxide saturated. As there is no more light energy for more rapid photosynthesis, raising the carbon dioxide concentration cannot increase the rate. If the light intensity is increased, however, the rate of photosynthesis will continue to increase until light becomes once more the limiting factor, when again the increase in rate stops abruptly and the rate becomes constant. This sequence of events is represented graphically in Fig. 17.5a.

Subsequent investigators, using improved techniques not available to Blackman, have shown that Blackman's principle is an oversimplification. When curves relating the rate of photosynthesis to increasing carbon dioxide concentration at different light intensities are determined by experiment, they differ from the theoretical curves in two respects. (1) The curve for any given light intensity does not show the sharp break implicit in Blackman's concept but bends over smoothly (Fig. 17.5b). This indicates that carbon dioxide is not suddenly replaced by light as the limiting factor, but that in the region where the curve bends over both carbon dioxide concentration and light intensity are simultaneously controlling the rate. (2) The curves for different light intensities are distinct from one another throughout their length, even though they converge as zero carbon dioxide concentration is approached. This indicates that carbon dioxide and light interact not only over a limited range of carbon dioxide concentration but over the whole

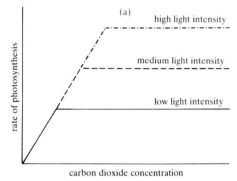

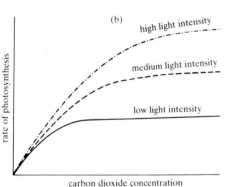

Fig. 17.5 Limiting factor curves. (a) Curves illustrating Blackman's theoretical interpretation of the operation of a limiting factor. (b) Typical curves obtained when relation between rate of photosynthesis and CO_2 concentration at different light intensities is investigated experimentally. Compare shape of curves with those of (a).

range from zero concentration to carbon dioxide saturation. It will be clear that, due to the interaction of factors, a given factor can be regarded as rate-limiting only when further increases in the other controlling factors have no effect on the rate.

Blackman's principle has been illustrated by considering the relationship of carbon dioxide concentration to photosynthetic rate, but it could equally well have been illustrated by describing photosynthetic rate as a function of light intensity at different carbon dioxide concentrations. Here the situation would be reversed, photosynthesis being light-saturated when carbon dioxide becomes the limiting factor.

Although Blackman's principle is now known to be an oversimplification, it nevertheless remains of fundamental importance because it was instrumental in formalizing the type of technique that has become standard practice in experimental physiology. His principle made it clear for the first time that to study the effect of one factor the intensity of the other controllable factors must be enough for their effects to be non-limiting.

Factors affecting the rate of photosynthesis

From the summary equation for photosynthesis it is possible to predict five factors that should influence the rate of the process, namely: (1) carbon dioxide concentration; (2) water supply; (3) light intensity; (4) chlorophyll content; and (5) accumulation of the products of photosynthesis. Because, as will be seen later, photosynthesis is not simply a photochemical reaction but involves strictly chemical stages as well, one can add another controlling factor, namely (6) temperature. If it is also remembered that photosynthesis takes place inside a living green leaf, it will be appreciated that the carbon dioxide has to diffuse from the external atmosphere to the site of the reaction. From this it follows that any features in the structure of the leaf which afford a resistance to the inward diffusion of carbon dioxide will affect the ultimate rate of the process. One can thus list yet another factor influencing the rate, namely (7) resistances in the leaf to free gaseous diffusion.

All the above factors have been derived from a knowledge of the basic facts of photosynthesis, but there is definite evidence that (8) protoplasmic factors are also involved. These factors are not well understood, although they are known to be associated with the activity of certain enzymes essential for photosynthesis.

These eight factors affecting the rate of photosynthesis may be conveniently divided into two categories, *external* (environmental) and *internal* (due to the constitution of the plant), although the division is not always clear-cut. The external factors are carbon dioxide concentration, water supply, light intensity, and temperature. The internal factors are chlorophyll content, protoplasmic factors, resistance to free gaseous diffusion, and accumulation of the products of photosynthesis.

External factors

Of the four external factors that could theoretically influence the rate of photosynthesis, the effect of water supply cannot be investigated directly because the photosynthetic apparatus is located in the watery medium of the cytoplasm. In view, however, of the fact that only a very small proportion of the water which passes through the plant is used in photosynthesis, it

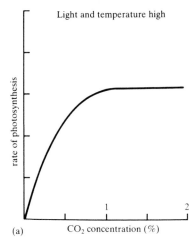

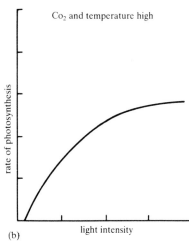

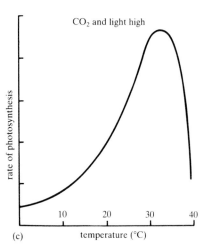

Fig. 17.6 The effect on the rate of photosynthesis of increasing (a) CO_2 concentration, (b) light intensity, and (c) temperature.

would appear that water in its role as one of the reactants in the process is seldom, if ever, a limiting factor. It is, nevertheless, true that an inadequate supply of water to a plant results in a drastic reduction in the rate of photosynthesis. Since a water deficit in the leaves causes the stomata to close, it used to be thought that stomatal closure was the main reason for this reduction in photosynthetic rate. There is abundant evidence, however, that photosynthesis is retarded in water-deficient leaves long before the stomata close. It now appears that a shortage of water, by causing dehydration of the protoplasm and thus a change in its colloidal structure, affects all or most of the metabolic processes of the plant. The reduced rate of photosynthesis due to an inadequate supply of water is thus more likely to be merely one of many indirect effects resulting from the dehydration of the protoplasm.

By contrast to water supply, the three other external factors controlling photosynthesis lend themselves readily to experimental variation. These factors are often spoken of as the three primary variables of photosynthesis because, in nature, they are the factors which are likely to be responsible for the daily variations in rate.

Carbon dioxide concentration With light intensity and temperature high enough to be non-limiting, it is found that at low carbon dioxide concentrations the rate of photosynthesis is nearly proportional to carbon dioxide concentration, indicating that at low concentrations carbon dioxide is almost entirely controlling the rate of the process. As the carbon dioxide concentration is increased, however, the increase in rate falls off progressively until a maximum rate is reached at about 1 per cent concentration, above which the rate remains constant over a wide range of carbon dioxide concentration (Fig. 17.6a).

Because the carbon dioxide content of the atmosphere averages only 0.03 per cent (i.e. 3 parts in 10 000 parts of air) it might be expected that carbon dioxide concentration would be the chief limiting factor of photosynthesis under field conditions. Broadly speaking this is true, and it can be shown experimentally that raising the carbon dioxide concentration of the atmosphere to about 1 per cent causes a rise in the photosynthetic rate. The possibility of enriching the atmosphere with carbon dioxide as a means of increasing the yield of a crop (which amounts to 'fertilizing' the crop with carbon dioxide) is of course feasible only in glasshouses where, because they are totally enclosed, the environmental conditions can be controlled. This has been done on a commercial scale for certain glasshouse crops, e.g. tomatoes (*Lycopersicon esculentum*), and the yields have been substantially increased but only at a high cost.

Although the concentration of carbon dioxide in the air is always rather low as compared with that of oxygen, it is worth noting that considerable deviations from the average value of 0.03 per cent can sometimes occur. Thus over dense vegetation, such as a forest canopy or a field of maize, the carbon dioxide concentration can drop significantly during the day as a result of photosynthesis. At night, when photosynthesis ceases but respiration continues, the carbon dioxide concentration can rise rapidly if there is no wind. For instance, at 1 m above ground level in a maize field the carbon dioxide concentration was shown to vary between an average maximum of 0.068 per cent at night and an average minimum of 0.025 per cent during the day.

Light intensity With carbon dioxide concentration and temperature high enough to be non-limiting, the relation between light intensity and photosynthesis is similar to that between carbon dioxide concentration and photosynthesis. At very low light intensities, however, no photosynthesis is detectable by the standard methods of gas analysis because, under such conditions, the gaseous exchanges of photosynthesis are smaller than those of respiration. Above the compensation point (i.e. the light intensity at which the photosynthetic intake and respiratory output of carbon dioxide exactly balance one another) increase in light intensity at first causes a proportional increase in photosynthetic rate, but at moderate light intensities the increase in rate begins to fall off until at high intensities the rate becomes constant. At such high intensities the plant is said to be light-saturated (Fig. 17.6b).

The light intensity at which a plant becomes light-saturated clearly depends on the level of the other controlling factors but, even when these are non-limiting, most plants are light-saturated at intensities well below full sunlight. This means that only in the early morning and on dull days will light intensity be the factor which is chiefly controlling the photosynthetic rate in nature. Because plants depend on photosynthesis for their growth and survival, it is reasonable to suppose that in the course of millions of years of evolution they would have developed a photosynthetic mechanism which uses light more efficiently than appears to be the case. The reason why they have not done so seems to be tied up with the low concentration of carbon dioxide in the atmosphere. The reaction by which carbon dioxide enters into the photosynthetic mechanism operating in most plants requires a relatively high concentration of carbon dioxide (i.e. the affinity of the enzyme for carbon dioxide is low), and consequently most plants can only reduce the carbon dioxide content of the air inside the leaf from the normal atmospheric value of about 0.03 per cent to 0.01 per cent. In tropical grasses such as maize and sugar cane, and certain other plants, Hatch and Slack have recently discovered a different biochemical pathway of carbon dioxide fixation (see p. 218) which has a far higher affinity for carbon dioxide. These grasses can extract virtually all the carbon dioxide entering the leaf from the air, and consequently their photosynthetic rate continues to increase with light intensity up to full sunlight. This important discovery of Hatch and Slack has some practical bearing because it opens up the possibility of being able to increase the yield of plants by modifying the photosynthetic apparatus itself rather than by making the growing conditions as favourable as possible to photosynthesis.

Temperature The range of temperature within which photosynthesis can take place varies widely with different plants, but for most plants of tropical regions the range is approximately 5–40 °C.

With light intensity and carbon dioxide concentration non-limiting, the rate of photosynthesis in most tropical plants increases from a minimum temperature, say 5 °C, until a temperature of about 35 °C is reached above which it begins to decrease (Fig. 17.6c). Over this range the photosynthetic rate is approximately doubled for every 10 °C rise in temperature. Between about 5 °C and 35 °C, therefore, photosynthesis has a Q_{10} of about 2, a value characteristic of strictly chemical reactions (Van't Hoff's rule). Above about 35 °C temperature causes temporary or permanent damage to the protoplasm, with a consequent decrease in photosynthetic rate. The higher the temperature and the more prolonged the exposure to high temperature, the more rapid is the decline in rate.

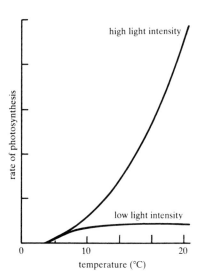

Fig. 17.7 The effect of temperature on the rate of photosynthesis varies according to whether the light intensity is high or low.

It will be evident, therefore, that it is impossible to specify an optimum temperature for photosynthesis (i.e. a temperature at which photosynthesis takes place most rapidly when no other factors are limiting the process) without stating the length of time that elapses between the start of the experiment and the measurement of the rate. If only a short time is allowed to elapse before the rate is measured, a higher optimum will normally be found than if the measurements are made after longer time intervals. From this standpoint the optimum temperature might be defined as the highest temperature at which photosynthesis can be maintained at a steady rate over a 'relatively' long period of time, but this is not the commonly accepted meaning of the word optimum.

Interaction of external factors Although it is axiomatic that the effect of any one factor on the rate of a physiological process must be studied under conditions where the levels of the other controllable factors are non-limiting, investigation of the way in which factors interact with one another involves varying the levels of more than one factor at a time. When the interactions between any two of the three controllable external factors of photosynthesis are investigated, the interaction between temperature and light is found to be more complicated than the other interactions because the pattern of response to variation in temperature is entirely different at low and high light intensities (Fig. 17.7). When the light intensity is low, increases in temperature do not increase the photosynthetic rate, i.e. the Q_{10} is unity. At higher light intensities, on the other hand, the photosynthetic rate is markedly increased by temperature increments, the Q_{10} now being about 2. The fact that the effect of temperature depends on whether or not light is also limiting, gives an important clue about the mechanism of photosynthesis. The interpretation of this differential response to temperature will not be considered until later in the chapter (see p. 211).

Internal factors

Although chlorophyll concentration and protoplasmic factors have been shown to be closely associated with photosynthetic performance, it would seem that they are limiting factors only while the machinery of the process is being built. In mature organs the level of these factors is considered to be more than adequate to meet normal requirements, and therefore only the internal factors that could be expected to influence the rate of photosynthesis of mature plants will be considered here. They are: (1) any structural features of the leaf which afford a resistance to the diffusion of carbon dioxide from the atmosphere to the chloroplast surface; and (2) the accumulation of large amounts of photosynthetic products within the chloroplasts. The effects of internal factors are not well understood, partly because they are so complex and partly because of the experimental difficulties encountered in their investigation.

Resistances in the leaf to free gaseous diffusion Since the cuticle covering the aerial parts of a plant is almost impervious to gases, the carbon dioxide used in photosynthesis must enter through the stomata before it can reach the chloroplasts where photosynthesis takes place. It follows therefore that any feature in the path which impedes gaseous diffusion from the atmosphere to the site of carbon dioxide fixation will influence the rate of photosynthesis by controlling the supply of carbon dioxide to the chloroplast surface. It is generally assumed that little resistance is offered to free gaseous diffusion by the intercellular air spaces, but that considerable resistance is offered by the water-saturated walls and cytoplasm which separate the intercellular spaces from the chloroplasts. Nevertheless, provided the stomata are open, the total resistance path is considered to be small because that part of the path where the carbon dioxide diffuses in solution is very short. Except for the stomata, the resistances offered in a given leaf by the various characters of the path are more or less constant because they depend upon the anatomical structure of the leaf. To all intents and purposes, therefore, the stomata constitute the only resistance factors that can effectively modify the rate of photosynthesis in a plant. When the stomata are closed inward diffusion of carbon dioxide is virtually prevented, and when they are nearly shut they offer such a large resistance as to slow down the entry of carbon dioxide into the leaf. All factors which influence stomatal movement will therefore have an indirect effect on the rate of photosynthesis.

Accumulation of the products of photosynthesis According to the law of mass action the rate of a chemical reaction falls off as the products accumulate, and it might be expected, therefore, that active photosynthesis would be inhibited when its products begin to accumulate. There are, however, two processes accompanying photosynthesis which tend to prevent such a build-up. Firstly, there is a continual translocation of soluble photosynthetic products out of the leaves to the rest of the plant, and, secondly, excess products are often stored temporarily as insoluble molecules (e.g. starch) within the chloroplasts. Quite apart from any effects resulting from the removal and temporary storage of photosynthetic products, the rate of photosynthesis, like that of all metabolic processes, is regulated by 'feedbacks' at certain points in the biochemical pathways of the process. These built-in metabolic controls are almost certainly far more important than any possible mass action effects.

The leaf as a photosynthetic organ

Knowing what the raw materials of photosynthesis are and what conditions are necessary for it to take place, it is possible to draw up a list of specifications for the process and then to design a structural model of an organ which meets these specifications. A design admirably suited for one function may however be quite inappropriate for another essential function. For example, an ideal photosynthetic organ should allow free access to carbon dioxide, but such free access might clash with the necessity to conserve water for an entirely different reason. A structure deliberately designed to meet the specifications for photosynthesis, but providing compromise solutions to the problems inherent in the specifications, would probably take the form shown in Fig. 17.8, which in fact resembles the structure of a typical foliage leaf. Such a theoretical attempt to explain leaf form in terms of its primary function leads to the concept of a single ideal photosynthetic organ, whereas it is a readily observable fact that leaves show a wide range of modification in form and structure. Nevertheless, practically all foliage leaves share certain basic features, such as a thin plate-like form and an elaborate system of intercellular spaces, which correspond to the requirements of photosynthesis. It can be said therefore that foliage leaves appear to be well adapted to their primary function of photosynthesis.

The cells of a photosynthetic organ must:

1. Contain chlorophyll

2. Be separated by intercellular spaces to allow access to CO_2

chloroplasts
intercellular air spaces

3. Be arranged to capture light efficiently

cells placed side by side to form a thin plate-like organ

4. Be provided with a transport system to supply water and remove soluble end products

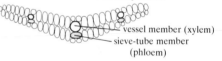
vessel member (xylem)
sieve-tube member (phloem)

5. Be protected from desiccation

epidermis with cuticle

6. Yet still be exposed to atmospheric CO_2

stomata

7. Be mechanically supported

vessel members with thickened walls

Fig. 17.8 The specifications for a photosynthetic organ.

For photosynthesis to be efficient, adequate levels of carbon dioxide, light energy and water must be supplied to the leaf and an adequate rate of removal of the photosynthetic products must be maintained from the leaf. The vascular system, which permeates the mesophyll, is responsible for conducting water to, and removing photosynthetic products from, the photosynthetic tissue. It remains therefore to consider leaves as carbon-dioxide-absorbing and light-absorbing organs.

Absorption of carbon dioxide

It has already been mentioned that the main path of entry of carbon dioxide into the leaf is through the stomata. Although there may be many thousands of stomata per square centimetre of leaf surface, it has been estimated that when fully open they never occupy more than 3 per cent of the area of the leaf epidermis. Despite this fact, leaves do absorb large quantities of carbon dioxide during photosynthesis. This apparent paradox can be explained by the fact that gases are able to diffuse through a thin barrier pierced by numerous suitably spaced small pores almost as rapidly as across an equal area with no obstructions.

The facility with which gases diffuse through multiperforate septa was first demonstrated experimentally in 1900 by Brown and Escombe. These workers showed that, whereas the rate of diffusion over a large surface is proportional to area, diffusion rates through small pores are more nearly proportional to the perimeter than to the area of the pores through which the molecules pass. The explanation of this phenomenon is as follows. Where diffusion is taking place over a large surface, the molecules can diffuse only at right angles to the surface because, except at the edge of the surface, any tendency to diverge from this course is impeded by neighbouring molecules (Fig. 17.9a). On the other hand, where diffusion is occurring through a small pore in an otherwise impervious barrier, the molecules around the edge of the pore are less impeded by their neighbours than those at the centre of the pore, because they are able to fan out sideways to fill the space, above the barrier, which surrounds the pore and is not already occupied by molecules of their own kind. This means that the rate of diffusion around the edge is greater than that at the centre of the pore, and the lines of diffusive flow therefore radiate outwards so that a hemispherical 'diffusion shell' is formed (Fig. 17.9b). Since the majority of molecules diffusing through a small pore slip around the edge, as it were, the mean rate of diffusion through the pore area is faster than through a comparable area of a continuous surface. As the size of a circular pore gets smaller, the area (πr^2) decreases as the square of the radius whereas the perimeter ($2\pi r$) decreases in direct proportion to the radius. In other words, the smaller a pore becomes, the larger the perimeter becomes in relation to the area of the pore. (N.B. It is instructive to compare this area/perimeter relationship with the area/volume relationship discussed (p. 13) in connection with cell size.) In terms of gaseous diffusion through a single pore this means that the magnitude of the fanwise diffusion taking place around the perimeter, relative to the perpendicular diffusion occurring through the centre of the pore, increases as the pore size becomes smaller. Eventually a pore size is reached, at a diameter of about 2 mm, below which diffusion is more nearly proportional to the perimeter of the pore than to its area. The overall effect is that the smaller a pore becomes, the faster is the rate of diffusion per unit area of

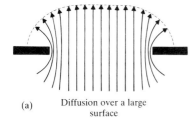

(a)   Diffusion over a large surface

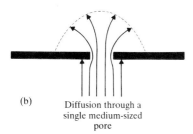

(b)   Diffusion through a single medium-sized pore

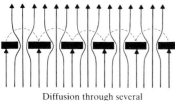

(c)   Diffusion through several small pores spaced so that the diffusion shells do not interfere with one another

Fig. 17.9 Gaseous diffusion through openings of various sizes under a uniform diffusion gradient.

pore. Thus if a number of sufficiently small pores are spaced just far enough apart so that their diffusion shells do not interfere with one another, a gas may diffuse almost as rapidly through a multiperforate septum as through an open surface equal in area to the septum (Fig. 17.9c).

Since stomatal pores are minute (on the average about 10 μm diameter when fully open) and they are generally spaced about 10 diameters apart, their high diffusive capacity in proportion to the small area which they occupy is fully explicable in terms of the physical principles of gaseous diffusion through multiperforate septa.

Chloroplasts and light absorption

Light reflected or transmitted by a leaf cannot be effective in photosynthesis, because to produce chemical change light must first be absorbed. It has long been known that only the green parts of a plant carry out photosynthesis, and it is reasonable to suppose that it is the green pigments in the chloroplasts which absorb the light needed for the process. Conclusive evidence that the chloroplasts contain the necessary pigments was obtained in 1954 by Arnon and his co-workers, who demonstrated that isolated chloroplasts can carry out the entire process of photosynthesis outside the living cell.

Sunlight, although appearing colourless, is really a mixture of light of many different wavelengths or 'colours'. The human eye is sensitive to wavelengths ranging from about 700 nanometres (1 nanometre (nm) = 0.001 microns (μm)) at the red end of the spectrum to about 400 nm at the violet end; the correspondence between wavelength and colour within the visible range of the spectrum is shown at the bottom of Fig. 17.10. Substances are white or colourless only when they reflect or transmit all the light which falls on them. Coloured substances absorb light of different wavelengths to different extents, their colour being that of the light least absorbed.

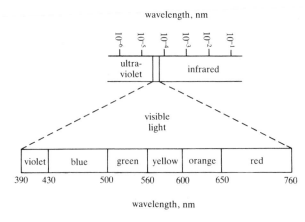

Fig. 17.10 Wavelengths of the different colours of visible light.

The green colour of the chloroplasts is due to the presence in them of four main types of pigment, *chlorophylls a and b*, which are green because they absorb violet and red light strongly and transmit green light, and *xanthophylls* and *carotenes*, which are yellow to orange-yellow in colour because they absorb blue and violet light more strongly than light of other col-

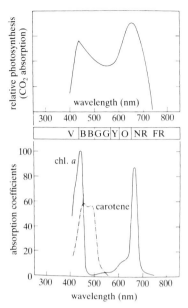

Fig. 17.11 An action spectrum for photosynthesis in wheat (*Triticum aestivum*) leaves, compared with absorption spectra for chlorophyll a and carotene. The peaks coincide in the blue and the near-red regions of the spectrum. (From Coult 1973, p. 84)

ours. In flowering plants the two chlorophylls are found only in the chloroplasts, whereas the orange-yellow pigments sometimes also occur in non-green parts of the plant body where they can play no part in photosynthesis.

The proportion of each wavelength of the incident light that is absorbed by a coloured material can be measured accurately with an instrument called a spectrophotometer. These measurements can then be plotted as a graph which relates absorption to wavelength of light; such a graph is called an absorption spectrum. For any particular coloured substance the shape of the absorption spectrum, that is the wavelengths at which absorption peaks occur and the relative heights of different peaks, is extremely characteristic; those for chlorophyll a and carotene are shown in Fig. 17.11. By means of this technique it is found that the absorption spectrum of a living green leaf can be largely accounted for as the sum of the absorption spectra of the four pigments present in the chloroplasts.

It now remains to show that the light absorbed by the chloroplast pigments is used in photosynthesis. This can be done by determining what is called the action spectrum of photosynthesis, that is, by determining the relative capacities of light of different wavelengths for photosynthesis. If equally intense beams of monochromatic light of different wavelengths are shone on a green leaf and the photosynthetic rate is measured at each wavelength, it is found that blue light and red light are the most effective and green light the least effective in carrying out photosynthesis. The similarity between the absorption spectrum of a green leaf and the action spectrum of photosynthesis by the same leaf can only mean that it is the light absorbed by the chloroplast pigments which provides the light energy needed for photosynthesis. From the pattern of the action spectrum it is clear that the light absorbed by the xanthophylls and carotenes is also functional in photosynthesis. Since carotenoids cannot operate photosynthetically in the absence of chlorophylls, it is believed that the light energy absorbed by the orange-yellow pigments is transferred to the chlorophylls. The chlorophylls then convert this energy into chemical energy just as if they had absorbed the light themselves. For this reason the orange-yellow pigments are sometimes described as accessory pigments, their apparent function being to increase the efficiency of photosynthesis.

The mechanism of photosynthesis

The overall process of photosynthesis is chemically an oxidation–reduction reaction of the type which can be represented as follows:

$$A + H_2B \rightarrow H_2A + B$$

A	H$_2$B	H$_2$A	B
(electron acceptor)	(electron donor)	(reduced acceptor)	(oxidized donor)

In photosynthesis, water is the electron donor and carbon dioxide the electron acceptor. In essence, therefore, photosynthesis is the photoreduction of carbon dioxide.

Photosynthesis is not a simple reaction taking place in one step, and the conventional equation merely indicates the raw materials and end products of the process. Since about 1950 the essential features of the intervening reactions have been determined mainly as a result of the combined use of paper chromatography and radioactive isotopes as tracers. The development of the modern understanding of photosynthesis is an outstanding example of the way in which scientists tackle a complex biological problem, and it is primarily from this point of view that the mechanism of photosynthesis will be discussed.

Evidence for the existence of 'light' and 'dark' reactions

The first real evidence that photosynthesis could not be a simple process, but must take place in at least two distinct steps, was provided early in this century by F. F. Blackman (who has already been mentioned in this chapter in connection with his principle of limiting factors). While studying the rate of photosynthesis in various conditions of carbon dioxide concentration, light intensity, and temperature, Blackman noticed that when light intensity was low but carbon dioxide concentration was high, the Q_{10} of photosynthesis was approximately 1 (i.e. the rate was practically independent of temperature), whereas when both light intensity and carbon dioxide concentration were high, the Q_{10} was approximately 2 (see Fig. 17.7). Since photochemical reactions characteristically have a Q_{10} of 1, and strictly chemical reactions have a Q_{10} of about 2, Blackman interpreted this difference in temperature sensitivity as indicating that under conditions of low light intensity the rate of photosynthesis is limited by a photochemical reaction, whereas under conditions of high light intensity the rate is limited by a purely chemical reaction. On this basis he suggested that photosynthesis involves two distinct stages, a light-sensitive temperature-independent stage (which he called the 'light' reaction) and a light-insensitive temperature-dependent stage (which he called the 'dark' reaction).

Blackman's suggestion was subsequently supported by the results of experiments in which plants were exposed to continuous and intermittent light of the same intensity. When plants are grown in continuous illumination, with the other prerequisite conditions such as carbon dioxide concentration and temperature favourable for a high rate of photosynthesis, it is found that the amount of photosynthesis is proportional to the amount of light falling on the plants, i.e. it is proportional to the product of light intensity and duration of illumination. If, however, plants are exposed alternately to very short periods (fractions of a second) of light and darkness of equal duration, so that the total amount of light reaching the plants is halved, the amount of photosynthesis is more than half the amount which occurs in continuous illumination under similar conditions of carbon dioxide concentration and temperature. This observation can be explained by assuming that during a continuous period of illumination the products of a light reaction are formed faster than they can be used up by a relatively slower chemical reaction, which is independent of light. If the light is switched off, the chemical reaction continues until the excess photoproducts are depleted. Under a regime of equally short periods of alternating light and darkness, therefore, the chemical reaction has time to catch up on the faster light reaction during the intervening dark periods, so that the photosynthetic yield (i.e. moles of CO_2 fixed per unit of light energy absorbed) for a given period of illumination is increased.

These two lines of evidence leave no doubt that photosynthesis consists of at least two distinct stages, one of which is photochemical and the other chemical in nature. Some of the work that has been done to elucidate these two stages will now be considered.

The photochemical stage

All attempts to carry out photosynthesis *in vitro* with chlorophyll extracted from leaves have failed, because chlorophyll is not the only component of the chloroplast which is essential to the process. It is, however, possible to remove intact chloroplasts from a cell, and in 1937 R. Hill showed that when an aqueous suspension of isolated chloroplasts is illuminated oxygen is freely evolved, provided that a ferric salt or other suitable electron acceptor is present in the suspension. He found, furthermore, that the amount of oxygen evolved was equivalent to the amount of ferric iron reduced to the ferrous form. This reaction, which involves the photochemical release of electrons from water to bring about the reduction, is known as the *Hill reaction* and can be represented by the following summary equation:

$$2H_2O + 2A \xrightarrow[\text{chlorophyll}]{\text{light}} 2H_2A + O_2$$

where A stands for the electron acceptor. Although Hill was not able to demonstrate the reduction of carbon dioxide by this technique, the light-dependent evolution of oxygen in both the Hill reaction and in photosynthesis strongly suggests that the Hill reaction is an essential part of the photosynthetic process. If this suggestion is correct, then it would appear that the light reaction of photosynthesis consists, at least in part, of the decomposition of water with the evolution of oxygen and the addition of hydrogen to some cellular electron acceptor, which thereby becomes a reducing agent.

This interpretation, which requires that the oxygen evolved in photosynthesis must come exclusively from water molecules, has been shown to be correct by using the isotope ^{18}O. According to the summary equation that we have been using to represent the overall process of photosynthesis, namely:

$$CO_2 + H_2O \xrightarrow[\text{chloroplasts}]{\text{light}} (CH_2O) + O_2$$

two atoms of oxygen are evolved but, of these, only one can come from the water molecule and so the other one must come from the carbon dioxide molecule. In fact, however, it has been found that the oxygen evolved comes exclusively from the water. If a photosynthesizing plant is supplied with water containing the isotope ^{18}O as a means of labelling the oxygen, then the oxygen evolved contains the isotopic marker. When, however, ^{18}O is supplied in the carbon dioxide molecule, no ^{18}O is found in the oxygen evolved. To meet the requirement that all the oxygen is derived from water, our generalized equation must therefore be rewritten as follows, where $^{*}O$ indicates labelled oxygen.

$$CO_2 + 2H_2{}^{*}O \xrightarrow[\text{chlorophyll}]{\text{light energy}} (CH_2O) + H_2O + {}^{*}O_2$$

The above evidence, although it shows that the photochemical stage consumes one of the reactants and forms one of the products of the overall process, provides no real understanding of the essential feature of photosynthesis which is the conversion of light energy into potential chemical energy.

As a result of research carried out since 1950 it is now known that illuminated chloroplasts can use light energy to synthesize ATP from ADP and inorganic phosphate. This is called *photosynthetic phosphorylation* and can be represented thus:

$$ADP + P_i \xrightarrow[\text{chlorophyll}]{\text{light energy}} ATP + H_2O$$

By means of photosynthetic phosphorylation green cells are provided with a source of chemical energy for driving energy-consuming reactions.

It is also known that the chemical substance in the chloroplasts which is the final acceptor of electrons from the splitting of water molecules (i.e. the natural counterpart of the ferric ion in the Hill reaction) is NADP, nicotinamide-adenine-dinucleotide-phosphate. Reduced NADP is the electron carrier which is formed in the photochemical stage, and subsequently used for the reduction of carbon dioxide in the chemical stage. If illuminated chloroplasts are suspended in a suitable medium containing NADP, ADP and inorganic phosphate, they can use light energy to reduce carbon dioxide to sugar. In other words it is now possible to duplicate the entire process of photosynthesis outside the living cell.

Although there is no doubt that ATP and reduced NADP are the essential products of the photochemical stage, the stages which occur between the absorption of light energy by chlorophyll and the formation of these two products have not yet been completely worked out. The sequence of events involves the photoexcitation of chlorophyll molecules.

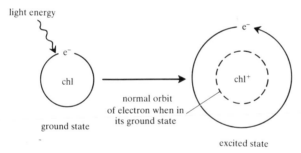

Fig. 17.12 The photoexcitation of a chlorophyll molecule.

Electrons spinning in their normal orbits around the nucleus of an atom are said to be at ground state. An input of light energy can force an electron from its normal orbit into a new orbit at a higher energy level, a process known as photoexcitation (Fig. 17.12). However, an excited electron, owing to its high energy state, represents an unstable condition and has a natural tendency to return rapidly (within a fraction of a second) to its original orbit, releasing the energy of photoexcitation as it returns to the ground state. When a chlorophyll molecule absorbs light energy during photosynthesis, an electron is raised to such a high energy level that it escapes from the chlorophyll molecule, which therefore acquires a positive charge (Chl $\rightarrow$ Chl$^+$ + e$^-$). The electron is immediately captured by an iron-containing electron acceptor called ferredoxin, which thereby becomes reduced, and this reduction represents the initial conversion of light into chemical energy. Once ferredoxin has been reduced, it is immediately re-oxidized by losing an electron to other electron acceptors. Along one route, the high-energy electron is passed along a series of electron carriers (including cytochromes), gradually losing some of its energy as it does so, back to an electron-deficient molecule of chlorophyll, which thus reverts to the ground state ($Chl^+ + e^- \rightarrow Chl$). As the electron flows down the electron-carrier chain, some of the energy of photoexcitation is conserved by coupling the exergonic oxidation–reduction reactions with the endergonic formation of ATP just as happens in the respiratory electron-transport chain (see p. 193). Along another route, the electron lost by ferredoxin is used to reduce $NADP^+$ to NADPH. The fate of the photoexcited electrons along this route necessitates an alternative source of electrons to convert the chlorophyll ion back to its ground state, because this condition cannot be reestablished with the same electron as was originally removed from the molecule. This alternative source is provided by the photochemical splitting of water, with the simultaneous production of oxygen and additional ATP. The reactions involved in the photochemical stage of photosynthesis may thus be summarized.

$$ADP + P_i \xrightarrow{\text{light}} ATP + H_2O$$
$$NADP^+ + H^+ \xrightarrow{\text{light}} NADPH$$
$$H_2O \xrightarrow{2e^- + 2H^+} \tfrac{1}{2}O_2$$

The chemical stage

Since the photochemical stage consists of the decomposition of water with the production of ATP, NADPH and oxygen, the chemical stage resolves itself into a consideration of how carbon dioxide is reduced to organic compounds such as sugars. The reactions involved in carbon dioxide reduction take place in the stroma of the chloroplast, are light-independent, enzyme-controlled, and have a Q_{10} of about 2.

Techniques of study

A knowledge of how CO_2 becomes incorporated into organic molecules has been obtained by the use of the long-lived radioactive isotope ^{14}C as a means of tracing the path of carbon in photosynthesis, coupled with the techniques of paper chromatography and autoradiography.

The combined techniques were developed by Calvin and his associates at the University of California and, like so many other significant advances in science, the experiments are simple in principle. Since it is more convenient to use unicellular organisms than flowering plants for investigating photosynthesis, the early experiments were carried out with unicellular green algae such as *Chlorella* and *Scenedesmus*. However, later studies with higher plants have shown that the pathway of carbon dioxide reduction is basically similar in most green plants. The experimental plants are allowed to photosynthesize for a suitable time in an atmosphere of carbon dioxide enriched with radioactive ^{14}C. The radioactive carbon behaves chemically like the ordinary isotope ^{12}C, but is readily distinguishable from it in analysis on the basis of its radioactivity. During photosynthesis the ^{14}C of the labelled carbon dioxide becomes incorporated in the first intermediate compound, then in the next intermediate, and so on throughout the entire process. The reaction may be stopped at any time by immersing the plant sample in boiling alcohol, which immediately stops metabolism and extracts any photosynthetic products formed. Chromatograms are then made of the alcoholic extract; the various compounds present in the extract can be identified from the position they occupy on the chromatogram, and those showing radioactivity can be detected by exposing the chromatogram to a sensitive photographic film. It is necessary to carry out a control experiment with plant material in the dark, because carbon dioxide is known to be fixed by many biochemical reactions other than those which are light dependent.

From the results of numerous experiments it is found that, if photosynthesis in an atmosphere containing labelled carbon dioxide is carried on for periods of 5 minutes or more, then radioactivity appears in a very large number of compounds including starch and soluble sugars, which are often the end products of the process. With progressive reduction in the time allowed for photosynthetic $^{14}CO_2$ fixation, however, the number of labelled compounds decreases. For the majority of plants when periods of the $^{14}CO_2$ fixation are as short as 2 seconds approximately 90 per cent of the radioactive ^{14}C appears in the 3–carbon compound phosphoglyceric acid. This compound does not become labelled to any appreciable extent when radioactive carbon dioxide is supplied to plants in the dark, and therefore in most plants it is considered to be the first stable intermediate formed in photosynthesis. Phosphoglyceric acid has the following formula, where Ⓟ is a convenient abbreviation for the phosphate group.

$$CH_2O\ Ⓟ$$
$$|$$
$$CHOH$$
$$|$$
$$*COOH$$

The radioactive carbon in phosphoglyceric acid is located in the terminal acidic group, $-*COOH$, which might suggest that the initial reaction of carbon dioxide fixation is the addition of carbon dioxide to a pre-existing carbon dioxide acceptor which contains two carbon atoms. However, it has been shown that the carbon dioxide acceptor is the 5–carbon compound ribulose diphosphate, which, by the addition of carbon dioxide, becomes converted into an unstable 6–carbon compound. This unstable intermediate then splits into two molecules of the 3–carbon compound phosphoglyceric acid. If photosynthesis is allowed to continue for considerably longer than 2 seconds, radioactive carbon also appears in the two remaining carbon atoms of the phosphoglyceric acid molecule, suggesting that the acceptor molecule, ribulose diphosphate, must itself be formed from carbon dioxide that has been fixed. It is therefore evident that carbon dioxide is incorporated into a 'carboxylation' cycle in which the initial carbon dioxide acceptor molecule is continually regenerated.

The Calvin cycle

Phosphoglyceric acid is known to be one of the 3–carbon compounds formed when a hexose sugar is broken down in respiration. Two molecules of phosphoglyceric acid could therefore be synthesized into a hexose molecule by a reversal of the pathway by which sugars are converted into phosphoglyceric acid during respiration. The results of experiments in which photosynthesis in the presence of $^{14}CO_2$ is allowed to continue for periods up to 2 minutes suggest that this is the case. After such intervals it is found that radioactivity is concentrated in the middle two carbon atoms (i.e. C–3 and C–4) of the hexose units formed, as would be expected if they were formed by the union of two phosphoglyceric acid molecules labelled in the carboxyl group.

The carbon dioxide fixed in the form of the terminal carboxyl group of phosphoglyceric acid is still in a highly oxidized condition, and before two phosphoglyceric acid molecules can combine to form a hexose sugar they must first be reduced. The electrons required to effect this reduction come from the photochemical stage of photosynthesis. The specific electron carrier is in fact NADP which accepts the electrons produced by the photochemical stage to become NADPH. This reduced coenzyme is the compound most frequently involved in reductive biosynthesis, and it is used in the Calvin cycle to reduce phosphoglyceric acid to phosphoglyceraldehyde. Not all the phosphoglyceraldehyde formed by the reduction of phosphoglyceric acid is converted into hexose units. Most of it is diverted through a series of reactions which 'shuffle' carbon atoms between various triose, tetrose, pentose, hexose and heptose sugar phosphates until eventually ribulose diphosphate molecules are re-formed and become available again as carbon dioxide acceptors. The chemical details involved in the regeneration are unimportant for the present purpose; it is only necessary to appreciate the overall result. Each time a molecule of ribulose diphosphate picks up a molecule of carbon dioxide two molecules of phosphoglyceric acid are formed. If this is repeated six times twelve molecules of phosphoglyceric acid will be formed, but of these only ten are necessary to regenerate the original six molecules of ribulose diphosphate used up. The net result is that six molecules of carbon dioxide have been incorporated into two phosphoglyceric acid molecules, which can be converted into a hexose molecule. This can be represented in the following way.

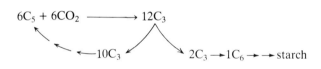

$$6C_5 + 6CO_2 \longrightarrow 12C_3$$
$$\longleftarrow 10C_3 \qquad 2C_3 \longrightarrow 1C_6 \longrightarrow \longrightarrow starch$$

The cycle is thus self-sufficient, ribulose diphosphate being re-formed as constantly as it is used up and carbon dioxide meanwhile becoming incorporated into phosphoglyceric acid molecules.

To drive the carboxylation cycle, energy and a reductant must be supplied. The only reductive step is in the formation of phosphoglyceraldehyde, which utilizes NADPH. There are two energy-requiring steps, one in the reductive formation of glyceraldehyde, and one in the reaction leading to the formation of ribulose diphosphate. The energy is provided by the ATP and NADPH molecules produced by photosynthetic phosphorylation in the photochemical stage. Some of the energy put into the cycle is 'drained off' and conserved as the chemical energy contained in the molecules of the products of photosynthesis, such as hexoses.

$$6CO_2 + 18ATP + 12NADPH + 12H^+ \rightarrow C_6H_{12}O_6 + 18ADP + 18P_i + 12NADP^+ + 6H_2O$$

It is this fraction of the energy supplied to the cycle which ultimately represents the radiant energy of light that is stored by photosynthesis as the potential chemical energy of a stable organic compound.

The end product of photosynthesis

At one time it was customary to regard carbohydrates as the end product of photosynthesis. With the recognition that phosphoglyceric acid is the first stable intermediate, it is now realized that no group of organic compounds can be singled out as the end product of photosynthesis. Phosphoglyceric acid occupies a central position in plant metabolism and can be readily converted to proteins or fats as well as to carbohydrates. Thus the products of photosynthesis can differ widely in different cells and at different times in the same cell.

It should also be pointed out that none of the reactions of the chemical stage involves light or chlorophyll directly. The photochemical stage provides electrons (as reduced NADP) for the reduction of phosphoglyceric acid, and chemical energy in the form of ATP molecules. If these requirements can be met in other ways the cycle should operate in the darkness. That it can do so has been shown by the production, in the dark, of sugar from carbon dioxide in a mixture consisting of NADPH, ATP, and the necessary enzymes from the stroma of the chloroplast.

Having described separately the reactions involved in the photochemical and the chemical stages of photosynthesis, it is now possible to summarize the whole process in the form of a diagram (Fig. 17.13).

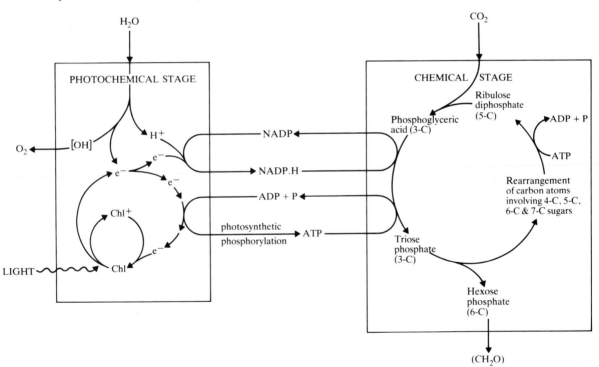

Fig. 17.13 A simplified scheme showing the reactions of the photochemical and chemical stages of photosynthesis.

Other photosynthetic pathways

C$_4$ photosynthesis

In the 1960s, about 10 years after the Calvin cycle had been elucidated, it became clear that sugar cane (*Saccharum officinarum*), maize (*Zea mays*) and certain other plants do not fix carbon dioxide directly by means of this cycle. In these plants the first compound formed after very short periods of photosynthesis is not the 3-carbon compound phosphoglyceric acid (PGA) but the 4-carbon compound oxaloacetic acid (OAA). Oxaloacetic acid is formed when carbon dioxide is added to the 3-carbon compound phosphoenolpyruvate (PEP), a reaction mediated by the enzyme PEP carboxylase. Phosphoenolpyruvate is a high-energy compound and its high-energy phosphate group is split off when it is carboxylated. The enzyme PEP carboxylase therefore has a high affinity for carbon dioxide, i.e. it fixes it very efficiently. This alternative method of photosynthetic carbon dioxide fixation is usually called the Hatch – Slack pathway after the two plant physiologists who clarified the steps involved. Plants using this pathway are called C_4, or 4-carbon, plants to distinguish them from those plants which carry out the usual type of photosynthesis in which the initial carbon dioxide product is the 3-carbon compound PGA, i.e. C_3 plants.

Oxaloacetic acid is readily converted into other 4-carbon acids, especially malic and aspartic acids, which in effect transfer the fixed carbon dioxide to enzymes of the Calvin cycle. The carbon dioxide initially fixed by PEP carboxylase is released by decarboxylation and combined with the 5-carbon compound ribulose diphosphate (RuDP) to give two molecules of PGA.

Experiments using $^{14}CO_2$ indicate that the PGA is converted into sugars by participating in the Calvin cycle. As a result of being decarboxylated the C_4 acids are themselves converted into pyruvic acid which, after being phosphorylated by ATP, is used to regenerate PEP, the initial carbon dioxide acceptor of this pathway. Thus it is only during the initial steps of carbon dioxide fixation that C_4 photosynthesis differs from the more familiar C_3 photosynthesis.

Leaf anatomy of C_4 plants

As well as differing biochemically, C_4 plants differ from C_3 plants in their leaf anatomy (Fig. 17.14). Generally speaking, leaves of C_4 plants can be distinguished by having small intercellular spaces, frequent veins, and large bundle-sheath cells which contain abundant chloroplasts. In C_3 plants chloroplasts are present in all the mesophyll cells, which each contain the same complement of photosynthetic enzymes and independently fix some of the carbon dioxide that has diffused into the leaf. In C_4 plants, by contrast,

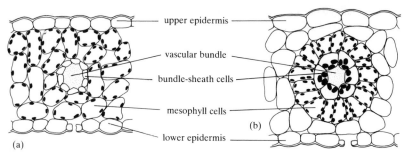

upper epidermis

vascular bundle

bundle-sheath cells

mesophyll cells

(b)

lower epidermis

(a)

Fig. 17.14 Leaf anatomy of two species of *Atriplex*. (a) *A. patula*, which has C_3 photosynthesis. (b) *A. rosea*, which has C_4 photosynthesis. Note the prominent bundle-sheath cells with extra large chloroplasts in the leaf of *A. rosea*.

there are two types of photosynthetic cells, the large bundle-sheath cells around the veins, and the mesophyll cells around the bundle sheaths. The chloroplasts in the bundle-sheath cells contain abundant starch grains and are agranal, i.e. the internal lamellae are not organized into grana and intergranal regions. The chloroplasts of the mesophyll cells usually lack starch grains and have grana. The fixation of carbon dioxide by the action of PEP carboxylase occurs in the mesophyll cells, and the C_4 acids so formed are translocated via plasmodesmata to the bundle-sheath cells where they are decarboxylated to pyruvic acid. The carbon dioxide formed in this reaction enters the Calvin cycle to produce carbohydrate, and the pyruvic acid returns to the mesophyll cells where it is converted into PEP, and the cycle can begin again. There is thus a spatial separation of the biochemical events between the mesophyll cells and the bundle-sheath cells.

Crassulacean acid metabolism (CAM) photosynthesis

A variation of the Hatch–Slack pathway is found in several types of succulent plants, including members of the cactus (Cactaceae), century plant (Agavaceae), stonecrop and houseleek (Crassulaceae) families, which inhabit very dry regions. Contrary to the usual pattern of stomatal movement, the stomata of these plants open during the night but close during the day. In plants belonging to the Crassulaceae (stonecrops and houseleeks) it has long been known that organic acids, particularly malic and isocitric acids, accumulate in the leaves at night but disappear during the day. The accompanying shift in pH (low pH at night and higher pH during the day) can be readily demonstrated by titrating cell sap which has been extracted from the fleshy leaves. Because the accumulation of malate and other organic acids causes such a spectacular increase in acidity and the process has been found to occur in all species of Crassulaceae tested, the phenomenon was called crassulacean acid metabolism (CAM).

At night, when conditions are least conducive to transpiration, the stomata of CAM plants open, carbon dioxide diffuses into the leaf and is fixed by the PEP carboxylation system to form OAA and malate. The malate is transferred from the cytoplasm to the central vacuoles of the mesophyll cells where it accumulates in large amounts. During the day the stomata close, thereby reducing water loss, and the accumulated malate and other organic acids are decarboxylated to provide a supply of carbon dioxide which the cells immediately re-fix via the Calvin cycle (i.e. the carbon dioxide is accepted by RuDP). The biochemical events of the CAM pathway are thus similar to those of the Hatch–Slack pathway, but the PEP and Calvin systems are separated *temporally* (i.e. by time) rather than *spatially* (i.e. in different cells) as in the Hatch–Slack pathway. A diagrammatic comparison of the Hatch–Slack and CAM pathways (Fig. 17.15) serves to emphasize the similarities and differences between these two types of carbon dioxide fixation.

Photorespiration

In the course of studies on plant productivity, it was discovered that a considerable proportion of the photosynthetic products was being oxidized to carbon dioxide and water without the production of metabolically useful energy. This was traced to a type of respiration which occurs in small organelles called peroxisomes and, because it occurs only in the light, is called *photorespiration*. In contrast to 'normal' respiration which occurs in the mitochondria, no ATP or reduced NAD is produced by photorespiration, the energy released being lost as heat. Under conditions of low carbon dioxide and high oxygen concentrations, which are normal atmospheric conditions, as much as 30–50 per cent of the photosynthetically fixed carbon may be lost as carbon dioxide by photorespiration, and the energy that was expended to fix it is wasted to the plant. The loss of carbon is due to the dual activity of the enzyme RuDP carboxylase, which can act not only as a carboxylase (which is its biologically useful role in the Calvin cycle) but also as an oxygenase, i.e. an enzyme which catalyses the direct insertion of molecular oxygen into an organic substrate. In a photosynthetic cell both oxygen and carbon dioxide compete for RuDP as a substrate, and the relative rates of carboxylation and oxygenation depend upon the ratio of carbon dioxide to oxygen concentrations in the immediate vicinity of the enzyme. Carboxylation of RuDP is the first reaction of the Calvin cycle and gives two molecules of PGA. Oxygenation of RuDP results in the formation of one molecule of phosphoglycolic acid (C_2) and one molecule of PGA (C_3), instead of the two molecules of PGA that are formed when carbon dioxide is fixed.

$$
\begin{array}{c}
CH_2OP \\
| \\
CO \\
| \\
H-C-OH \\
| \\
H-C-OH \\
| \\
CH_2OP \\
RuDP
\end{array}
\quad + \ O_2 \quad
\xrightarrow[\text{(oxygenase)}]{\text{RuDP carboxylase}}
\quad
\begin{array}{c}
CH_2OP \\
| \\
COOH \\
\text{phosphoglycolic acid} \\
+ \\
COOH \\
| \\
CHOH \\
| \\
CH_2OP \\
PGA
\end{array}
$$

The PGA can be channelled into the Calvin cycle to give sugars and thereby conserve carbon. The phosphoglycolic acid is fed into a C_2 cycle, the glycolic oxidase system which is located in the peroxisomes, and is oxidized to carbon dioxide although during the reaction sequence some of the carbon in the phosphoglycolic acid is converted into PGA and therefore conserved.

The diversion of phosphoglycolic acid from the Calvin cycle appears to be an unavoidable side-effect of the operation of RuDP carboxylase in air. The C_2 cycle of photorespiration functions as a kind of salvage mechanism which recovers at least some of the carbon lost from the Calvin cycle. Photorespiration is costly to the plant in terms of energy consumption, but the de-

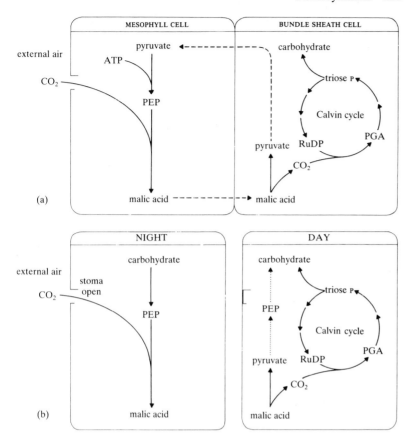

Fig. 17.15 Comparison of the (a) Hatch-Slack and (b) CAM pathways of carbon dioxide fixation.

crease in net productivity is less than would be the case if the C_2 scavenging cycle did not take place.

The significance of the Hatch–Slack and CAM pathways

Plants which fix carbon dioxide by the CAM pathway are generally succulent or semi-succulent, and are often to be found growing in arid conditions. Many of these plants possess obvious anatomical features which may be looked upon as adaptations to the dry habitat, e.g. fleshy leaves and stems in which substantial amounts of water may be stored. The CAM mechanism of carbon dioxide fixation, in which large amounts of atmospheric carbon dioxide are fixed at night when the stomata are fully open, enables these plants to carry out photosynthesis during the day with the stomata closed. In this way CAM plants can photosynthesize without losing an excessive amount of water by stomatal transpiration. Under conditions of severe drought CAM plants grow extremely slowly, but at least they can survive when other plants would die.

The significance of the Hatch–Slack pathway is not so obvious. It may seem pointless for C_4 plants to fix atmospheric carbon dioxide as OAA in one type of cell, and then decarboxylate and re-fix it in another type of cell. However, it has been demonstrated that, in conditions of high light intensity and high temperature, the leaves of C_4 plants fix carbon dioxide much more efficiently than those of C_3 plants. C_4 plants are capable of reducing the carbon dioxide concentration of the surrounding atmosphere to almost zero, whereas the leaves of C_3 plants can only lower it to approximately 80 ppm. In other words, the net photosynthesis of C_3 plants ceases when the carbon dioxide concentration of the atmosphere falls to about one-quarter of its normal level, but the net photosynthesis of C_4 plants continues at carbon dioxide concentrations down to 10 ppm or even lower.

The greater photosynthetic efficiency of C_4 plants is partly due to the fact that the enzyme PEP carboxylase, which occurs at higher concentrations in the leaves of C_4 plants than in those of C_3 plants, has a much greater affinity for carbon dioxide than the enzyme RuDP carboxylase of the Calvin cycle. By having their photosynthetic mechanism divided into two spatially separated cycles, the mesophyll cells of C_4 plants are able to absorb carbon dioxide efficiently and then to transfer it into the bundle-sheath cells where it accumulates and can be fixed by the Calvin cycle. Another reason why C_4 photosynthesis is highly efficient is that C_4 plants show little or no photorespiration. Because the bundle-sheath cells are completely surrounded by mesophyll cells, it is thought that any carbon dioxide released by the photorespiration of the bundle-sheath cells is immediately recovered by the carbon dioxide fixing mechanism of the mesophyll cells. In this way the leaves of C_4 plants apparently circumvent the problem of photorespiration.

The fact that C_4 plants can fix carbon dioxide very efficiently means that at high light intensities and high temperatures they are able to carry out photosynthesis with their stomata almost closed. The leaves of C_4 plants can therefore photosynthesize efficiently while at the same time reduce water loss, and will grow rapidly under environmental conditions which would be harmful to C_3 plants. Generally speaking, C_4 plants are characteristic of tropical climates with extremes of light, temperature and dryness, whereas C_3 plants are abundant in all climates.

Although the Hatch–Slack pathway was discovered in tropical grasses, it has now been shown to occur in over 100 genera, belonging to both monocotyledons and dicotyledons. At least twelve of these genera, e.g. *Atriplex*, contain both C_3 and C_4 species. The possibility of introducing the desirable characteristics of C_4 photosynthesis into C_3 crop plants is now being actively investigated.

Chapter 18

Nitrogen metabolism

Plants are capable of converting inorganic nitrogenous compounds, such as nitrates and ammonia, into amino acids which are the building blocks for proteins. The synthesis of proteins from amino acids is closely associated with the action of the genes and is therefore more conveniently discussed in Chapter 39. In this chapter the study of nitrogen metabolism will deal mainly with the conversion of inorganic nitrogenous compounds into amino acids.

The formation of amino acids

Reduction of nitrates to ammonia

Although the atmosphere contains an abundant supply of molecular nitrogen, flowering plants are unable to tap this source directly and must obtain their nitrogen in a combined form. Apart from about 400 species of insectivorous plants which capture insects and appear to be able to absorb amino acids directly through their modified leaves, plants obtain their nitrogen by absorbing the nitrate or ammonium ions present in the soil solution. Although ammonia is the main nitrogen compound set free by the decay of organic material, in most soils it is immediately oxidized to nitrate by bacteria and hence nitrate ions are normally the chief source of nitrogen available to plants. However, once they have been absorbed, the nitrate ions must be reduced back to ammonia before their nitrogen component can be incorporated into amino acids and other organic nitrogenous compounds. The reduction of nitrate to ammonia in plants takes place in three stages by the following pathway:

$$NO_3^- \xrightarrow{\text{nitrate reductase}} NO_2^- \xrightarrow{\text{nitrite reductase}} NH_2OH \xrightarrow{\text{hydroxylamine reductase}} NH_3$$

(nitrate) (nitrite) (hydroxylamine) (ammonia)

The main evidence for the existence of this pathway is that the reductase enzymes necessary for all three stages have been shown to be present in flowering plants.

The reactions involved in the reduction of nitrate are all endergonic, and therefore require not only electron donors but also a source of ATP. Both requirements can be supplied by respiration (N.B. nitrate reduction can take place in the dark, e.g. in roots), but there is evidence that they can also be supplied by the 'light' (i.e. photochemical) reactions of photosynthesis. In green leaves, for instance, nitrate reduction is accelerated in the light, suggesting that nitrate as well as carbon dioxide can be reduced by the 'dark' (i.e. strictly chemical) reactions of photosynthesis. Furthermore, the three reductases necessary for the reduction of nitrate to ammonia have all been shown to be present in chloroplasts.

Synthesis of amino acids

In flowering plants the incorporation of the nitrogen in ammonia into amino-acid molecules apparently takes place in only one way, namely by the initial formation of glutamic acid. This is accomplished by the reductive amination of α-ketoglutaric acid (which is one of the acids produced in the TCA cycle):

$$\alpha\text{-ketoglutaric acid} + NH_3 + 2(H) \rightleftharpoons \text{glutamic acid} + H_2O$$

As in nitrate reduction, the hydrogen donor (which is NADH or NADPH) can be provided either by respiration or by photosynthesis.

Since glutamic acid seems to be the only amino acid that can be formed directly from ammonia, it is reasonable to suppose that it should occupy a key position in nitrogen metabolism. That it does so is indicated by the rapidity with which glutamic acid becomes labelled with the heavy isotope of nitrogen (^{15}N) when ammonium sulphate labelled with this isotope is fed to plants. Furthermore, glutamic dehydrogenase (the enzyme catalysing its formation) is found to be almost universally present in plants.

Once glutamic acid has been formed it can function as the precursor for the synthesis of other amino acids by the process of transamination (i.e. the transfer of an amino group, $-NH_2$, from one molecule to another without the formation of ammonia). This process is catalysed by the transaminase (or, as they are sometimes called, aminotransferase) enzymes which accept an amino group from an amino acid and transfer it to an α-keto acid which is thereby converted into the corresponding amino acid. For example, transfer of the amino group from glutamic acid to oxaloacetic acid results in the formation of α-ketoglutaric acid (the α-keto acid corresponding to glutamic acid) and aspartic acid (the amino acid corresponding to oxaloacetic acid).

glutamic acid + oxaloacetic acid $\rightleftharpoons$ α-ketoglutaric acid + aspartic acid
  (amino acid)   (α-keto acid)   (α-keto acid)   (amino acid)

Since a wide variety of α-keto acids occurs in plant tissues, it seems likely that many of the naturally occurring amino acids are formed by transamination of glutamic acid, or an amino acid derived from it, with the appropriate α-keto acid. Such transaminations are ultimately dependent on the reductive amination of α-ketoglutaric acid because this is the only α-keto acid capable of reacting with ammonia directly.

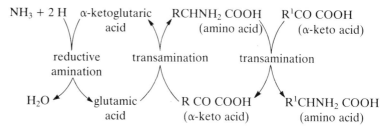

Storage and translocation of nitrogen

The amount of ammonia in a plant is often in excess of immediate requirements. This may either happen in the normal course of growth, for example during germination when the reserve proteins of the seed are being broken down, or result from the liberal application of nitrogenous fertilizers to the soil. Under conditions of excess ammonia, both glutamic and aspartic acids are able to react with more ammonia to form the amides glutamine and asparagine respectively. ATP is required for the reactions to take place.

$$
\begin{array}{lllll}
\text{COOH} & + \quad \text{NH}_3 & \rightleftharpoons & \text{CO NH}_2 & + \quad \text{H}_2\text{O} \\
| & & & | & \\
\text{CH}_2 & & & \text{CH}_2 & \\
| & & & | & \\
\text{CH}_2 & & & \text{CH}_2 & \\
| & & & | & \\
\text{CHNH}_2 & & & \text{CHNH}_2 & \\
| & & & | & \\
\text{COOH} & & & \text{COOH} & \\
\text{(glutamic acid)} & \text{(ammonia)} & & \text{(glutamine)} & \text{(water)} \\
\text{with one} & & & \text{with two} & \\
\text{amino group} & & & \text{amino groups} &
\end{array}
$$

Because excess ammonia is toxic to plant cells, the formation of glutamine and asparagine may be regarded as a 'safety device' whereby nitrogen can be temporarily stored in a form which is non-toxic but yet readily available for synthesis when the occasion arises.

Besides functioning as a temporary nitrogen store, glutamine and asparagine may also act as a means of translocating amino groups when storage protein is mobilized for growth. This applies particularly to asparagine which often accumulates in enormous amounts in germinating seedlings. The protein in the storage region is first hydrolysed into its component amino acids which undergo transamination with oxaloacetic acid to give aspartic acid. This then combines with ammonia to form asparagine which is translocated in the xylem to the growing regions. Here, because there is a supply of α-keto acids produced by photosynthesis, the reverse series of reactions takes place and new protein is formed. The translocatory function of amides, using asparagine as the example, is shown in Fig. 18.1.

Biological fixation of nitrogen

Although flowering plants must obtain their nitrogen in a combined form, some organisms (notably certain bacteria, many blue-green algae, and a few fungi) are able to fix free nitrogen, i.e. to use the molecular nitrogen of the atmosphere as the starting point in the synthesis of proteins and other nitrogenous organic compounds. The end product of this fixation process is apparently ammonia, which is then brought into organic combination by the metabolic pathways already described for flowering plants. Some of the organisms capable of fixing nitrogen are free-living, others exist in a symbiotic association with another organism. (Symbiosis is an association of two organisms which are physiologically interdependent, and in which neither reacts defensively against the other.) The free-living organisms include bac-

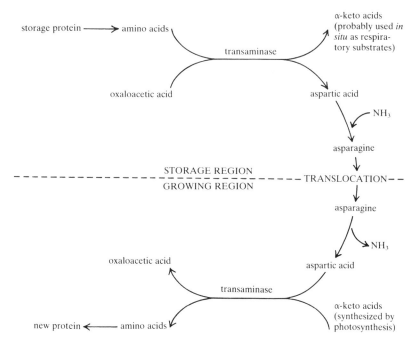

storage protein ⟶ amino acids

transaminase

oxaloacetic acid

aspartic acid

α-keto acids (probably used *in situ* as respiratory substrates)

NH₃

asparagine

STORAGE REGION
- TRANSLOCATION - -
GROWING REGION

asparagine

NH₃

oxaloacetic acid

aspartic acid

transaminase

new protein ⟵ amino acids

α-keto acids (synthesized by photosynthesis)

Fig. 18.1 Amides (e.g. asparagine) as a means of translocating amino groups from storage to growing regions.

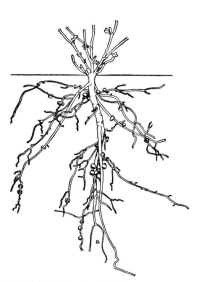

Fig. 18.2 Root system of the legume *Tetragonolobus siliquosus* showing root nodules. (From Strasburger 1976, p. 294)

teria of the genus *Beijerinckia* (which is widely distributed in most tropical soils) and blue-green algae of the genera *Nostoc* and *Anabaena* (which are important fixers of nitrogen in certain tropical soils, especially rice paddies). The most familiar examples of symbiotic associations are those between bacteria and flowering plants belonging to the family Leguminosae (the pod-bearing family.) The ability of leguminous plants to restore soil fertility after it has been reduced by the growth of a cereal crop has been known for nearly 2000 years and has formed the basis of crop rotation for several centuries. This knowledge was purely empirical until Boussingault, in 1838, produced evidence to show that the value of leguminous plants lay in their apparent ability to fix free nitrogen. His evidence was based on the demonstration that the nitrogen content of clover plants, when grown in sand, increased whereas that of wheat, under similar conditions, did not. Subsequently Hellriegel and Wilfarth, in 1888, related the ability of leguminous plants to fix nitrogen with the presence on their roots of localized swellings, called nodules, caused by the invasion of bacteria (Fig. 18.2). When legumes were grown in sterile soil they did not develop root nodules, and their growth could be maintained only by the addition of an external supply of nitrogenous fertilizer. If, however, they were grown in non-sterile soil and subsequently developed root nodules, they grew normally in the absence of any external supply of nitrogenous fertilizer. These results showed that association of the legume with a bacterium is necessary for nitrogen fixation. It is now known that soil-inhabiting bacteria of the genus *Rhizobium* infect the root hairs and penetrate into the root cortex, multiplying rapidly as they do so. The cortex cells react to the presence of the bacteria by dividing to form a mass of tissue, which projects from the root surface as a pink-coloured nodule. The centre of each nodule is occupied by enlarged cells containing numerous bacteria in their cytoplasm. Ultimately the nodule becomes soft and brown, its interior is digested, and it eventually peels off, returning the bacteria to the soil. The genus *Rhizobium* is divided into several groups or 'species' according to the genera of legume which they can infect. The division, however, is not a sharp one because many legumes can be infected with bacteria from more than one species. Furthermore, within each *Rhizobium* species there are strains which differ from one another in their ability to fix nitrogen. Some strains may fix nitrogen in association with one legume species but not with another closely related species of the same genus. Other strains, though readily producing nodules, fix only a small amount of nitrogen.

The amount of nitrogen fixed by a nodule is correlated with the extent and persistence of the bacterial tissue in its centre. In *ineffective* strains of *Rhizobium*, which cause nodulation but little or no fixation, this bacterial tissue is small and short-lived. In *effective* strains, where the bacterial tissue is extensive and persists for perhaps several months, a symbiotic relationship is established. In this relationship the legume supplies the bacterium with carbohydrates, and the bacterium provides the legume with fixed nitrogen. However, the relationship is delicately balanced because *Rhizobium*, although it grows readily in pure culture if supplied with ammonium salts, will not fix nitrogen except under very specialized conditions. In fact, for about 80 years the bacterium was thought to be incapable of fixing nitrogen unless in association with legume roots, but in 1975 it was discovered that a combination of a very low oxygen concentration with a source of dicarboxylic acids enabled several *Rhizobium* spp. to fix nitrogen independently of any legume. Furthermore, the application of nitrogenous fertilizers to legumes is known to suppress nodule development and so reduce fixation. This has been established by supplying a solution of ammonium sulphate labelled with the isotope ^{15}N to the roots of legumes, and then estimating how much of the total nitrogen content of the plants is contributed by the labelled nitrogen absorbed from the soil and how much by the unlabelled nitrogen fixed by the bacteria in the nodules. The results of many experiments suggest that the size and number of the nodules formed on the roots is largely determined by the carbohydrate–nitrogen balance of the plant.

The mechanism of nitrogen fixation is far from understood, but the responsible enzymes have now been isolated in cell-free systems and this will greatly assist its investigation. The mechanism is known to be associated with the red pigment responsible for the pink colour of nodules produced by effective strains of *Rhizobium*. The pigment belongs to the haemoglobin group of respiratory pigments, which also occur in the red blood cells of vertebrates and some other animals. It appears that neither the root cells nor the root-nodule bacteria are capable of forming haemoglobin by themselves, and it is only after the root cells have been invaded by the bacteria that haemoglobin is formed. The pink colour of nodules is a reliable indication of active nitrogen fixation, and spectrometric estimation of the haemoglobin content of nodules has been used as a means of determining the efficiency of *Rhizobium* strains.

The beneficial effect on soil fertility of growing legumes is not entirely due to the increased nitrogen content of the roots and aerial parts of the leguminous crop which, by being left to rot or by being ploughed into the soil, increase the soil's store of combined nitrogen. Leguminous crops, while still growing, sometimes excrete part (often 10–20 per cent) of the nitrogen fixed in their nodules into the surrounding soil. This occurs under conditions when the rate of photosynthesis is such that the plant cannot utilize all the nitrogen fixed, and the excess is excreted. It has been a common practice for many centuries and in many countries to grow mixed crops of legumes and non-legumes (particularly grasses and cereals). The success of this practice is now known to be due to the absorption by the non-leguminous crop of nitrogen excreted by the leguminous crop.

Not all legumes appear to be capable of forming root nodules. There are about 13 000 species of Leguminosae, and of a sample of 1200 species investigated only 88 per cent were found to have nodules. There are also a number of non-leguminous plants (e.g. cycads, podocarps, and *Casuarina* spp.) which have root nodules containing symbiotic nitrogen-fixing micro-organisms. It is probable that such plants, like legumes, play a significant part in maintaining soil fertility in nature.

In the tropics, bacterial symbiosis is also found in the leaves of certain plants belonging to the Rubiaceae (e.g. *Pavetta*, *Psychotria* and *Ixora*) and Myrsinaceae (e.g. *Ardisia*). In plants with leaf nodules the bacterium occurs at the growing tip, and each new leaf is infected as it unfolds. The ovary and the seeds also become infected, thus enabling the bacterium to be passed to the next generation. Another type of association between bacteria and non-leguminous plants has recently been discovered in South America. For example, the nitrogen-fixing bacterium *Azospirillum lipoferum* grows on the root surface of tropical grasses belonging to the genus *Digitaria* and in association with the roots of maize plants. This discovery immediately led to the suggestion that, if *Azospirillum* were inoculated into soil in which maize is growing, the amount of nitrogenous fertilizer required for economic yields would be substantially reduced. Unfortunately, the increase in nitrogen fixation of *Azospirillum*-inoculated soil proved so small as to be of no practical significance, and scientists are now trying to find out the reasons for this unexpected result.

The nitrogen cycle

Apart from the relatively few plants (legumes and a few non-leguminous genera) which can use atmospheric nitrogen by means of the symbiotic nitrogen-fixing bacteria in their root nodules, plants obtain their nitrogen by absorption of nitrate or ammonium ions from the soil. Once inside the plant, this inorganic nitrogen is converted into amino acids and then into plant proteins. These may either remain within the plant or, if the plant is eaten by an animal, they may be broken down into their amino-acid components and reconstituted as animal proteins. In either case the nitrogen is 'locked up' within organisms. The continued existence of plant and animal life therefore requires that this element be returned to the environment so that it can be made available again to plants and thence to animals. The pathways by which atoms of nitrogen are circulated in nature are known as the *nitrogen cycle*. Since the pathways are cyclic it is possible to start at any point, but it is convenient to begin with the organic nitrogen present as proteins within the bodies of plants and animals.

When plants and animals die their bodies are acted upon by *putrefying* or decay-causing bacteria, which are found chiefly in soil and in the mud at the bottom of lakes, rivers and oceans. These bacteria are saprophytes which require dead organisms as their source of food. By their activity the putrefying bacteria break down plant and animal proteins into ammonia, which is released into the surrounding soil or water. During the lifetime of many animals some nitrogen is returned to the environment by the excretion of ammonia or of urea (which, for simplicity, can be regarded as ammonia in a combined form).

The ammonia produced by putrefaction and excretion is then acted upon by a second group of bacteria, the *nitrifying* bacteria, which are also present in soil and water. There are only two known genera of nitrifying bacteria, *Nitrosomonas* which oxidizes ammonia into nitrites, and *Nitrobacter* which oxidizes nitrites into nitrates. These oxidation processes supply the bacteria with the energy they need for carbon dioxide fixation. The result of the combined activities of the nitrifying bacteria is that nitrates are again made available to plants for synthesis into plant proteins.

This cycle is complicated by the fact that there is a 'leak' caused by a third group of soil bacteria, the *denitrifying* bacteria. These convert either nitrates or ammonia into molecular nitrogen which, being gaseous, escapes into the atmosphere. However, this loss is not permanent because there are several methods whereby molecular nitrogen is converted into nitrogenous compounds which are fed back into the main cycle. Any process by which free nitrogen combines chemically with other elements to form a nitrogenous compound is called *nitrogen fixation*. One important method is through the activity of free-living and symbiotic organisms capable of converting atmospheric nitrogen into ammonia (i.e. the reverse of denitrification). These nitrogen-fixing organisms have already been considered in the preceding section (Biological fixation of nitrogen). Some atmospheric nitrogen is also fixed in the form of nitrogen oxides by electrical discharges (lightning) and conveyed by rain to the soil, where they form nitrites and nitrates. In addition to these naturally occurring processes, man has discovered how to use atmospheric nitrogen for synthesizing ammonia, which he applies to the soil as ammonium salts (mainly ammonium sulphate). The artificial synthesis of ammonia has become one of the largest chemical industries in the world, because, wherever intensive agriculture is practised, the demand for nitrogen as a plant nutrient always exceeds the potential supply produced by natural methods.

The complete nitrogen cycle thus involves a number of processes (putrefaction, nitrification, denitrification, nitrogen fixation, and nitrogen assimilation) and a variety of organisms. If one could follow the fate of a single atom of nitrogen it would enter into the composition of many different compounds as the result of the activities of different living organisms. The nitrogen cycle is, in fact, not one cycle but a group of cycles all interacting with one another. The major pathways are summarized in Fig. 18.3. In this diagram ammonia (and not nitrate) is represented as the end product of the biological fixation of nitrogen, because it is the final inorganic product of this process and the immediate precursor of organic nitrogen. Nitrogen assimilation is shown as occurring from nitrate as well as from ammonia because, although assimilation of nitrate involves its reduction to ammonia, this occurs within the plant and there is no release of nitrites into the environment where they could be acted upon by *Nitrosomonas*.

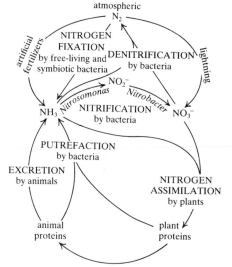

Fig. 18.3 The nitrogen cycle. (The names of the main naturally occurring processes are shown in capital letters.)

Chapter 19

Mineral nutrition and salt uptake

Most of the fresh weight of a herbaceous plant is due to water. The proportion of water is found by heating a weighed quantity of fresh plant material in an oven 98°C until there is no further loss of weight; the weight of the dry matter that is left after the water has been removed is known as the dry weight. Table 19.1 gives an analysis of the various chemical elements in the dry matter of whole maize plants. It will be seen that the three elements carbon, hydrogen and oxygen make up just over 94 per cent of the total dry weight. The reason is that most of the dry matter is composed of organic compounds produced by photosynthesis from the raw materials carbon dioxide and water. Some of these organic compounds consist entirely of carbon, hydrogen and oxygen. This is true for cellulose which is the most abundant constituent of the dry matter, accounting for about one-third of the dry weight. Other organic compounds have a carbon skeleton composed of these three elements, although one or more other elements occur in the rest of the molecule.

The fourth most abundant element in Table 19.1 is nitrogen. This is a constituent of all proteins, the main component of the dry matter derived from the protoplasmic material of the plant. The remaining elements are all present in only very small amounts; they form the ash which is left when plant material is burnt and the organic compounds are lost in the volatile products of combustion.

Table 19.1 An analysis of the chemical elements in the dry matter of five maize plants. (Data of Latshaw and Miller, 1924)

Element	Percentage of total dry weight	Relative number of atoms
Oxygen	44.43	4640
Carbon	43.57	6060
Hydrogen	6.24	10 440
Nitrogen	1.46	174
Phosphorus	0.20	11
Potassium	0.92	39
Calcium	0.23	9
Magnesium	0.18	12
Sulphur	0.17	9
Iron	0.08	2
Silicon	1.17	70
Aluminium	0.11	7
Chlorine	0.14	7
Manganese	0.03	1
Undetermined elements	0.93	

With the exception of carbon and part of the oxygen, both of which are derived from the carbon dioxide of the atmosphere, the chemical elements composing plants are all normally absorbed from the soil by the roots. Hydrogen, which in terms of number of atoms is the most abundant element, is derived exclusively from water, which also supplies the oxygen not derived from the atmosphere. All the remaining elements are absorbed as inorganic salts and hence are called *mineral elements*. Except for nitrogen, they are derived exclusively by the chemical dissolution of the parent rock giving rise to soil. The inorganic salts of nitrogen come partly from rock dissolution, but mainly from the atmosphere as a result of the processes of nitrogen fixation.

The presence of a mineral element in a plant does not necessarily mean that it is essential for growth; thus silicon is often present in appreciable amounts but most plants can grow normally when it is deliberately excluded from their environment. The most widely employed method of establishing which elements are essential for a plant to grow to maturity is to grow the plant in water culture. This means that the roots, instead of growing in soil, are suspended in a large vessel containing distilled water to which has been added an appropriate mixture of inorganic salts in known proportions. To provide the suspended roots with an adequate supply of oxygen for their respiration, it is always desirable and sometimes essential to aerate the solution by passing a current of air through it. The test plant is also exposed to light conditions suitable for photosynthesis. If the culture solution contains all the essential elements in sufficient amounts the plant will develop normally and complete its life cycle. If, however, the omission of a particular element from the salts used in the culture solution causes the plant to develop abnormally, the element in question is considered to be essential. By the use of this technique it was established before the end of the last century that seven elements must be present in the solution supplied to plant roots for healthy growth to be maintained; these are nitrogen, phosphorus, potassium, calcium, sulphur, magnesium and iron. Nitrogen is required in considerably larger quantities, and iron in considerably smaller quantities, than the others. Of these elements nitrogen is obtained as nitrate (NO_3^-) or ammonium (NH_4^+) ions, phosphorus as orthophosphate ($H_2PO_4^-$) ions, and sulphur as sulphate (SO_4^{2-}) ions; the remaining four elements are absorbed as cations (K^+, Ca^{2+}, Mg^{2+}, Fe^{2+} or Fe^{3+}).

These seven elements are required by growing plants in quantities large enough for an absence or deficiency of any one of them to become apparent in a short time (often after only a few days) by the development of symptoms of unhealthy growth. Because deficiency of each element tends to produce its own characteristic symptoms, it is usually possible to recognize which element is lacking from the external appearance of the plants. For example, lack of nitrogen causes stunted growth, yellowing of the leaves and excessive woodiness; lack of potassium (a mobile element, see p. 244) causes the older leaves to wither and die while the tip of the shoot continues to grow; by contrast, lack of calcium (a non-mobile element) is manifest first in the young leaves which turn pale green and eventually die at the tips and along the margins. The interpretation of deficiency symptoms becomes difficult when more than one element is lacking.

No addition was made to the above list of seven essential elements until the second decade of this century when inorganic salts of a very high degree of purity became commercially available. It was then found that several other elements were essential but only in very low concentrations. These additional elements, which are present in most soils in minute amounts, include boron, manganese, zinc, copper, molybdenum, chlorine and possibly others. They are called either *micronutrient* or *trace* elements, to distinguish them from the macronutrient elements which are required in relatively large amounts. Proof of the indispensability of micronutrient elements presents considerable experimental difficulty, and this is the reason why they were not detected by the earlier workers. The concentrations required may be (as in the case of molybdenum) of the order of only one part in a thousand million of solution, so that mere contaminating traces are often sufficient to supply the needs of plants when impure chemicals are used to make up culture solutions. Hence, establishing the requirement for a micronutrient element necessitates scrupulous avoidance of impurities not only in the solution and container but also from dust in the surrounding air. In certain plants, especially those with large seeds (e.g. beans), it is also necessary to remove the cotyledons, because these may contain sufficient stored quantities of certain essential elements for the plant to reach maturity without an external supply being necessary. If the cotyledons are not removed from the young seedling, it may take several generations of seed-to-seed growth in culture solutions to establish a requirement for particular micronutrient elements.

Typical amounts of the various elements required to maintain growth of plants in water culture solutions are shown in Table 19.2. The amount of iron is only about twice as much as that of boron, which is why iron is now usually listed (as in Table 19.2) as one of the micronutrient elements, even

Table 19.2 Mineral elements required by higher plants

Element	Approx. amount needed*	Role in plant
MACRONUTRIENTS		
Nitrogen (N)	15	Constituent of amino acids, proteins, nucleic acids, etc.
Potassium (K)	5	Activator for one or more enzymes; involved in the control of the osmotic potential of cells, particularly guard cells
Calcium (Ca)	3	Structure and permeability properties of cell membranes; structure of the middle lamella
Phosphorus (P)	2	Component of ATP, nucleic acids and many metabolic substrates; cofactor for several enzymes
Sulphur (S)	1	Constituent of certain proteins (−SH group essential to many enzymes)
Magnesium (Mg)	1	Constituent of chlorophyll; cofactor for many enzymes
MICRONUTRIENTS		
Iron (Fe)	0.1	Associated with electron transport (cytochrome oxidase system)
Boron (B)	0.05	Uncertain, but possibly essential for translocation of sugars across cell membranes
Manganese (Mn)	0.01	Cofactor for one or more enzymes (e.g. arginase)
Zinc (Zn)	0.001	Cofactor for one or more enzymes (e.g. carbonic anhydrase)
Copper (Cu)	0.0003	Associated with certain oxidase systems, and reduction of nitrite to ammonia
Molybdenum (Mo)	0.0001	Essential for reduction of nitrate
Chlorine (Cl)	0.05	Essential for the photosynthetic reactions in which oxygen is produced

* Milligram atoms per litre of nutrient solution.

though the requirement for it is easy to detect and hence did not escape the notice of earlier workers.

In addition to the above elements for which an absolute requirement has been established, there are other elements which are required by certain plants only. Thus small amounts of sodium have been found necessary for *Atriplex*, a shrub which grows naturally in saline soils (hence its common name 'saltbush') and accumulates large quantities of this element in its tissues. It may be that sodium is required in trace amounts by all plants but the difficulty of establishing its necessity is that it is not yet possible to reduce contamination to a sufficiently low level to induce disease symptoms. (Until recently a similar difficulty existed for the chloride ion, and it was only after a technique for avoiding contamination was evolved that chlorine was proved to be a micronutrient.) A specific requirement for cobalt has also been shown for those leguminous plants which have nitrogen-fixing nodules on their roots.

The role of mineral elements in plants

A general principle that has emerged from the study of mineral requirements is that the need for a particular element is almost invariably quite specific; thus sodium cannot substitute for potassium, even though the two are chemically very similar. For some elements this specificity can be explained because they are components of organic molecules essential in metabolism. Table 19.2 summarizes the specific roles that mineral elements are known to serve in plants.

Macronutrients

Nitrogen, the macronutrient which is required in the greatest amount, is a constituent of all amino acids and hence of all proteins; it is also a constituent of nucleic acids, chlorophylls, and many other metabolically important compounds. Sulphur, like nitrogen, is also involved in protein metabolism because it occurs in proteins which are capable of forming disulphide linkages both within and between polypeptide chains. Magnesium enters into the structure of the chlorophylls and is therefore essential for photosynthesis. It is, however, required by all plants, whether green or not, because it plays an essential role in many enzymic reactions. Phosphorus is associated with the biochemical mechanisms by which energy is stored and transferred in living cells. Apart from the synthesis of ATP from ADP and inorganic phosphate, it therefore participates in the phosphorylation of many intermediate compounds of photosynthesis and respiration. Phosphorus is also present in all nucleic acids, and in so many other compounds that a detailed listing would be pointless. Calcium has no known metabolic role; as the insoluble salts of pectic acids it may be an essential component of the middle lamella of the cells of higher plants. There is also evidence that it is important in maintaining the structure and properties of cell membranes.

In addition to being incorporated into organic molecules, some mineral elements also exist as inorganic ions within the protoplasm and central vacuole. The element potassium occurs largely, if not entirely, in this form in the living plant. Although potassium is an element for which it is easy to demonstrate a requirement, it has proved extremely difficult to assign a specific role to it. There is evidence that it functions as an activator for at least one enzyme in glycolysis, but since it is required in relatively large amounts it seems unlikely that this is its only role. Lack of potassium affects the rates of photosynthesis, protein synthesis and respiration, and therefore potassium presumably plays a part in all these metabolic pathways. It also seems probable that potassium is an important factor in regulating the osmotic potential of cells; in the case of the guard cells of leaves it may play a key role in the mechanism of stomatal movement (see p. 84). It is certain that potassium plays a very important role in plants, but the ways in which it functions are not yet understood.

Micronutrients

In general, micronutrients are essential because they are involved in some way with specific enzymic reactions. This explains why they are necessary only in small amounts. The elements iron, copper and molybdenum are all associated with enzymes catalysing oxidation–reduction reactions. They are able to function in this way by virtue of valency changes in their ions. Thus iron enters into the molecular structure of the respiratory pigments known as cytochromes, which negotiate the transfer of electrons from a reduced hydrogen acceptor to molecular oxygen during the terminal stages of aerobic respiration. In this chain of reactions the iron attached to the cytochrome molecules alternates between the ferric and ferrous state according to whether it is acting as an oxidant (by the ferric state gaining an electron) or a reductant (by the ferrous state donating an electron). Copper behaves in a similar way in the ascorbic acid oxidase system which, like the cytochrome oxidase system, is a pathway by which hydrogen can be transferred to molecular oxygen. This system depends on the interconversion of cuprous and cupric ions attached to the carrier molecules. In the case of molybdenum, the metal is part of the enzyme nitrate reductase, which catalyses the reduction of nitrate to nitrite. It is probable that molybdenum also acts by valency changes.

Of the four remaining micronutrient elements which are thought to be essential to all flowering plants, a specific role can be assigned to zinc and manganese because each is a constituent of a particular enzyme. (The reactions catalysed by these enzymes are too complex to be considered here.) It is by no means certain that the known biochemical roles of zinc and manganese account entirely for their requirement because they may play other, as yet unknown, roles. For boron and chlorine (respectively the earliest and the most recent of the micronutrient elements to be discovered) no clearly defined role has been established, although there is evidence of an association between boron and sugar translocation, and between chlorine and photosynthesis. Depriving plants of either of these elements leads to severe disease or death, so it seems likely that each is involved in at least one enzymic reaction whose chemical nature is not yet known.

Practical aspects of mineral nutrition

Nitrogen, phosphorus and potassium are the elements in shortest supply in most agricultural soils because they are removed by crops in greater amounts than the other essential elements. To increase the yield of plantation crops it has long been standard practice to add these three macronutrients to the soil in the form of artificial fertilizers. However, the continual removal of crops from soils to which nitrogenous, phosphatic and potassic fertilizers alone have been applied results in a general decrease in the amounts of the other essential elements required in smaller quantities. If economic yields are to be maintained it becomes necessary sooner or later to supply some of these other elements also. On certain soils even micronutrients may eventually need to be applied; in fact, it was the surprising cure of certain diseases of unknown cause by the application of micronutrients that stimulated the scientific investigation of the role played by trace elements in plant nutrition.

Among the better known disorders now recognized as deficiency diseases may be mentioned 'yellows' of tea (sulphur deficiency), 'little leaf' of oil palms (boron deficiency), 'die back' of citrus (copper deficiency), and many others which do not have common names, including zinc deficiency of coffee and magnesium deficiency of cocoa.

Deficiencies other than those of nitrogen, phosphorus and potassium are generally remedied by spraying the foliage with an aqueous solution of the sulphate of the deficient element, or borax (sodium borate) in the case of boron deficiency. Soil application is often not effective owing to the element being rendered insoluble, and therefore unavailable to the plant, by reaction with other substances already present in the soil.

In intensive agriculture rapid methods of diagnosing nutrient deficiencies are essential. Recognition of deficiency symptoms by visual inspection is clearly the quickest method, but unfortunately the symptoms of a particular deficiency may vary somewhat from one crop to another. For many of the major tropical crops like sugar cane and bananas, colour photographs of plants suffering from known deficiencies have been published. While useful in diagnosis, the value of such photographs is limited by the fact that a deficiency of one element may completely mask that of another; for example, deficiency of nitrogen alone often cannot be distinguished from deficiency involving nitrogen and phosphorus, or nitrogen and potassium. Hence visual diagnosis is being replaced to an ever-increasing extent by diagnosis based on chemical analysis of leaf samples taken at standard distances from the stem apex. With modern spectrographic techniques only a very small amount of plant material is needed for a very detailed assessment of its nutrient status to be made.

Ion uptake by plant cells

Mineral salts are taken up from the soil as ions, and it is appropriate to consider the process of ion uptake by individual cells before discussing the absorption of mineral salts by plant roots.

It is now well established that plant cells are capable of absorbing and retaining inorganic ions at concentrations far above those in the solution which surrounds them. Any ion (or indeed any solute) which is more concentrated inside the cell than in the external medium is said to *accumulate* within the cell. The phenomenon of accumulation has been demonstrated very convincingly by comparing the ionic composition of the vacuolar sap of certain algae (e.g. freshwater algae of the genus *Nitella* and marine algae of the genus *Valonia*) with that of the medium in which they are growing. These algae have exceptionally large cells and their vacuolar sap can be extracted easily and in sufficient quantity for gross chemical analysis. For example, in the case of *Nitella* it has been shown that potassium ions can be accumulated to concentrations as much as a 1000 times greater than those in the surrounding pond water. However, the cell sap must not be regarded merely as a concentrated form of the solution outside, because some ions are absorbed much more readily than others. There is, in fact, a selective accumulation of some ions and a partial or complete exclusion of others, so that the ionic composition of the vacuole differs, often markedly, from that of the external medium (Table 19.3). For example, the concentration of sodium ions in sea-water is about forty times that of potassium ions, yet in the vacuolar sap of *Valonia* spp. the concentration of potassium ions is often five to six times, and can be as much as twelve times, that of sodium ions. These marine algae therefore discriminate very effectively between the cations of two such chemically related elements as sodium and potassium. The ability to accumulate ions is a property of living cells because, when death occurs, ions are rapidly lost into the surrounding solution until a concentration equilibrium is established.

Table 19.3 The relative concentrations of certain ions inside and outside the cells of a fresh-water alga (*Nitella*) and a marine alga (*Valonia*). (Adapted from Höber 1945)

	Nitella clavata				Valonia macrophysa		
Ion	Ionic concentration (millimoles per litre) of:		Ratio of internal/ external concentration	Ion	Ionic concentration (moles per litre) of:		Ratio of internal/ external concentration
	cell contents	pond water			cell contents	sea water	
K^+	54.3	0.051	1065	K^+	0.50	0.012	42
Na^+	10.0	0.217	46	Na^+	0.09	0.498	0.18
Ca^{2+}	10.2	0.775	13	Ca^{2+}	0.0017	0.012	0.14
Cl^-	90.8	0.903	100	Cl^-	0.597	0.580	1.0

The mechanism by which ions are absorbed is not a simple process. When a plant tissue is immersed in a relatively concentrated solution of an inorganic salt, there is usually an initial rapid uptake of ions followed by a slower steady uptake. Whereas the rate of the initial rapid uptake is affected only slightly by temperature, that of the subsequent slower uptake has a temperature coefficient (Q_{10}) of between 2 and 3 (Fig. 19.1). This differential response to temperature indicates that the total uptake of ions can be divided into two stages, a short initial phase which is not under metabolic control (*passive ion uptake*) followed by a prolonged phase in which ions continue to enter by a process linked to metabolic activity (*active ion uptake*).

Passive ion uptake

When a plant tissue that has been immersed in a relatively concentrated salt solution is transferred to a dilute salt solution, some of the ions already absorbed diffuse rapidly out into the external solution. If the tissue is then replaced in the concentrated salt solution ions will diffuse rapidly back into the tissue. It is thus evident that diffusion of ions can occur freely between the external solution and a certain part of the tissue which is called the 'free space'. Although there have been claims that the free space includes at least part of the cytoplasm, it is now generally held that the free space of a tissue consists only of the volume not occupied by the protoplasts, i.e. the cell walls, the intercellular spaces, and any dead cells that may be present. Be-

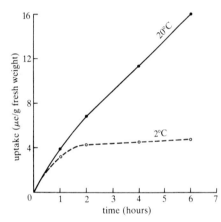

Fig. 19.1 Uptake of K^+ ions by discs of carrot (*Daucus carota*) tissue over a 6-hour period at 2 °C and 20 °C. Note that the rapid uptake during the first hour is affected only slightly by temperature. This represents the phase of passive ion uptake. (From J. F. Sutcliffe (1962), *Mineral salts absorption in plants*, p. 48. Pergamon Press, London)

cause the intercellular spaces are usually filled almost entirely with air, they are not considered to form a significant part of the free space.

By preventing the occurrence of active ion uptake (e.g. by using metabolic inhibitors or by low temperature) it is possible to study the passive uptake of ions into the free space of a plant tissue as an isolated process. When, under these conditions, a tissue is immersed in a salt solution of known concentration and allowed to come to equilibrium, it is sometimes found that the amount of salt absorbed by the tissue is greater than would be expected on the basis of the estimated volume of the free space. This means that an accumulation of ions against the concentration gradient must have occurred during passive uptake. How can this be explained without invoking metabolic energy? One way in which such an accumulation can occur is through ion exchange. By this mechanism ions already absorbed *either* on or within the cell wall *or* on the outer surface of the plasmalemma may exchange with ions carrying the same charge in the surrounding solution. For example, a hydrogen ion, H^+, adsorbed on the cell wall could be displaced by a cation, say a potassium ion, K^+. The hydrogen ion would pass into the external solution and the potassium ion would then become adsorbed on the cell wall where it would be rendered osmotically inactive. In a similar way anions could exchange with adsorbed hydroxyl ions. The operation of ion exchange mechanisms in no way invalidates the assumption that the equilibrium concentration of osmotically active ions will be the same both in the free space of the tissue and in the surrounding solution, but what it does imply is that accumulation of ions can occur against an *apparent* concentration gradient.

Active ion uptake

Although the rapid exchange of ions by passive uptake can account for salt accumulation within plant tissues, it is not adequate to explain the total accumulation that occurs. Many ions continue to enter the tissue, at a considerably slower rate, long after this initial phase of free diffusion is over. Such ions enter the cytoplasm and accumulate in the vacuole, from which they cannot readily pass back again into the external solution. This slow steady accumulation of ions against a concentration gradient would seem to require metabolic energy because, when the metabolic activity of the tissue is inhibited, the rate of ion uptake is also inhibited. This phase of ion uptake, which has a Q_{10} of between 2 and 3, is called active ion uptake because the cells must do work to move the ions. There is extensive evidence that the necessary energy is derived ultimately from respiration (i.e. that ATP energy is used). Most of the experiments have been done with thin discs of tissue cut from excised cylinders of storage organs, such as potato (*Solanum tuberosum*) tubers and carrot (*Daucus carota*) roots, which contain abundant starch grains in their cells. When freshly cut discs are immersed in a solution of an inorganic salt, their rates of respiration and salt uptake are low. If, however, the solution is aerated by passing a current of air through it, the starch within the cells begins to disappear and the rates of respiration and salt uptake simultaneously increase. By varying the percentage of oxygen in the gas stream aerating the solution, it can be shown that there is a positive correlation between the rate of respiration and that of ion uptake (Fig. 19.2), i.e. the greater the rate of respiration the faster is the accumulation of inorganic ions. The dependence of salt uptake on respira-

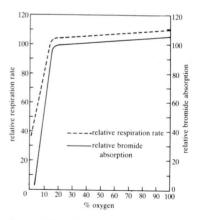

Fig. 19.2 The effect of varying the oxygen concentration in the aeration stream on (i) respiration rate and (ii) uptake of Br^- ions, by discs of potato (*Solanum tuberosum*) tissue. Note the close correlation between the rates of respiration and bromide uptake. (From F. C. Steward (1933), *Protoplasma*, **18**, 208–42)

tion is supported by the fact that if respiration is inhibited by respiratory poisons, such as cyanide, salt uptake is also inhibited.

Since ions, once inside the central vacuole, cannot readily pass back into the free space of the cell, it is reasonable to assume that the cytoplasm forms a permeability barrier across which ions must be 'pushed', using energy supplied by respiration. The mechanism by which ions pass through the cytoplasm and accumulate in the vacuole is incompletely understood, but most theories postulate the operation of so-called *carriers*. These may be envisaged as enzyme-like substances located in or on the two cell membranes, the plasmalemma and the tonoplast. A carrier, it is assumed, combines on the outer surface of the membrane with an ion in the external solution, transports it across the membrane and then releases it on the inner surface (Fig. 19.3). The ion, once it has been transported across the membrane, cannot pass back because the membrane is permeable only to the carrier – ion complex. The carrier, however, returns to the outer surface of the membrane where it is available to repeat the cycle. If it is further assumed that a given carrier combines only with certain ions, then the carrier concept provides a possible basis for explaining the selective absorption of different ions. To account for the dependence of active ion uptake on respiration, it is postulated that the carrier is able to act only during active metabolism. One theory even supposes that the carriers themselves are intermediates involved in respiration. According to this theory it is assumed that the carrier exists in the oxidized condition (i.e. loses an electron) at the outer surface of the membrane and is thus capable of accepting a negative ion, but that when the carrier – ion complex reaches the inner surface of the membrane it is reduced and so releases the negative ion. If this is so, then the energy needed for ion accumulation against a concentration gradient comes directly from respiration, but whether the link is such a direct one remains to be proved.

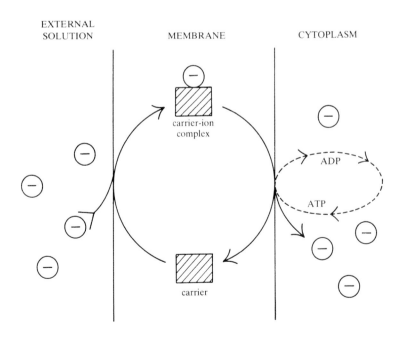

Fig. 19.3 Diagram to explain the concept of a carrier mechanism in the passage of ions across the plasmalemma into the cytoplasm.

Absorption of mineral salts by plant roots

It was formerly believed that mineral salts are taken into plants with the water that enters via the root hairs. It is now known that the part of the root most actively engaged in salt absorption is not the root-hair region but the region of elongating cells just behind the root tip. This fact alone provides strong circumstantial evidence that water absorption and salt uptake are dependent on different mechanisms. Numerous experiments have also failed to show a consistent correlation between the rates of the two processes, which would be the relationship expected if salt uptake were merely an accompaniment of water absorption. Nevertheless mineral salts, once absorbed by the root, are distributed to rest of the plant in the transpiration stream, and hence the rate of water absorption can sometimes have an indirect influence on the rate of salt uptake. It has been shown by carefully controlled experiments that, if the concentration of salts in the external solution is low, the rate of water absorption has little effect on the rate of salt uptake. If, however, the external concentration is high, then rates of the two processes run approximately parallel to one another. These results can be explained by assuming that, when the concentration of salts supplied to the root is low, the amount of salts absorbed is small and can be removed by the transpiration stream even when the rate of water absorption is slow. Under these conditions wide variations in the rate of water absorption are without effect on salt uptake because the rate of the latter is limited by the supply of salts available to the root. If, on the other hand, the external concentration is high, there is an accumulation of salts in the xylem and, unless these are rapidly removed, further uptake in impeded. In this case the rate of salt uptake is limited by the rate at which the salts are moved up the xylem, and this is related to the rate of water absorption. The existence, under these conditions, of a relationship between salt uptake and water absorption does not invalidate the postulate that the mechanisms of the two processes are different.

In order for mineral salts to be transported to the shoot system, the ions absorbed by the root from the soil solution must cross the cortex and then enter the xylem. The mechanism of this radial movement across the cortex is far from being understood, but it appears to involve the same type of active transport as is used by an individual cell to accumulate ions within its vacuole. For example, under certain conditions the concentration of mineral ions within the xylem is known to be many times greater than that in the soil solution. Accumulation in the xylem is also inhibited by respiratory poisons or by depriving the roots of oxygen.

If the concept of free space (see previous section) is correct, then it is possible to regard the root, or any other plant organ, as consisting of two regions, the *apoplast* and the *symplast*. The apoplast corresponds to the system of free spaces where free diffusion of water and dissolved ions is possible. It includes the cell walls, the intercellular spaces, and the water-filled cavities of the dead xylem elements. The symplast represents the volume of the root occupied by the protoplasts of the living cells which, because they are in contact with one another by means of the plasmodesmata in their cell walls, form a continuous system. Figure 19.4 shows the distribution of these regions in a transverse section of a root. When root structure is interpreted in this way, it is clear that ions could move across the cortex either in the symplast or in the apoplast or in both. If the symplast route (pathway B) is

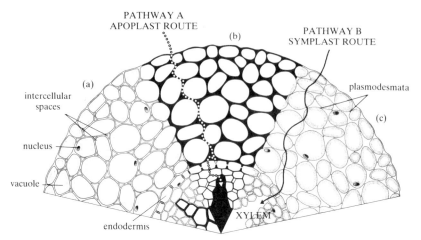

Fig. 19.4 Diagrammatic cross-section of a root showing alternative pathways for the movement of water and mineral ions across the cortex. (a) Anatomical drawing of sector of root, (b) Similar sector with cell walls and intercellular spaces (*apoplast*) shaded in black; note break in the apoplast route due to Casparian strip in cell walls of endodermis. (c) Similar sector with living cells and plasmodesmata (*symplast*) shown stippled.

postulated, then all the tissue between the soil solution and the xylem sap can be compared to the cytoplasmic region of a single cell. There is an outer permeability barrier (corresponding to the plasmalemma of an individual cell) located in the epidermal cells, and an inner permeability barrier (corresponding to the tonoplast) located in the cells abutting on the dead conducting elements of the xylem. To account for the accumulation of ions within the xylem it is postulated that active transport occurs at one or both of the permeability barriers, i.e. where ions enter the symplast from the soil solution and/or where they leave the symplast to enter the xylem. The ions are thought to move through the intervening region of the symplast either passively by diffusion or actively by protoplasmic streaming. Some ions will be diverted along the route into the vacuoles of cortical cells, where they will be retained.

If the apoplast route (pathway A) is postulated, then it is suggested that ions diffuse across the cortex mainly in the anticlinal walls of the cortical cells without entering the adjacent protoplasts. At the endodermis, however, there is a break in the continuity of the system of free spaces (though not of the apoplast itself) due to the peculiar structure of the cells forming this layer. Endodermal cells fit tightly together, but their radial and transverse walls are impregnated with the continuous band of suberin called the Casparian strip (see p. 45). Since suberin is impervious to water, it would seem that water and solutes cannot pass from one side of the endodermis to the other except through the protoplasts of the endodermal cells. If it is assumed that the cytoplasm of the endodermal cells offers a high resistance to the free diffusion of ions, then any leakage of ions already accumulated in the xylem would be effectively prevented. Because the endodermis appears to be the only permeability barrier along the apoplast route between the vascular cylinder and the soil solution, it may also be supposed that it is the layer responsible for the active transport of ions into the xylem. In other words, it is the one layer where those who advocate the apoplast route are obliged to invoke metabolic energy to explain salt uptake by roots. It remains to be shown whether or not there is any justification for singling out, from all the cells of the symplast, the endodermis as the layer that plays the major role in salt uptake.

There is no general agreement among plant physiologists about whether ions cross the cortex primarily in the symplast or in the apoplast, although recent evidence favours the apoplast route as far as the endodermis. The reason for this uncertainty is that it is very difficult to demonstrate where the main permeability barriers in the root are located. It is therefore appropriate to emphasize that in salt absorption, is in many other physiological processes, one should distinguish clearly between phraseology which merely describes what may happen and a proven explanation. The following quotation by the eminent plant physiologist F. C. Steward is an apt comment about salt absorption by roots: 'One does not understand an obscure mechanism, dependent on cellular organization, by merely giving it a name!'

Chapter 20

Translocation of solutes

The flowering plant is a highly specialized organism showing division of labour between the separate organs and between the cells within any one organ. Since every living cell of the plant requires a supply of inorganic and organic nutrients for its maintenance and growth, there must be a continual movement of dissolved nutrients from the points of supply (*sources*) to the places where they are used (*sinks*). Some nutrients, notably inorganic salts, are absorbed from the soil by the roots whereas others, notably carbohydrates, are formed in the foliage leaves. The conduction of any dissolved substance from any part of a plant to any other part is called *translocation*. For movements over microscopically short distances simple diffusion of solutes from cell to cell might possibly maintain adequate rates, but this process is far too slow to account for the conduction of nutrients from the roots to the leaves or from the leaves to the growing points of the root and shoot. Such long-distance translocation is effected through the specialized conducting tissues, xylem and phloem, which extend as a continuous vascular system throughout the plant.

Conduction in the xylem

Soil nutrients move upwards from the roots to the aerial parts in the tracheary elements (i.e. tracheids or vessel members) of the xylem. The inorganic solutes absorbed from the soil by the living root cells are passed into the water-filled lumina of the tracheary elements (see p. 239), and are then carried upwards in the transpiration stream set up by the evaporation of water from the leaves and other aerial parts. Some inorganic solutes may be absorbed along this pathway by the living cells adjacent to the tracheary elements, but most of the water and dissolved solutes reach the leaves. Experiments with fluorescent dyes (which are detectable in very low concentrations and therefore can be used as marker solutes in tissues) have shown that inside the leaves the solutes pass out of the xylem into the solution that permeates and covers the walls of the mesophyll cells. Each mesophyll cell can thus be envisaged as being bathed in a dilute salt solution from which it can absorb ions. Any salts which are not absorbed may eventually find their way to the surface of the leaf where, if they accumulate in high concentrations, they may be deposited as a crust when the water in which they are dissolved evaporates.

In woody plants the capacity of the xylem for this upward transport can be shown by 'ringing' the main stem, i.e. removing a continuous strip of bark (all tissues external to the vascular cambium) from the circumference

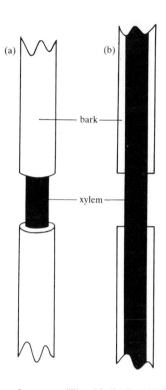
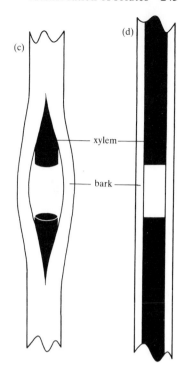

Fig. 20.1 Methods of interrupting the longitudinal continuity of the xylem and bark (containing the phloem) in translocation studies. (a) Ringed stem. (b) Longitudinal section of (a) showing cut phloem. (c) Stem with short length of xylem removed. (d) Longitudinal section of (c) showing break in continuity of the xylem.

of a stem (Fig. 20.1). By this means the continuity of the phloem is completely broken but the central core of xylem is left intact, apart from superficial damage. In such ringed plants mineral salts continue for a time to accumulate in the stem above the ring at rates similar to those found in control plants. The rate of accumulation is not maintained because the roots become starved by being cut off from their carbohydrate supply in the leaves (see later) and as a result their uptake of soil nutrients diminishes. If, however, a short length of the central core of xylem is cut out, leaving the bark continuous, the normal upward movement of salts stops. The conclusion that mineral nutrients are carried up mainly in the xylem has been confirmed by more refined experiments using radioactive tracers. When, for example, a solution of potassium phosphate labelled with radioactive potassium or phosphorus is supplied to the roots of a woody plant growing in sand culture, the radioisotope can be detected some hours later in both the xylem and bark of the stem. If, however, a longitudinal slit about 20 cm long is made in the bark so that a strip of wax paper, or some other suitable material impervious to water, can be inserted between the xylem and the bark, high radioactivity appears in the xylem above, below and along the incision whereas in the bark it appears only where the bark is in contact with the xylem (Fig. 20.2). This indicates that in the intact stem the xylem is the main channel for the upward conduction of inorganic solutes, although lateral transport from the xylem into the bark also occurs.

Since mineral salts are translocated in the transpiration stream, it might be expected that the amounts of solute transported upwards would increase with the rate of transpiration. In fact, because salt uptake by the root depends on a different mechanism from that of water absorption (see p. 239), the rate of transpiration is not directly related to the upward translocation

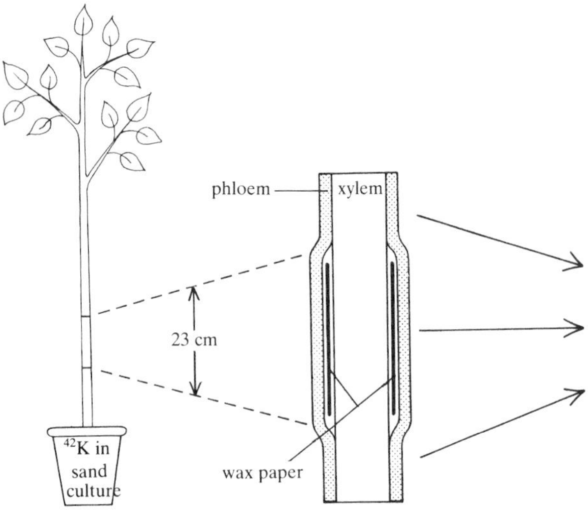

	Concentration of ^{42}K (ppm) after 5 hours of uptake			
	Stem with xylem and phloem separated by strip of wax paper		Control stem	
	phloem	xylem	phloem	xylem
Above strip	53	47	64	56
Region of strip top	11.6	119		
middle	0.7	112	87	69
bottom	20	113		
Below strip	84	58	74	67

Fig. 20.2 Method of separating xylem and phloem by wax paper to demonstrate that the xylem is the pathway for the upward translocation of mineral ions. A table of results giving the distribution of radioactive potassium in the stem of *Salix* after an absorption period of 5 hours is shown at the right of the diagram. (Data from Stout, P. R. and Hoagland, D. R. (1939), *American Journal of Botany*, **26**,320)

of mineral salts; usually faster flow of water is offset by a more dilute xylem sap.

The xylem sap also contains dissolved organic substances but, since these are normally present only in relatively low concentrations, the xylem is unlikely to be the main channel by which they are transported.

Conduction in the phloem

The supply of nutrients from the leaves to the rest of the plant takes place through the sieve tubes of the phloem. The principal export from leaves is carbohydrate (almost exclusively in the form of sucrose) produced by photosynthesis. Since some inorganic solutes, notably nitrates, are incorporated into organic molecules when they reach a leaf, and other solutes do not remain in the leaves to which they are first translocated, there is also a continual export of organic nitrogen compounds and inorganic salts out of leaves. Such re-export of mineral elements also occurs through the phloem. In mature leaves the content of total nitrogen and of inorganic elements is maintained at a more or less constant level as a result of continual import via the xylem, but in senescent leaves export of nitrogen, potassium, phosphorus, magnesium and other *mobile elements* usually exceeds import, so that prior to leaf fall considerable amounts of these elements are translocated back into the stem via the phloem. The older leaves on starved plants show a similar net loss, the elements exported commonly moving to the younger leaves and the growing points of the root and shoot. Calcium and a few other elements are not remobilized in this way because, having once been absorbed by a leaf, they remain in it permanently.

When nutrient materials from the leaves reach the stem they may move either upwards or downwards, but the movement is often predominantly downwards because the roots constitute the most massive non-photosynthesizing part of the plant and therefore require large amounts of carbohydrate from elsewhere. Downward translocation of organic substances through the phloem can be shown by ringing experiments similar to those used for showing that the xylem is the main channel for the upward transport of mineral salts. If a woody stem is ringed to remove the bark *below* the foliage-bearing region, there is an accumulation of carbohydrates, organic nitrogen compounds and mobile elements such as phosphorus and potassium in the phloem above the ring, whereas the tissue below the ring becomes depleted of these substances (Fig. 20.3). Confirmatory results have been obtained by the use of radioactive tracers. If a single leaf is enclosed in an illuminated chamber while still attached to the stem of the plant, and radioactive carbon dioxide is supplied to the chamber, the leaf forms labelled products of photosynthesis which are translocated both upwards and downwards in the stem from the point of attachment of the treated leaf. Radioactivity is found to be localized in the phloem of the stem, and in some cases can be seen to be confined to the sieve-tube members.

To determine what substances are conducted in the phloem it is necessary to sample the contents of sieve tubes. A technique for obtaining sieve-tube sap takes advantage of the fact that aphids, during feeding, insert their tubular mouthparts into a sieve tube of the host plant (Fig. 20.4). A feeding aphid is anaesthetized with a gentle stream of carbon dioxide, and its body is cut off leaving the mouthparts in position as an inert tube. Sieve-tube sap then exudes for hours, and sometimes for days, from the cut stump, and the exudate can be collected in a micropipette in quantities sufficient for a chemical analysis to be made. Analysis of the exudates from a large number of plants has shown that the solute content of sieve-tube sap consists mainly of sucrose (commonly 5–15 per cent by volume), with relatively low concentrations of amino acids and mineral ions. Hexose sugars such as glucose and fructose are either absent or present only in traces.

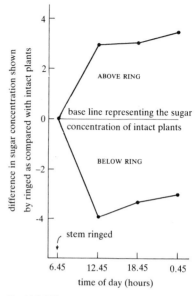

Fig. 20.3 Effect of ringing on sugar concentration in the phloem sap of Sea Island cotton (*Gossypium barbadense*). Since there are diurnal fluctuations in the sugar content of the stem tissues of intact plants, the accumulation or depletion shown by the ringed as compared with the normal plant is expressed in terms of the diurnal variation shown by the phloem of intact plants. (From Mason, T. G. and Maskell, E. J. (1928), *Annals of Botany* **42**, 189–253)

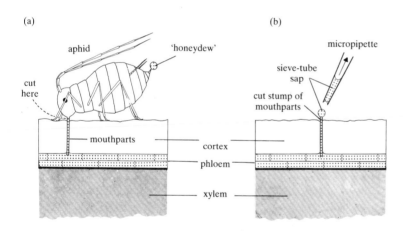

(a)

(b)

Fig. 20.4 Aphid mouthparts technique for obtaining sieve-tube sap. In a feeding aphid (a) the phloem sap is forced into the gut and, after being chemically altered, emerges as 'honeydew'. When the mouthparts are cut off (b) pure sieve-tube sap exudes and can be collected in a micropipette. (Adapted from Ray 1972, p. 116)

Whereas xylem transport is always upwards from the roots (i.e. polar), the direction of phloem translocation is variable (i.e. non-polar). Nevertheless the direction of flow is always from regions of supply towards regions where organic nutrients are being used for growth or being rendered insoluble for storage reserves. Taking an annual plant as an example (Fig. 20.5), the direction of flow in the germinating seed is from the storage tissues to the growing regions. At the seedling stage most of the carbohydrate exported from the mature leaves is moved downwards to the rapidly growing root system, although some is passed upwards to the growing shoot apex and the younger still-developing leaves. As the plant gets larger the uppermost mature leaves transport primarily to the developing leaves directly above them and to the shoot apices, the lowermost leaves predominantly to

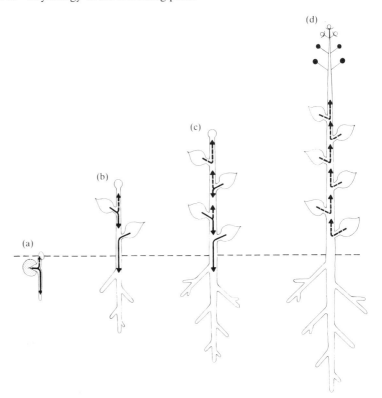

Fig. 20.5 Diagrams showing the changing pattern of phloem translocation at successive stages in the life of an annual plant. (a) The germinating seed. (b) The seedling. (c) The mature vegetative plant. (d) The flowering and fruiting stage.

the roots, and the leaves in between to both the roots and the shoot apices. When the reproductive stage is reached, the direction of flow is almost entirely towards the flowers and developing fruits; this movement includes not only organic nutrients but also mineral elements which at this stage are being re-exported from the senescent leaves. Similar patterns of reversible translocation in the phloem operate in perennial plants.

It is thus clear that the plant is able to regulate the direction in which phloem transport occurs, but the mechanism of this control is not understood. It is often said that the direction is determined by the relative positions of 'the source and the sink', but this is a statement of fact rather than an explanation. There is circumstantial evidence that, in shoots, the ability of an organ to compete for organic nutrients from the supply available may be correlated with its content of growth substances, particularly indole-acetic acid (IAA). This would agree with the fact that old leaves, which have a low content of IAA, are not supplied with organic nutrients, whereas shoot apices and young leaves, which have the highest concentrations of IAA, attract the flow of organic nutrients. The way in which IAA might attract sucrose and other translocates towards the tissues in which it occurs is unknown.

Mechanism of phloem transport

Unlike xylem transport, which takes place along a track of dead cells and requires merely a difference in water potential between the two ends of the track, phloem transport requires active living cells all along the track and cannot take place through a section of stem which has been killed by heat or poisons. This suggests that metabolic processes may be involved in the mechanism.

Rates of phloem transport were originally measured by correlating the rate at which starch accumulates in a developing tuber (e.g. yam) or fruit (e.g. pumpkin) with the cross-sectional area of the phloem in the organ (stolon or fruit stalk) supplying the tuber or fruit. Assuming phloem sap to be a 10 per cent sucrose solution, it was calculated that sucrose would need to flow through the sieve tubes at rates ranging from 50 to 100 cm per hour in order to maintain the observed rates of starch accumulation. More recently, actual rates of phloem transport have been measured by feeding radioactive carbon dioxide to mature leaves and then determining the rate at which the radioactivity spreads. The observed rates of movement are of the same order of magnitude as the calculated ones. These rates rule out diffusion as the mechanism of phloem transport because they are some 20 000 to 40 000 times faster than could be achieved by diffusion alone.

The exudation of sieve-tube sap from the cut stump of aphid mouthparts indicates that the sap moves under a positive pressure, but the nature of the forces creating this pressure is still largely unknown. Probably the most widely held hypothesis is due to the German botanist Münch who, in 1930, suggested that there is a mass flow (bulk flow) of sap along the phloem track caused by a gradient in osmotic potential between the phloem cells located at the opposite ends of the track. The principle of Münch's *mass-flow hypothesis* can be explained by reference to the physical model shown in Fig. 20.6. Two semi-permeable reservoirs, one (labelled A) containing a concentrated sucrose solution and the other (labelled C) containing a dilute sucrose solution, are joined by a connecting tube (labelled B). If the two reservoirs are then placed into the arms of a U-shaped vessel (labelled D) containing water, osmosis will occur but the uptake of water into A will be faster than into C. There will thus be a mass flow of sucrose solution from A to C through the connecting tube B and, provided the force of this mass flow is greater than the osmotic potential of the sucrose solution in C, water will be forced out of C into the water of the surrounding vessel. Because the arms of the vessel are exposed to the atmosphere, water will also flow from the arm containing C (where water is being forced out) into the arm containing A (where water is being absorbed). There will thus be a circulation of water around the system until the sucrose concentrations in A and C become equal. The circulation could theoretically be made continuous by a regular addition of sucrose to A and a regular removal of it from C. The analogy between such a continuous flow system and the living plant is as follows (Fig. 20.7). The leaves correspond to reservoir A, the roots to reservoir C, the sieve tubes of the phloem to the connecting tube B, and the xylem sap of the tracheary elements to the water in the U-shaped vessel D. In the leaves, where photosynthesis occurs, sucrose is 'loaded' into the sieve tubes by active metabolic processes, and water is therefore absorbed from the xylem by osmosis. In the roots the sucrose concentration in the sieve tubes is low because it is continually being removed for growth or storage. It is assumed that the difference between the osmotic potential of the concentrated sap in the leaves and that of the dilute sap in the roots will result in a mass flow of water (carrying any dissolved substances along with it) through the phloem from the leaves to the roots. Since the phloem is always closely associated with the xylem, in which water is moving from the roots

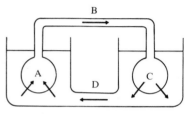

Fig. 20.6 A physical model demonstrating the principle of Münch's mass-flow hypothesis.

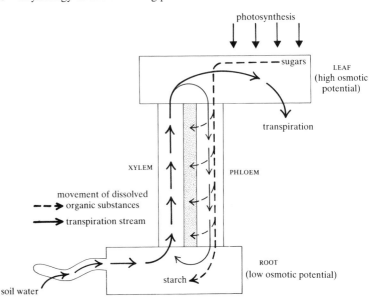

Fig. 20.7 Application of Münch's mass-flow hypothesis to the living plant.

through the stem to the leaves, this latter tissue will provide for the necessary movement of water in the opposite direction so as to maintain a circulatory system comparable to that of the physical model.

Acceptance of Münch's mass-flow hypothesis requires, among other things, the demonstration that sucrose concentration decreases in the direction of phloem transport. By simultaneous, or nearly simultaneous, sampling of the contents of the phloem at many points along the trunk of a tree, there is evidence that such gradients exist in many species of tree during the growing season when phloem translocation is active. However, there is a serious objection to the hypothesis. According to Münch's explanation the mass flow of phloem sap in one direction implies a mass flow of xylem sap in the opposite direction. Because the direction of the transpiration stream in the xylem is always up the stem, this would mean that phloem translocation should always be down the stem whereas, in fact, it is known that the direction may be either upwards or downwards. This feature of phloem translocation is incompatible with Münch's explanation of the process, but it does not rule out the possibility that dissolved substances in the phloem could be carried along by a mass flow of the sieve-tube sap. A more modern hypothesis of mass flow explains reversible movement in the phloem by postulating that the sieve tubes are merely passive channels through which sap is pumped as a result of the metabolic activities of the adjacent companion cells. On this basis the direction of mass flow is supposed to vary according to the prevailing relation between supply and demand for sucrose.

There are, however, two criticisms which apply to any hypothesis involving the mass flow of sieve-tube sap. Firstly, there is convincing evidence that different solutes can travel in opposite directions at the same time in the phloem. The significance of this observation is uncertain, because simultaneous bidirectional flow could be explained by separate sieve tubes transporting in opposite directions. Obviously if proof were obtained that different solutes can move independently in the same sieve tube, the concept of mass flow would have to be abandoned. Secondly, it is extremely difficult to reconcile any form of mass flow with the structure of the sieve tubes. In phloem translocation sieve-tube sap must pass from one sieve-tube member to the next through the minute pores in the intervening sieve plates. The presence of these pores would seem to offer a resistance that would effectively prevent mass flow at the surprisingly high rates observed.

In view of these basic uncertainties it is not surprising that alternative hypotheses for phloem translocation have been put forward. Since the essential feature of phloem translocation seems to be an acceleration in the movement of any solute that gains entry into the sieve tubes, renewed attention is being given to the possibility that the sieve tubes themselves may play some active role in the translocation process. Despite earlier reports to the contrary, protoplasmic streaming has now been demonstrated both within individual sieve-tube members and through the pores in the sieve plates separating adjacent sieve-tube members (Fig. 20.8). It would be premature on the basis of these observations to discard hypotheses involving mass flow in favour of one based on protoplasmic streaming. There is still considerable controversy about the detailed structure of phloem, and clearly any proposed mechanism of phloem translocation must be consistent with the actual structure of this tissue. The advent of the electron microscope stimulated a renewed interest in the structure of phloem, and current work on the fine structure of this tissue is likely to play an important part in the formulation of a generally acceptable theory of phloem translocation.

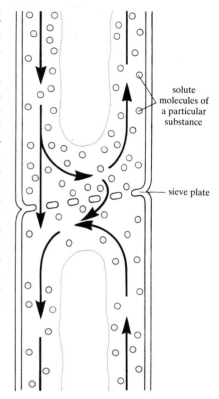

solute molecules of a particular substance

sieve plate

Fig. 20.8 The protoplasmic-streaming hypothesis of phloem translocation. All materials in the sieve tubes are kept in circulation by protoplasmic streaming, represented by the arrows. The direction in which a given solute moves is dependent upon the concentration gradient for that particular substance.

Chapter 21

Plant growth substances

In addition to inorganic raw materials (water, carbon dioxide and mineral salts) which the plant obtains from its environment, there are certain organic compounds, called *hormones*, which are necessary for plant growth. When the word hormone was first introduced it was intended to denote any substance which is produced in one part of an organism but is translocated to another part where it exerts a profound effect on metabolism and growth, and thus functions as a chemical messenger. Since its introduction, two further notions have become attached to the word. The first is that only minute amounts of the substance are needed to influence a physiological process, hormones almost always acting at less than one-thousandth molar and often in the one-millionth molar range. The second notion is that they are naturally occurring substances. With plant hormones, unlike animal hormones which have their effect on specific 'target' organs, it is often difficult to separate the sites of production from the sites of action, and furthermore different hormones often interact in complex ways to control a particular physiological process. Use of the phrase 'plant hormone' has also been unnecessarily complicated by the fact that many synthetic chemicals, which modify growth, were formerly called hormones. These synthetic chemicals with hormone-like action are best referred to as plant regulators, with the term sometimes qualified by the process affected, for example, plant growth regulators. This does not mean that synthetic and natural chemicals other than hormones cannot act as chemical messengers but rather that certain criteria additional to the promotion of a physiological action must be satisfied if an organic substance is to qualify as a hormone.

The first real evidence for the existence of chemical messengers in plants was provided by Charles Darwin (famous for his theory of organic evolution) who, in the course of experiments on the bending of plant organs to light, found that grass seedlings could be prevented from bending if the extreme tip of the coleoptile (the tubular structure, enclosing the true leaves, which is the first part of a grass seedling to emerge above ground at germination) were covered with an opaque hat, even though the region which bends towards light lies some distance behind the tip. There must, therefore, be some controlling 'influence' transmitted from the region of perception of the light stimulus (the coleoptile tip) to the region of response (the bending region). This influence is now known to be a hormone which is produced in the tip and passes downwards into the growing region where it causes curvature.

Plant growth substances, as hormones and plant regulators may collectively be called, are classified primarily by the influence which they have on certain physiological processes. This is implied by the names of some of the groups; for example, abscisins regulate abscission, auxins regulate cell elongation, cytokinins influence cytokinesis and hence are involved in the regulation of cell division, and 'florigen' is involved in inducing flower formation. Since these substances are grouped on the basis of a physiological action, a single chemical, or closely allied group of chemicals, can sometimes be classified in several ways because it may influence more than one process. For example, ethylene, a simple gaseous hydrocarbon, which has been known for years to have a drastic effect on plants, is a fruit-ripening hormone for certain types of fruit, an abscisin because of its action on abscission, and a growth inhibitor because of its interactions with auxins in the growth process. The correlation of the chemical and physiological classifications depends on precise knowledge of the chemical identity, and mode of action, of the various growth substances.

In common with higher animals, flowering plants have several different types of hormones which regulate many aspects of growth, development and metabolism. Since plants lack a nervous system, the hormonal control system is essential for communication between cells. Just as a machine needs both accelerator and brake for effective and precise control, so also the growth systems of plants need growth-promoting substances and inhibitors. There are five main groups of plant growth substances, three of them promoters and two of them inhibitors.

Growth promoters

Auxins

The earliest known and most studied plant hormones belong to this group. They are produced in growing shoot apices and, after migrating to other parts or organs, produce a variety of different effects, among which the regulation of cell enlargement is the most characteristic. In the presence of auxins the cellulose wall becomes plastic and is extended by the osmotic potential of the cell sap. The interweaving of the cellulose fibrils composing the framework of the wall is relaxed or loosened, and this allows new cellulose fibrils to be added. Auxins are translocated away from the site of synthesis by a special polarized transport mechanism which requires metabolic energy and which moves auxins in only one direction. This direction is always away from the tips of the shoot, and is thus anatomically determined.

In investigations of auxin action the coleoptile of seedling oats (*Avena sativa*) has been used as the standard test material. The great advantage of *Avena* coleoptiles for this type of work is that their growth from 1 cm to a final length of about 6 cm is accomplished almost entirely by cell elongation. If the tip of an *Avena* coleoptile is cut off, elongation of the stump is greatly retarded as compared with an intact coleoptile. If the removed tip is replaced, elongation of the coleoptile takes place at nearly the normal rate, showing that the influence originating in the tip can pass across a cut in the living tissues. Furthermore, if the tip is placed, with the cut surface downwards, on a small block of gelatine or agar gel for several hours, and the block minus the tip then transferred to the cut surface of a decapitated coleoptile, this coleoptile will elongate almost as fast as an intact coleoptile. A block of gelatine or agar gel which has not been in contact with cut-off tips has no such effect. All these results can be interpreted in terms of an auxin

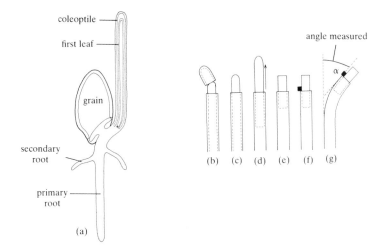

Fig. 21.1 Went's *Avena* coleoptile method for the bioassay of auxin. (a) *Avena* seedling in vertical section (diagrammatic). (b) and (c) Tip of the coleoptile removed. (d) First leaf pulled loose so that its elongation will not dislodge agar block. (e) Tip of first leaf cut off. (f) Agar block placed unilaterally on coleoptile tip. (g) The curvature produced is measured.

produced in the tip which passes longitudinally down the coleptile and stimulates the extension of the adjoining tissues.

Using the agar block technique F. W. Went developed a method for detecting and measuring the concentration of auxins in a plant tissue. If an agar block containing an auxin is placed on one side of the stump of a decapitated coleoptile, extension is greater on the side directly below the block than elsewhere, so that the coleoptile bends. By allowing the auxin present in one agar block to diffuse into other agar blocks, Went was able to obtain auxin in a series of known dilutions and to show that the curvature (measured as the angle α in Fig. 21.1) produced when the blocks were placed unilaterally on decapitated coleoptiles was proportional, within limits, to the concentration of auxin in the blocks. This relation between auxin concentration and coleoptile curvature made it possible to use the bending of oat coleoptiles for measuring the concentration of auxin in a plant tissue or extract. A technique which uses the response of an organism to determine the concentration of a chemical in an extract is called a bioassay. The development of Went's bioassay method opened the way for the chemical identification of auxin by enabling techniques for its purification to be worked out.

Several substances having auxin activity have been isolated in pure form from plant tissues, but the most common and important is the relatively simple substance indole-acetic acid (abbreviated to IAA).

IAA is active at very low concentrations, 10^{-6} mg being sufficient to produce a conspicuous curvature if applied to one side of a decapitated coleoptile.

In addition to naturally occurring auxins, there are many synthetic compounds, similar in chemical structure to IAA, which are extremely active as growth promoters. They differ from natural auxins in that they are highly toxic if applied even in slight excess. Furthermore the threshold concentration at which they become toxic varies widely from species to species of plant, thus allowing the use of these compounds as extremely efficient selective weedkillers. An example of such a compound that is now used extensively all over the world is 2,4-D (2,4-dichlorophenoxy-acetic acid) which is used as a weedkiller of broad-leaved species growing among grass crops (narrow leaved) such as cereals and lawns. Besides controlling the growth in length of shoots, including their response to unilateral light which will be discussed in the next chapter, auxins play many other roles in plant development. They are formed not only in growing shoots but also in several other regions, including some involved in the reproductive phase, for example pollen, fruits and seeds. One well known phenomenon which is mediated, at least in part, by auxins is 'apical dominance'. When the apical bud of many normally unbranched plants is removed some of the axillary buds, which normally remain dormant, grow out. If the decapitated bud is replaced immediately by a source of auxins the growth inhibition is reimposed and the lateral buds fail to grow out. Presumably the dominance of the intact apex is due to auxins produced in the bud and transmitted down the stem.

Lateral roots, like lateral buds, are also affected by auxins, and external applications of these substances greatly promotes the formation of lateral roots. A very important practical application of this phenomenon is in the promotion of root formation in plant propagation by cuttings. This was in fact the first of the many commercial uses to which auxins are now put.

One final example is the role played by auxins present in pollen grains. One of the main results of the pollination of flowers is an increase in the auxin content of the ovary and this initiates fruit growth and brings about fruit set by preventing abscission of the flower stalk. External applications of synthetic auxins have long been known to induce the same processes without pollination, and produce seedless fruit.

Gibberellins

Gibberellins are named after the fungus *Gibberella fujikuroi* which causes the 'bakanae' disease of rice plants, the infected plants becoming long and spindly and yellow in colour. As a result of the pioneering work of E. Kurosawa in Japan, who showed that the increased growth of the plant was due to a chemical secreted by the fungus, a new group of plant hormones was revealed by P. W. Brian in England and F. H. Stodola in the USA. These hormones are now known to be widely distributed in higher plants where they produce growth responses which are more dramatic than those of auxins. The striking responses promoted by gibberellins are generally accentuations of normal growth patterns rather than alterations into abnormal forms. For example, dwarf varieties of many species, such as maize and coffee, can be induced to grow as tall as the corresponding tall varieties; other aspects of the induced growth are normal. Rosette plants, in which little or no stem growth occurs during the first year normally require exposure to a particular day-length or some other environmental signal before stem elongation takes place. Treatment with gibberellins can usually produce the same effects as these signals, apparently by duplicating the action of natural gibberellins whose formation is initiated by the signals normally required.

There are over thirty natural gibberellins, all having a specific chemical configuration (a gibbane skeleton), but the one most frequently detected is gibberellic acid, GA_3, and many of the physiological effects ascribed to gibberellins as a group have been obtained only with this compound.

Cytokinins

These are compounds derived from the nitrogen-containing compound, adenine. They were discovered when plant cells in culture enlarged but failed to divide in the presence of auxin alone. Cell division was found to be stimulated by a variety of natural extracts whose activity is now known to be due to cytokinins.

Cytokinins are synthesized in roots from which they move upwards in the xylem and pass into the leaves and fruit. Required for normal growth and differentiation, cytokinins act, in conjunction with auxins, to promote cell division and to retard senescence which, at least in its early stages, is an organized phase of metabolism and not just a breakdown of tissue. An example of senescence is the yellowing of isolated leaves, which occurs as proteins are broken down and chlorophyll is destroyed. Cytokinins prevent yellowing by stabilizing the content of protein and chlorophyll in the leaf. They are used commercially in the storage of green vegetables.

Growth inhibitors

Although in the past a number of inhibitory substances have been isolated, their true role as naturally occurring growth regulators remained suspect. However, one growth inhibitor has been unequivocally established as a category of growth substances equal to auxins, gibberellins and cytokinins. This substance is abscisic acid.

Abscisic acid

The name abscisic acid (ABA) is derived from the ability of this substance to promote abscission, a discovery made by F. T. Addicott working on the abscission of cotton bolls. Independent studies by P. F. Wareing on the endogenous inhibitor relations in bud dormancy in deciduous trees led to the isolation and characterization of a potent growth inhibitor that was called dormin. However, it was soon shown to be identical with abscisic acid which is the accepted name for this inhibitor.

Abscisic acid (ABA) seems to be widely distributed in the plant kingdom and, like auxin, it is synthesized in fruits and leaves. It has a wide spectrum of activities in growth and development, but in all its roles it appears to be inhibitory. It may be that the relative balance, which is environmentally controlled, between ABA and the three categories of growth promoters determines the type of response it evokes.

Ethylene

Ethylene is a rather different type of hormone from the four previous categories in that it is a gas. It is released from most plant organs in varying concentrations, most obviously from ripening fruits. It probably triggers ripening and is responsible for the accompanying 'climacteric' rise in respiration rate of some fruits, for example avocado. Fruit ripening can be hastened by ethylene at concentrations as low as one part per million or even less. Ethylene is used commercially in the fumigation of the holds of banana boats at a suitable time before docking, thus guaranteeing that fruit picked green will have started ripening before sale.

Hormonal interaction

The discovery that auxin is involved in controlling growth activities as widely separated in time and space as the extension growth of the coleoptile and
the formation of fruit after pollination, led for a time to the idea that IAA
was the 'master hormone' of the plant. Different responses were attributed
to different sensitivities of the target tissues. Research work based on this
concept of auxin action brought confusion rather than clarity to the interpretation of growth phenomena, and the concept was eventually abandoned
when study of plant tissue cultures highlighted two situations incompatible
with its basic assumption. The first of these incompatibilities was that the
application of auxin at different concentrations could invoke quite different
responses in the same cultured tissue. The second was that the maintenance
of tissue growth required the presence of a complex of growth factors, at
that time largely unidentified but available in such natural sources as the nutrient endosperm of coconuts. From this work emerged the current concept
of hormone control as the outcome of a delicate balance of at least five major types of hormones, all interacting with each other in a bewildering complexity of antagonisms and synergisms. An antagonism is the interaction of
two substances acting in the same system in such a way that one partially or
completely inhibits the effect of the other. By contrast, a synergism is the
interaction of two or more substances having a similar effect in a given system to produce an effect which is greater than the sum of the effects of the
substances acting in isolation. This complexity can be illustrated by the following example. It has long been known that the cambium of trees, or at
any rate those of temperate regions, is most active when the buds open and
the shoots extend. Cell division of the cambium begins near the bud in each
shoot and then spreads away from it. The terminal bud stimulates the cambium to divide rapidly through the action of two groups of hormones, auxins and gibberellins. Application of either hormone alone stimulates cambial activity, but the effect of them both together is many times greater than
the sum of the responses of the two individually. The composition of the
vascular tissue produced by the cambium also depends on the hormone
treatments. With IAA by itself fully differentiated xylem is produced, with
GA_3 alone cambial products on the xylem side do not differentiate into mature xylem. With various combinations of the two hormones varying degrees of differentiation are obtained, the maximum production of fully
differentiated xylem occurring with a GA_3/IAA ratio of 100/500 ppm. On
the phloem side no differentiation takes place with IAA alone, but GA_3
alone considerably promotes it. IAA in low concentrations increases the stimulatory effect of GA_3, but higher concentrations are antagonistic, maximum phloem production occurring at a GA_3/IAA ratio of 500/100 ppm,
which is the reciprocal of the fraction promoting xylem production. It is
from discoveries such as these that an insight into the normal working of the
plant is being slowly gained.

Chapter 22

The response of plants to external stimuli

The factors which control plant growth can be divided into three groups. First are the inherent genetic factors, the genes, carried by all cells, which set the limits within which each cell, organ and, ultimately, the whole plant can grow. These factors can be altered only by breeding, and once the egg cell has been fertilized the genetical potentialities of the resultant plant are fixed.

Second are the internal factors, which integrate the individual cells, the tissues and the organs into a single structural and functional unit, the organism. These factors have been discussed in the previous chapter on plant growth substances.

The third group includes the external factors of the environment. Apart from the immediate effects on the nutrient (e.g. photosynthesis) and water (e.g. transpiration) status of the plant, plants may respond to external factors by movement or by morphogenetic changes. In this chapter consideration will be given to plant movements and to photoperiodism, which is the morphogenetic response of plants to the relative length of day and night.

One of the obvious features distinguishing plants from animals is that, generally speaking, plants are rooted in one place whereas animals move around. However, the ability to move is not restricted to animals, and flowering plants can respond to certain stimuli in their surroundings by moving some part of their body. Often such movements are important in the normal development of the plant, because they enable it to take up a position favourable for the gathering of raw materials and energy from the environment. Plant movements are not usually spectacular because most of them occur too slowly to be noticed by casual observation. Nevertheless their reality can be strikingly shown by time-lapse photography, whereby pictures of the moving part are taken at relatively long intervals and then projected at such a speed that the movements are greatly accelerated.

Plant movements may be brought about in two ways, firstly, by one side of the organ growing faster than the others, or secondly, by a change in turgor of particular cells causing a change in position of the responding organ. In the first type the response can only take place in those parts of the plant which are still growing or capable of resuming growth, whereas movements of the second type are shown only by mature organs. An example of this second type of movement is provided by the 'sleep movements' of the leaves of leguminous trees such as *Samanea saman* (rain tree).

Irrespective of how the movement takes place, plant movements are conventionally classified into two broad categories on the basis of the relation between the direction of movement and the direction from which the stimulus comes. Where the direction of the movement is determined by the direction from which the stimulus originates the response is called a *tropism*, whereas if the direction of movement is not related to the direction of the stimulus but depends on the structure of the responding organ, the response is called a *nastic movement*. Although it would be logical to study tropisms and nastic movements together under the umbrella topic of plant movements, an understanding of the mechanism of nyctinasty, or the sleep movements of leaves, involves information about the plant pigment, phytochrome, which is discussed in connection with photoperiodism. For this reason, nyctinasty will be considered after photoperiodism instead of with tropisms.

Tropisms

A tropism may be defined as a bending movement produced by unequal, or differential, growth on opposite sides of an extending organ, in response to a directional external stimulus. There are several tropisms known but, since unilateral light and gravity are the two directional stimuli which largely determine the direction of extension growth, and hence the orientation, of shoots and roots, only phototropism and geotropism will be considered here. In any case, the phenomena of phototropism and geotropism are by far the best studied examples of tropisms.

Phototropism

When a plant, such as a potted plant standing in a window, is illuminated more brightly on one side than on the others, the stems usually bend towards the brighter light and the leaves tend to orientate themselves at right angles to the direction of the light source. These movements are examples of phototropism. Since stems bend towards the light source they are said to be positively phototropic, in contrast to most leaves, which align themselves at right angles to the direction of light (diaphototropic). Because roots normally grow in the opposite direction from stems, one might expect roots to be negatively phototropic. In fact, there are only a few plants whose roots are known to respond to light.

Phototropism is particularly well demonstrated by seedlings, and the coleoptiles of seedling oats (*Avena sativa*) have been extensively used as experimental material for investigating the phenomenon. As long ago as 1880 Charles Darwin showed that, although it is the tip of a growing shoot which is sensitive to one-sided illumination, the bending of the stem towards the source of light occurs behind the tip in the region of cell elongation. This observation, as already mentioned (p. 250), was the starting point of the investigations which eventually led to the discovery of auxins.

The positive phototropic curvature of a growing shoot is caused by the cells on the shaded side elongating more than those on the illuminated side. This curvature towards light closely resembles the curvature produced by the unilateral application of an agar block containing IAA to a decapitated *Avena* coleoptile. It is now known that the difference in rates of elongation on the two sides of a curving stem is correlated with a greater concentration of IAA on the shaded than on the illuminated side. If the coleoptile tip of an *Avena* seedling that has been exposed to unilateral illumination is removed and bisected longitudinally at right angles to the direction of the

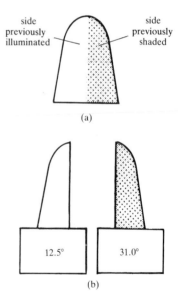

side previously illuminated

side previously shaded

(a)

12.5°   31.0°

(b)

Fig. 22.1 Lateral movement of auxin in coleoptile tip of *Avena* in response to unilateral illumination. (a) Coleoptile tip removed from seedling after exposure to unilateral light. (b) Coleoptile tip bisected and halves stood on separate agar blocks. The figures indicate the amount of auxin which, in one experiment, diffused into each block, expressed as degrees of curvature measured by Went's bioassay method.

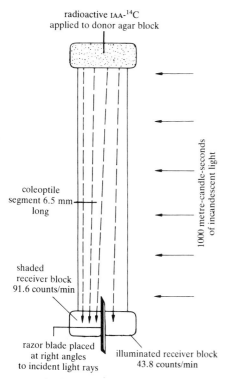

radioactive IAA-^{14}C
applied to donor agar block

coleoptile
segment 6.5 mm
long

shaded
receiver block
91.6 counts/min

1000 metre-candle-seconds
of incandescent light

razor blade placed
at right angles
to incident light rays

illuminated receiver block
43.8 counts/min

(A shaded/illuminated ratio of 71.4:28.6)

Fig. 22.2 Lateral transport of radioactive
auxin (IAA −^{14}C) in a segment of an *Avena*
coleoptile tip, as a result of unilateral
illumination. An agar block containing
radioactive auxin is placed on the apical end of
a coleoptile segment, and the amount of
radioactive auxin diffusing into the two
receiver blocks is expressed as counts per
minute above background. Radioactivity
appearing in the two receiver blocks is equal in
the dark, but unequal in unilateral
illumination. (Data of Pickard, B. G. and
Thimann, K. V. (1964), *Plant Physiology*, **39**,
341–50. Figures are the mean of nine
experiments)

incident light (Fig. 22.1), the IAA from the illuminated and shaded halves
can be collected in separate agar blocks and assayed by Went's bioassay
method. It is found that the shaded side always contains more IAA than the
illuminated side.

Having established that there is an unequal distribution of IAA under the
influence of unilateral light, the next problem is to explain how this is
brought about. The obvious possibilities that IAA is either destroyed on the
illuminated side or produced in greater quantity on the shaded side would
seem to be ruled out, because the total amount of IAA in a coleoptile tip
has been shown to be the same irrespective of whether or not it is exposed
to lateral light. This would suggest that the IAA in the tip migrates from the
more strongly illuminated side to the shaded side, and experiments using
radioactively labelled IAA have confirmed that such a lateral migration
takes place (Fig. 22.2). This redistribution of IAA occurs at the extreme
apex of the coleoptile, but the mechanism of how it is achieved is still unknown. It is reasonable to suppose that some pigment is involved because
light must be absorbed before it can produce a chemical effect. By determining the action spectrum for phototropism (i.e. the relative efficiencies of
light of different wavelengths in eliciting phototropic curvatures), it is found
that only blue light is effective and so the postulated pigment is presumably
yellow because this is the colour which absorbs blue light. Two yellow substances which are present in coleoptiles, β-carotene and riboflavin, have
been suggested as photoreceptive pigments, but the absorption spectrum of
neither of them exactly matches the action spectrum of phototropism
(Fig. 22.3). The spectrum for β-carotene does not have a peak in the range
of ultraviolet light, while that for riboflavin fits the ultraviolet peak but
does not match the two peaks in the range of blue light. This difficulty
might be met by assuming that both substances function in the phototropic
reaction, but there are additional reasons for doubting whether either of
them, particularly β-carotene, is involved in phototropism. The objection to
β-carotene is that, although it is present along the length of the coleoptile, it
is often absent from the extreme tip which is the photoreceptive region;
furthermore, albino mutants of various plants, in which β-carotene is completely absent, show normal phototropic responses. The case for riboflavin,
which does occur at the extreme tip, is based primarily on the fact that it
accelerates the enzymic breakdown of IAA in the presence of light, and
thus its presence offers a possible explanation for the unequal distribution
of IAA as a result of its destruction on the lighted side. However, such a
reaction mechanism is inconsistent with the facts (1) that unilateral light
does not decrease the total amount of IAA in a coleoptile tip, and (2) that
the unequal distribution of IAA is known to result from lateral transport at
the extreme tip rather than from its destruction on the lighted side. Since
neither β-carotene nor riboflavin meets the necessary requirements, it must
be concluded that the identity of the photoreceptive pigment in phototropism has not yet been resolved. Although the mechanism for the lateral
transport of IAA at the coleoptile tip is unknown, there can be no doubt
that, because the downward translocation of IAA is almost strictly longitudinal, the cells below the tip on the shaded side receive more IAA, and
hence elongate more, than those on the illuminated side, with the result
that the coleoptile bends towards the light source.

Geotropism

Geotropism is the response of plants to the stimulus of gravity. If a seedling growing in diffuse light (i.e. in light conditions where the possibility of phototropic effects can be ruled out) is placed in a horizontal position, the main root bends down until pointing vertically downwards and the main stem bends up until pointing vertically upwards. Primary roots and main stems are therefore said to be positively and negatively geotropic respectively. Leaves, lateral roots and lateral branches are all plagiogeotropic, maintaining a constant angle with the vertical, not necessarily at right angles to it (e.g. lateral roots often make an acute angle with the vertical). Some organs change their responses to gravity as they develop. For example, the flower stalk of *Arachis hypogaea* (groundnut) becomes positively geotropic after the fertilization of the ovules and grows actively downwards, not just bending under its own weight, so that it buries the developing fruits in the ground.

That the bending of main stems and roots, when placed horizontally, is due to gravity can be demonstrated by means of a klinostat. This instrument slowly rotates a plant about the horizontal axis, so that each flank of the plant is in turn stimulated and the unidirectional effect of gravity is eliminated. When a plant is mounted in a horizontal position on a klinostat it no longer responds to gravity and grows on horizontally. This behaviour indicates that the normal behaviour is due to a differential response to gravity by the upper and lower sides of the stimulated plant.

The perception of gravity by shoots and coleoptiles occurs in the extreme tip; when the tip is removed the organ fails to respond to gravity, but, if the tip is replaced, the organ again responds as if the tip had not been cut off. In roots the site of gravity perception is the root cap; in maize (*Zea mays*) and barley (*Hordeum vulgare*) it is possible to remove the intact root cap cleanly from the rest of the root tip and such decapped roots fail to respond to gravity, even though their extension growth is unimpaired. Replacement of the cap restores the ability to respond.

The bending of main stems and roots in response to gravity is due, as in phototropism, to differential growth in the elongating region behind the apex. In a horizontally placed stem or coleoptile the elongation is greater on the lower than on the upper side whereas in a root the reverse happens, thus giving the upward and downward bending respectively.

In stems and coleoptiles IAA is involved in geotropic curvatures as it is in phototropic curvatures. This can be shown by removing the coleoptile tip of a seedling which has been kept horizontal for a sufficient time to produce a geotropic response. If the cut surface of the tip is then immediately placed in the centre of a vertical face of an agar block which has been separated into upper and lower halves with a barrier, it is found that more IAA accumulates in the lower half of the block (Fig. 22.4). Since IAA stimulates the elongation of coleoptile cells it is not surprising to find that the lower side of a horizontal coleoptile elongates more than the upper side, so that the coleoptile bends upwards.

Until recently IAA was also invoked to explain the geotropic curvature of roots. Since elongation of root cells is retarded by concentrations of IAA which stimulate the elongation of stem cells (p. 257) it seemed reasonable to suppose that in the root a higher concentration of IAA on the lower side should retard elongation, so that the root bends downwards instead of up-

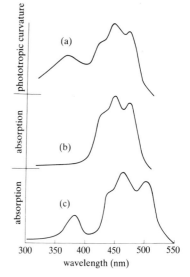

Fig. 22.3   The action spectrum for phototropism of *Avena* coleoptile (a) compared with the absorption spectra of carotene (b) and riboflavin (c). Neither of the two pigments exactly qualifies as the phototropic photoreceptor. (From Bara, M. and Galston, A. W. (1968), *Physiologia Plantarum* **21**, 109–18).

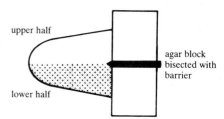

Fig. 22.4   Coleoptile tip mounted centrally on vertical face of a divided agar block, for demonstrating the effect of gravity on the distribution of auxin. More auxin accumulates in the bottom half of the agar block.

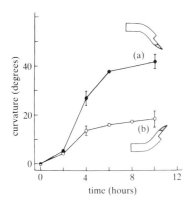

Fig. 22.5 Curvature of *Zea mays* roots, from which one half of the cap has been removed, and orientated so that the remaining half cap is on the lower (a) or upper (b) half of a horizontal root. Vertical lines extending from one side of the points show the standard error of the mean. (From Wilkins, M. B., Gibbons, G. S. B. and Shaw, S. (1972) in D. J. Carr (Ed.) *Plant growth substances 1970*, p. 721, Springer-Verlag)

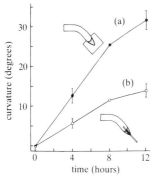

Fig. 22.6 Comparison of the curvatures developed by horizontal *Zea mays* roots with an intact cap but into the apex of which a barrier has been inserted either vertically (a) or horizontally (b). Root caps dotted. Vertical lines on one side of the points show the standard error of the mean. (From Wilkins, M. B., Gibbons, G. S. B. and Shaw, S. (1972) in D. J. Carr (Ed.) *Plant growth substances 1970*, p. 723, Springer-Verlag)

wards as in the stem. However, it is doubtful whether IAA exists in root apices, and recent evidence suggests that the positive geotropic response of the primary root of *Zea mays* depends upon a growth inhibitor produced in the root cap moving backwards into the region of elongation. On this basis the downward bending results from the growth of the lower half being depressed more than that of the upper half. In support of this hypothesis it is found that, when one half of the root cap is removed and the root is then placed in the horizontal position, the root always bends towards the remaining half-cap regardless of whether this is on the upper or the lower side of the root (Fig. 22.5). The downward curvature of a horizontal root with the remaining half-cap on the lower side is nearly twice as much as the upward curvature developed by a horizontal root with the remaining half-cap on the upper side. There is no reason why a lower half-cap should produce more inhibitor than an upper half-cap, and the simplest explanation for this difference in curvature is that a downward movement of inhibitor occurs in the root cap, in a way comparable to the downward movement of IAA in a coleoptile tip. If this is the case, a steeper concentration gradient of inhibitor between the upper and lower halves of the elongating region would be expected when the remaining half-cap is on the lower side than when it is on the upper side. Further evidence for this interpretation of the positive geotropism of roots is obtained by inserting vertical and horizontal barriers across the root tip of a horizontally placed root (Fig. 22.6). When the barrier is placed vertically it is found that the downward curvature is much greater than when the barrier is placed horizontally. This can be explained by postulating that the horizontal barrier impedes the downward migration of the inhibitor produced by the top half of the root cap, so that less inhibitor reaches the lower side of the elongating region. The experimental evidence for a growth inhibitor being involved in the geotropic responses of roots is convincing, but so far the nature of the inhibitor (if it exists) is unknown.

The mechanism of how gravity causes the redistribution of IAA (in the tips of shoots and coleoptiles) or a growth inhibitor (in root caps) is not understood because it is difficult to conceive how gravity, unlike light, could be detected and bring about a direct chemical effect. Most theories of geotropism propose the displacement by gravity of conspicuous starch grains, called *statoliths* (Fig. 22.7), which are present in the cells of most organs sensitive to gravity. It is supposed that when the orientation of the gravity-perceptive organ is changed the statoliths shift their position within the cells and, presumably by pressing against another region of the protoplasm, they cause distributional changes in the appropriate growth-controlling substance. Studies with the electron microscope have shown that statoliths change their position under the influence of gravity at a rate which is consistent with the presentation time (i.e. the time for which an organ must be continually subjected to a given intensity of stimulus for a visible response to follow inevitably), but they have so far failed to reveal any organelles which might be sensitive to the pressure of statoliths. To test whether statoliths are an essential part of the gravity-perceiving mechanism, coleoptiles have been destarched (i.e. by incubating them in solutions which promote the hydrolysis of starch to sugar) and their subsequent response to gravity studied. Some workers claim that the disappearance of statolith starch results in loss of geotropic response, but other workers find that destarched coleoptiles respond to gravity in much the same way as do the controls, which contain

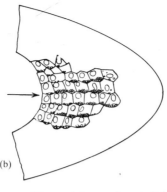

Fig. 22.7 Distribution of statoliths in the root tip of a vertically (a) and horizontally (b) orientated root. The arrows indicate the axis of the root.

statoliths. In view of the conflicting evidence on this crucial experiment, it is clear that further research must be done to determine the means by which the gravitational stimulus is perceived.

Photoperiodism

It is well known that certain plants normally flower only at a particular time of year; for instance, in the subtropics of the Northern Hemisphere, poinsettia (*Euphorbia pulcherrima*) comes into flower about December and, in some countries, is a traditional flower of the Christmas season. In 1920, two American physiologists, W. W. Garner and H. A. Allard, discovered that the timing of flowering is controlled primarily by the relative lengths of day and night. At the equator day-length is approximately 12 hours throughout the year, but in mid-temperate latitudes it varies between 8 and 16 hours according to the time of year. The length of the daily period of light is called the *photoperiod*, and the reaction of plants to photoperiods of different lengths is termed *photoperiodism*.

On the basis of their day-length requirements for flowering, most plants can be separated into three major groups:

1. *Short-day plants* which flower only when the day-length is *less than* a certain critical period, e.g. poinsettia (*Euphorbia pulcherrima*), sweet potato (*Ipomoea batatas*), pineapple (*Ananas comosus*) and rice (*Oryza sativa*).
2. *Long-day plants* which flower only when the day-length is *more than* a certain critical period, eg castor oil plant (*Ricinus communis*) and English potato (*Solanum tuberosum*).
3. *Day-neutral plants* which flower independently of day-length producing flowers at any time of the year, usually when the plant has reached a particular size, e.g. maize (*Zea mays*) and many tropical garden plants, including *Zinnia* spp.

The separation of plants into these three groups has nothing to do with their taxonomic classification into genera and families. However, the geographical distribution of many plants, both wild and cultivated, is very definitely related to the length of the photoperiod. In the tropics most plants are either short-day or day-neutral types because, with a few exceptions, only these are able to flower and thus perpetuate themselves by sexual reproduction. Similarly in temperate regions it is long-day or day-neutral plants which flower during the summer months when the day-lengths are of the order of 15–16 hours. It is clear, therefore, that before attempting to

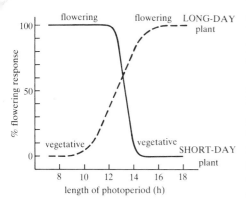

Fig. 22.8 The effect of day-length on the flowering response of a short-day plant with a high critical photoperiod and a long-day plant with a low critical photoperiod. Note that a daily light period of about 13 hours induces flowering in both types of plant.

cultivate a plant in another country at a different latitude its photoperiodic behaviour should be taken into account. It is now known that some of the past attempts to introduce plants into other countries have failed mainly because of unfavourable day-lengths.

The day-length requirement for flowering is characteristic for a given plant, but the critical values of both short-day and long-day plants differ widely from species to species and even among varieties of the same species. Generally speaking, critical day-lengths for short-day plants vary between 11 and 15 hours (i.e. day-length must be less than 11–15 hours before flowering will occur) and for long-day plants between 12 and 14 hours (i.e. day-length must be more than 12–14 hours before flowering will occur). There is thus a considerable overlap in the day-length requirements of short-day and long-day plants, so that a photoperiod of, say, 13 hours will induce flowering in both short-day plants with a high critical photoperiod and long-day plants with a low critical photoperiod (Fig. 22.8). It should be stressed that the concept of short-day and long-day plants refers not to the absolute length of day but to whether flowering occurs at day-lengths *shorter than or longer than* a certain critical value. Many short-day plants, for example, flower at a time when the days are considerably longer than the nights, contrary to what would seem to be implied by the term short-day plant.

In the natural state the alternation of light and dark is based on a 24-hour cycle, so that variation in the light period is inevitably accompanied by variation in the dark period, but under experimental conditions it is possible to vary the light period while keeping the dark period constant, and vice versa. By varying the duration of the light and dark periods in this way it is possible to assess the relative importance of the light and dark periods in inducing flowering. It is found that the important factor is not the length of the light period (nor the ratio of light to dark) but the length of the dark period: short-day plants will flower in long days if long nights are also given, and long-day plants will flower in short days with short nights (Fig. 22.9). Hence, although the names short-day and long-day are used to describe the photoperiodic requirements of plants for flowering, it would be more accurate to regard short-day plants as long-night plants, and long-day plants as short-night plants. The word photoperiodism, although strictly incorrect, is nevertheless retained because it has become so well established. Short-day plants flower because the dark periods exceed a certain length, not because the light periods are short; the minimum dark period must be given, irrespective of the length of the intervening light period. Because exposure to a light period of some length is essential for photosynthesis, short-day plants require alternating dark and light periods if flowers are to be formed.

Fig. 22.9 Responses of typical short-day and long-day plants to variations in length of exposure to light and dark. The results show that flowering is controlled by the length of the dark period rather than by the length of the light period.

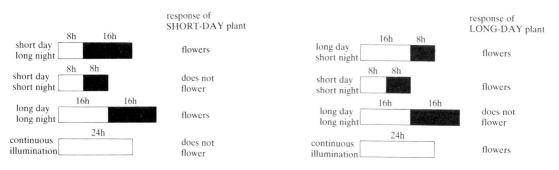

In the Biloxi variety of soybean (*Glycine soja* var. *Biloxi*), for example, a dark period of at least 10 hours (although it can be considerably longer) must alternate with light periods which may be anything from 4 to 18 hours (but optimally from 10 to 12 hours) if flowers are to be produced. Long-day plants, on the other hand, will not flower if the dark period exceeds a certain critical length, but will flower under any shorter dark period, up to and including continuous light. Long-day plants do not require a dark period for flowering, so that the ability to flower in continuous light provides a useful criterion for distinguishing long-day from short-day plants. For example, *Xanthium pennsylvanicum* (a plant which has been much used in photoperiodic studies), will flower on a day-length of 15.5 hours and therefore might be assumed to be a long-day plant, yet it is nevertheless a short-day plant because it fails to flower in continuous light. In this plant a dark period is essential for flowering, even though this is as short as 8.5 hours.

To induce flowering it is not necessary for plants to be exposed throughout the period of growth to photoperiods favourable to flowering. For example, if short-day Biloxi soybeans growing under long-day conditions are exposed to short-day conditions for two consecutive days and then returned to long days, flowering will still occur. This 'triggering-off' of flowering is known as *photoperiodic induction*, and has been demonstrated for a number of both short-day and long-day plants. An extreme case of induction is found in *Xanthium pennsylvanicum*, the short-day plant already mentioned. When kept in continuous light this plant remains vegetative indefinitely but, after exposure to only one dark period of not less than 8.5 hours, it will flower about a fortnight later. Once the potential for flowering has been induced, the plant will continue to flower even when maintained in continuous light. Biloxi soybeans and *Xanthium pennsylvanicum* are examples of plants which respond very rapidly to a light regime suitable for flowering, but other plants may require several days or weeks of appropriate photoperiods before photoperiodic induction will occur.

The importance of the dark period for the induction of flowering is also shown by the results of experiments in which the dark period is interrupted with a brief exposure to light. If a short-day plant is put on a short-day cycle, such as 8 hours of light alternating with 16 hours of dark, it will flower. If, however, the dark period is interrupted at somewhere near its mid-point by a short period of relatively low-intensity light, flowering is completely prevented, just as if the plant were being kept under a long-day cycle (Fig. 22.10). An interruption of only 1 minute's exposure to a 25 watt bulb is enough to prevent flowering in some short-day plants. When a long-day plant is kept under the same short-day cycle, interruption of the dark period has the opposite effect and induces flowering. In both short-day and long-day plants the treatment reverses the usual photoperiodic reaction, indicating that in both cases the plant appears to be measuring the length of the dark period. It would thus seem that a similar timing mechanism is operating in both short-day and long-day plants, even though the effect of interrupting the dark period differs in the two types of plants.

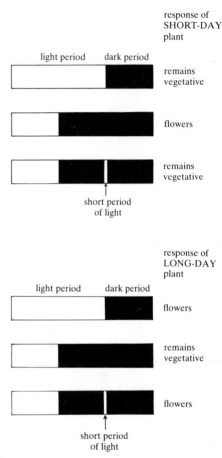

Fig. 22.10 The effect of interrupting the middle of a long dark period with a brief (say 15 minute) 'light break' on the flowering of short-day and long-day plants. The responses indicate that plants measure the absolute length of the dark period.

Mechanism of photoperiodism

For both short-day and long-day plants it has been shown that the leaves are the receptors of the photoperiodic stimulus. When a plant, growing in an appropriate light regime for flowering, is stripped of all its leaves it will not normally flower, but if a single young leaf (or sometimes even only a small part of a leaf) is allowed to remain the plant will flower. Alternatively, when a single leaf of a short-day plant is enclosed in a light-proof box and given short photoperiods suitable for flowering, while the rest of the plant is exposed to long photoperiods or even to continuous illumination, flowering takes place. Since the effect of the treatment (i.e. the production of flowers) occurs in the growing meristem at the shoot apex which is separated in space from the treated leaf, it would seem that some hormone which is produced by the leaf in response to short days, moves to the apical meristem where it induces the initiation of flowers rather than of leaves and axillary buds. Grafting experiments point to the same conclusion. If a short-day plant, the receptor, growing in long days (i.e. under conditions unsuitable for flowering) is grafted on to a plant, the donor, of the same species that has been photoperiodically induced by previous exposure to short days, the receptor will flower even though the grafted plant is retained under a long-day regime (Fig. 22.11). Some sort of stimulus must therefore pass from the photo-induced donor to the receptor. Provided that a graft can be established, the receptor and donor in this type of experiment can belong to different species, or even be a short-day plant and a long-day plant, so long as the donor has previously been photo-induced. This suggests that the chemical factor inducing flowering is at least functionally equivalent, if not chemically identical, in different plants. The transmission of the flowering stimulus can be prevented by girdling the petiole or otherwise interrupting phloem transport. This clearly indicates that the chemical factor is translocated through the phloem to the apical meristem.

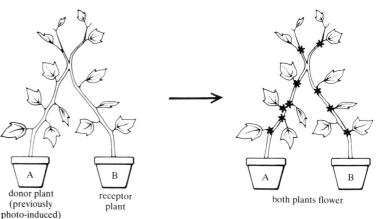

Fig. 22.11 Evidence for the existence of a flower-inducing hormone from grafting experiments. For explanation see text. (From Adams, Baker and Allen 1970, p. 324)

donor plant (previously photo-induced)   receptor plant   both plants flower

The name *'florigen'* has been proposed for this hypothetical flowering hormone, but efforts to isolate such a substance have so far been unsuccessful. There appears to be some connection between florigen (if it exists) and a gibberellin, because application of gibberellins to many long-day plants causes them to flower under short photoperiods normally unsuitable for flowering. Nevertheless the relation between florigen and a gibberellin would seem to be indirect, because gibberellins do not promote flowering in short-day plants even though grafting experiments indicate that the flowering hormone is of the same nature in both short-day and long-day plants. Furthermore, they often induce stem elongation but not flowering in some long-day plants.

The effect on flowering of interrupting the middle of a long dark period with a brief exposure to light provides a convenient method for determining the action spectrum of photoperiodism (i.e. the relative efficiencies of different wavelengths of light in destroying the effect of a continuous dark period). It is found that red light, of around 660 nm wavelength, is the most effective in preventing flowering of short-day plants and in promoting flowering of long-day plants (Fig. 22.12). Since flowering is controlled in both types of plants by light of the same wavelength range, it would appear that the basic photochemical reactions are the same even though in short-day plants they suppress and in long-day plants promote flowering.

A surprising feature of the effect of red light is that it is nullified by exposure to light in the 'far red' region, of wavelength around 730 nm. Thus, a short exposure of red light given to short-day plants in the middle of a long dark period inhibits flowering, but if this treatment is followed immediately by a short exposure to far-red light flowering is promoted as if the dark period had not been interrupted. This switching on and off of the capacity to flower can be repeated many times, the subsequent response always depending on the kind of light given last (Fig. 22.13).

An examination of the action spectrum of photoperiodism (Fig. 22.12) suggests that plant cells must contain a blue-coloured pigment which absorbs the wavelengths of light effective in the control of flowering. Reversal of the action of red by far-red light, and vice versa, indicates the presence of a pigment in two forms, a red-absorbing form and a far-red-absorbing form, which are interconvertible. A blue-coloured pigment, which shows different absorption spectra when exposed to red and far-red light (Fig. 22.14), has now been detected in many different plants, ranging from algae to angiosperms; it is called *phytochrome*. After illumination with red light phytochrome exists in the form which absorbs maximally in the far-red region of the spectrum at 730 nm, and for this reason is referred to as the P_{730} or P_{fr} form. Illumination with far-red light, or alternatively a long uninterrupted dark period, converts P_{730} into the red-absorbing form with maximum absorption at 660 nm, referred to as the P_{660} or P_r form. The conversion of P_{660} into P_{730} requires considerably less energy than the reverse reaction, so that in light containing both red and far-red wavelengths (e.g. sunlight) phytochrome exists mainly in the far-red-absorbing form (i.e. P_{730}). Hence it is the red-sensitive response which is normally regarded as the light response, the far-red response being the one that occurs in darkness. Since the dark period is the important factor in inducing flowering, it can be inferred that P_{730} is the biologically active form which is responsible for inhibiting flowering in short-day plants but promoting it in long-day plants. Although phytochrome is clearly involved in the mechanism of photoperiodism, and in many other developmental processes, the way in which it is instrumental in effecting these responses is still undiscovered.

Sleep movements of leaves

The leaves of a considerable number of tropical plants, particularly members of the Leguminosae, take up different positions during the day and at night; this process is called nyctinasty (literally, night movement). In most cases of nyctinasty the leaf surfaces are horizontal and face the sun during the day, but become folded together in a vertical position at sunset (Fig. 22.15). The folding is either upwards or downwards according to the

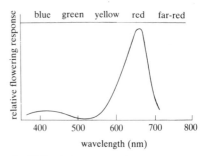

Fig. 22.12 Action spectrum for photoperiodism.

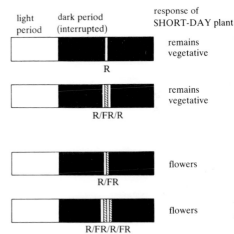

Fig. 22.13 The effect of interrupting the middle of a long dark period by flashes of red (R) and far-red (FR) light on the flowering of a short-day plant. Note that only when far-red light is given last does flowering subsequently occur.

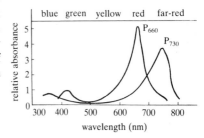

Fig. 22.14 The absorption spectra of the two forms of phytochrome. The two forms are usually called P_{660} and P_{730} to indicate the wavelength of maximum absorption by each form. (From Black and Edelman 1970, p. 119)

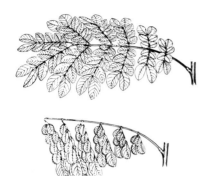

Fig. 22.15 Bipinnate leaf of *Samanea saman* (rain tree) in the open day-time position (above) and the closed night-time position (below). (From Brown 1935, p. 113)

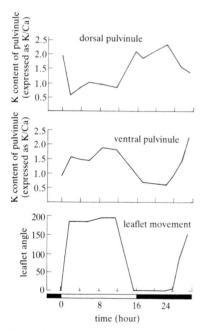

Fig. 22.16 Diurnal variations in leaflet angle and in the potassium content of the motor cells on the dorsal and ventral sides of the pulvinule of *Albizia julibrissin*. In order to compensate for changes in cell volume during leaflet movement, changes in potassium content per cell are expressed as changes in the ratio of K to the immobile element Ca which is largely bound in the cell walls. (Redrawn from Satter, R. L. and Galston, A. W. (1971), *Science*, **174**, 518–20)

species. It is often assumed that the function of nyctinasty is to enable plants to capture sunlight during the day, but reduce heat loss or the absorption of moonlight (which, if too intense, would interfere with the floral induction of short-day plants) during the night.

The petiolar base of all leaves showing sleep movements is swollen to form a small organ, the *pulvinus*, which is the region where the movement takes place. In the pulvinus the vascular tissue is concentrated in a flat plane at right angles to the direction of movement, and is surrounded by a wide cortex of thin walled parenchyma. Under the electron microscope the outermost cortical cells on both the dorsal and ventral sides of the pulvinus are seen to differ from the other cortical cells in having their cytoplasm filled with numerous small vacuoles. These specialized cells are called *motor cells* because they show alternate contraction and expansion. Leaf movement depends on the relative turgor of the motor cells on opposite sides of the pulvinus. For example, in *Albizia julibrissin* and *Samanea saman* whose leaflets hang downwards at night, the upward movement of the leaflets in the morning results from an increased turgor in the ventral motor cells and a decrease in the dorsal ones. The reverse changes occur when the leaflets fold downwards at sunset.

Since nyctinastic leaves usually open at dawn and close at dusk, it is reasonable to suppose that the occurrence of alternating periods of light and darkness controls the movement. However, under experimental conditions, the leaflets of *Albizia julibrissin*, on which this account is based, continue to open and close each day for several days when kept in constant intensity light or uninterrupted darkness. Furthermore, if *Albizia* plants are grown under a 16 hour photoperiod the leaflets open rapidly on transfer from dark to light, begin to close after 12 hours, and are usually completely closed before darkness. They remain closed during the 8 hour dark period, but will open without illumination if the dark period is extended by 2 or 3 hours. It is thus clear that, even though the movement has been shown to be affected by the light–dark transition, nyctinasty in *Albizia* depends also on a built-in endogenous rhythm which is even more decisive than the light stimulus.

Like stomatal movement, which also depends on turgor changes, nyctinastic movements are accompanied by potassium (K^+) fluxes into and out of the motor cells. The K^+ concentration in the *Albizia* pulvinule (the name given to the pulvinus of a leaflet) is about 0.5 M and K^+ is the only cation present in sufficiently high concentration for its movement to exert a significant osmotic effect. The possibility of a relationship between K^+ flux and leaflet movement has been investigated by measuring the angle between paired leaflets at specified times during the 24 hour cycle, and determining the K^+ level in the motor cells of the pulvinules at the same times. The K^+ level is measured with an electron probe, an instrument which enables the concentration of several different elements to be measured simultaneously within a very small volume, such as several cells of a tissue. The curves for leaflet movement and for K^+ flux in the ventral motor cells are almost parallel, whereas the curve for K^+ flux in the dorsal cells is almost their mirror image (Fig. 22.16). Although these results do not establish whether K^+ flux is the cause or the consequence of water movement, there is little or no evidence for active transport of water in other plant species. Thus, changes in K^+ provide an explanation for turgor changes in the motor cells in term of osmosis.

Since it has been shown that the total amount of K^+ in a pulvinule remains more or less constant, the fact that the dorsal cells are turgid when the ventral cells are flaccid, and vice versa, suggests that K^+ ions shuttle from contracting cells on one side of the pulvinule to expanding cells on the other. Experimental measurements of K^+ flux during closure, however, show that intake into expanding dorsal cells is completed before efflux from contracting ventral cells has stopped. This implies that dorsal and ventral motor cells have different rhythms, and both exchange K^+ ions with a common reservoir which seems to be the central region of the pulvinule.

The part played by white light in leaflet movement is also revealed by studying the effects it has on the K^+ fluxes of dorsal and ventral motor cells. It is found that white light is not necessary for K^+ flux into ventral motor cells, but that it does control the K^+ flux out of dorsal cells. The K^+ content of these cells decreases rapidly when leaflets are transferred from darkness to light, and increases rapidly when leaflets are returned to darkness. Blue light is the most effective region of the spectrum in promoting leaflet opening and keeping them open.

Studies of nyctinasty started in 1966 have established that the regulatory pigment phytochrome is involved in controlling the closure of leaflets. If leaflets that have recently opened in white light are briefly exposed to red or far-red light and then transferred to darkness, effects are evident within 10 minutes. If leaflets have been exposed to red light (i.e. phytochrome exists in the P_{fr} form), K^+ moves out of the ventral motor cells and the leaflets close, whereas if leaflets have been irradiated with far-red light (i.e. phytochrome exists in the P_r form) K^+ moves into the ventral motor cells but the leaflets scarcely move because these cells are already turgid. By contrast, the movement of K^+ into the dorsal motor cells is not controlled by phytochrome. Since nyctinastic closure is inhibited by anaerobic conditions, or by compounds such as dinitro-phenol (DNP) and sodium azide (NaN_3) which interfere with ATP production, it appears that during closure K^+ ions are pumped out of the ventral motor cells by a P_{fr}–dependent mechanism. The continued presence of P_{fr} is required throughout the closure period, because leaflets exposed to red light stop closing within 10 minutes if subsequently irradiated with far-red light.

In an attempt to integrate the results from the numerous experiments made on nyctinasty, Satter and Galston have proposed a mechanism for this process, with K^+ flux in relation to membrane properties as its basis (Fig. 22.17). In view of the dominant role of endogenous rhythmic behaviour in nyctinasty it is suggested that there are rhythmic changes in the state of the motor-cell membranes, which are expressed as rhythmic changes in the dynamic equilibrium between active transport and passive diffusion of K^+ ions through motor-cell membranes. The equilibrium favours active transport of K^+ ions into ventral and out of dorsal motor cells when the leaflets are open, and passive diffusion in the opposite directions when the leaflets are closed. Superimposed on this basic rhythm is the requirement of the P_{fr} form of phytochrome for normal leaflet closure. There is considerable evidence that phytochrome is located in the membranes of the ventral motor cells adjacent to or at the sites through which K^+ ions move. The movement across the membrane involves the expenditure of ATP. It is therefore suggested that phytochrome influences movement by activating a K^+ pump present in the membrane. The phase of the rhythm can be modified by white (or blue) light which is thought to increase membrane integrity. The presence of white light will thus maintain low K^+ in

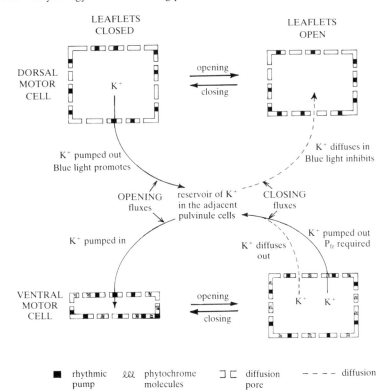

LEAFLETS
CLOSED

LEAFLETS
OPEN

DORSAL
MOTOR
CELL

K⁺

opening

closing

K⁺ pumped out
Blue light promotes

K⁺ diffuses in
Blue light inhibits

OPENING
fluxes

reservoir of K⁺
in the adjacent
pulvinule cells

CLOSING
fluxes

K⁺ pumped in

K⁺ diffuses
out

K⁺ pumped out
P_{fr} required

VENTRAL
MOTOR
CELL

opening

closing

K⁺   K⁺

Fig. 22.17 The K⁺ fluxes involved in the nyctinastic movements of the leaflets of *Albizia julibrissin.* (Based on Satter, R. L. and Galston, A. W. (1973), *Bioscience* (*Washington*), **23**, 407–16)

■ rhythmic pump   ℓℓℓ phytochrome molecules   ⊐⊏ diffusion pore   ---- diffusion

dorsal motor cells and high K⁺ in the pulvinule reservoir cells. When leaflets are transferred from white light to darkness the membranes of all cells become leaky, enabling K⁺ ions to diffuse into the dorsal cells from the reservoir where they are stored when the leaflets are in white light.

Nyctinasty is not a subject with any obvious economic applications, but the study of how leaf movements occur is of considerable scientific importance. Endogenous rhythms and phytochrome are involved not only in leaflet movements but also in such widely different processes of plant development as seed germination and flower induction. The advantage of using leaf movements for investigating the mode of action of phytochrome is that the phytochrome-mediated responses are measurable shortly after the stimulus is applied. Thus an understanding of the mechanism of interaction of rhythms and phytochrome in pulvinules can be expected to contribute to an understanding of many important questions in plant biology, some of which are certainly of importance in crop production.

Suggestions for further reading

Elementary books

Baker, J. J. and Allen, G. E. (1974) *Matter, energy and life*, 2nd edn. Addison-Wesley.
  Intended for students who have little or no knowledge of chemistry or physics, this book presents the basic chemico–physical principles necessary for understanding such biological topics as enzyme action, respiration, photosynthesis and protein synthesis.
Barker, G. R. (1968) *Understanding the chemistry of the cell.* Edward Arnold.
  This booklet presents the methods used in studying the chemistry of living systems in such a way that cellular reactions are seen not as the consequence of living processes but as the means whereby such processes operate. This booklet achieves the aim expressed in the title.
Lehninger, A. L. (1971) *Bioenergetics: the molecular basis of biological transformations,* 2nd edn. W. A. Benjamin.
  The classic, semipopular account of the flow of energy in cells, including a discussion of the functions of both mitochondria and chloroplasts.
McElroy, W. D. (1971) *Cellular physiology and biochemistry*, 3rd edn. Prentice-Hall.
  A brief introduction to the basic principles of cell physiology and biochemistry, especially useful for its clear account of the exchanges and uses of energy within cells.

General textbooks of plant physiology

Bidwell, R. G. S. (1979) *Plant physiology,* 2nd edn. Collier Macmillan.
  An advanced book which is particularly valuable because it discusses the physiology of a number of important topics not usually included in textbooks of plant physiology. Such topics include parasites and disease, symbiosis, plant distribution, and other 'bridge' areas within biology.
Coult, D. A. (1973) *The working plant.* Longman.
  A short book which is noteworthy because it stresses the thesis that structure and function are always intimately related and should be studied together.
Devlin, R. M. (1975) *Plant physiology,* 3rd edn. Van Nostrand Reinhold.
  A standard introductory textbook covering the entire field of plant physiology at the level of the 'whole plant'.
Galston, A. W. (1964) *The life of the green plant,* 2nd edn. Prentice-Hall.
  A concise, very readable summary of the nutrition, growth and morphogenesis of the flowering plant by one of the leading workers in the field.

Kumar, H. D. and Singh, H. N. (1980) *Plant metabolism*. Macmillan.
One of the special features of this introductory text is the emphasis it gives to certain metabolic features of tropical plants, such as the Hatch-Slack pathway of carbon dioxide fixation.

Ray, P. M. (1972) *The living plant,* 2nd edn. Holt, Rinehart & Winston.
A brief but outstanding text on the growth and development of the flowering plant.

Steward, F. C. (1964) *Plants at work*. Addison-Wesley.
This brief volume gives a masterly survey of the entire field of plant physiology by one of the outstanding experimentalists of the subject. Although it has not been revised, it should still be useful to all students of biology.

Special topics

Conn, E. E. and Stumpf, P. K. (1976) *Outlines of biochemistry,* 4th edn. John Wiley and Sons, Inc.
An excellent introductory textbook of biochemistry, designed for the student with a limited background of chemistry.

Hill, T. A. (1980) *Endogenous plant growth substances,* 2nd edn. Edward Arnold.
An outline of current knowledge on plant hormones, with special reference to the control of growth and differentiation.

Longman, K. A. and Jenik, J. (1974) *Tropical forest and its environment.* Longman.
An introduction to the ecology of tropical forests, noteworthy for the useful data it contains on the effects of day-length on tropical plants.

Richardson, M. (1975) *Translocation in plants,* 2nd edn. Edward Arnold.
This well-written booklet reviews the old and new techniques used for studying the problems of translocation, and discusses the conclusions so far reached.

Rutter, A. J. (1972) *Transpiration* (Oxford/Carolina Biology Reader, no. 24). Packard Publishing Ltd.
Apart from giving the standard information about transpiration, this slim booklet emphasizes the significance of the process in the productivity of crops and the hydrology of the land surfaces of the earth.

Sutcliffe, J. (1979) *Plants and water,* 2nd edn. Edward Arnold.
A very clear exposition of the subject, which goes beyond an introductory level.

Villiers, T. A (1975) *Dormancy and survival of plants*. Edward Arnold.
An interesting account of just one area in the province of the physiological ecology of plants. Dormancy is here interpreted in terms of physiological responses to seasonal changes in the environment.

Wareing, P. F. and Phillips, I. D. J. (1981) *Growth and differentiation in plants*, 3rd edn. Pergamon Press.
A thorough and readable treatise on the controls of plant development.

Yudkin, M. and Offord, R. (1980) *A guidebook to biochemistry,* 4th edn. Cambrige University Press.
The important features of biochemistry are introduced by discussing some of the key concepts of the subject. Well worth careful study.

Part III

The plant kingdom

Chapter 23

Introduction to the plant kingdom

The beginnings of plant life

Nearly all biologists agree that life originated in the sea and subsequently emerged on to the land. The flowering plants which dominate most of the world's vegetation today show no obvious sign of an aquatic ancestry, but certain features in the reproduction of both flowering plants and their non-flowering ancestors clearly indicate an aquatic origin. Detailed study of fossil and living plants has provided a general outline of the development of plant life after its emergence on to land, but the way in which plant life originated in the sea is more obscure. Nevertheless recent work, particularly on the borderline of genetics and biochemistry, has enabled a tentative account to be given of the beginnings of life.

Early stages of the earth

According to the available evidence the earth originated about 4500–5000 million years ago. It probably started as a whirling globe of hot gases which, due to the high temperature, were present as free atoms. These eventually became sorted out according to mass, so that heavy atoms, such as iron, gravitated towards the centre of the earth, where they are still present to-day. The lightest atoms, such as hydrogen and carbon, formed an outer layer, while atoms of intermediate mass collected as a middle layer. As the earth cooled in the cold conditions of outer space, the temperature became low enough to allow the formation of molecules. Hydrogen, carbon, nitrogen and oxygen were the four most abundant elements in the outer layer. Hydrogen, the most reactive of the four, combines more readily with the other three than these combine with one another, so that three types of molecules must at this stage have appeared in the earth's outer layer.

$$2H_2 + O_2 \rightarrow 2H_2O \quad \text{Water}$$
$$3H_2 + N_2 \rightarrow 2NH_3 \quad \text{Ammonia}$$
$$2H_2 + C \rightarrow CH_4 \quad \text{Methane}$$

Temperatures were still so hot that these three compounds were present in the form of gases.

In time, as the earth continued to cool, some of the gases turned first into liquids and then into solids, so that a crust was gradually formed. As this thickened and cooled it wrinkled and folded to give rise to the first mountain ranges. When the earth's crust became cool enough, water no longer inevitably existed as steam but condensed into rain over long periods of geological time. In this way the rivers and oceans were formed. At first the oceans contained fresh water, but they became increasingly saline as the rivers eroded the mountain sides and dissolved them away. Also dissolved in these oceans were some of the atmospheric ammonia and methane, compounds which remain as gases at temperatures at which water is liquid.

The formation firstly of water, ammonia and methane, and secondly of large bodies of liquid water containing ammonia, methane and many dissolved salts were key events which made the subsequent origin of life possible.

The origin of organic compounds

The element carbon displays a tremendous capacity for linking directly with other carbon atoms, so as to form molecules containing long chains of carbon atoms. Carbon-to-carbon linkages therefore render possible the formation of extremely complex and varied molecules. Carbon compounds are called organic compounds because they scarcely occur in nature today except as the product of living organisms. Organic compounds consist mostly of carbon, hydrogen, oxygen and nitrogen, the same four elements that comprise the molecules of water, ammonia and methane.

It has been shown that organic substances can be formed by non-living processes such as the action of ultra-violet light on mixtures of simple inorganic substances. Therefore under the influence of cosmic ultra-violet light the methane in the oceans of the early earth must have reacted with other methane molecules, with water, with ammonia and many of the other simple compounds present, so that there came into existence a large variety of molecules with shorter or longer chains of linked carbon atoms. In this way it is thought that the basic chemical units that were destined to form living matter were created. These basic units belong to six groups of chemical substances: sugars, glycerol, fatty acids, amino acids, pyrimidines and purines. When these substances become available today, they are immediately absorbed or broken down by bacteria or other micro-organisms, but in the primeval seas, before life had originated, it is likely that such substances would have accumulated. Indeed it is probable that these basic chemical units combined with one another in various ways, resulting in the formation of such complex molecules as polysaccharides, fats, proteins and nucleotides.

Polysaccharides and fats were to become the main sources of chemical energy for living systems, whereas proteins became the main constructional material. The development of proteins was also important because some of them took on the function of catalysing the synthesis of other complex molecules, and became enzymes. The appearance of enzymes, therefore, must have speeded up the whole tempo of chemical changes in the early seas.

The purines and pyrimidines also played a vital part in the ultimate formation of living matter. By combining with sugar and phosphate they gave rise to substances called nucleotides. Hundreds and thousands of different nucleotides then combined with one another to produce the exceedingly complex giant molecules known as nucleic acids. These were the first and, as far as we know, still are the only kind of molecule to be endowed with the capacity to make exact replicas of themselves; that is, they can reproduce. This property of self-duplication, remarkable though it is, is no more or no less than one of the properties inherent in matter made from atoms. The fact that nucleic acids have this property is strictly a consequence of their particular structural make-up and atomic complexity. On account of being able to duplicate themselves, it is not surprising that nucleic acids became the molecules which store the information that is passed on from generation to generation. In other words, they came to house the 'blueprint' of instructions which informs each new individual how to develop and function as a mature adult. In assuming this 'informational' role nucleic acids combined with proteins to form nucleoproteins, which are the largest and most complex molecules known.

The origin of the living cell

All the events which led up to the formation of polymers such as proteins and nucleic acids, and ultimately to nucleoproteins, may be described as chemical evolution. The subsequent organization of such complex substances to form the earliest living organisms may, by contrast, be regarded as biological evolution, because these units came to display the characteristics and properties associated with life. The crucial step which marked the transition from non-life to life was undoubtedly the enclosure of the basic components of living matter within a surface membrane, but it is precisely here that there is least evidence on which to propose hypotheses. Granted, however, that such a crucial step did take place, the next important milestones in the evolution of life were the origin of photosynthesis and of aerobic respiration.

The origin of plants and animals

Most modern organisms are dependent, directly or indirectly, on the two closely linked metabolic processes, photosynthesis and aerobic respiration. As previously discussed (p. 173), organisms which are energetically self-supporting are known as autotrophs (e.g. green plants), whereas those which have no primary source of energy and depend on an outside supply of organic substances are known as heterotrophs (e.g. animals and fungi).

In the world of today, the heterotrophs depend on green plants for their continued existence, and this is the reason why it was previously thought that the earliest forms of life were autotrophic. If, however, the above hypothesis for the origin of life is correct, the first cells must have been heterotrophic, even though the organic substances they used had originated from non-living processes and not from green plants, as in modern heterotrophic organisms.

The first living organisms probably existed in a medium where the chemical polymers, which contemporary organisms have to synthesize for themselves, were freely available. Under such conditions the earliest organisms must have increased very rapidly in numbers, but such increases could not go on indefinitely. As the foodstuffs available in the primeval seas became depleted, some of the primitive organisms began to perform in their own bodies some of the syntheses which up till then had been performed on molecules outside themselves. This development had far-reaching consequences because it led to the evolution of photosynthesis and aerobic respiration. The following five stages are thought to have occurred in this evolution.

The first true organisms probably obtained energy by the partial oxidation of organic substances synthesized in the environment by non-biological means. They evolved into organisms that used gaseous hydrogen for the reduction of atmospheric carbon dioxide to produce carbohydrates which, by partial oxidation, provided them with energy (*anaerobic heterotrophs*). The first primitive type of photosynthetic system arose when anaerobic heterotrophs started using porphyrins, a group of nitrogenous pigments present in their cytoplasm, for trapping light energy, as well as for partially oxidizing hydrocarbons formed chemically (*anaerobic photoheterotrophs*). Anaerobic photoheterotrophy was followed by the evolution of anaerobic organisms whose method of synthesizing organic molecules was obligately dependent on light energy (*anaerobic phototrophs*). The crucial steps leading to the evolution of photosynthesis as practised by plants today were taken when anaerobic phototrophs used light energy to split water so as to obtain hydrogen for the reduction of carbon dioxide to produce carbohydrates (*the photolysis of water*). One far-reaching consequence of using the photolysis of water to synthesize carbohydrates was that elemental oxygen was released into the atmosphere for the first time. The final important stage, which made possible the evolution of plant life as it exists today, occurred when organisms developed the capacity to use atmospheric oxygen for completely oxidizing carbohydrates, thus yielding much more energy than was possible by the earlier fermentation processes (*aerobic phototrophs and aerobic heterotrophs*). Although opinion differs about the time scale involved, most biologists agree that the above sequence of stages leading to the first primitive plants and animals fits the evidence available from chemical fossils (i.e. the presence of organic molecules of presumed biological origin in geological strata, especially of pre-Cambrian age) and from geochemical data on the conditions present during rock formation. When primitive organisms had acquired the ability to photosynthesize and respire aerobically, they had reached a stage of evolutionary development beyond which some very simple plants alive today have apparently not progressed.

The origin of sexual reproduction

In the previous section an account was given of how life could have originated from non-living matter in the physical and chemical conditions which are thought to have prevailed during the early stages of the earth's history. These conditions have not persisted, and it is unlikely that living organisms have arisen from non-living matter since that time. The earliest organisms are envisaged as single cells surrounded by a semi-permeable membrane, and containing a shell of gelatinous material around a central aggregate of nucleoproteins. When such a cell divides, the nucleoprotein molecules of the central aggregate divide so that two nucleoprotein aggregates, each a replica of the original one, are formed. Apparently to facilitate this division into two identical halves, natural selection has favoured the joining together of the nucleoproteins into threads along which they are arranged in a definite sequence. When the cell divides, each thread divides lengthwise into two similar daughter threads, which then separate so that both daughter cells contain an identical set of threads. These threads are the chromosomes found in the nuclei of present-day cells, and their nucleoprotein components are called genes. The process by which each chromosome splits lengthwise and the daughter threads separate into two groups, prior to cell division, is called mitosis.

Although nucleoproteins are stable compounds, small abrupt alterations in their chemical structure can, and do, occur from time to time. Such alterations, known as *gene mutations*, are inherited by the offspring. The gradual changes which occur in living organisms over long periods of time are apparently caused by natural selection acting on populations of individuals which differ among themselves as a result of gene mutations. Natural selection would favour individuals in which there had occurred mutations that were advantageous in the particular environment of the organism.

It is likely that different favourable mutations would occur in different individuals of a species. If two individuals of a primitive species of unicellular organism happened to fuse together, a cell might have resulted which possessed the advantageous characteristics of both, and this would doubtless be favoured by natural selection to a greater extent than either of the original types. It is evident that every time such a fusion took place the number of chromosomes in the resulting cell would be doubled, until ultimately an impossible state of affairs would occur. For fusion to take place repeatedly there must therefore be a corresponding halving of the chromosome number. If the biological advantage of doubling the chromosome number is to be retained, the genes would have to be reshuffled when the chromosome number is halved, because the advantage would disappear if the chromosomes simply separated into the two sets that originally fused.

A process of this kind has in fact been developed by natural selection, and now occurs in almost all groups of living organisms. The process is called *meiosis*, and it probably arose as a modification of mitosis. Meiosis, which can take place only in nuclei containing two sets of chromosomes, consists of two successive nuclear divisions and results in four nuclei, each with only one set of chromosomes. The two sets of chromosomes in the original nucleus are generally derived from two different individuals and consequently they are likely to differ from one another in a number of their genes. As a result of the reshuffling of the genes that takes place during meiosis, the four resulting nuclei will usually each have a different gene constitution, thus increasing the possibility of the appearance of advantageous gene combinations.

Once meiosis had been evolved, fusion of cells could become a regular process, provided that meiosis occurred before the next fusion. Such an alternation of nuclear fusion and meiosis constitutes the essential feature of sexual reproduction. Until about 50 years ago it was usual to regard the fusion of sex cells, or gametes, as the important feature of the sexual process; where the sex cells are all alike, the condition was regarded as a lower grade of sexuality and less beneficial to the species. With the recognition that the primary function of sexual reproduction lies in combining in one individual the favourable attributes of separate individuals, the emphasis was transferred from the fusing cells to the fusing nuclei. The presence of morphologically differentiated sex cells or sex organs (to which the names male and female can be applied) is now seen to be of secondary importance in sexual reproduction. During the early stages of the evolution of sex there were probably no separate sexes. The origin of the difference between the sexes was the result of a divergence whereby one type of fusing cell retained its power of locomotion and became a male gamete, while the other type accumulated food products in its body, grew larger, lost its power of locomotion and became the egg or female gamete. The chief advantage of this differentiation of gametes into two sexes appears to be that the provision of a store of food within the egg confers survival value on the resulting fusion product, or zygote. It is evident that the differences between the sexes were superimposed on the basic nuclear cycle, and it is therefore advisable to draw a clear distinction (as C.D. Darlington advocated as long ago as 1937) between sexual reproduction and sexual differentiation. Whereas sexual reproduction is the regular alternation of nuclear fusion and meiosis in the life cycle, sexual differentiation may be defined as the morphological differentiation of the sex cells into larger female gametes and smaller male gametes. The former is the essential criterion of sexual reproduction, while the latter is not.

There can be no doubt that the rate of evolution was enormously increased when sexual reproduction appeared, because the shuffling of genes between different individuals would give rise to many new gene combinations, some of which would be more highly favoured by selection. The fact that sexual reproduction occurs almost universally throughout the plant and animal kingdoms is very strong circumstantial evidence for its fundamental importance.

In many plants both asexual and sexual reproduction occur side by side, but this does not imply that both methods of reproduction are equally advantageous. Sexual reproduction, with its shuffling of the genes between different individuals, undoubtedly renders a species more readily adaptable to new conditions of life, and hence is extremely important from a long-term point of view. Asexual reproduction, on the other hand, provides little opportunity for genetic variation and hence for adaptation, but it does allow rapid multiplication to occur without the complications of nuclear fusion and meiosis, and it may therefore be advantageous from a short-term point of view. The fact that asexual reproduction often occurs side by side with sexual reproduction is thus readily understandable on the grounds that these two methods of reproduction have different functions to perform.

Life cycles of plants

Since a regular alternation of nuclear fusion and meiosis is the essential feature of sexual reproduction, it follows that the life history of any sexually reproducing organism consists of a cycle of two alternating phases: one which begins with meiosis and ends with nuclear fusion, during which there will be a single set of chromosomes in each nucleus, and the other from nuclear fusion to the subsequent meiosis, during which the nuclei will contain a double set of chromosomes. Cells or nuclei in the former condition are said to be *haploid* and in the latter *diploid*. The essential features of the life cycle of any sexually reproducing organism can be represented as follows:

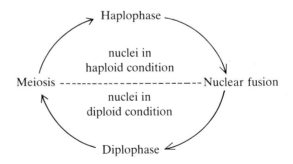

The two haploid cells which undergo nuclear fusion in a sexual cycle are the *gametes*, and the cell or organ in which they are produced is a *gametangium* (literally, gamete box). The terms *isogametes* and *isogametangia* are applied to gametes and gametangia which are morphologically indistinguish-able and, correspondingly, *heterogametes* and *heterogametangia* to gametes and gametangia which are morphologically different. Commonly in hetero-gamy (i.e. sexual reproduction involving heterogametes) the smaller gamete is motile and is called a male gamete, while the larger gamete is sta-tionary and is referred to as the female gamete or egg. The gametangia pro-ducing male gametes are often known as *antheridia*, those producing eggs as either *oogonia* (where the gametangium is unicellular as in algae and certain fungi) or *archegonia* (where the gametangium is surrounded by a jacket of sterile cells). The diploid cell resulting from the fusion of two gametes is called the *zygote*.

The simplest possible life cycle consists of an alternation of a one-celled haplophase with a one-celled diplophase. Such a life cycle can occur only in unicellular organisms (e.g. *Chlamydomonas*), and in multicellular plants there are three basic types of life cycle (Fig. 23.1) depending on whether cell multiplication occurs in one or other or both phases.

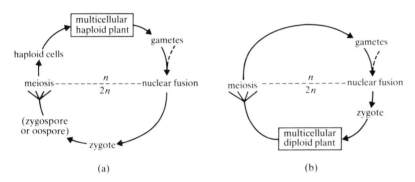

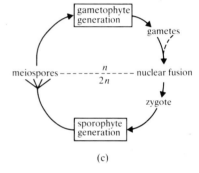

(a)   (b)   (c)

Fig. 23.1   The three basic types of life cycle in plants. (a) Haploid life cycle. (b) Diploid life cycle. (c) Alternation of generations.

1. In the haploid life cycle the haploid cells divide by mitosis and develop into multicellular haploid plants, which eventually produce gametes. The diploid stage is represented only by the zygote, which undergoes meiosis to form four haploid cells that start the cycle again. In haploid life cycles meiosis is frequently delayed until after the zygote has passed through a resting stage. It is common practice to refer to the resting zygote formed from the fusion of isogametes as a *zygospore* and to the resting zygote formed from the fusion of heterogametes as an *oospore*. Haploid life cy-cles occur in certain algae.

2. In the diploid life cycle the diploid zygote divides by mitosis and pro-duces a multicellular plant in which meiosis, when it occurs, gives rise directly to gametes. The only haploid stage in a diploid life cycle is thus the gametes themselves. This diploid life cycle, with nuclear fusion fol-lowing directly after meiosis, is the exact opposite of the haploid type of life cycle. Diploid life cycles occur in certain algae, and are almost uni-versal in the animal kingdom.

3. In the third type of life cycle the diploid zygote divides by mitosis and gives rise, as in the diploid life cycle, to a multicellular diploid plant, but at maturity this produces a sporangium (literally, spore box) containing haploid spores formed by meiosis; such spores may be referred to as *meiospores*. Each meiospore then develops into a haploid phase (a free-living plant or merely a tissue) which eventually forms gametes. The life cycle thus includes two alternating phases, or so-called generations,

which differ in chromosome complement and in type of reproductive cell produced. Since the diploid phase produces spores (meiospores) it is called the *sporophyte generation*, whereas the haploid generation, because it produces gametes, is called the *gametophyte generation*. This type of life cycle is described as an *alternation of generations*, and is found in the majority of plants. There is, however, considerable variation in the relative size and duration of the two generations. In ferns and seed-bearing plants the diploid sporophyte generation predominates, but in mosses and liverworts the converse is true.

Survey of plants and plant-like organisms

The development and extensive use of the electron microscope during the last 30 years has resulted in a greatly increased understanding of the microstructure of a wide range of organisms. This new knowledge has had important implications in our concept of the living world. Formerly, organisms were divided into two groups, the plant and animal kingdoms, because people only had the knowledge to think in terms of the two most obvious forms of life — plants and animals. Any organisms that were not unmistakably animals were regarded as plants. So long as the plant kingdom was regarded in this negative fashion as the 'not-animal' kingdom, it was extremely difficult to make meaningful generalizations about 'plants' because this term covered a very heterogeneous collection of unrelated organisms. Any general statement about plants had to be qualified with all sorts of exceptions.

Modern systems of classifying organisms accept the animal kingdom more or less unchanged but divide the plant kingdom into four distinct groups – viruses, bacteria and blue-green algae, fungi, and true plants. These four groups are probably as different from one another as they are from animals, and the simplest way (which will be followed in this book) of recognizing these fundamental differences is to classify each group as a separate kingdom in its own right. There are many reasons to justify this drastic revision of the plant kingdom (in the broad 'not-animal' sense), but the following are the most important ones.

Viruses stand apart from all other forms of life, both plant and animal, in that they are not organized into cells, and can sometimes be precipitated in the form of pure crystals which have a constant chemical composition. Individual virus particles are too small to be visible with a light microscope, but they can be seen when examined with an electron microscope. A virus crystal may contain millions of such particles. Although viruses can grow and reproduce inside living cells, they cannot do so on their own and it is therefore an open question whether or not they should be regarded as organisms at all. They can be considered as a half-way house between inanimate macro-molecules and true organisms. Viruses will not be discussed in this book except to illustrate certain points of genetics (Chapter 39).

Bacteria and blue-green algae are undoubtedly organisms because they have a cellular structure and can grow and reproduce without the assistance of other living cells. They are either unicellular or colonies of cells which are not joined together by protoplasmic connections. However, their cells do not contain any membrane-bound organelles (e.g. nuclei, mitochondria and plastids), a feature shown by all other organisms. Most biologists now regard the presence or absence of membrane-bound organelles as indicative of a fundamental difference between groups of organisms. Those organisms whose cells lack membrane-bound organelles are called 'prokaryotic' (literally, pre-nucleate), whereas those that have them are termed 'eukaryotic' (literally, with a proper nucleus). On this basis, bacteria and blue-green algae are grouped together as the Prokaryota, while all other true organisms are grouped as the Eukaryota. If bacteria and blue-green algae are given the status of a kingdom, they are said to belong to the kingdom *Monera*.

The case for excluding the fungi from the plant kingdom is based primarily on the fact that their cells lack plastids, and in this respect resemble animals more closely than plants. Since there are only three main types of membrane-bound organelles (nuclei, mitochondria and plastids) present in eukaryotic cells, the presence or absence of plastids is considered to reflect an important bifurcation in the course of evolution. Because fungi and animals lack plastids (which house the photosynthetic pigments when these are present) they are unable to photosynthesize and therefore depend for their energy on an external supply of organic compounds. In other words, fungi and animals are heterotrophs whereas green plants, which are capable of photosynthesis, are autotrophs. The fact that fungi and animals both lack plastids does not imply that the two groups are closely related, and any attempt to argue a close relationship must recognize that fungi resemble plants more closely than animals in every respect except their nutrition. Traditionally fungi have always been classified as plants because they have cell walls, but the chemical composition of fungal cell walls is nevertheless quite different from that of the walls of plant cells with plastids. Also, fungi never accumulate starch which is the typical carbohydrate reserve of organisms with plastids. From such evidence as this, it is difficult to avoid the conclusion that fungi are neither animals nor plants, and merit classification as a distinct kingdom.

Having placed viruses, bacteria and blue-green algae, and fungi in three separate kingdoms, we are left with the eukaryotic organisms which have plastids. These comprise the various groups of green plants (including algae other than blue-green algae), which collectively form the plant kingdom as it is now understood. The plant kingdom, defined in this narrow sense as eukaryotic organisms with plastids, becomes a relatively homogeneous group which can be meaningfully compared with the animal kingdom, composed of eukaryotic organisms without plastids or cell walls.

Having settled for a five-kingdom system of classifying organisms, it must not be concluded that there are always hard and fast dividing lines between them. It must not be forgotten that all organisms alive today have evolved from one or a few ancestral types that arose after the earth cooled down some 3500 million years ago. As time passed some of the original organisms became extinct, others diverged as they evolved along independent lines, and others survived more or less unchanged. It is not surprising, therefore, that any attempt to pigeonhole organisms into man-made categories (i.e. to classify them) is likely to be complicated by the existence of intermediate forms. There is, in fact, no absolute dividing line between the plant and animal kingdoms. While most eukaryotes which are not fungi can readily be allocated to either the plant or the animal kingdom on the basis of whether or not they have plastids, there are many unicellular organisms which can function as either autotrophs or as heterotrophs, according to the conditions prevailing in their environment. There are also unicellular organisms which are regularly heterotrophic but nevertheless resemble in detailed structure and life cycle other organisms that have plastids. Such organisms are often studied by both botanists and zoologists. The existence of these 'plant/ animals' is understandable in terms of the occurrence of evolution. It is meaningless to try and classify them as either plants or animals, because they do not fit logically into these man-made categories. It is common practice to adopt a compromise by using the term Protista to cover all unicellular plants and animals. Some biologists consider that the Protista should be treated as yet another kingdom, but such a separation will not be made here because the protistans with which we shall be concerned can be classified unmistakably as unicellular plants. Figure 23.2 summarizes the criteria on which the five-kingdom system of classification is based.

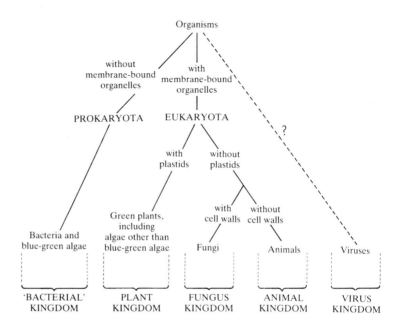

Fig. 23.2 The classification of organisms into five kingdoms.

Classification of the plant kingdom

During the last 20 to 30 years several new classifications of the plant kingdom have been proposed. All of them are prompted by a desire to emphasize some particular aspect of plant structure or evolution or both. Practically every botanist has his own ideas about how plants should be classified, but there never can be a 'best' way because the most appropriate classification depends on the viewpoint from which plants are being studied. A fossil botanist, for example, is likely to favour a different classification from one preferred by a chemotaxonomist. It is, however, essential to have some system of classifying the plant kingdom in order to have a framework on which to hang information about the levels of organization and possible relationships of different types of plants. The classification shown in Table 23.1 is a simple scheme which incorporates modern ideas about the phylogeny of plants but at the same time retains the traditional status of certain groupings. This compromise scheme is adopted because, in the author's view, some of the traditional groupings (e.g. the seed plants) draw attention to certain important concepts (e.g. the seed habit) more readily than the groupings of some of the modern classifications.

Table 23.1 The plant kingdom – an outline arrangement of the main groups

Level of organization	Division	Subdivision	Class	Common name
	(CYANOPHYTA – belong to the kingdom Monera)			Blue-green algae
	EUGLENOPHYTA			Euglenoids
	CHLOROPHYTA			Green algae
Algae	CHRYSOPHYTA			Diatoms
	PHAEOPHYTA			Brown algae
	RHODOPHYTA			Red algae
Bryophytes	BRYOPHYTA		Hepaticae	Liverworts
			Musci	Mosses
	TRACHEOPHYTA	LYCOPSIDA		Club mosses
Pteridophytes		SPHENOPSIDA		Horsetails
		PTEROPSIDA (in its restricted sense)		Ferns
Seed plants		SPERMOPSIDA	Gymnospermae	Cycads and conifers
			Angiospermae	Flowering plants

Of the divisions listed in Table 23.1 those which are collectively grouped as algae are essentially simple aquatic plants, whereas the two divisions Bryophyta and Tracheophyta are, by comparison, complex land plants. The Tracheophyta, as the name implies, are characterized by the possession of the specialized conducting tissues, xylem and phloem, for the transport of water and soluble foods around the plant body. The sporophyte is always the dominant generation and the one in which the vascular tissues are developed. The term Tracheophyta was introduced to emphasize the phylogenetic as well as the physiological importance of the vascular system. The Tracheophyta are also often called the *vascular plants*.

Whereas the various divisions of the algae have little in common with one another beyond the fact that they are at about the same level of structural organization, it is an interesting fact that the bryophytes and tracheophytes have two fundamental characters in common. This would suggest that these two groups, although apparently so diverse, are descended from a common ancestral stock which possessed these characters. The two characters possessed by both bryophytes and tracheophytes are as follows:

1. The possession of a type of female sex organ called an *archegonium* (see Fig. 25.7d). An archegonium is a flask-shaped structure in which the oosphere or egg cell occupies the main body of the flask, the *venter*, which is surrounded by a wall one-cell thick. The tubular *neck* of the archegonium, also one-cell thick, surrounds a central column of neck-canal cells which break down at maturity to provide a liquid medium through which the male gametes (antherozoids) can pass.

2. The possession of wind-dispersed meiospores. Both bryophytes and tracheophytes have an alternation of generations in which the haploid meiospores produced by the sporophyte generation are dispersed by wind. It is possible that this feature was important in the successful colonization of the land by plants, because the development of numerous minute spores adapted for dispersal by the wind would have been an immense advantage to the primary colonists.

Chapter 24

Algae

Although the word 'algae' is no longer accepted as the name of a formal taxonomic group within the plant kingdom, it is still widely used as a convenient term to include all the simpler photosynthetic plants. There is ample justification for studying the algal groups together because, despite their diversity, they have certain important characters in common. Their reproductive structures are nearly always unicellular but, if multicellular, every component cell forms a reproductive unit (either zoospore or gamete) and there is no outer layer of sterile cells. Another distinguishing feature is that algae never produce embryos, i.e. the zygotes never develop into multicellular young plants while still enclosed within the female sex organ. By contrast, plants above the level of algae produce multicellular sex organs with a sterile jacket of cells in which multicellular embryo plants develop.

In addition to these technical distinctions, however, there are more obvious features which are common to this heterogeneous assemblage of about 20 000 known species.

General characteristics of algae

Habitats

Most algae are aquatic, living in either the sea or fresh water, although some forms are able to exist in the films of water in soil, on tree bark and similar habitats. Indeed, algae will grow in almost any place where there is enough moisture and enough light for photosynthesis. One of the most extreme habitats colonized by algae is the tissue of animals. Some sea anemones and corals, for instance, regularly contain algal cells among their tissues, and both partners gain some advantage from the symbiotic association.

Most species of algae are anchored to a substratum, but many of the very minute forms float freely in water where, together with other free-floating forms of life, they comprise the plankton. Such algae are important as *primary producers*, forming the ultimate source of food for most aquatic animals. This fact is expressed by the saying, 'All flesh is grass, and all fish is diatom' – diatoms being a particularly abundant group of algae in the plankton.

Range of form

In all algae the plant body is called a *thallus*, which is the term used to describe any plant body not differentiated into roots, stems and leaves.

Although they lack these organs, algae range in size from microscopic unicellular plants to highly complex multicellular seaweeds that sometimes
reach 70 m in length. In so wide a range of forms it is convenient to distinguish several levels of organization.

Unicells   Unicellular algae are found in all the algal divisions of the plant
kingdom, except the Phaeophyta (brown algae). They are considered to be
the basic cell types from which, in the course of evolution, algae with larger
and more complex bodies developed. There are motile and non-motile
types of unicell, and both represent starting points for the evolution of more
elaborate types.

Colonies   Colonial algae are formed when the daughter cells resulting from
the division of a mother cell remain together. The association may be loose
and easily broken without detriment to the component cells or, at the other
extreme, it may be highly organized with a specific number of cells per colony, the individual cells sometimes being interconnected by thin protoplasmic strands. Colonies may be motile or non-motile according to the type of
unicellular ancestor from which they are derived.

Filaments   Filamentous algae develop from single cells in which division
occurs entirely, or almost entirely, in one dimension and is not followed by
separation of the daughter cells. If the division is strictly in one dimension
the result is an unbranched filament, but if some of the cells develop lateral
bulges which continue to grow and divide the result is a branching filament.
A more specialized type of thallus, called *heterotrichous*, is produced when
branching filaments form both a prostrate attachment system and an upright
photosynthetic system.

Coenocytes (siphonaceous thalli)   The coenocytic habit is achieved by repeated nuclear division without the formation of cross-walls. This type is
most highly developed in tropical species.

Parenchymatous thalli   This type of thallus is formed when the cells of a
filament divide in three planes. The simplest type is a flat plate two cells
thick, while the most complex types, found in some of the brown seaweeds,
have a thallus with many similarities to the body of a flowering plant.

Basic structure of algal cells

Although there are variations in structure between the unicells of the separate algal divisions, there are many general similarities. A description of
these common features will serve as a useful introduction to the cellular
structure of algae in general.

Cell wall   The walls of most algae consist primarily of cellulose, but those
of many algae also contain a gelatinous component which is also a polysaccharide. This component apppears to have different functions in different
algae; in colonial types it helps to hold the cells together, whereas in algae
which have colonized habitats liable to seasonal drying-out it probably helps
to retard desiccation.

Chloroplasts   The possession of one to many chloroplasts is a conspicuous feature of algal cells. Chloroplasts vary widely in shape (from minute lens-shaped bodies to large cup-shaped structures) and are not always green, the colour of their chlorophyll component being masked by the presence of other pigments. The chloroplasts of algae, with a double membrane on the outside and a system of internal thylakoids, are essentially similar in structure to those of flowering plants.

The chloroplasts of many algae contain one or more specialized proteinaceous bodies called *pyrenoids*. Pyrenoids in algae belonging to the division Chlorophyta are associated with the synthesis of starch which, in these algae, is formed within the matrix of the chloroplast. Pyrenoids are also common in the chloroplasts of algae belonging to other divisions but their function is unknown, especially since it has been established that in these algae the polysaccharide and other food reserves are synthesized outside the chloroplasts.

Flagella   Since algae are essentially an aquatic group it is not surprising that most of them are either motile organisms or else have motile reproductive stages. Motility is due to the activity of one or more whip-like organs called flagella which are anteriorly or laterally inserted. Each flagellum arises from a basal granule within the cytoplasm and extends through a pore in the cell wall to the outside.

Studies with the electron microscope have shown that the flagellum is a complex structure with two central fibrils surrounded by a ring of nine double fibrils (Fig. 24.1a). This so-called nine plus two flagellar structure is found throughout the plant kingdom. The electron microscope has also revealed that there are two types of flagella, whiplash and tinsel. Tinsel flagella have lateral hair-like appendages, whereas whiplash flagella are devoid of them (Fig. 24.1b).

The motile cells of many algae are capable of reacting to variations in light intensity. Depending on the intensity of the light source they may swim towards or away from it (positive or negative phototaxis respectively). In order to react, however, cells must absorb light and most flagellate cells possess an eyespot, which is a modified part of the chloroplast containing a red or orange carotenoid pigment. The eyespot is involved in the phototactic responses of algae, but the mechanism of its involvement has been clearly shown in only one organism, namely *Euglena*.

Contractile vacuoles and other organelles   The flagellate cells of most freshwater algae, but only a few marine algae, also contain one or more contractile vacuoles which function as osmotic regulators. Since the concentration of dissolved substances in fresh water is less than that within the cell, water tends to enter the cell by osmosis, but the excess is removed by the activity of these vacuoles.

Mitochondria and nuclei are present in the cells of all algae except the Cyanophyta, but need no special mention because they are structurally and functionally similar to those of other plants.

Reproduction

Algae may reproduce asexually or sexually. Vegetative reproduction, which includes any process whereby portions of the plant body become separated

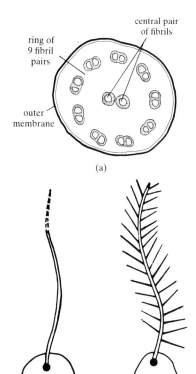

Fig. 24.1   Structure of flagella. (a) T.S. of an algal flagellum, showing the characteristic 9+2 fibrillar organization. (b) The whiplash (left) and tinsel (right) types of flagella.

off and produce new individuals without the formation of special reproductive cells, is treated here as a type of asexual reproduction. In unicellular algae like *Euglena* this can happen by binary fission, which is a simple division of the whole organism into two equal parts which then grow and form new individuals. Filamentous algae can reproduce vegetatively by becoming fragmented, either through chance breakage or by separation at specific points of weakness between cells, into two or more pieces which grow into independent plants. Asexual reproduction, apart from vegetative methods, involves the reorganization of the contents of certain cells to produce one or more spores, which are then liberated from the parental cell. Although there are many different kinds of asexually produced spores, motile zoospores are the most common in algae. Zoospores are flagellate, self-propelling spores which frequently have chloroplasts and an eyespot. Liberation of zoospores may be by the rupture of the surrounding wall, or through a pore in the wall.

Sexual reproduction involves the fusion of the contents of two cells (gametes), produced within ordinary vegetative cells or within cells specially modified as gametangia. Where the two gametes which fuse are motile and equal in size they are called isogametes, or + and − gametes. Where both gametes are motile but one is smaller and more active than the other, the gametes are called anisogametes. Since anisogamy represents an early stage in the evolution of sexual differentiation, the smaller gamete of the two is described as the male gamete, and the larger as the female. Further sexual differentiation leads to oogamy which is the fusion of a small motile male gamete with a larger, non-motile female gamete, the egg or oosphere (hence the name oogamy). These gametes are often produced in highly specialized reproductive structures. Sexual fusion (whether isogamous, anisogamous or oogamous) usually takes place in the liquid medium in which the alga is growing.

Classification

The algae are a very diverse group and represent several divisions or phyla which all have a similar range of forms, suggesting that the evolution of elaborate types from simple ones has occurred independently within each separate division. Nowadays it is customary to classify the algae into seven or eight divisions which are considered to represent parallel evolutionary lines. One of these divisions, the Cyanophyta, should strictly speaking be excluded from the plant kingdom because its members, unlike those of the other divisions, are prokaryotic in their organization. Another division, the Euglenophyta, is on the borderline between the plant and animal kingdoms but in this book will be considered as belonging to the plant kingdom, even though it is often combined with the animal phylum Protozoa to form the kingdom Protista. Of the other divisions of algae, the three main ones (Chlorophyta, Phaeophyta and Rhodophyta) are definitely members of the plant kingdom, and all include multicellular types.

The algal divisions are separated primarily on (1) the chemical nature of their photosynthetic pigments, including both chlorophylls and accessory pigments, (2) the chemical composition of their food reserves, and (3) the number and type of flagella on any motile cells in their life cycle (Table 24.1). Additional characters such as the chemistry and structure of the cell wall, types of reproduction, and morphology of reproductive organs are also used in classification, especially of individual divisions.

Table 24.1 Characteristic features of the four algal divisions discussed in this book

Division	Common name	Chief photosynthetic pigments	Food reserves	Flagella
CYANOPHYTA	Blue-green algae	Chlorophyll a Phycocyanin Phycoerythrin	Cyanophycean starch	None
EUGLENOPHYTA	Euglenoids	Chlorophylls a and b	Paramylon	One or more tinsel
CHLOROPHYTA	Green algae	Chlorophylls a and b	Starch	Two (or more) whiplash
PHAEOPHYTA	Brown algae	Chlorophylls a and c Fucoxanthin	Laminarin	One whiplash and one tinsel

All the divisions of algae are photosynthetic and contain chlorophyll a and b-carotene, but each division also has its own characteristic combination of accessory pigments. These include other chlorophylls, carotenoid pigments such as xanthophylls, and phycobilin pigments, which are structurally similar to the bile pigments of animals. Indeed, the common names of some of the algal divisions are based on the colour of the predominant accessory pigments which mask the green colour of the chlorophylls.

Most of the algal divisions also have their own particular carbohydrate reserve. The Chlorophyta, like the Bryophyta and Tracheophyta, store their carbohydrates as starch, whereas in the Phaeophyta laminarin, another polymer of glucose, takes the place of starch. Paramylon and floridean starch are the distinctive carbohydrate reserves of the Euglenophyta and Rhodophyta respectively.

The general characteristics of the algal divisions discussed in this book are as follows:

Cyanophyta   In the Cyanophyta (blue-green algae) two accessory pigments, a blue phycocyanin and a red phycoerythrin, mask the green of chlorophyll and result in various colours, including blue-green from which the division gets its common name. The 'plants' are small and have a prokaryotic organization. The food reserve is cyanophycean starch. There are no flagellate stages, and sexual reproduction is unknown.

Euglenophyta   In the Euglenophyta the distinctive green colour due to chlorophylls a and b is not masked by other pigments. All the species are microscopic and most inhabit fresh water. Food reserves include a carbohydrate, paramylon.

Chlorophyta   In the Chlorophyta (green algae), like the Euglenophyta, the distinctive green colour of chlorophylls a and b is not masked by other pigments. The plants vary from minute unicelluar forms to plants of moderate size. The food reserve is starch.

Phaeophyta   In the Phaeophyta (brown algae) an additional brown pigment, fucoxanthin, masks the green of chlorophylls a and c. The plants are nearly all marine, and some attain enormous size. The chief food reserve is a carbohydrate, laminarin.

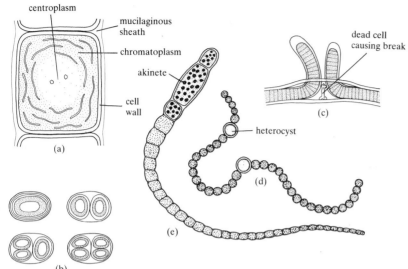

Fig. 24.2 Blue-green algae. (a) Hypothetical cell of a filamentous blue-green alga, showing prokaryotic organization. (b) Single cell and two-, three- and four-celled colonies of *Gloeocapsa*. (c) False branching in *Scytonema*. (d) Filament of *Nostoc* showing heterocysts. (e) Filament of *Rivularia* with akinetes.

Cyanophyta – the blue-green algae

Distribution

Blue-green algae are inconspicuous organisms but are very widely distributed. Although blue-green algae are a common component of the phytoplankton in both salt and fresh water where, under certain conditions, they may become so abundant as to discolour the water to form what are called 'blooms', they are most frequently seen on damp soil and on surfaces like plant pots, concrete walls or rock faces which are constantly moist. A feature involved in the distribution of blue-green algae is that they are able to withstand extremely adverse conditions. Certain species are found in hot springs, others can live in areas where the daily temperature range is from −60 °C to +15 °C. Blue-green algae also grow in full sunlight, in complete darkness, and in water with as much as 27 per cent salt content. Some live in symbiotic association with both plants and animals.

Range of form

Blue-green algae are essentially unicellular and reproduce by binary fission (Fig. 24.2b). They may occur as individual cells or, more usually, as loosely organized colonies. Each cell is typically surrounded by a mucilaginous sheath, and colonies arise when daughter cells remain held together in a jelly-like mass. Such colonies may be irregular masses of a few to many cells, or may have characteristic shapes ranging from filaments, through flat plates and hollow spheres, to cubes.

The most common type of body is the filament, the cells of which share a common outer wall. Increase in the number of cells in a filament is achieved by circular in growth of wall material from the outer wall. Filaments may be unbranched (e.g. *Nostoc* and *Oscillatoria*), or apparently branched as a result of a break between two cells which is followed by the growth of one or both of the broken ends through the parent sheath (e.g. *Scytonema* Fig. 24.2c). In some species this 'false branching' occurs adjacent to special cells which develop at intervals along the filament. Such cells, which are called *heterocysts* (Fig. 24.2d), differ from normal vegetative cells in having transparent contents and thick walls with a characteristic thickening at each end of the cell. True branching, due to the lateral division of certain cells in a filament, is shown by only a few genera.

Structure of the cell

All blue-green algae can be recognized by their cellular structure which is prokaryotic and therefore quite different from that of other algae (Fig. 24.2a). When a single cell is examined under the high power of a light microscope, all that can be seen is a wall enclosing a mass of granular protoplasm, which is blue-green in some species but may vary from greyish through yellow, green and blue to red. The wall is quite unlike that of a typical plant cell, but shows certain similarities to that of many bacteria. It is covered by an extracellular cell sheath, which is a rigid structure in some species but forms an extensive gelatinous envelope in others. The wall itself has a complex structure, and can only with difficulty be distinguished from the plasmalemma.

The protoplasm of the cell is not divided into cytoplasm and nucleus, nor are there any recognizable organelles. Under the electron microscope, however, it is possible to recognize two regions. The outer one is called the chromatoplasm and contains numerous flattened photosynthetic membranes calles chromatophores, while the inner region is called the centroplasm. The protoplasm itself appears to have the consistency of a stiff jelly. Protoplasmic streaming has never been observed, and there is no fluid-filled central vacuole such as occurs in most plant cells. Scattered small vacuoles are sometimes present, but these are filled with gas – a rare phenomenon which is found elsewhere only in certain bacteria and protozoa.

Although there are no recognizable chloroplasts in the cell, the essentials of the photosynthetic machinery are present in the chromatophores which can be considered structurally and functionally equivalent to the thylakoids of a chloroplast. Chlorophyll molecules are a component of both the chloroplast thylakoids and the chromatophore. Only one kind of chlorophyll is present (chlorophyll a), but other pigments involved in photosynthesis are also present. Two of these are structurally similar to the bile pigments of animals and are hence called phycobilins. One of them is blue (phycocyanin) and the other is red (phycoerythrin). These two pigments, which absorb light in regions of the spectrum different from chlorophyll molecules, can apparently transfer this energy to chlorophyll a for use in photosynthesis. The product of photosynthesis is stored as a glycogen-like polysaccharide which is best called cyanophycean starch, until its chemistry is better understood.

Although there is no recognizable nucleus, DNA material is present in the centroplasm but is not organized into chromosomes. In the absence of chromosomes, mitosis does not occur, and the way in which the DNA material is divided between two daughter cells is not known.

Nutrition

All blue-green algae contain chlorophyll and can photosynthesize, but some are also able to exist indefinitely using ready-made organic compounds. This is the essential feature of *heterotrophic* nutrition characteristic of fungi and animals. It is thought that the blue-greens which are found in caves or under stones embedded in mud can live there only because of their ability to adopt a heterotrophic mode of life.

Blue-green algae are also unique among green plants in their nitrogen metabolism, because some of them can 'fix' atmospheric nitrogen in a manner similar to that found in certain bacteria. This capacity is thought to be correlated with the presence of heterocysts because most of the blue-greens which have so far been shown to fix nitrogen possess heterocysts. Blue-green algae are abundant in some tropical soils where they play a role comparable to that of the nitrogen–fixers on the nodules of leguminous plants. In the rice fields of India and Pakistan, for example, rice has been grown continuously for years without the need to add nitrogenous fertilizers. The reason is that the waterlogging of the rice paddies during the rainy season induces a luxuriant growth of *Nostoc* and this replenishes the nitrogen which has been removed by the growth of the rice crop.

Reproduction

Blue-green algae reproduce vegetatively by cell division which, in practice, means binary fission in unicellular forms and fragmentation in filamentous forms. The development of heterocysts in the filaments of some blue-green algae creates points of weakness and facilitates the fragmentation of the filament into shorter lengths. The death of single cells at intervals along the length of a filament achieves the same result. A few blue-green algae (e.g. *Anabaena* and *Rivularia*) form asexual spores called *akinetes* (Fig. 24.2e). These are thick-walled, non-motile cells containing copious food reserves, which tide the alga over conditions unfavourable to the vegetative cells. There is no sexual reproduction in the life cycle.

Euglenophyta

General characteristics

The euglenoids are a small group of aquatic organisms which combine the characteristics of plants and animals to such an extent that it is difficult to classify them as either plants or animals. Their major features are:

1. They are unicellular organisms with eukaryotic cellular structure.
2. The cell is not surrounded by a cellulose wall but by a proteinaceous *pellicle*, occurring within the plasmalemma. In most euglenoids the pellicle is pliable and allows the cell to change its shape, but in some species it is rigid and the cell has a fixed shape.
3. The anterior end of the euglenoid cell is invaginated to form a canal which is dilated at its inner end into a more or less spherical cavity, the reservoir. The canal and reservoir, although suggestive of a channel by which solid food particles might pass into the cell, are probably not involved in the ingestion of solid food by colourless euglenoids.
4. Some euglenoids are photosynthetic, others are not. The pigmented members have chloroplasts containing chlorophylls a and b. The photosynthate is stored as paramylon, a glucose polymer, which occurs as granules in the cytoplasm.
5. Euglenoids basically have two flagella of the tinsel type, with the lateral hair-like projections occurring in a single row along the length of the flagellum.

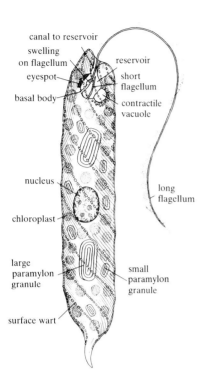

canal to reservoir
swelling
on flagellum
eyespot
basal body
reservoir
short
flagellum
contractile
vacuole
nucleus
long
flagellum
chloroplast
large
paramylon
granule
small
paramylon
granule
surface wart

Fig. 24.3 Euglena spirogyra, a species with two very large paramylon granules in the cytoplasm. (From Vickerman and Cox 1967, p. 15)

6. Sexual reproduction probably does not occur or, if it does, is extremely rare.

As a group euglenoids favour fresh-water habitats, particularly small bodies of water rich in organic nitrogenous compounds. *Euglena*, the type genus with probably more than 150 different species, is selected as an example of this division.

Euglena

Euglena is common in stagnant water and, if conditions for growth are particularly favourable, it may become so abundant as to colour the water green. Like all euglenoids, *Euglena* is unicellular and surrounded by a pellicle. Most species have a long cylindrical body, blunt at the front end but somewhat pointed behind (Fig. 24.3). The smallest species are less than 10 μm long but the largest are about 50 times longer, that is about 0.5 mm long, and therefore visible to the unaided eye. Two flagella arise at the base of the reservoir, one is long and emerges from the canal as the locomotory flagellum, the other is short and confined to the reservoir. This short flagellum is often in contact with the shaft of the long flagellum, and this was responsible for the earlier, erroneous belief that there was a single flagellum which forked at its base. Alongside the reservoir is a plate-like organelle, the eyespot, which is red in colour because it contains carotenoid pigments. This is used in conjunction with a light-receptive swelling which is situated on the locomotory flagellum at a point opposite the eyespot. When *Euglena* is swimming freely, the eyespot periodically masks the photoreceptor, but movement of the flagellum re-orientates the organism until the photoreceptor is continuously exposed. This results in a phototactic response to unidirectional light which is characteristic of *Euglena* and other motile photosynthetic organisms.

Another organelle associated with the reservoir is the contractile vacuole which periodically empties its contents into the reservoir by disruption of the intervening membrane. The contractile vacuole is filled from several smaller accessory vacuoles which radiate from it.

The cell of *Euglena* contains one or more chloroplasts, the shape of which varies according to the species and thus provides a useful diagnostic feature. The electron microscope has shown that the chloroplasts are much simpler in their internal structure than those of higher plants in that their thylakoids are not divided into grana and intergranal lamellae. *Euglena* contains chlorophylls a and b and, apart from a need for vitamin B_{12}, the organism is autotrophic in sunlight. However, in the dark it has the power to become heterotrophic. If kept in the dark for two weeks the chloroplasts bleach, but normally the organism recovers its green colour when replaced in the light. While living as a heterotroph, it almost certainly absorbs its food as organic molecules in the medium, and no solid food is taken into the reservoir. The chief food reserve is paramylon.

Euglena reproduces asexually by binary fission. Following division of the nucleus by an unusual method of mitosis, the whole cell divides by splitting longitudinally from the anterior end. There is no reliable evidence that sexual reproduction takes place.

Chlorophyta – the green algae

The Chlorophyta, or green algae, have been selected for more detailed study than any other algal division because their photosynthetic pigments (chlorophylls a and b, etc.), main food reserve (starch), cell-wall composition (cellulose), and ultrastructure have so much in common with those of higher plants. It is among the green algae, therefore, that evolutionary trends leading to increased specialization of body form and mode of reproduction can be profitably sought. Because of its common occurrence, *Chlamydomonas* is used as the standard example to illustrate the unicellular state from which more elaborate types must have evolved.

Chlamydomonas

Chlamydomonas is a microscopic unicellular green alga which is very common in stagnant fresh water (e.g. in pools, rain-water tanks, etc.), and, like *Euglena,* may be one of the reasons why stagnant water sometimes turns green. A single cell is ovoid or sometimes pear-shaped in outline, and surrounded by a thin cellulose wall (Fig. 24.4). From the anterior end of the cell project two equal flagella which are about the same length as the cell. The movement of the flagella in locomotion is similar to the movement of a person's arms when swimming the breast stroke. The greater part of the inside of the cell is occupied by a thick cup-shaped chloroplast which is pressed against the wall and gives the organism its overall green colour. Embedded in the thickened basal part of the chloroplast is a conspicuous organelle, the pyrenoid, which might easily be mistaken for the nucleus except for its position. The front end of the cell and the cavity of the chloroplast are filled with colourless cytoplasm. The cytoplasm inside the cavity of the chloroplast contains the nucleus, mitochondria and other organelles. Just below the points of attachment of the flagella there are two contractile vacuoles, which alternate with one another in showing a slow expansion followed by a sudden contraction. These vacuoles doubtless serve to get rid of excess water and soluble waste products. Inside the chloroplast, usually near the rim, is a red eyespot, which is sensitive to light.

    Chlamydomonas has both asexual and sexual means of reproduction. In asexual reproduction the motile cell comes to rest, the flagella are with-

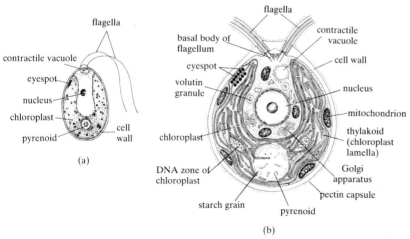

Fig. 24.4   Chlamydomonas. (a) Vegetative cell as seen under the high power of a light microscope. (b) Section of a vegetative cell to show the detailed structure revealed by the electron microscope. (b from Vickerman and Cox 1967, p. 20)

drawn, and the protoplast divides into two, four or eight daughter cells (Fig. 24.5). Each of these cells develops into a biflagellate zoospore which resembles the parent cell but is initially smaller. The zoospores swim away when the wall of the parent cell breaks down.

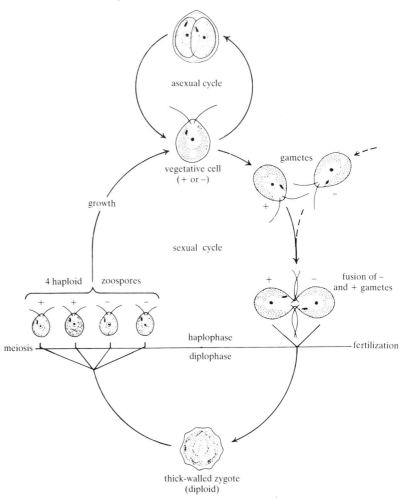

Fig. 24.5   Life cycle of *Chlamydomonas*.

Sexual reproduction takes place when two individuals of opposite mating types come into contact with one another. The protoplast of each individual divides into 16, 32 or even 64 daughter cells which function as gametes when they are released from the parental cell. The gametes of the two fusing individuals are at first clumped together in a single mass, but they break away from it in pairs. The gametes of a pair then fuse at their flagellar ends to form zygotes. The wall of the zygote becomes thickened, and the zygote is converted into a zygospore. After a resting period, the contents of the zygospore divide by meiosis to form four haploid zoospores, which are freed by the splitting of the zygospore wall. The zygote is the only diploid stage in the life cycle.

In most species of *Chlamydomonas* the gametes look exactly alike, and are therefore described as isogamous. They are, however, physiologically different, and fusion will only take place between gametes of opposite mating types. Because it is impossible to refer to one gamete as male and the other as female, the two mating types are called + and −. Mating type is genetically determined. Since the zygote is formed by the fusion of + and – gametes, its nucleus inevitably contains the alleles for both + and – mating type. These alleles are segregated at meiosis so that, of the four zoospores formed, two are + and two are −. The condition in which gametes from the same individual or strain of individuals cannot fuse is described as *heterothallic*.

Wherease in most species of *Chlamydomonas* the fusing gametes are similar in size (isogamous), in some they differ in size according to the number of divisions which took place before they were formed (anisogamous). In these anisogamous species it is possible to regard the gametes as sexually differentiated, and to call the smaller of the two motile gametes the male gamete. In at least one species of *Chlamydomonas* (*C. coccifera*) sexual differentiation of the gametes is carried to the limit where one entire individual functions as a non-motile female gamete, which is fertilized by a smaller motile male gamete produced by mitotic divisions of the protoplast of another individual. This type of sexual reproduction is called oogamy, because a gamete which loses its motility is referred to as an oosphere or egg.

Colonial green algae and the origin of the 'soma'

A unicellular, free-swimming, green alga such as *Chlamydomonas* may be considered to represent the most primitive type of plant organization from which the plant kingdom has evolved. To bridge the gap between the unicellular condition and the more elaborate types of organization found in multicellular plants, it is necessary to postulate certain evolutionary changes. The precise manner in which such changes took place must always remain in doubt, but it is possible to visualize how they may have happened by comparing present-day forms which appear to represent different stages in the evolution of such characters as complexity of bodily structure, division of labour between vegetative and reproductive functions, and differentiation of sex. By arranging selected living forms in sequence, certain trends can be recognized which are probably similar to those which took place in the past.

The concept of the 'soma' The body of any unicellular organism not only feeds and grows, but also reproduces after reaching a certain size. Reproduction in a unicellular organism is essentially a continuation of the growth process because the protoplasm of which the organism is composed continues to increase in bulk, even though it can only increase beyond a certain limiting size by dividing to form two or more new individuals. So long as conditions are favourable, growth and reproduction continue indefinitely, and therefore unicellular forms are potentially immortal in the sense that they do not die of 'old age'. In the majority of higher plants and animals, by contrast, the functions of feeding and growing are separated from the function of reproduction. In these organisms there is a definite mortal body or 'soma' which dies of old age, whether the environmental conditions continue to be favourable or not, and the only cells that are potentially immortal are the reproductive cells because these alone have the capacity to grow into new individuals of the species.

The soma in both plants and animals originates when certain cells are set apart solely for nutrition and growth (somatic cells), leaving the rest to be entirely concerned with reproduction (germ cells). The distinction between somatic and germ cells is much more marked in higher animals than in higher plants, because at least some of the somatic cells of most plants have retained the power to reproduce the species by various methods of so-called vegetative reproduction. This difference is undoubtedly related to the fact that plant cells are on the whole far less highly specialized for particular functions than animal cells.

The Gonium–Volvox series

One series of flagellate, colonial green algae, beginning with *Gonium* and ending with *Volvox*, illustrates very strikingly the way in which the soma may have originated. The simplest members of the series have no soma, but they are connected by transitional forms with the most complex member, *Volvox*, which has a very well-developed soma. The same series of algae also illustrates the differentiation of sex between the gametes.

All members of this series consist of cells which are held together in a common layer of mucilage to form a colony consisting of a definite number of cells arranged in a definite pattern. Such a colony, which is called a *coenobium*, behaves as a single organism, moving through the water by the coordinated beating of the flagella of all the cells.

Gonium   This genus represents a very simple type of colonial aggregation. *Gonium pectorale*, a typical species, consists of sixteen cells embedded in a flat plate of mucilage (Fig. 24.6a, b). Each cell has the same essential structure as a single *Chlamydomonas* cell, and the flagella protrude outwards through the mucilage.

Gonium reproduces asexually and sexually. In asexual reproduction each cell of the coenobium simultaneously divides into sixteen daughter cells, each group constituting a new coenobium. The sixteen young coenobia escape as a result of the disintegration of the parent colony, and then grow to full size. In sexual reproduction all the cells of a colony participate in the formation of gametes. Each cell divides to form a coenobium-like mass of sixteen gametes, which separate from one another shortly after leaving the parent colony. Two gametes, usually from separate colonies, fuse to form a zygote which immediately surrounds itself with a thick membrane and becomes a zygospore. Zygospore germination, during which meiosis occurs, results in the formation of a four-celled colony, each cell of which subsequently gives rise by asexual reproduction to a new sixteen-celled colony.

Pandorina   Thus is a colony of sixteen chlamydomonad (i.e. *Chlamydomonas*-like) cells arranged in the form of a sphere. The cells are pressed tightly together so that each cell is cone-shaped with the point facing inwards (Fig. 24.6c). As in all coenobia, the cells are held together by a common mucilaginous envelope, from which the two flagella of each cell project. In spite of its spherical shape, the coenobium has a definite polarity which is manifested by the fact that the same side of the sphere is always directed forwards during movement.

Asexual reproduction is similar to that in *Gonium*, each cell of the colony dividing to form a daughter coenobium. Sexual reproduction is also similar except that either sixteen larger or thirty-two smaller gametes are produced.

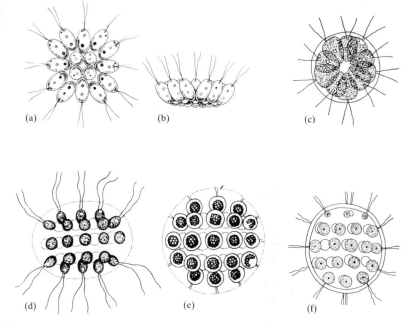

Fig. 24.6 Motile colonial green algae. (a) *Gonium pectorale*, surface view of colony. (b) *Gonium pectorale*, side view of colony. (c) *Pandorina* sp., mature colony. (d) *Eudorina elegans*, a spherical colony of 32 cells. (e) *Eudorina elegans*, showing formation of daughter colonies. (f) *Eudorina illinoiensis*, with four smaller cells at one end forming a soma. (a and b from Delevoryas 1966, p. 26; d and e from Brook, A. J. (1964), *The living plant*, p. 358, Edinburgh University Press)

Fusion occurs without reference to size, so that the fusing gametes are sometimes isogametes and sometimes anisogametes. The zygote gives rise to a new sixteen-celled coenobium by cell division.

Eudorina   The coenobia of *Eudorina* are spherical or ellipsoidal, and contain basically thirty-two chlamydomonad cells arranged in a single layer on the surface of a hollow sphere (Fig. 24.6d). Like all hollow spherical colonies, *Eudorina* begins development as a flat plate which curves round to form a sphere, leaving a small pore where the closure is not quite complete. The polarity of the coenobia is emphasized in the ellipsoidal types by the fact that the cells frequently lie in regular rows at right angles to the long axis of the ellipsoid.

The reproduction of *Eudorina* is noteworthy because different species show a progressive specialization in the cells of the colony. In *asexual reproduction* individual cells of the coenobium divide to form daughter coenobia but, whereas in *Eudorina elegans* all thirty-two cells are capable of division, in *Eudorina illinoiensis* four cells at the anterior end, which are slightly smaller that the rest, do not divide. These four cells form a soma or mortal body, because they die whenever the other twenty-eight cells reproduce (Fig. 24.6e, f).

In *sexual reproduction* the same tendency for certain cells to remain purely vegetative is shown, and there is also a well-marked differentiation into male and female gametes. The male and female gametes may be formed either in the same (monoecious) or in different (dioecious) colonies according to the species. In the former case the male gametes develop at the anterior end of the colony. The formation of the male gametes involves division of a vegetative cell into sixty-four slender flagellate cells which swim as a unit until they reach a female gamete when they break apart as separate male gametes. The female gametes are similar to vegetative cells except that

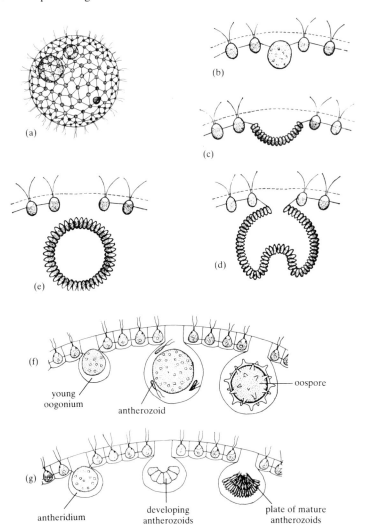

Fig. 24.7 Volvox. (a) Mature colony with three daughter colonies at various stages of development. (b)–(e) Stages in asexual reproduction. (f) Sexual reproduction showing a developing egg, fertilization, and an oospore. (g) Sexual reproduction, stages in the formation of the male gametes or antherozoids. (b–e from Lowson 1973, p. 502; f and g from Smith 1955, vol. I, p. 41)

they are larger and function, without division, as reproductive cells. A male gamete fuses with a female gamete to form a thick-walled zygote which, on germination, releases a motile zoospore. This, after swimming for a time, comes to rest and divides to give a new colony.

Volvox   This, the last member of this series of motile coenobia, consists of a very large number of chlamydomonad cells arranged in a single layer around the periphery of a fluid-filled sphere (Fig. 24.7a). The actual number of cells varies according to the species, from as few as 500 to as many as 60 000. Each colony may grow to about 1 mm in diameter, which probably represents the upper limit of mechanical stability for this type of colonial construction. In some species the individual cells are connected with each other by prominent protoplasmic strands, in others by fine protoplasmic threads.

Nearly all the cells of a *Volvox* colony are exclusively vegetative, and only a few in the rear half of the coenobium are capable of giving rise to daughter colonies or gametes. *Volvox* thus has a very well-developed soma, as do all higher plants and animals. As a colony approaches maturity, the reproductive cells become very much larger than the vegetative cells, and lose their flagella.

In *asexual reproduction* (Fig. 24.7b–e) each reproductive cell divides repeatedly to produce a new colony, in a manner analogous to the formation of daughter colonies by *Pandorina* and *Eudorina*. The daughter colonies become quite large and protrude into the cavity of the parent colony. Eventually they break loose and swim about inside the parent colony until the latter dies and disintegrates. This may be delayed so long that the daughter colonies may themselves contain granddaughter colonies before being liberated. In *Volvox africana* as many as four generations may be formed in one original parent colony.

In *sexual reproduction* (Fig. 24.7f, g) there are marked differences of size, activity and structure between the male and female gametes. Colonies are monoecious or dioecious according to the species. The female reproductive cells, the *oogonia*, do not divide and remain in protoplasmic contact with the neighbouring somatic cells, from which they derive nourishment. On account of their large size, the oogonia project into the interior of the colony. Each oogonium produces a single female gamete which is large and inactive, and densely filled with food reserves. Because the female gamete is non-motile it is called an egg. Fertilization of the egg occurs within the oogonium. The male reproductive cells, the *antheridia*, divide repeatedly to form a flat plate containing numerous slender, biflagellate antherozoids, very similar in appearance to the male gametes of *Eudorina*. The plate of antherozoids is set free in the interior of the parent colony, where it either breaks up into separate antherozoids or continues to swim as a unit until it approaches an egg. If the colony is monoecious, the antherozoids are attracted chemotactically to the oogonia where they effect fertilization. In dioecious species the antherozoids have to escape from the interior of a male colony via the pore in the surface layer, before they can fertilize the eggs of a female colony. The antherozoids are chemotactically attracted to a female colony, which they are thought to enter through the surface pore and then fertilize the eggs from the inside.

After fertilization the zygote forms a thick-walled resting spore which is called an oospore instead of a zygospore, because the female gamete is an egg. The oospore is set free by the break-up of the parent coenobium, and eventually germinates to liberate one functional zoospore which is produced by meiosis. The zoospore divides and gives rise to a spherical colony consisting of only a few cells. This small colony reproduces asexually, forming a somewhat larger colony, and the process is repeated until the mature size characteristic of the species is reached.

Filamentous and parenchymatous green algae

Ulothrix

Ulothrix is a genus of filamentous green algae which grow on rocks or stones in streams and fresh-water lakes, although there are a few marine species. A single plant of *Ulothrix* consists of an unbranched filament of short cylindrical cells which are wider than they are long (Fig. 24.8a). All the cells of a filament are similar in shape and capable of division and reproduction, except for the basal cell which attaches the plant to its substratum. This cell,

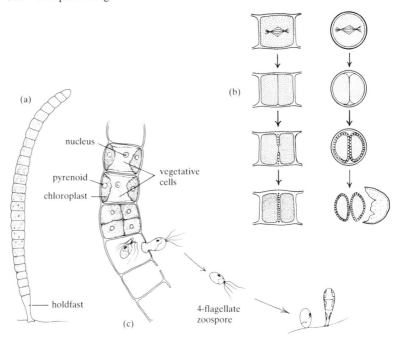

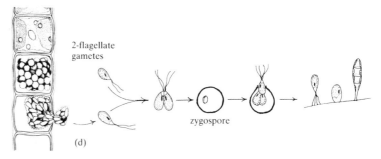

Fig. 24.8   Ulothrix. (a) Single filament
attached to a substratum. (b) Method of wall
formation in *Ulothrix* and a unicellular alga.
New wall material is shown by dotted lines. (c)
Stages in asexual reproduction. (d) Stages in
sexual reproduction. (a and c from
Alexopoulos and Bold 1967, p. 37)

known as the holdfast, is longer and narrower than the others and its lower
surface, conforming to irregularities on the surface of the substratum, anchors
the entire filament. Each vegetative cell possesses a characteristically
shaped chloroplast which forms an incomplete ring just inside the cellulose
cell wall (Fig. 24.8c). Within the chloroplast there is a central vacuole in
which a single nucleus is suspended by fine strands of protoplasm.

The filament increases in length by cell division. The method of cell division is different from that found in any alga studied so far, but very similar
to what is found in higher plants. When a *Ulothrix* cell divides, the protoplast divides into two daughter protoplasts in the usual way but, whereas in
Chlamydomonas (for example) new walls are formed around each daughter
protoplast and the wall of the mother cell eventually disintegrates, in
Ulothrix the daughter protoplasts become separated from each other by the
formation of a cross-wall which grows in from the persistent wall of the
mother cell. The two daughter cells are therefore enclosed partly by a new
partition wall and partly by the persistent wall of the parental cell

(Fig. 24.8b). In *Ulothrix* the partition walls are always formed in the same plane, and this results in the formation of a filamentous plant body.

Asexual reproduction i.e. multiplication of filaments as opposed to the increase in their length, takes place in two ways:
1. Fragmentation This is very common at times of active growth.
2. Zoospore production This can occur in all the cells of a filament except the basal cell.

When zoospore formation begins the vegetative cells become darker green and their contents begin to shrink away from the cell wall. Each protoplast then divides one or more times, so that eventually each cell contains anything from two to sixteen (to thirty-two) tiny protoplasts, the number being constant for any given filament (Fig. 24.8c). These protoplasts develop into zoospores, each with a chloroplast, pyrenoid, eyespot, nucleus, two contractile vacuoles and four flagella. Apart from being four-flagellate the zoospores are very similar to *Chlamydomonas* cells. The zoospores escape through a pore which develops in the lateral wall of the cell in which they are formed, and, after swimming for a time, ultimately settle down, flagellar end first, on a suitable surface. Each zoospore then resorbs or loses its flagella, elongates and then divides repeatedly to give a new filament. This sequence of development, starting with a motile unicell and ending in an unbranched filament, perhaps indicates one method by which filamentous organisms could have evolved.

Ulothrix reproduces sexually by means of biflagellate gametes (Fig. 24.8d). These are produced in the same way as zoospores, but division of the protoplast continues until there are thirty-two or sixty-four daughter protoplasts per cell. The gametes are essentially similar to the zoospores but they are biflagellate and thus even more like *Chlamydomonas* than the zoospores. They are apparently incapable of germinating into new filaments, which is not surprising in view of their small size. As with zoospores, the gametes are released through a pore in the wall of the parental cell into the surrounding water, where they fuse in pairs. All species of *Ulothrix* are physiologically heterothallic and isogamous. The diploid zygote continues to swim about for a short time but then settles down, and turns into a zygospore by secreting a thick wall around itself. Upon the advent of favourable conditions, meiosis occurs and four haploid four-flagellate zoospores emerge. Each of these zoospores, like the asexually produced zoospores, gives rise to a new vegetative filament. Thus *Ulothrix*, although a non-motile filament, reproduces by motile zoospores and gametes resembling *Chlamydomonas* cells, suggesting an ancestry from *Chlamydomonas*-like organisms.

Ulva

Ulva, or sea lettuce as it is sometimes called, is a marine alga that often grows in profusion on rocky shores between tide marks. It is particularly abundant where fresh water runs into the sea, especially when the fresh water has a high nitrogen content. The thallus (Fig. 24.9a) is a thin, crinkled, slimy sheet, up to 30 cm across and bright green in colour. It is attached to the rocky substratum by a multicellular holdfast.

The leaf-like thallus of *Ulva* is two cells thick, and thus represents the simplest possible type of parenchymatous thallus. In surface view the cells are seen to have very thick walls (Fig. 24.9b), and this may be associated with the severe desiccation to which the plants are subjected at low tide.

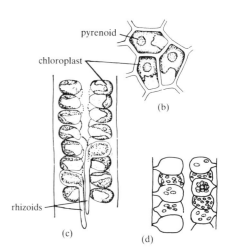

Fig. 24.9   Ulva. (a) Habit. (b) Surface view of vegetative cells of thallus. (c) V.S. basal portion of thallus showing formation of rhizoids. (d) V.S. thallus with cells producing zoospores. (a and d from Bell and Woodcock 1971, p. 35; b and c from Brook, A. J. (1964), *The living plant*, p. 363, Edinburgh University Press)

Each cell has a single nucleus and chloroplast. The holdfast is formed from hundreds of tubular filaments which grow down between the two sheets of vegetative cells towards the base of the thallus (Fig. 24.9c).

Ulva belongs to the same order (Ulotrichales) as *Ulothrix* and its methods of reproduction are similar to those of *Ulothrix*. *Ulva* reproduces asexually by four-flagellate zoospores which are formed in ordinary vegetative cells near the margin of the thallus. The protoplasts of cells functioning as zoosporangia divide by meiosis, followed by mitosis, into usually thirty-two uninucleate protoplasts which metamorphose into four-flagellate zoospores. These escape through a pore which develops on the surface of the zoosporangium (Fig. 24.9d). When the zoospores come to rest on a suitable substratum, the flagella are withdrawn and the cell proceeds to divide. At first the divisions of the thallus are all in the transverse plane so that a uniseriate filament results, as in *Ulothrix*, but subsequent divisions give rise to the sheet-like thallus of the adult plant.

Sexual reproduction is similar to asexual reproduction, but the motile cells are biflagellate and function as gametes. *Ulva* is heterothallic so that only gametes derived from thalli of opposite mating types will fuse to form zygotes. The four-flagellate zygote, after swimming for a time, settles down, withdraws its flagella and secretes a cell wall. Germination follows almost immediately to form first a filament and then a two-layered thallus.

It is an observed fact that a given plant of *Ulva* reproduces either sexually or asexually, or in other words gives rise to either zoospores or gametes but never to both. The reason is that, although all plants look alike, some are haploid giving rise to gametes, while others are diploid giving rise to zoospores. There is therefore an alternation of generations in the life cycle and, since *Ulva* is heterothallic, there are three morphologically similar but cytologically dissimilar types of thallus (Fig. 24.10). This type of life cycle in which the haploid and diploid generations are similar in form is called an isomorphic alternation of generations.

Coenocytic green algae

This series includes the green algae whose thalli are either completely coenocytic or semi-coenocytic, i.e. their multinucleate thalli are either without cross-walls or with cross-walls which separate the thallus into multi-

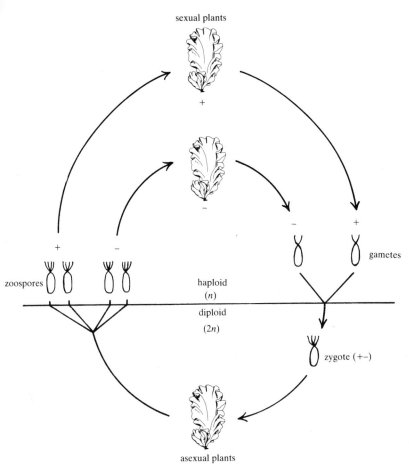

Fig. 24.10 Life cycle of *Ulva*, showing isomorphic alternation of generations.

nucleate portions. The completely coenocytic construction is best illustrated by the *siphonaceous algae*, the adjective siphonaceous (from the Greek word siphon, meaning a tube) referring to the non-cellular tubular structure of their thallus. The semi-coenocytic condition is shown by *Cladophora*, which will be described first because its construction forms an interesting link between filamentous forms like *Ulothrix* and the siphonaceous algae.

Cladophora

Cladophora is a 'filamentous' alga found in fresh and salt water, growing attached to rocks, vegetation or other submerged objects. The plant body is heterotrichous (i.e. consists of two systems of branching filaments), having an erect system which is bright green and divided into elongate units, and a prostrate system of colourless filaments divided into cubical units. Although the units look like cells they are in fact multinucleate coenocytes because the formation of cross-walls bears no relation to nuclear division.

The wall of *Cladophora* is thick and stratified into three layers, the inner and middle of which are composed of cellulose but differ in that the microfibrils run almost at right angles to each other. The outer layer contains chitin, and this gives the alga a characteristic 'crisp' feel when touched. Within the wall of each unit of the erect portion of the thallus there is a complex reticulate chloroplast with numerous embedded pyrenoids

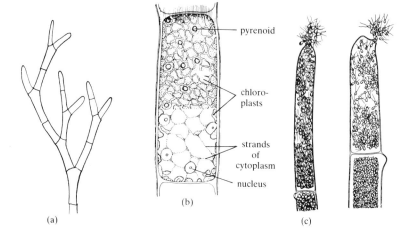

Fig. 24.11 Cladophora. (a) Part of plant,
showing pattern of branching. (b) Cell
structure, as seen in surface view (top half)
and median optical section (bottom half). (c)
Zoosporangia with four-flagellate zoospores.
(d) Gametangia with two-flagellate gametes.
(b from Bold, H. C. (1973), *Morphology of
plants*, 3rd edn, p. 44, Harper & Row; c and d
from Brown 1935, p. 458)

(Fig. 24.11b top half). The central region is filled with a highly vacuolate
cytoplasm in which numerous nuclei are suspended (Fig. 24.11b bottom
half). In surface view under the microscope it is difficult to distinguish between nuclei and pyrenoids.

The branching of the thallus occurs in a very characteristic manner. At
the anterior end of a cell, just below the septum, a small protrusion is
formed which continues to elongate and eventually becomes cut off by a
cross-wall. The development of a branch displaces the original septum, so
that where a branch occurs there is an inverted V-shaped septum, and this
provides a useful diagnostic feature of the genus (Fig. 24.11a).

Increase in length of the filament is noteworthy in that it is more or less
restricted to the apices. The terminal units of every branch are filled with
dense contents, and increase in length until they eventually divide into two
segments by the formation of a cross-wall. The subterminal segment then
enlarges to the normal size but undergoes no further division, although it
may subsequently form a branch by the method already described.

Asexual and sexual reproduction are superficially similar because in both
cases the coenocytic units of the erect portion of the thallus divide into
numerous motile reproductive cells. If these are four-flagellate they function as asexual zoospores, whereas if they are two-flagellate they function as
sexual gametes (Fig. 24.11c, d). The life-cycle pattern varies according to
the species. In some it is strictly comparable with that in *Ulva*, having an
isomorphic alternation of generations, whereas in others the plants are all
diploid and produce both zoospores and gametes.

Siphonaceous algae

One of the simplest types of siphonaceous algae is *Protosiphon*, a genus of
small algae which grow on the surface of soil. The body of *Protosiphon* is
divided into a globose or ovoid green portion, about 1 mm in diameter,
growing above the ground, and a narrow, colourless rhizoidal tube which
penetrates the soil (Fig. 24.12a). The aerial, green part contains one large
chloroplast with many pyrenoids, and throughout the cell are scattered
several nuclei. There are several methods by which *Protosiphon* reproduces,
and the one used depends on the amount of available moisture. When the
soil on which the alga grows is flooded, sexual reproduction by biflagellate
isogametes takes place. Unfertilized gametes can function asexually as zoo-

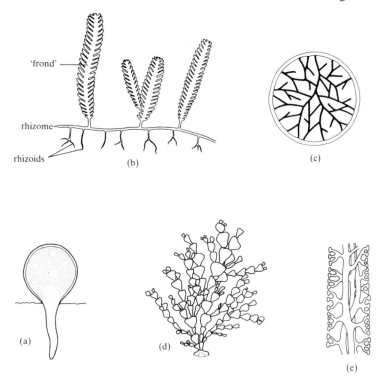

'frond'

rhizome

rhizoids

(b)

(c)

(a)

(d)

(e)

Fig. 24.12 Siphonaceous green algae. (a) *Protosiphon*. (b) *Caulerpa*, portion of plant. (c) T.S. of rhizome, showing cross-supports. (d) *Halimeda*, habit sketch. (e) Section through one of the flattened lobes showing central strands with branches forming photosynthetic layer at surface.

spores. There are also asexual methods by which the alga can withstand the drying out of the soil.

Caulerpa This is a far larger and more complex siphonaceous alga than is *Protosiphon*. It grows in tropical seas and combines a highly differentiated external form with an internal structure of extreme simplicity. It consists of a creeping and branching 'stem' portion from which arise erect leaf-like or tubular branches of various shapes, and the whole is anchored by root-like structures which adhere to, or penetrate, the substrate (Fig. 24.12b). The upright portion of some species may be as much as 10–30 cm high. The form of the thallus is maintained by turgor, but this is reinforced by a system of ingrowths which traverse the cavity of the coenocyte (Fig. 24.12c). Although *Caulerpa* has an apparently complex structure it would be unable to grow as large as it does if it were not buoyed up by a watery medium.

In addition to its superficial similarity, the resemblance of *Caulerpa* to a higher plant is also shown at the microscopic level. Although internally there are no boundaries in the form of cross-walls, the nuclei are nevertheless concentrated at the growing apices, being more widely dispersed in the older regions behind the growing points. Exactly the same arrangement of the nuclei would have been encountered if the growing apex had been organized on a cellular basis like the apical meristem of a higher plant. The coenocytic structure of *Caulerpa* thus shows several parallels in its vegetative organization to the cellular body of a higher plant.

Reproduction is largely vegetative, by the fragmentation of the thallus. Sexual reproduction is by means of biflagellate anisogametes formed in the 'leafy' area of the plant.

Halimeda   This is a very complex coenocyte, which grows on coral reefs. Its external appearance resembles that of a small prickly pear (Fig. 24.12d). This highly complex plant body is constructed of branching filaments which are intricately woven together. The central region of the thallus is occupied by a system of large, longitudinal filaments. These give rise to numerous small branchlets which are massed together at the surface (Fig. 24.12e). The chlorophyll of the plant is concentrated in these superficial branches, the larger, more central filaments being used for food storage. The whole plant is covered with a deposit of lime, which gives it added rigidity.

The siphonaceous mode of construction precludes the formation of a large thallus, because it lacks the rigidity which is given by internal cross-walls. In addition, the absence of cellular construction makes specialization of parts more difficult, and true division of labour impossible. The siphonaceous habit thus represents an evolutionary 'dead end'.

Phaeophyta – the brown algae

The Phaeophyta, or brown algae, are formally characterized by the three characters listed in Table 24.1, namely: (1) the possession of a brown pigment, fucoxanthin, which masks the green of their photosynthetic pigments, chlorophylls a and c; (2) the photosynthate is stored as laminarin, named after the genus *Laminaria*; and (3) the possession by the biflagellate male gametes, the only motile bodies in the life cycle, of one whiplash and one tinsel flagellum, both emerging from the side of the cell.

The group is almost exclusively marine, and its members range from simple branching filaments to complex seaweeds, such as *Sargassum*, whose complexity of organization exceeds that of other algae and, indeed, that of many land plants. Unicellular and colonial types are unknown. A brief account of *Sargassum* will be given to illustrate the parenchymatous type of thallus shown by the brown seaweeds, which represent the most complex structural organization reached by any algae.

Sargassum is a genus of brown seaweeds which grows on rocky coastlines around the world mainly within the tropical zone. A vegetative plant of *Sargassum* (Fig. 24.13), which represents the diploid generation, has many features in common with land plants in that it has a stem and leaf-like structures, which are actually flattened branches, arising from the stem. This high degree of external differentiation is matched by an equally high degree of internal differentiation. In transverse sections of the thallus two distinct tissues can be recognized, a central *medulla* composed of elongate colourless cells, and a peripheral *cortex* of more or less isodiametric cells of which the more superficial ones contain the chloroplasts. *Sargassum* is attached to the surface of a rock by a specialized organ, known as the *holdfast*, which is morphologically the expanded base of the stem. The rest of the thallus also shows a marked division of labour between the various parts. Some of the branches in the axils of the 'leaves' form special structures called air-bladders, which contain gas and have the function of enabling the plant to be buoyed up when it is covered by the tide. Other branches are fertile and bear the sexual reproductive organs which, in surface view, look like little pits. In most species the male and female sex organs are both borne on the same branch, and both shed their gametes into the sea where fertilization takes place. The female gamete is large and non-motile, and the male is small and motile. The zygote germinates directly into a new plant. The life

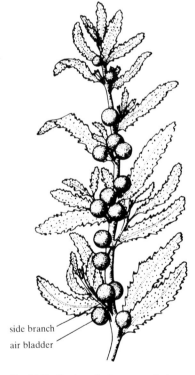

side branch
air bladder

Fig. 24.13 Portion of a *Sargassum* plant. (From Stone and Cozens 1969, p. 110)

cycle of *Sargassum*, in which the haploid phase is represented only by the gametes, is thus an example of a diploid life cycle.

Sargassum gives its name to the Sargasso Sea, where several species of *Sargassum* grow in great sheets extending over thousands of square kilometres, floating by means of their air-bladders. *Sargassum* is the only genus of seaweeds that includes members which habitually thrive adrift on the surface of the sea. The species of *Sargassum* in the Sargasso Sea are sterile, but maintain themselves vegetatively by increasing in length in front and dying off behind. How the plants first came to be in this region of the north Atlantic Ocean is unknown, but it is possible that plants growing along the coasts of the Guineas and the West Indies sometimes drift out to sea and are carried north-eastwards by the Gulf Stream to the Sargasso Sea. The Sargasso Sea is the biggest stretch of uniform vegetation in the world, and has probably existed for millions of years. Living on and in the floating mats of *Sargassum* is a whole range of animals (shrimps, crabs, molluscs and small fish) which feed on the weed and each other as well as on the encrusting hydroids and polyzoa that grow on the weed.

Although the brown algae have a more complex bodily organization than any other algae, they are not the stock which successfully colonized the dry land. The reasons for this conclusion are that the combination of their photosynthetic pigments, the nature of their carbohydrate reserve and the flagellation of their reproductive stages are all different from those found in any of the groups of land plants.

Chapter 25

Bryophytes

The bryophytes (Bryophyta) are a well-defined division of land plants with no close relationships to the rest of the plant kingdom. The group includes the mosses (Musci) and the liverworts (Hepaticae). Most bryophytes are small in size, the smallest scarcely visible without the aid of a hand lens and the largest seldom more than 50 cm in height or length. They are common on trees, rocks, logs and soil in every part of the world and in almost every habitat except the sea. They grow best under moist conditions, and are particularly abundant in tropical forests and in the damp woodlands of temperate regions. Despite their preference for moist habitats, bryophytes are nevertheless essentially land plants and those which grow in fresh water are adapted secondarily to an aquatic life. This is shown by the fact that aquatic bryophytes have retained many features typical of land plants, including cutinized spores dispersed by wind.

All mosses and most liverworts (the leafy liverworts) have a stem which bears leaves, but in some liverworts (the thalloid liverworts) the plant body is a dorsiventrally flattened thallus without leafy appendages. No bryophytes have true roots, although they do have hair-like outgrowths called rhizoids which anchor the plant to the substratum and may absorb water and dissolved substances. Internal differentiation of tissues is simple by comparison with the tissue systems of higher land-plants. Some bryophytes possess a simple central conducting system in the stem, but lignified conducting elements (tracheids and vessels) and true vascular tissues (xylem and phloem) are completely absent. In fact, most bryophytes rely to a considerable extent on capillary conduction of water in the narrow spaces between overlapping leaves. This poor development of internal conducting tissues may be one reason why bryophytes grow best in moist habitats.

Although bryophytes can always be recognized by their structure, they are also distinguished from all other land plants by their life cycle. The life cycle of bryophytes, like that of most plants, involves an alternation of generations between a sexual or gametophyte generation which reproduces sexually (and sometimes vegetatively as well), and an asexual or sporophyte generation which reproduces by spores. The cells of the gametophyte have the haploid (n) number of chromosomes, those of the sporophyte the diploid (2n) number. The characteristic feature of bryophytes is that the gametophyte lives longer, is physically larger and more conspicuous, is nutritionally independent, and in general represents the familiar moss or liverwort 'plant'. By contrast, the sporophyte is a spore capsule, usually stalked, which bears no leaves or branches and remains permanently attached to the gametophyte, on which it is to a large extent nutritionally dependent. The alternation of a dominant, independent gametophyte with an attached, de-

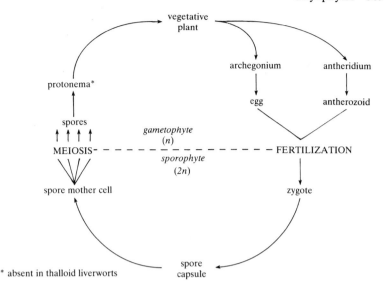

* absent in thalloid liverworts

Fig. 25.1 The life cycle of bryophytes.

pendent sporophyte is such a constant feature of bryophytes that it can be singled out as the one truly diagnostic character of the group (Fig. 25.1).

Bryophytes have probably existed on the earth for a very long time, but they are rarely preserved as fossils except in deposits of comparatively recent date. This may be because the plants are small and their structure (particularly the absence of lignified tissues) does not lend itself to preservation. The evolutionary origin of bryophytes is unknown, but there are nevertheless good reasons for believing that the mosses and liverworts alive today are survivors of an aquatic group which colonized the land at some very remote geological period.

Despite their small size and relatively simple structure, bryophytes are undoubtedly a successful group of plants as indicated by the vast number (at least 20 000) of recognized species. However, because they have never developed an efficient conducting system, they have not been able to achieve the large size necessary for becoming the dominant vegetation on land. Thus bryophytes, although they have successfully colonized numerous niches on land, appear to have been a blind alley in plant evolution.

Liverworts

Liverworts differ fundamentally from mosses in the structure of the sporophyte generation, but they also differ in the bigger and more conspicuous gametophyte generation. All mosses are leafy, but liverworts are of two types, leafy and thalloid. Although superficially resembling mosses, leafy liverworts can always be distinguished from them in having leaves with no midrib; they are also almost always prostrate in habit, whereas most mosses are erect. Leafy liverworts are the more common of the two types of liverworts, but the thalloid liverworts are the more conspicuous. The thalloid types have a ribbon-like leafless thallus, which creeps along the surface on which it is growing. The word 'liverwort' refers to a supposed resemblance between the lobes of thalloid liverworts and those of the human liver. Because thalloid liverworts are easier to recognize and more suitable for examination than leafy liverworts, a thalloid type is selected to illustrate the group.

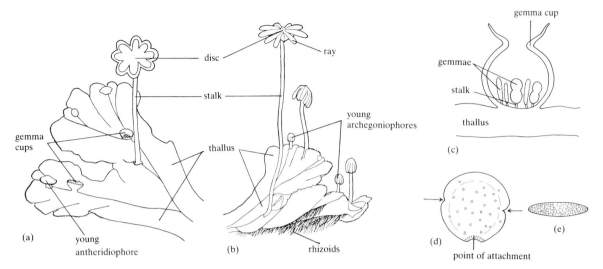

Fig. 25.2 The gametophytes of *Marchantia polymorpha*. (a) Terminal portion of a male gametophyte showing antheridiophores and gemma cups. (b) Terminal portion of a female gametophyte showing archegoniophores. (c) V.S. gemma cup showing attachment of gemmae to thallus. (d) Detached gemma in surface view. Meristematic cells occur in the notches (indicated by arrows) on each side. (e) T.S. gemma taken in plane shown in diagram (d). (a and b from Holman, R. M. and Robbins, W. W. (1933), *Elements of Botany*, p. 296, John Wiley & Sons).

Marchantia

Liverworts belonging to the genus *Marchantia* (named in honour of Marchant, a French botanist) are common in tropical countries, where they can be found 'in the hills' on moist shaded banks of streams, roadside embankments, and similar places. Since the plants usually grow together in large numbers, they may cover considerable areas of ground. *Marchantia* can be easily recognized by its curious, umbrella-shaped, reproductive structures (Fig. 25.2) which are produced in great profusion at certain times of the year.

In the vegetative state (i.e. when it is not producing any reproductive structures) a *Marchantia* plant consists of a green, ribbon-like thallus about 1–1.5 cm wide and several centimetres long. Except at the tip the central part of the thallus is thickened on the under side to form a distinct midrib, which is visible on the upper side as a darkly pigmented line. The thallus grows along the surface of the soil and repeatedly branches in a dichotomous fashion, i.e. it forks repeatedly into two equally developed parts. At the tip of each branch is a small depression in which lies a group of meristematic cells. It is by the division and subsequent differentiation of the cells derived from these apical initials that the tissues of the wings and midrib of the mature thallus are formed. From time to time the apical group of meristematic cells divides into two separate groups, each of which continues to grow independently. This is the reason for the characteristic forking of the thallus.

Structure of the thallus

The thallus is many cells thick and is stratified into three distinct regions, an upper photosynthetic layer, a middle storage layer, and a lower absorptive and anchoring layer (Fig. 25.3a). When the upper surface of the thallus is examined under a hand lens, it is seen to be divided by a regular network of faint lines into numerous, small, polygonal areas. In the centre of each such area is a dot representing a small pore which leads into an air chamber beneath the epidermis (Fig. 25.3b, c). The lines seen on the surface corres-

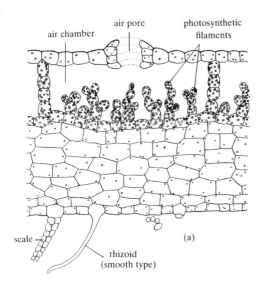

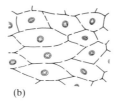

(b)

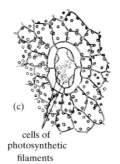

(c)

scale

(a)

rhizoid
(smooth type)

cells of
photosynthetic
filaments

Fig. 25.3 Vegetative structure of *Marchantia polymorpha*. (a) V.S. thallus, showing colourless tissue (below) and photosynthetic filaments projecting into air chamber (above). (b) Surface view of thallus to show 'hexagonal' pattern of air chambers. (c) Surface view of single air pore. (a from Coulter and Dittmer 1964, p. 251)

pond in position with vertical walls that separate neighbouring air chambers, so that each polygonal area marks the limits of an underlying air chamber. A vertical section through the upper part of the thallus shows that the pores are in fact short tubes which pierce the epidermis like minute ventilating shafts; each tube looks like a barrel without top or bottom, and consists of several superimposed tiers of concentric rings of cells. Arising from the floor of each air chamber are numerous, irregularly branching, chains of cells which are densely packed with chloroplasts, and collectively form the main phtosynthetic tissue of the plant.

The tissue below the air chambers consists of closely packed cells with few or no chloroplasts. Most of the cells are thin-walled storage cells of various types (e.g. starch-containing cells and oil-containing cells), but here and there may be found thick-walled cells which probably have a mechanical function.

The ventral surface of the thallus bears rhizoids and scales. The rhizoids, each of which is a single thread-like cell, emerge mainly from the midrib region and may penetrate the soil to a depth of over 2 cm. They are of two types, smooth rhizoids which have evenly thickened walls, and tuberculate rhizoids which have frequent peg-like thickenings of the wall projecting into the cell cavity. The two types of rhizoids may have different functions but, if so, the difference is not understood. The other outgrowths of the lower surface, the ventral scales, are less conspicuous than the rhizoids and consist of delicate, overlapping plates of tissue only one cell thick. They are arranged in rows near the ends of the branches, and possibly serve to protect the delicate apical cells against desiccation and injury.

The complex thallus of *Marchantia* with its pores. air chambers and photosynthetic filaments, bears some resemblance to the leaf of a higher plant with its stomata, intercellular air spaces and palisade cells. The similarity is, however, more apparent than real; the pores of *Marchantia* cannot open and close like stomata and the photosynthetic filaments are only superficially similar to the palisade cells of a leaf. The resemblance between the two organs is a striking example of convergent evolution, similar structures having been evolved separately by different types of plants to carry out similar functions.

Asexual reproduction

There are two methods of asexual reproduction in *Marchantia*. The first is common to all liverworts with dichotomously branched thalli, and depends upon the fact that the older parts of a thallus die off behind as fast as new branches are produced in front. When this dying back process reaches a dichotomy, each branch continues life independently and becomes a separate plant.

The second method involves the production of special asexual reproductive bodies called *gemmae*, which are produced on the dorsal side of the thallus in cup-shaped organs, the *gemma cups* (Fig. 25.2c). Each gemma is a biconvex multicellular body (Fig. 25.2d) with a growing point in a small notch in the margin on each side. When mature the gemmae become detached from the short stalk on which they are borne, and are readily dispersed by splashing raindrops or by insects and other small animals. If a gemma comes to rest on damp soil it grows into a double-ended thallus, from which two plants develop by the rotting away of the central portion of the developing thallus. The production of gemmae is a very effective method of reproduction, as shown by the numerous plants arising from gemmae which spring up in a very short time around a vegetative plant of *Marchantia*.

Sexual reproduction

The onset of sexual maturity in *Marchantia* is indicated by the development of special branches which grow vertically upwards and bear the sex organs above the general level of the thallus. *Marchantia* is dioecious, an individual plant being either male or female. In a male plant the reproductive branches are called *antheridiophores* and look like small umbrellas, the top of the 'umbrella' forming a disc with a wavy margin (Fig. 25.2a) The antheridia are sunk into deep ovoid cavities which open on the upper surface of the disc; a single large antheridium is attached by a short stalk to the base of each cavity (Fig. 25.4a). An antheridium arises from a single superficial cell which develops into a stalked ovoid body consisting of a central mass of antherozoid mother cells surrounded by a wall one cell thick. Each antherozoid mother-cell matures into a single biflagellate antherozoid. In a female plant the reproductive branches, or *archegoniophores*, resemble miniature palm trees with about ten finger-like lobes (rays) radiating from a small central disc (Fig. 25.4b). The archegonia are borne in groups on the underside of the disc between each pair of rays (Fig. 25.4c). The archegonia are actually formed on the top side of the young disc but, as the disc develops, the tissue between the rays grows downwards and inwards in such a way as to carry the archegonia to the underside. At maturity each archegonium is a flask-shaped organ with a long neck and a swollen basal portion, the venter (Fig. 25.5a). When the archegonium is ready for fertilization the column of neck-canal cells in the centre of the neck disintegrates, creating a mucilage-filled channel through the neck which provides free access from the outside to the egg inside the venter.

A careful study of the development and mature structure of both the antheridiophores and archegoniophores shows them to be branches of the vegetative thallus which have become modified for bearing the sex organs. On both kinds of branches air chambers open by pores on the upper surface of the disc, and rhizoids, similar to those on the underside of the midrib of

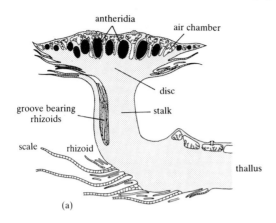

(a)

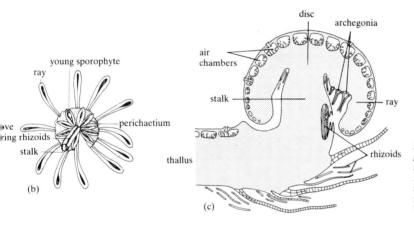

(b)

(c)

Fig. 25.4 Sex organs of *Marchantia polymorpha*. (a) V.S. antheridiophore and adjoining portion of male gametophyte. (b) Archegoniophore seen from below. (c) V. S. archegoniophore and adjoining portion of female gametophyte. (b from Bell and Woodcock 1971, p. 108)

the vegetative thallus, are present in two grooves which run the length of the apparently solid stalk.

When fertilization occurs the stalk of the archegoniophore is very short so that the archegonia, which at this stage in their development are present on the upper surface of the disc, are only slightly raised above the level of the female thallus. In wet weather, when the surface of the antheridiophore is covered with a thin film of water, the antherozoids are released from the antheridia and exude from the cavity containing them on to the surface of the disc. Although the antherozoids can swim in water, it is unlikely that they would be able to swim far enough to effect fertilization. It is thought that their transference to a female plant depends largely on splashing rain-drops. If a raindrop makes a direct hit on a ripe male disc, the water splashed off will contain antherozoids and could easily land on a female disc several centimetres away. Once they have landed in the water film on a female disc, the antherozoids can swim the relatively small distance neces-sary to reach the archegonia. The antherozoids are chemotactically attracted to the archegonia, and fertilization occurs when an antherozoid swims down the neck canal and fuses with the egg inside the venter to form the zygote, the first cell of the sporophyte generation.

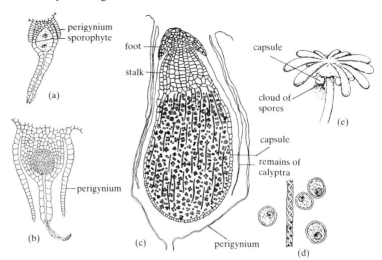

Fig. 25.5 The sporophyte of *Marchantia polymorpha*. (a) L. S. two-celled sporophyte inside archegonium. (b) A somewhat older stage than in diagram (a). (c) L.S. nearly mature sporophyte. (d) Spores and portion of an elater. (e) Side view of archegoniophore with spores being discharged from one of the capsules. (a–d from Smith 1955, vol. 2, pp. 54–5; e from Brook, A. J (1964). *The Living Plant*, p. 428, Edinburgh University Press).

The sporophyte

After fertilization, the stalk of the archegoniophore elongates and, by differential growth of the central disc, the archegonia come to lie on the under surface of the disc. Within the archegonium, the zygote begins to develop into a spore-bearing structure, the *sporogonium*. The first division of the zygote is by a transverse wall (Fig. 25.5a), successive transverse and longitudinal divisions result in the formation of a spherical mass of diploid cells enclosed in the venter (Fig. 25.5b). With further development the embryo sporogonium differentiates into three regions, a terminal *capsule* in which the spores are subsequently formed, a short stout *stalk*, which is prolonged into a *foot* (Fig. 25.5c). The foot anchors the sporophyte in the gametophytic tissue of the archegoniophore, from which it absorbs water and nutrients for the development of the sporogonium. The capsule and stalk for a long time remain enclosed in the archegonial wall, which increases in size to keep pace with the growth of the developing capsule and remains as a covering layer, the *calyptra*. At this stage the young sporogonium is also enclosed in a cylindrical sheath, the perigynium which develops from the cells around the base of the archegonium (Fig. 25.5a–c). The cells inside the capsule divide by mitosis until spore mother cells are formed, when meiosis takes place to give rise to tetrads of haploid spores. Not all the cells in the capsule become spores; some elongate and develop into long thread-like cells with a marked spiral thickening of the wall. These are the *elaters* which subsequently assist in the dispersal of the spores (Fig. 25.5d).

When the spores are ripe, the stalk elongates greatly and the capsule bursts through the calyptra and perigynium and projects freely downwards into the air. Not all the archegonia which are fertilized develop mature sporogonia, but a large archegoniophore may have twenty or more spore capsules hanging beneath it (Fig. 25.5e). Once exposed to the air the capsule wall, which is only one cell thick, splits into from four to eight petal-like flaps, exposing a yellowish brown mass of spores mixed with elaters. The liberation of the spores is assisted by the elaters which have hygroscopic properties. As they dry out, the elaters coil and twist vigorously and so loosen the spore mass; thus the ripe spores, instead of sticking together in a mass, are distributed singly or in small groups which are readily blown away by wind. A spore, if it alights on a suitable damp substrate, germinates rapidly into a new gametophytic thallus, thus completing the life cycle.

Mosses

Mosses (Musci) form the larger of the two classes into which the Bryophyta are divided. Many plants which are called 'mosses' in common speech are not mosses according to the botanical meaning of the word; thus club mosses are relatives of ferns, and Spanish moss may be a seed plant belonging to the pineapple family (Bromeliaceae). The true mosses include a large number of species which, apart from one or two notable exceptions such as the bog mosses (*Sphagnum spp.*), are remarkably similar in their general plan of construction, although within the general plan there is enormous variation in detail. *Funaria hygrometrica* is one of the few mosses with an almost cosmopolitan distribution and, since it is typical of many mosses, provides a convenient example to illustrate the features of the group.

Funaria hygrometrica

This moss forms compact, bright yellowish green tufts on soils rich in mineral salts. For this reason it is particularly common in areas which have recently been burnt and where the ground has become enriched by minerals from the ash of the fire. At such sites large colonies of *Funaria* often cover many square metres with an almost continuous carpet, in which may be found plants at all stages of development.

On being broken apart, a tuft of *Funaria* will be seen to be a dense aggregate of short, erect, slender stems bearing spirally arranged leaves (Fig. 25.6a). The upper part of the stem and the upper leaves are bright green, but the lower part of the shoot is brown as a result of being cut off from the light. The leaves are oblong-ovate in shape, with an acute apex and a broad base of attachment to the stem. Along the midline of each leaf there is a distinct vein or midrib which, unlike the veins of most leaves, is not branched. Leaf size increases towards the tip of the stem, and in mature plants the terminal leaves surround the sex organs which develop at the apex. The stems are sparsely branched but, since the older portion of a shoot system dies away as the younger portion continues to grow, axes which originated as branches often seem to be completely separate plants.

There are no roots but the base of each stem is anchored to the ground by branched multicellular filaments, called rhizoids, with oblique cross-walls. If the soil is carefully removed from around the basal portion of a young healthy *Funaria* plant, the rhizoid system will be found to be surprisingly extensive. Furthermore, it consists of branches of at least three orders of magnitude. From the base of the stem a number of stout brown rhizoids grow almost vertically downwards into the soil for a distance of about 1 cm. Along the length of these main anchoring strands arise much finer, freely branching laterals which in their turn bear even finer branches. The ultimate rhizoid branchlets are in close contact with soil particles and, like the root hairs of higher plants, assume all sorts of distorted shapes according to the contact their walls make with the soil particles.

The growth of the leafy shoot results from the activity of a single apical cell situated at the extreme tip of the stem (Fig. 25.6b). This cell is a four-sided structure rather like an inverted pyramid, of which the curved base is

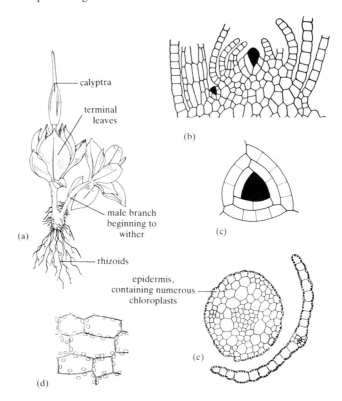

Fig. 25.6 Vegetative structure of *Funaria hygrometrica*. (a) Whole (gametophyte) plant after fertilization, with the terminal leaves of the female shoot clasping the base of a young sporophyte. The male shoot, now withering, is shown on the right. (b) Median L.S. apex of gametophyte, showing the apical cells (shaded black) of the main axis and of an embryonic lateral branch. (c) T.S. apex of gametophyte, showing the apical cell (shaded black) and the derivative cells cut off in regular succession from its three sides. (d) Surface view of a few leaf cells, each containing numerous chloroplasts. (e) T.S. stem and leaf of gametophytic plant. (a from E. V. Watson (1957), Famous plants – 6. Funaria, *New Biology*, **22**, 106, Penguin Books, England; b and e from Smith 1955, vol. 2, pp. 110–11)

exposed upwards while the three sides are wedged in the surrounding tissue. Mitotic division in the apical cell is followed by the formation of a new cell wall parallel to one of the three downwardly pointing sides. At each division a new flat cell is formed and the apical cell then grows to its former size. In rapid and regular succession cells are cut off parallel to each of the three sides in the manner indicated in Fig. 25.6c. These derivative cells divide both anticlinally and periclinally to form the tissues of the entire shoot system. The leaves originate as outgrowths from superficial cells just behind the apical cell and are therefore formed in three ranks. During subsequent elongation of the stem this three-ranked arrangement is soon lost and the leaves assume the spiral arrangement characteristic of the mature shoot.

Anatomy of the leafy shoot

The leaves of *Funaria* are thin flat plates of tissue only one cell thick except for the thickened midrib (Fig. 25.6c). The thin part of the leaf consists of large, thin-walled cells which are filled with conspicuous, ovoid chloroplasts; a single detached leaf of *Funaria* is, in fact, a very suitable object for the microscopic observation of chloroplasts (Fig. 25.6d). The midrib is several cells thick and quite complex in structure. In transverse section two types of cells can be recognized, one thick walled and one thin walled. The thick-walled cells, which are far more common, undoubtedly provide mechanical support which, in a delicate leaf like that of *Funaria*, must be an important function of the midrib. The thin-walled cells, which tend to form a group in the centre of the midrib, are usually regarded as a simple type of conducting strand, although in *Funaria* there appears to be little experimental evidence for this suggestion. No perceptible cuticle is present over the surface of the leaf which can very readily absorb or lose water.

The stem of *Funaria* is differentiated into three anatomical regions: a superficial epidermis, a thick cortex and a central strand (Fig. 25.6e). The epidermis is one cell thick, green and without stomata. The cortex is composed of several layers of parenchyma cells which, when young, contain chloroplasts. The outer cells of the cortex usually have much thicker walls than the inner cells. The central strand consists of narrow, elongated, thin-walled cells in which the protoplasm degenerates during development. These cells, in contrast to the similar thin-walled cells in the midribs of the leaves, are known to have a conducting function. Using fluorescent dyes to trace the flow of water, it has been shown that *Funaria hygrometrica* obtains its water by a combination of internal and external pathways. Internal conduction through the rhizoids was found to be extremely slow (only ten cells traversed in 40 hours) and the stem base seems to be supplied with water mainly from outside by capillary action. Once inside the stem, however, water travels readily up the central strand but it is not conducted into the leaves, which are supplied by external conduction in capillary channels formed by the overlapping leaves. In *Funaria* the differentiated tissues of the leaf midrib do not link up with the central strand of the stem but end blindly in the outer cortex, and hence the efficiency of the internal conducting system may be regarded as limited. In other genera of mosses the continuity is complete and water is conducted up the stems directly into the leaves.

Sexual reproduction

The vegetative plant just described is the gametophyte or sexual generation which produces the sex cells or gametes. The gametangia (male antheridia and female archegonia) are produced in groups at the tips of the stems, where they are surrounded by the enlarged terminal leaves of the vegetative shoot. In *Funaria* antheridia and archegonia are borne on the same plant but on separate branches; the tips of the fertile branches are called antheridial and archegonial cups (Fig. 25.7). On any one plant the antheridia typically ripen first and are borne at the extremity of the main stem. The archegonia arise on a lateral branch but this subsequently elongates and assumes a terminal position because, by the time an archegonial branch is mature, the antheridial shoot has died and withered.

At the apex of a male shoot about thirty antheridia develop in succession so that both young and mature stages can usually be found in a single antheridial cup (Fig. 25.7b). The antheridia are similar in structure to those of liverworts. At maturity the apical cell of the antheridial wall is larger than the other wall cells and functions as a lid which ultimately ruptures to release the antherozoids. The antheridia are interspersed with sterile hair-like structures, called *paraphyses*, each of which ends in an almost globose cell rich in chloroplasts. Apart from being active in photosynthesis the paraphyses have no known function but, since their swollen tips meet one another above the level of the antheridia, they may serve for water retention.

At the extremity of a lateral female branch a small group of stalked archegonia (Fig. 25.7d) develop from single superficial cells. Apart from being stalked, the archegonia are similar in structure to those of liverworts. Like the antheridia, the archegonia are interspersed with sterile paraphyses, but these are much smaller than those which occur in the antheridial cups.

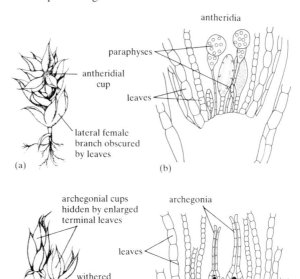

Fig. 25.7 Sex organs of *Funaria hygrometrica*. (a) Male shoot at fertilization. (b) L.S. antheridial cup. (c) Two female branches after fertilization, with withered male shoot displaced to one side. (d) L.S. archegonial cup. (a and c from Bold, H. C. (1973), *Morphology of Plants*, 3rd edn, p. 302. Harper & Row).

Fertilization in *Funaria*, as in all bryophytes, depends upon the presence of free water. It is the uptake of water by the mature antheridium that causes the 'lid cell' to rupture and the antherozoids to be exuded as a mucilaginous mass through the aperture thus created. The individual antherozoids do not separate from one another until the mucilaginous mass comes into contact with the water in the antheridial cup. It is thought that splashing raindrops transfer the antherozoids to the archegonial cups where fertilization takes place, in a manner similar to that described for *Marchantia* (see p. 313).

The sporophyte

The zygote of *Funaria* does not develop directly into another leafy moss plant but into a spore-producing structure, the sporophyte, consisting of a spore capsule on a stalk (Fig. 25.8); this stalk is always unbranched and leafless. The sporophyte, which remains attached throughout its life to the parent gametophyte plant, produces haploid spores and these give rise eventually to new moss plants.

The zygote develops within the venter to form an ellipsoidal embryo sporophyte with a growing point at each end. The lower end bores through the base of the venter and penetrates the tip of the stem on which the archegonium is borne. This dagger-like prolongation, called the *foot*, anchors the sporophyte and absorbs food and water from the tissues of the parent gametophyte (Fig. 25.8a). The upper end elongates to produce the spore capsule and its supporting stalk or *seta*. For a time the archegonial venter enlarges to keep pace with the elongating sporophyte, but eventually it ruptures round the base. The upper part of the archegonium is then carried upwards as a conical cap, the *calyptra* (Fig. 25.8b,c), on top of the distal end of the

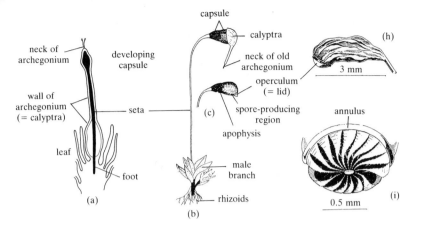

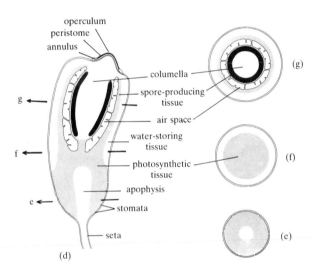

Fig. 25.8 The sporophyte of *Funaria hygrometrica*. (a) L. S. archegonial cup after fertilization, showing young sporophyte (shaded black) with 'dagger-like' foot. Note that the distal part of the archegonium forms the calyptra covering the sporophyte. (b) The leafy gametophytic plant with the leafless sporophyte growing from the top of it. (c) Capsule after removal of calyptra. (d) Diagrammatic L.S. capsule, fully swollen but still immature. (e) – (g) Diagrammatic T.S.s of capsule taken at the levels indicated in diagram (d). (h) Dry mature capsule with lid still attached. (i) Mouth of capsule after the lid has fallen off, showing the peristome teeth. Note the outer teeth attached to the central disc; only the tips of the inner teeth are visible. (b and c from Scott, D.H. and Ingold, C.T. (1955), *Flowerless plants*, p. 82, published by A. & C. Black, London; h and i from E. V. Watson (1957), Famous plants – 6. Funaria, *New Biology*, **22**, 108, Penguin Books, England)

sporophyte, where it remains until the capsule swells to achieve its mature size.

When fully grown the capsule hangs down from the end of the erect wiry seta, which is about 2 cm high and forms a strong supporting column. The seta supplies the capsule with water and mineral salts by means of a strand of conducting tissue which runs up the centre. The capsule is a very complex structure with a high degree of internal tissue differentiation. For descriptive purposes it can be divided into three regions: a basal region of sterile photosynthetic tissue, an upper region containing both sterile and fertile cells, and an apical region which is modified for the dispersal of spores (Fig. 25.8d).

The basal region of the capsule, called the apophysis, is penetrated centrally by a strand of conducting tissue which is continuous with the central strand of the seta. Surrounding this central strand is a broad zone of chlorophyllous cells with conspicuous intercellular spaces; the epidermis of this region is pierced by functional stomata. It is evident from its anatomical structure that the basal region of the capsule is photosynthetic, so that the mature sporophyte is to some extent autotrophic. The upper portion of the capsule contains a cylinder of fertile spore-producing cells embedded in sterile tissue. The cells inside the cylinder form a central column of colour-less parenchyma (the *columella*) which extends downwards into the apophysis. The spore-producing region is separated from the multicellular capsule wall by a cylindrical air space crossed by strands of green photosynthetic cells. The fertile cells divide repeatedly before meiosis finally produces tetrads of immature spores, and the new gametophyte generation begins. While the spores are maturing the sterile tissues collapse throughout the capsule, which consequently dries out. This change is marked externally by the capsule changing in colour from green to reddish brown and becoming deeply furrowed longitudinally. When the capsule is fully ripe it consists essentially of the original wall enclosing a powdery mass of cellular debris intermingled with mature spores which have now separated from one another.

At the apex of the capsule is a cap of tissue (the lid or *operculum*) which, when the capsule is ripe, is thrown off by the disintegration of a ring of thin-walled cells (the *annulus*) inserted between the edge of the lid and the rest of the capsule wall (Fig. 25.8h). After the lid has been shed the spores are still not free to drop out of the capsule, even though it hangs downwards, because the cell layers beneath the lid have developed into a double row of triangular teeth (the *peristome*) which guard the mouth of the capsule. The teeth are sheets of thickened wall material which persist in certain predetermined areas despite the general breakdown of tissue that accompanies the drying out of the capsule. In *Funaria hygrometrica* there are two rings of peristome teeth, one above the other, with sixteen teeth in each ring (Fig. 25.8i). The outer peristome teeth are arranged like the spokes of a wheel, their broad bases attached to the rim of the capsule wall and their points joined at the centre to form a central disc. The inner peristome teeth are more delicate than the outer ones and are attached only at the rim of the capsule wall; they partly cover the gaps between the teeth of the outer peristome ring and so hinder the shedding of the spores. The outer peristome teeth are hygroscopic, and respond to changes in humidity by bending inwards or outwards as they take up or lose water. Consequently, when all the outer teeth bend inwards simultaneously they depress the central disc and lessen the spaces between them. Conversely when the teeth arch outwards they raise the disc and widen the spaces, thus allowing spores to escape when the seta is shaken. The inner peristome teeth do not exhibit bending movements and, because they do not completely cover the capsule mouth, are thought to act as a sieve allowing the spores to fall out gradually rather than in one mass. It is generally assumed that the elaborate structure and mechanism of the peristome ensure that the spores are liberated only in dry weather, when they are more likely to be dispersed by air currents.

The protonema

After being shed from the capsule the spores germinate rapidly if they come to rest on a damp surface. The germinating spore gives rise in the first place not to a young moss plant but to a juvenile stage called the *protonema* (Fig. 25.9). This consists of a system of branching filaments, some of which grow on the surface of the soil and are green, whereas others penetrate it and resemble the rhizoids of the mature plant in being colourless or brown and in having oblique transverse walls. Apart from the formation of bran-

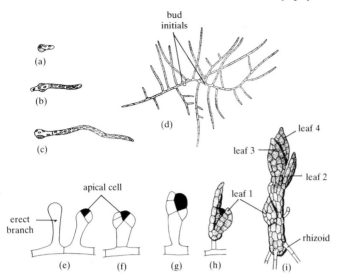

Fig. 25.9 Protonema of a moss (*Splachnum ovatum*). (a) – (d) Development of protonema from germinating spore to branching filament (as seen from above) with short erect branches which are bud initials. (e) – (i) Stages in development of young moss plant from formation of erect branch on protonema up to bud with first four leaves. (All from Smith 1955, vol. 2, pp. 126 and 128)

ches, growth of the protonema is apical (i.e. restricted to the terminal cells of the branches) so that the filaments are made up of single cells placed end to end. However, the protonema sooner or later gives rise to short lateral branches in which a pyramidal apical cell is created by the formation of three adjacent oblique walls. The division of such an apical cell with its three cutting faces soon results in the formation of a bud, which then grows into a young moss plant. Many buds are formed on a single protonema, and this explains why *Funaria* grows in dense carpets. The protonema dies once the young moss plants have developed their own rhizoids and become established as independent plants. The insertion of a protonema stage before the production of the typical gametophyte plant serves to distinguish the life cycle of mosses from that of thalloid liverworts. In other respects the sequence of their life cycles is very similar.

Chapter 26

Pteridophytes

The plants included in the pteridophyte level of organization show an alternation of dissimilar generations but, unlike the bryophytes, it is the diploid, sporophyte generation which predominates and constitutes the 'plant' as normally understood. The sporophyte has a well-developed vascular system with distinct xylem and phloem tissues, and is therefore potentially capable of reaching a much larger size than the gametophyte of bryophytes. Pteridophytes differ from plants at the next evolutionary level (i.e. the seed plants) in that they reproduce by spores and not by seeds.

The pteridophytes include three main groups, the ferns (subdivision Pteropsida, class Filicinae), the horsetails (subdivision Sphenopsida) and the club mosses (subdivision Lycopsida). There is no close relationship between these groups which, from fossil evidence, have been distinct for a very long time – at least since the Devonian period about 350 million years ago. Representatives from only two of the groups, the ferns and club mosses, will be considered in this book. Club mosses can be distinguished from ferns by two obvious features in their general appearance. Firstly, they have numerous, small, simple leaves instead of relatively large, usually compound ones. Secondly, the sporangia, instead of being borne in large numbers on the underside of the leaves, as in ferns, are borne singly on the upper side of specialized sporophylls which are closely packed round the ends of certain branches to form structures called *cones* or *strobili*.

Plants with a predominant haploid phase, such as the bryophytes discussed in the last chapter, depend on moist conditions for their reproduction, and are therefore severely limited in their ability to colonize dry land. A life cycle with a dominant diploid phase is capable of being modified so as to give rise to a type of plant in which the need for external water for fertilization is eliminated. The first step in the evolution of plants completely adapted for life on land therefore involves contracting the haploid phase of the life cycle, and this has been achieved by two changes. Firstly, the haploid phase, instead of being independent, is supplied with food material from the parental diploid phase, and so is able to develop rapidly without first having to build up its own reserve of food material by photosynthesis. This means that much of the non-reproductive part of the haploid phase is redundant and so allows considerable simplification. Secondly, the haploid phase, instead of bearing both antheridia and archegonia, is unisexual, its sex being controlled by the preceding diploid phase. This allows sexual differentiation to occur, so that those haploid plants which develop archegonia acquire a large food reserve, necessary to nourish the zygote until it establishes itself as an independent sporophyte, while those which develop antheridia acquire only a relatively small reserve of food material.

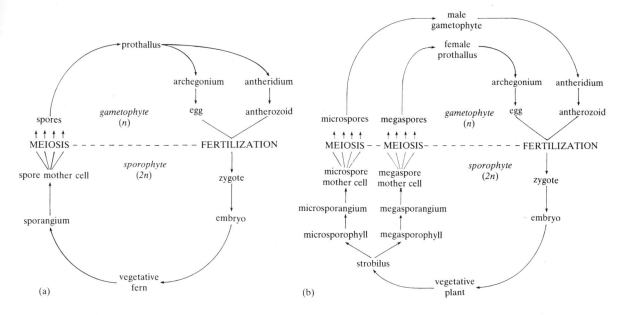

Fig. 26.1 Comparison of homosporous and heterosporous types of life cycles. (a) Homosporous life cycle of a typical terrestrial fern. (b) Heterosporous life cycle of *Selaginella* sp.

The characteristic feature of plants in which these two changes have occurred is the possession of two kinds of spores, which are restricted to different sporangia. Some sporangia contain a small number of large spores, which give rise to female haploid plants, while others contain a large number of small spores which develop into male haploid plants. The production of spores of two sizes by a plant is called *heterospory*.

Seed plants have been the dominant form of plant life on land for the last 250 million years, and the adoption of heterospory was undoubtedly one of the important factors which led to their success. However, heterospory has evolved independently in some members of all the three groups of plants at the pteridophyte level of organization. Apart from a few very unusual aquatic forms which are heterosporous, ferns are homosporous, but club mosses belonging to the genus *Selaginella* are heterosporous (Fig. 26.1). Seed plants have evolved from homosporous ancestors, and a knowledge of the similarities and differences between the homosporous type of life cycle, shown by practically all ferns, and the heterosporous type of life cycle, shown by *Selaginella*, is necessary for an understanding of the life cycle of angiosperms.

Ferns

Ferns are represented by less than 10 000 living species but, owing to their larger size and characteristic appearance, they are a more conspicuous component of the vegetation than liverworts and mosses, despite the greater abundance of species in the latter groups (about 20 000 species). Ferns are world-wide in distribution, but are most abundant in the humid tropics. They are also extensively grown in gardens and greenhouses for their attractive foliage. Most ferns have such a characteristic appearance that they cannot easily be mistaken for any other kind of plant. Part of this impression is due to the coiling of the young leaves which unfold as they mature; this almost

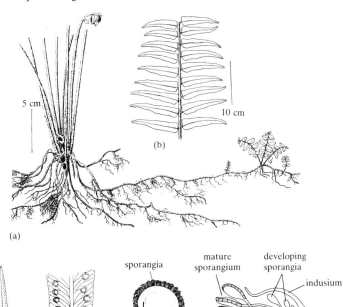

Fig. 26.2 *Nephrolepis biserrata*. (a) Base of old plant showing runners bearing new leafy stems and roots. (b) Part of a sterile frond. (c) A fertile pinna. (d) Part of a fertile pinna showing venation and sori. (e) A sorus with ripe sporangia. (f) Diagrammatic V.S. sorus to show attachment of sporangia to the pinna. (a–d from R. E. Holttum (1954), *A revised flora of Malaya*, Vol. 2, *Ferns of Malaya*, p. 373, Government Printing Office, Singapore; e from Ewer and Hall 1973, p. 173)

unique feature is called *circinate vernation* (not to be confused with venation), and is the result of the upper surface of the leaf growing more slowly than the lower during early development. Ferns vary widely in size and form, ranging from tree ferns which may reach a height of 5 m, to minute filmy ferns which have leaves only one cell thick and might easily be mistaken for mosses. Tree ferns look rather like small palm trees because they have a single trunk with a crown of leaves at the top of it. In addition to the numerous 'typical-looking' terrestrial species, many ferns (notably the 'nest-ferns') grow as epiphytes on trees and rocks. In Malaya and Singapore, for instance, more than half of the approximately 500 known species are epiphytic. In so diverse a group no one pattern of organization can be said to be representative of the group as a whole. For this reason, a general description of the sporophyte is used to illustrate the essential features of many common terrestrial ferns. By contrast, the inconspicuous gametophyte generation of all ferns, whether epiphytic or terrestrial, is remarkably similar.

The sporophyte

The essential vegetative parts of the fern sporophyte, like those of all vascular plants, are leaves, stems and roots. The leaves, or *fronds* as they are usually called, are the most conspicuous parts of a fern. The stalk of a frond is called a *stipe* to distinguish it from other kinds of stalk. It usually bears hairs or flat, sometimes elongated scales, the shape and colour of which may be useful in distinguishing different kinds of ferns. The flat part of the frond, often called the lamina, may be simple or variously divided into few or many separate leaflets. If the leaflets are arranged like the barbs of a feather the frond is called *pinnate*, each leaflet being called a *pinna*, and the axis bearing the pinnae is called the *rachis*. The frond of the fern *Nephrolepis biserrata*, illustrated in Fig. 26.2b, is an example of a pinnate frond. If each pinna is again pinnate, the whole frond is *bipinnate*, and each ultimate leaflet is called a *pinnule*. There also exist tripinnate and quadripinnate fronds but there are no special terms for the ultimate leaflets of these. The leaves, like the rest of the plant, are supplied with strands of xylem and phloem, which in leaves constitute the veins. These are usually easy to see, being either raised on the surface or visible when the frond is held up to the light. The veins of most fronds are branched; they may fork equally (dichotomously) or the branching may be pinnate, or a combination of the two. Veins are either free (or open), i.e. they branch but do not join again, or sometimes they form a network; in the latter case the veins are said to be reticulate.

The fronds are the conspicuous part of a fern; the rest of the plant (except in tree ferns, which have a massive stem) is colloquially called the 'root' but, in fact, is largely a stem. Every fern has a true stem from which arise the fronds and true roots. The stem of some ferns is quite short, in which case it is called a *stock*; it may be erect or horizontal. If the stock is erect it always bears a rosette of fronds at its apex, and roots radiate in all directions from it. In other ferns the stem is long and slender, and called a *rhizome*. *Nephrolepis biserrata* (Fig. 26.2a) has both a stock, bearing a group of erect stipes, and numerous radiating rhizomes. A rhizome is usually horizontal, growing at the surface of, or under, the ground but in epiphytic ferns it climbs upon the branches or trunk of a tree. Horizontal rhizomes and stocks usually bear fronds only on the upper surface, and roots only on the lower surface. There is no sharp distinction between a horizontal stock and a rhizome, and in practice all fern stems are often called rhizomes.

Although a primary root is present in the very young sporophyte of ferns, it is soon replaced by adventitious roots which develop along the entire length of the stem except near the growing apex.

Reproduction

The familiar fern plant belongs to the sporophyte generation because it reproduces asexually by means of haploid spores which result from meiosis. Morphologically, therefore, an individual fern plant is comparable to the spore capsule of a liverwort or moss. The spores are produced in spore boxes called sporangia. As in all tracheophytes, the sporangia are borne on leaves called *sporophylls*. In most ferns all the leaves of a mature plant are sporophylls which differ from the vegetative leaves produced by the immature plant only in that they bear sporangia. In a few ferns the sporangia are produced on specialized sporophylls which are morphologically distinct from the vegetative fronds.

Spore production and liberation

On the lower surface of a mature leaf of almost any common fern there are round or elongate rust-coloured patches which, when young, may be covered by a flap of tissue called an *indusium* (Fig. 26.1d–f). These rusty patches consist of numerous sporangia, and are called *sori* (singular, sorus), the word sorus being the technical term for a cluster of spore boxes. The shape of sori, their position in relation to the veins and edges of leaflets, and the presence and type of any indusium, are important characters in the classification of

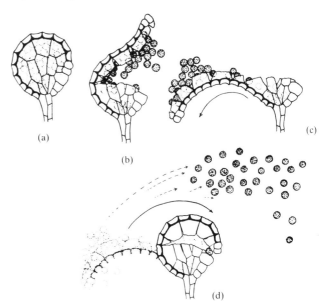

Fig. 26.3 Stages in the forcible discharge of spores from the sporangium of a fern. (a) The sporangium is closed; the spores are outlined by dotted lines. (b) The sporangium has opened at the stomium. (c) Later stage with the annulus bent backward. (d) Shortly after (c), the annulus has sprung forward to its original position; the cells of the annulus are now largely filled with gas, shown by shading. (From Coulter and Dittmer 1964, p. 318)

ferns. The sporangia are club-shaped structures, each having a slender sterile stalk and a flattened fertile head, rather like a biconvex lens. The cells of a developing sporangium are diploid, but as the sporangium approaches maturity some of the internal cells acquire dense contents and become spore mother cells. Each of these divides by meiosis to give four haploid spores.

The wall of the sporangium is one cell thick but along the rim of the lens there is a row of specialized cells, the *annulus*, extending from the stalk on one side, over the top, and half way back to the stalk on the other side (Fig. 26.3). Each annulus cell is heavily thickened on all but its outer wall. The line of cells along the remaining quarter of the rim is called the *stomium*. The walls of the stomium cells, like those of the cells composing the sides of the sporangium, are thin.

The spores are liberated from the sporangium by a type of mechanism which seems to have been evolved repeatedly in land plants, and will therefore be described in detail as an example of such a mechanism. Each annulus cell is at first filled with an aqueous solution but, under conditions favouring evaporation, the outer wall is sucked in and the side walls tend to bend inwards as if hinged at the base. When this decrease in volume happens in every cell of the annulus, a strain is set up in the thin-walled stomium, which eventually tears. Once the stomium is torn, the annulus starts to bend slowly backwards, and the slit in the stomium extends laterally across the side walls of the sporangium. This has the effect that the spherical wall of the sporangium is separated into two cups, united only by the annulus. As the cells lose more water, the whole annulus continues to bend backwards carrying with it the upper cup which contains most of the spores. The water inside the annulus cells at this stage is under considerable tension, and each cell is 'straining' to regain its original volume but is prevented from doing so by the cohesion of the water and its adhesion to the cell walls. Eventually a point is reached at which the tension is so great that the water can no longer adhere to the walls. At this point the water vaporizes and the cell becomes filled with gas. With the sudden release of tension the cell returns to its former

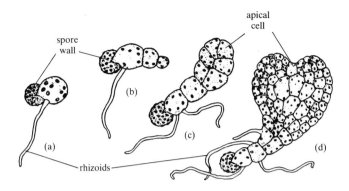

Fig. 26.4 Stages in the germination of a fern spore into a young prothallus. (From Coulter and Dittmer 1964, p. 319)

size and shape. The explosion in one cell appears to trigger off the others, with the result that the entire annulus springs back to its original position, slinging the spores from the upper sporangial cup into the air as it does so.

The gametophyte

The spores of ferns are sufficiently light to be easily airborne, and hence may be widely dispersed. Under dry conditions they retain their viability for several months, but when moistened at a suitable temperature they germinate (Fig. 26.4). The germinating spore gives rise to a small green plate of tissue, commonly about 5 mm across, which lies flat on the soil. This structure is called the *prothallus* (Fig. 26.5), and represents the gametophyte (sexual) phase in the life cycle. It contains chlorophyll and manufactures its own food. The wings of the prothallus are one cell thick, and even the stouter midrib region is only a few cells thick. The lower surface, particularly the central midrib region, bears abundant rhizoids, which attach the prothallus to the soil and absorb water and nutrients. The prothallus grows by means of a pyramidal cell which is located between the two lobes of the heart-shaped plate of tissue.

Although much simpler in structure than a liverwort or moss plant, the fern prothallus is the sexual stage (gametophyte) in the life cycle and bears on its lower surface archegonia and antheridia (Fig. 26.5b, c). The archegonia are found near the apical notch and look like small tubes projecting from the under surface. The antheridia form small round projections scattered among the rhizoids, somewhat behind the archegonia. The archegonia resemble those of bryophytes in their essential structure, but the venter is embedded in the tissue of the prothallus and the neck is proportionally shorter. The antheridium consists of a wall of three sterile cells surrounding a mass of cells that develop into coiled antherozoids bearing numerous flagella. Two of the wall cells are ring-shaped, like a motor tyre, while the third is a flat lid-cell. When a thin film of water covers the under surface of the prothallus, the antherozoids are set free by the opening of the lid-cell, and the neck of the ripe archegonium becomes an open tube. The antherozoids swim to the archegonia to which they are attracted chemically. Although several antherozoids may enter the neck of an archegonium only one of them fertilizes the egg cell which remains embedded in the prothallus. The resulting zygote divides into four quadrants, which form respectively the stem, leaf, root and 'foot' of the young sporophyte (Fig. 26.6). The foot is a tissue embedded in the prothallus and through it the developing sporophyte

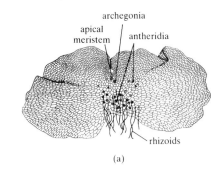

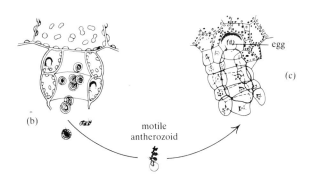

Fig. 26.5 The prothallus of a fern. (a) Ventral view of a mature prothallus of *Nephrolepis* sp. (b) L.S. antheridium in process of liberating antherozoids. (c) L.S. archegonium at time of fertilization. (a from Smith 1955, vol. 2, p. 353; b and c from Delevoryas 1977, p. 77)

absorbs food from the gametophyte. With the development of green leaves and roots the sporophyte rapidly becomes an independent plant in its own right, and the prothallus withers.

Because new fern plants can only grow from a prothallus, the distribution of ferns is limited to places where their prothalli can grow. Different species of fern grow best in different habitats; some flourish in rock crevices, while others are found only on the bark of trees. Most ferns depend on sexual reproduction for their spread, and if a species is to be successful there must be a correlation between the habitat requirements of the prothallus and those of the fern to which it gives rise.

Club mosses

Selaginella is a predominantly tropical genus of club mosses, which are usually found creeping along the ground in damp forests. Most species are rather inconspicuous plants which could easily be mistaken for large mosses because their stems bear numerous small leaves, and this is why they are called club mosses. However, the resemblance is superficial because, whereas in a true moss the leaves and stems belong to the haploid sexual generation, in *Selaginella* they form part of the diploid spore-producing generation.

There are several hundred species of *Selaginella* in the tropics, but the vast majority resemble one another in general appearance and belong to one section of the genus. The following description, though based on two typical species (*S. martensii* and *S. kraussiana*) of this section, will therefore apply to most species of *Selaginella*.

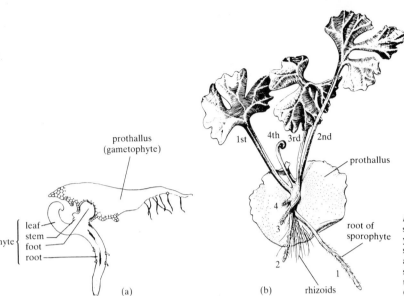

Fig. 26.6 Development of the young sporophyte of a fern. (a) L.S. prothallus with young sporophyte attached to the ventral surface. (b) Ventral view of prothallus with an attached sporophyte which is becoming self-supporting; note the circinate vernation of the youngest (4th) leaf. (b from Foster & Gifford 1974, p. 318).

Vegetative features

Selaginella has a prostrate, creeping stem which branches at frequent intervals. The branching is monopodial (i.e. the same growing point remains permanently in the lead) even though near the growing tips it appears to be dichotomous because the main stem and a newly formed branch grow with equal vigour for a time (Fig. 26.7). In the older portions of the plant, however, the main stem grows more vigorously than the branch so that the distinction between the two becomes evident. The stems bear numerous small leaves which are separated by distinct internodes on the older parts but are crowded together at the growing tips. The leaves are of two types and are arranged in four longitudinal rows. There are two rows of small dorsal leaves and two rows of larger ventral leaves. This arrangement imposes a marked dorsiventrality on the vegetative shoot. Close to the base of each leaf is a small, membranous, dorsal outgrowth called a *ligule*, which is best seen on very young leaves because it soon withers and disappears. The ligule has no known function but its existence can be traced back to the Carboniferous period (250–300 million years ago) where it occurs in the Lepidodendrales, a large group of fossil plants which possibly share a common ancestry with *Selaginella*.

The shoot system is attached to the soil by colourless leafless structures, called *rhizophores*, which resemble roots in appearance but have no root caps. The rhizophores arise singly or, more often, in pairs, at each fork of the stem and grow vertically downwards to the ground. On reaching the soil they give rise at their tips to a bunch of adventitious roots which have root caps and the anatomical features of true roots. The rhizophore is intermediate in structure between stem and root, and there is fossil evidence to indicate that it represents a type of organ which existed before roots and stems evolved independently.

Growth in *Selaginella*, as in all the land plants hitherto discussed, is entirely apical, i.e. every cell of the plant body is derived from the meri-

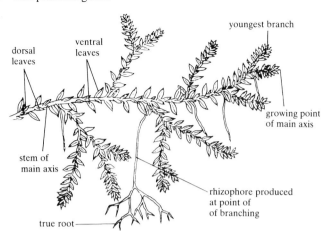

dorsal leaves

ventral leaves

youngest branch

growing point of main axis

stem of main axis

rhizophore produced at point of of branching

true root

Fig. 26.7 General view of *Selaginella kraussiana* showing the creeping stem and two types of leaves. (From N.S. Parihar (1962), *An introduction to the Embryophyta*, vol. 2, p. 67, Central Book Department, Allahabad)

stematic cells located at the tip of each shoot and root. The mature plant body, like that of ferns, has a conducting system of xylem and phloem.

Reproduction

The *Selaginella* plant described above is the sporophyte and sooner or later it produces short, erect branches which end in strobili (Fig. 26.8). These are radially symmetrical structures, in contrast to the dorsiventral vegetative shoots, and consist of a central axis with four vertical rows of fertile, over-lapping leaves, called sporophylls, radiating from it. Each sporophyll bears a single, stalked sporangium on its upper surface, between the stem and the ligule. The sporangia are of two kinds, *microsporangia* and *megasporangia*, and the sporophylls which bear them are therefore distinguished as *micro-sporophylls* and *megasporophylls* respectively. The prefixes micro- and mega- refer to the sizes of the spores contained within the sporangia and not to the sporangia themselves, which are about equal in size. The microspor-angia contain a large number of small spores, *microspores*, while the megas-porangia contain a small number of large spores, *megaspores*. Sporangia of both kinds occur in the same strobilus, the megasporangia usually occurring on the lower sporophylls. A plant which produces both microspores and megaspores is said to be *heterosporous*.

Each type of sporangium consists of a short stalk, a sporangial wall two cells thick, a single layer (the tapetum) of nutritive cells, and a central mass of spore mother cells that divide by meiosis to form haploid spores. In the microsporangium all, or nearly all, of the spore mother cells undergo meiosis to form a large number of microspores. In the megasporangium also there are initially many spore mother cells, but all except one degenerate. This surviving cell increases tremendously in size and divides by meiosis into a tetrad (i.e. a group of four) of large megaspores which, on account of the way in which they are packed in the tetrad, are roughly pyramid-shaped. The degeneration of megaspore mother cells within the megasporangium suggests one possible way by which the heterosporous condition could have evolved from the homosporous condition.

The gametophytes

In *Selaginella* each of the two kinds of spore germinates into a different type

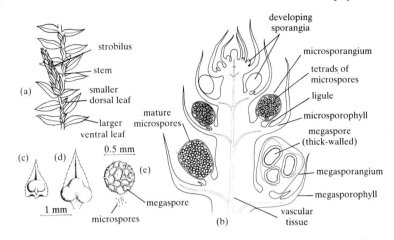

Fig. 26.8 The strobilus of *Selaginella*.
(a) Small portion of fertile shoot showing
young strobilus. (b) L.S. strobilus of *S.
kraussiana*. (c) Microsporophyll with
microsporangium. (d) Megasporophyll with
megasporangium. (e) Relative sizes of
microspores and megaspores. (a from N. S.
Parihar (1962), *An introduction to the
Embryophyta*, vol. 2, p. 67, Central Book
Department, Allahabad; c–e from Ewer and
Hall 1973, p. 176)

of gametophyte; the microspores give rise to male gametophytes while the
megaspores develop into female gametophytes. It is usual for both the
microspores and megaspores to start germinating before they are shed from
the sporangia containing them.

In the development of the male gametophyte (Fig. 26.9) the microspore
nucleus divides into two daughter nuclei, one of which takes up a position
at one side of the spore where it is walled off as a small cell and does not
divide again. This cell is called the *prothallial cell* because it is thought to
represent all that remains of the vegetative part of a prothallus. The other
cell is called the *antheridial cell* because it continues to divide inside the
microspore wall to form a mass of cells that differentiate into a single antheridium. The antheridium consists of an outer layer of sterile jacket cells and
an inner mass of biflagellate antherozoids, which swim freely inside the
microspore. At maturity the antherozoids are liberated during wet weather
by the rupture of the antheridial jacket and microspore wall. The development of the microspore into a male gametophyte thus takes place entirely
within the microspore wall until the antherozoids are mature.

The megaspore also starts to germinate inside the megaspore wall. The
megaspore nucleus divides many times, without wall formation, to give a
large number of nuclei which are concentrated at the apical end of the spore.
Later, walls are formed around these nuclei so that a patch of cellular tissue
develops inside the megaspore. At this stage in the development of the
female gametophyte the megaspore wall splits open along three lines of
weakness at the apical end of the spore, exposing part of the female prothallus (Fig. 26.10). On this exposed portion archegonia develop in considerable numbers. The archegonia are basically similar to those of ferns except

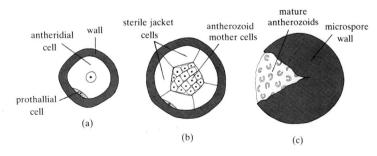

Fig. 26.9 Development of the male
gametophyte of *Selaginella*. (a) Section of
microspore, showing the small prothallial cell
which takes no further part in development.
(b) Section of microspore with cellular mass
that can be interpreted as an antheridium. (c)
Male gametophyte with mature antherozoids
about to be released.

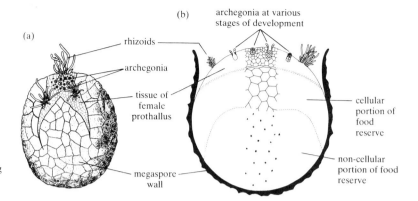

Fig. 26.10 Development of the female gametophyte of *Selaginella*. (a) Habit sketch of germinating megaspore, showing tissue of female prothallus between the splits in the megaspore wall. (b) V.S. germinating megaspore, showing female prothallus growing on top of large food reserve. (a from McLean and Cook 1962, vol. 1, p. 588)

that the necks only just protrude above the surface. In some species of *Selaginella* tufts of rhizoids may also develop on the exposed portion of the female prothallus. Even though part of its surface is freely exposed, the female prothallus remains permanently invested by the megaspore wall.

Fertilization and the development of the embryo

Fertilization usually takes place on the ground. In wet weather the free-swimming antherozoids are attracted to the archegonia which they enter through the neck and fertilize the egg cell. The resulting zygote, which is the first cell of the new sporophyte generation, starts to develop immediately into an embryo. The first division of the zygote is transverse to the long axis of the archegonium and produces an upper and a lower cell. The actual embryo develops from the lower of these two cells while the upper one (called the suspensor), by its subsequent growth, pushes the developing embryo into the tissues of the female prothallus (Fig. 26.11). The embryo soon becomes organized into a root–stem axis which is at first nourished by means of a temporary organ called the *foot*. This develops as an outgrowth of the embryo on the side opposite the suspensor, and buries into the lower part of the prothallus from which it absorbs food. As the embryo axis elongates it forces its way through the tissues of the prothallus until it projects through the cracks in the megaspore wall. From this point onwards the female prothallus degenerates and the embryo establishes itself as an independent young sporophyte.

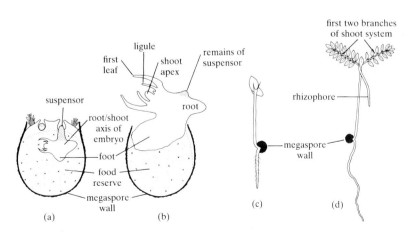

Fig. 26.11 Stages in development of the embryo sporophyte of *Selaginella*. (a) and (b) V.S. megaspores with developing embryo sporophytes. (c) Embryo with first leaves and root. (d) Embryo showing branched stem, leaves, and first rhizophore.

Chapter 27

Gymnosperms

All vascular plants at a higher level of organization than the pteridophytes form a group, the seed plants (Spermatophyta), which is characterized by the possession of modified megasporangia called *ovules*. In botanical usage the term *seed* has a precise meaning, and refers to the structure into which an ovule develops after fertilization. There are two essential features which distinguish an ovule from the megasporangium of a pteridophyte (Fig. 27.1). Firstly, a single functional megaspore is retained permanently *within* the megasporangium. Secondly, the megasporangium has one or more covering layers, called *integuments*, which completely surround the megasporangium except at its apex where there is a small opening, the *micropyle*. The integuments are morphologically part of the megasporangium and not outgrowths from the megasporophyll. Up to the stage of fertilization this composite structure, consisting of the megasporangium proper (or *nucellus* as it is called in seed plants) and one or more integuments, is termed an ovule which may therefore be defined as 'an indehiscent, integumented megasporangium containing a single functional megaspore'. After fertilization the ovule develops into a seed which, at maturity, is shed from the parent plant.

The reproductive structures of seed plants were named long before their homology with other structures found in the pteridophytes was understood, and for this reason they are referred to by a different set of terms. The following list is a comparison of the terminology used for homologous structures in the two groups.

Pteridophytes	*Seed plants*
megasporophyll	carpel (angiosperms only)
megasporangium	nucellus ⎱ ovule
–	integuments ⎰
megaspore	megaspore
female gametophyte	embryo sac (angiosperms only)
microsporophyll	stamen (angiosperms only)
microsporangium	pollen sac (usually restricted to angiosperms)
microspore	⎱ pollen grain
male gametophyte	⎰

Apart from the possession of ovules (and hence, subsequently, of seeds), there is no other absolute criterion which separates the seed plants from the pteridophytes. As a group the seed plants are structurally more complex and exhibit much greater diversity of form than the pteridophytes but, if

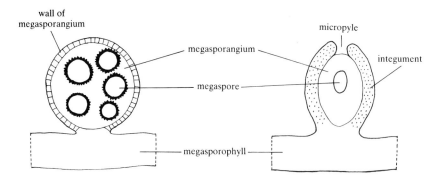

Fig. 27.1 Megasporangium of pteridophyte (left) compared with ovule of seed plant (right).

only vegetative features are considered, the two groups merge into one another. The distinction was less obvious in the geological past than at the present day, and several genera of fossil plants which had fronds indistinguishable from those of ferns reproduced by seeds.

The seed plants are divided into two main groups, the *gymnosperms* (meaning naked seeds) and the *angiosperms* (meaning vessel seeds), which differ in the way they carry the ovules. The gymnosperms bear their ovules externally on the surface of megasporophylls or analogous structures known as *ovuliferous scales*, which (with the notable exception of *Cycas* among living representatives of the group) are massed into woody strobili called *cones*. In angiosperms, on the other hand, the ovules occur within closed structures, called *carpels*, which may be regarded as being formed by the fusion, edge to edge, of the megasporophylls. The carpels are surrounded by other specialized organs to form the composite reproductive structures known as 'flowers'.

This chapter is concerned only with the gymnosperms. All gymnosperms are woody perennials, but they exhibit a wide range of life forms including tall trees (most conifers), palm-like trees (the cycads) as well as shrubs and climbers. In addition to having 'naked' seeds, gymnosperms are characterized by leathery, usually evergreen, leaves, and by the absence of vessels from the xylem and of companion cells from the phloem. At a previous stage in the earth's history the gymnosperms were the dominant type of vegetation, but with the advent of the angiosperms in the Jurassic, about 150 million years ago, they dwindled in importance and today are represented only by about 600 species as compared with about 250 000 species of angiosperms. The living gymnosperms belong to two main groups, a smaller group of palm-like trees called cycads and a much larger group, with distinctive cones, called conifers. Both groups illustrate features in the evolution of land plants, especially the reproductive modifications that took place when plants no longer had free access to water as a medium in which to liberate their gametes.

Cycads

Although cycads possess the advanced feature characteristic of all seed plants, namely the production of ovules, they also exhibit some very primitive features both in their morphology and in their life cycle. Fossils show that the cycads alive today are the survivors of a group which flourished during Mesozoic times. At the present day there are less than 100 species belonging to nine genera, one of which is *Cycas*, the type-genus from which the group takes its name. Having been almost world-wide in distribution in the past, cycads are now confined to the tropics and subtropics, where they are often grown as ornamental plants. Although the individual genera have a restricted distribution, the group as a whole is represented in three widely separated geographical regions (central America, southern Africa, and eastern Asia together with Australia). This pattern of discontinuous distribution is, in itself, evidence of an ancient group that was already widely distributed before the existing ocean barriers became established.

Vegetative features

In appearance most modern cycads resemble small palm trees (Fig. 27.2) and it is from this superficial resemblance, coupled with the fact that a type of sago can be obtained from some of them, that they are popularly called 'sago palms'. They are, however, not true palms for these belong to the other main group of seed plants, the angiosperms. In most cycads the stem is short, stout, unbranched (although there may be occasional adventitious branches), and bears a crown of large pinnate leaves just below the terminal bud. The leaves are arranged in a close spiral around the stem, and a new crown is produced at yearly or less frequent intervals. In some genera the pinnae of the young leaves are rolled up like a watch spring (*circinate vernation*) in the manner characteristic of ferns. Below the crown the stem is covered with an armour made up of the persistent bases of the leaves of previous seasons. The foliage leaves are protected in the bud by scale leaves, which are potential leaves that never develop leaf blades. As a consequence the armour of many cycads shows a distinct ribbing caused by the alternate production of large foliage leaves and small scale leaves. Underground there is typically a long, tuberous tap root.

Fig. 27.2 Cycad, typical habit.

Secondary thickening occurs in the stems of cycads but it is never very extensive, and the limited amount of secondary xylem which is produced is soft with very wide parenchymatous rays. The trunk therefore retains the soft texture of the unthickened stem, and its mechanical strength is due largely to the armour of persistent leaf bases. Growth rings, which form such a conspicuous feature of the secondary xylem of most woody plants, are present in cycad wood but they are not a reliable criterion for estimating the age of the stem; the differences in growth rate caused by the annual alternation of rainy and dry seasons are not sufficient to be marked internally by a distinct growth ring. In some species growth rings are formed every time a new crown of foliage leaves develops, but in others they mark the alternation of growing periods and prolonged resting periods. Prolonged resting periods occur at irregular intervals; for example, one individual of the cycad *Dioon edule* known to be about 100 years old showed only twenty growth rings, even though this plant had produced many more than twenty crowns of leaves during its lifetime.

Reproduction

All cycads are dioecious (i.e. the microspores and megaspores are produced on separate male and female plants). Generally speaking the sporangia are borne in dense cones; this is true for the microsporangia throughout the group, and for the megasporangia in all genera except *Cycas* itself. The cones

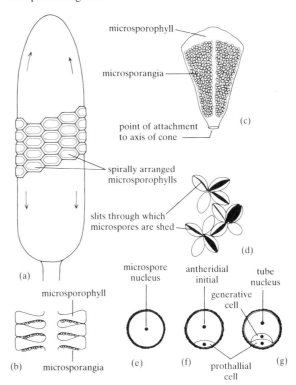

Fig. 27.3 Male cones of cycads. (a) Male cone of *Zamia floridana* (b) L.S. of (a) showing the microsporangia on the lower surface of the microsporophylls. (c) Microsporophyll, seen from below. (d) Microsporangia of *Cycas* sp. arranged in sori. (e)–(g) Stages in the development of the microspore.

develop from lateral buds at the apex and hence are borne at the top of the stem. The two types of cone are similar in external appearance except that the male cone is usually longer and thinner than the female cone of the same species. In the female plant of *Cycas* the megasporangia are not borne in cones but on special fertile leaves (megasporophylls) produced at the apex of the main stem in close spirals which alternate with the spirals of foliage leaves.

Male cones

The male cones of cycads (Fig. 27.3) are among the longest cones known, the cone *without* the stalk often reaching 30 cm or more in length. The cone consists of a central axis bearing many, closely packed, spirally arranged cone scales or microsporophylls. These are wedge-shaped, and the entire lower surface of each is covered with numerous (over 1000 in *Cycas*) microsporangia. It will be noted that, like the sporangia on the fronds of ferns, the microsporangia occur on the lower surface of the microsporophylls, and close inspection of a young microsporophyll will reveal that, also like ferns, they are arranged in small groups or sori (with three to six microsporangia in each group) arising from a central papilla. Each microsporangium consists of a wall several cells thick, and a central mass of numerous microspore mother cells which divide by meiosis to form tetrads of haploid microspores; these separate into individual microspores.

The microspores, like those of *Selaginella*, start to develop into the male gametophyte while still inside the microsporangium. The nucleus divides to form a smaller *prothallial* cell which is cut off at one side and does not divide again, and a larger *antheridial initial* which divides once more to give a

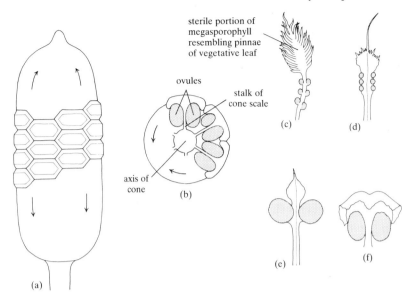

sterile portion of
megasporophyll
resembling pinnae
of vegetative leaf

ovules

stalk of
cone scale

axis of
cone

(a)   (b)   (c)   (d)   (e)   (f)

Fig. 27.4 Female cones and megasporophylls of cycads. (a) Female cone of *Zamia floridana*. (b) T.S. of (a) showing attachment of ovules to the cone scales. (c)–(f) Hypothetical stages in the evolution of a cone scale from a megasporophyll of *Cycas*. (c) *C. revoluta*. (d) *C. circinalis*. (e) *C. rumphii*. (f) Cone scale of *Encephalartos*.

smaller *generative* cell and a larger *tube* nucleus. The three-celled micro-spore, which from this stage onwards is called a pollen grain, is shed.

When the pollen grains are ripe the axis of the cone elongates slightly and each microsporangium opens by a longitudinal slit in its wall. The pollen grains thus have free access to the open air, and are dispersed by wind so that pollination of the ovules can take place.

Female cones, and the megasporophylls of Cycas

In all cycads except *Cycas* the megasporangia are borne in cones (Fig. 27.4), which are among the most massive known (e.g. those of *Macro-zamia* can weigh over 35 kg). As in the male cone the cone scales are arranged spirally around a central axis, but they are less numerous and different in shape from those of the male cone. Each cone scale bears two ovules which are seated one on each side of the stalk. The ovules are pro-tected by the ends of the cone scales which fit tightly together at the surface of the cone. Cones, like the flowers of angiosperms, are special branches on which the organs for sexual reproduction are borne. All living seed plants have such branches set apart for sexual reproduction with the one exception of the female plant of *Cycas*.

In *Cycas* the megasporangia are borne on leaf-like organs, megaspor-ophylls, which are much smaller than the foliage leaves but, like them, are produced in close spirals on the main stem. Throughout the life of a female plant of *Cycas* the apical meristem alternately produces foliage leaves and megasporophylls. When the megasporophylls have fulfilled their reproduc-tive function they drop off, so that the armour covering the stem of a fe-male *Cycas* plant consists of the persistent bases of three different kinds of leaves – foliage leaves, scale leaves and megasporophylls.

The megasporophylls of *Cycas* are flattened structures about 15–20 cm long and distinctly leaf-like in appearance. Each consists of a distal, pin-nately divided portion and a proximal, stalk-like portion with two to ten megasporangia (according to the species) borne on opposite sides of the stalk. The megasporangia apparently replace the pinnae on the proximal

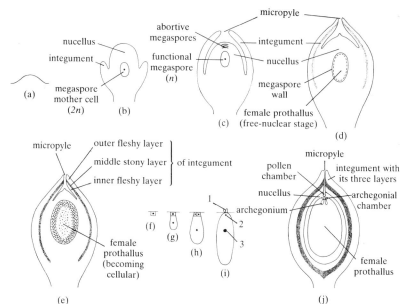

Fig. 27.5 Stages in the development of the ovule of cycads. (a) Ovule initial arising as a bump on the megasporophyll. (b) Very young ovule. (c) Young ovule in which meiosis has occurred. (d) Ovule with free-nuclear divisions in the female prothallus. (e) Ovule with female prothallus becoming cellular. (f)–(i) Stages in the development of the archegonium (1 = neck cells; 2 = ventral-canal nucleus; 3 = egg nucleus). (j) Mature ovule ready for pollination.

portion of the megasporophyll. At first sight a megasporophyll of *Cycas* bears little resemblance to a cone scale typical of the female cones of other cycad genera. However, the megasporophylls of different species of *Cycas* can be arranged in a series (Fig. 27.4c–f) showing a reduction in the number of ovules to two (e.g. *C. rumphii*), accompanied by a gradual suppression of the distal leaf-like portion of the megasporophyll. If the condition in *Cycas* is regarded as primitive, then the range of megasporophyll types within the genus is suggestive of the stages by which the female cone may have evolved, because the aggregation of reduced megasporophylls around a common axis would produce a recognizable cone. Unfortunately the fossil record is inconclusive on this point, and the sequence cannot be regarded as more than a possibility.

Ovule and female prothallus

In all cycads the structure and behaviour of the ovule is basically the same, irrespective of whether it is borne on a cone scale or a fertile leaf. The structure of the ovule can best be understood by following the various stages in its development (Fig. 27.5). The young ovule arises as a bump of meristematic tissue which, as it enlarges, becomes invested by a single integument; this grows around the nucellus (i.e. that part of the ovule which can be regarded as the equivalent of a megasporangium in a pteridophyte) except at the apex where it leaves a small opening, the micropyle, to admit the pollen grains. At an early stage in the development of the ovule, a single cell situated deeply within the nucellus enlarges to form a megaspore mother cell. This undergoes meiosis to produce four haploid megaspores, of which three degenerate and disappear. The single surviving megaspore enlarges and starts to develop into a female prothallus. As in all seed plants, the megaspore is permanently retained within the nucellus so that the germination of the female prothallus is internal. In other words, instead of the megaspore wall splitting open to expose a cushion of prothallial tissue as in

Selaginella, the female prothallus completes its development within the megaspore. It follows, therefore, that the female prothallus is entirely dependent on the parent sporophyte for its nourishment.

During the first phase of development of the female prothallus, the first and succeeding divisions of the megaspore nucleus are not accompanied by wall formation, so that numerous nuclei lie free within the cytoplasm of the megaspore. This method of division is described as *free nuclear division*, and it continues until over 1000 nuclei have been formed (the free-nuclear stage of the female prothallus). Cell walls are then laid down between the nuclei, a process which begins at the periphery of the megaspore and continues progressively inwards until the whole of the female prothallus becomes cellular.

During or following the process of wall formation, a few (usually three to five) superficial cells at the micropylar end of the female prothallus enlarge to become archegonial initials, which divide to form archegonia. The archegonia are considerably reduced compared with those of pteridophytes, but nevertheless still recognizable. Each typically consists of two neck cells, and a central cell containing an extremely large egg nucleus and a small ventral-canal nucleus. The ventral-canal nucleus appears to have no function and soon degenerates. When this happens the egg nucleus is ready for fertilization.

While the above changes have been taking place within the female prothallus, other changes have also been occurring elsewhere in the ovule. The integument differentiates into three layers, an outer fleshy layer, a middle stony (sclerenchymatous) layer and an inner fleshy layer. As the female prothallus increases in size the nucellus progressively diminishes, and some of the nucellar cells at the micropylar end break down to form a distinct cavity called the *pollen chamber*. At maturity the archegonia project into another cavity which develops between the micropylar end of the female prothallus and the overlying nucellus; this space is known as the *archegonial chamber* and is not to be confused with the pollen chamber.

Pollination and fertilization

The process of pollination consists of the transfer of pollen grains, in the three-celled condition already described, to the micropylar end of the ovules. The amount of pollen produced by cycads is enormous and, being a light dry powder, it is readily blown about by the wind. Some of the air-borne pollen grains alight on a drop of sticky fluid, the *pollination droplet*, which at this time exudes through the micropyle of the ovule. The pollination droplet subsequently dries up and, in doing so, draws any adhering pollen down into the pollen chamber. This is followed by the closure of the micropyle. With the arrival of pollen grains in the pollen chamber pollination is effected, but the subsequent process of fertilization does not occur until several months later.

Shortly after pollination the interrupted development of the male gametophyte is resumed (Fig. 27.6). On the side opposite the prothallial cell the pollen grain pushes out a short tubular outgrowth called a *pollen tube*, which penetrates the side of the pollen chamber and so anchors the developing pollen grain in the tissue of the adjacent nucellus. The function of the pollen tube seems to be the absorption of food material from the nucellus. The pollen-grain end of the male gametophyte then also elongates so that it hangs in the pollen chamber, suspended from the pollen tube. It may

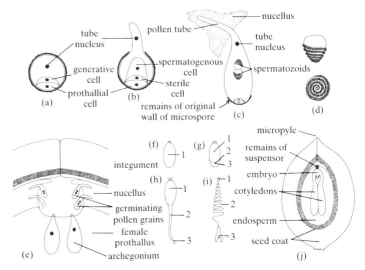

Fig. 27.6 Pollination and seed development in cycads. (a) Pollen grain at time of shedding. (b) Formation of pollen tube shortly after pollination. (c) Pollen tube hanging from the nucellus into the archegonial chamber. (d) Mature antherozoids. (e) Apical region of ovule at time of fertilization; four pollen grains have produced pollen tubes, one of which has discharged its contents. (f)–(i) Stages in development of the embryo (1=archegonium; 2=suspensor; 3=embryo proper). (j) Longitudinal section of mature seed, showing embryo with two cotyledons and two leaf primordia.

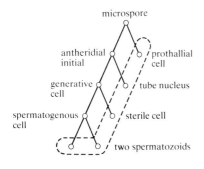

remain suspended for several months while the archegonia are maturing. Meanwhile the generative cell divides into two daughter cells, best termed the *sterile cell* and the *spermatogenous cell*, which stay at the pollen-grain end. As the name suggests, the sterile cell plays no further part in the development of the male gametophyte. Shortly before fertilization the spermatogenous cell divides to produce two pear-shaped *spermatozoids*, each with a spiral band of flagella. The formation of the spermatozoids completes the development of the male gametophyte which therefore consists of five cells. These are included inside the dotted line of the diagram on the left.

While the pollen tubes are growing laterally into the nucellus, the floor of the pollen chamber (i.e. the thin layer of nucellar tissue between the pollen chamber and the archegonial chamber) disintegrates so that the pollen chamber becomes continuous with the archegonial chamber. The pollen-grain end of the male gametophyte now swells and finally bursts, forcibly releasing into the archegonial chamber a drop of fluid containing the two spermatozoids. The spermatozoids are thus released in close proximity to the archegonia and, at the same time, are provided with a drop of fluid in which to swim. This fluid has such a high osmotic potential that when it comes into contact with the neck cells of the archegonium, they become plasmolysed and the spermatozoids can then enter the egg cell. More than one spermatozoid may enter an archegonium but only one fertilizes the nucleus of the egg cell.

It will be noted that, although cycads produce pollen tubes, they nevertheless have motile spermatozoids. Although the spermatozoids are strikingly different in size and shape from those of pteridophytes, the fact that they are produced at all suggests a connection with plants dependent on external water for fertilization. In cycads the pollen tube appears to function primarily as a means of supplying nourishment from the nucellus to the developing pollen grain, and is only indirectly concerned with bringing the male and female gametes together. By contrast, in conifers and angiosperms the pollen tube functions as the means of conveying the non-motile male gametes direct to the egg cell. The cycad method of fertilization is thus intermediate between the purely locomotory method shown by pteridophytes and the purely pollen-tube method shown by conifers and angiosperms.

The embryo and the seed

When fertilization has been effected, the zygote begins to develop into an embryo sporophyte and, because this marks the first stage in the formation of a new plant, the ovule is now called a seed. Since there are several archegonia within a single ovule more than one embryo may start to develop; this condition is known as *simple polyembryony* and is equivalent to the formation of fraternal (dizygotic) twins in animals. However, one of the embryos grows faster than the others which soon degenerate so that only one embryo reaches maturity in any one seed. This functional embryo first undergoes a period of free nuclear division similar to that of the female prothallus. When many nuclei have been formed wall formation begins and soon a conspicuous mass of cells develops at the base of the archegonium. The cells at the extreme tip of this cellular region become the embryo proper while those nearer the neck of the archegonium differentiate into a suspensor. The latter elongates very greatly and pushes the embryo through the wall of the archegonium into the centre of the female prothallus. From this stage onwards it is usual to refer to the female prothallus as the endosperm because it now functions as a nutritive tissue for the developing seed. The embryo, embedded in the endosperm, gradually becomes transformed into a miniature plant with recognizable root, stem and cotyledons (seed leaves).

While the embryo has been developing, the tissues that originally surrounded the fertilized egg cell have also been undergoing modification to form other parts of the seed. The endosperm continues to grow at the expense of the surrounding nucellus which gradually disappears until it is represented only by a dry, papery cap of tissue at the micropylar end of the mature seed. The integument of the ovule becomes the seed coat. The outer fleshy layer of the integument remains fleshy and forms a thick, brightly coloured (often red or orange) outer covering to the mature seed, the middle stony layer becomes so hard that it is difficult to cut with a pocket knife, and the inner fleshy layer is reduced to a thin papery membrane lining the inner surface of the stony layer. At maturity therefore the seed (Fig. 27.6j) consists of three main parts, the embryo sporophyte, the endosperm and the seed coat. The ripe seed looks rather like a mango, but the resemblance is superficial because, whereas the former is a seed, the latter is a fruit which is a structure found only in angiosperms.

Conifers

The conifers are the more important of the two main groups of living gymnosperms and, although more abundant in the Mesozoic era before the angiosperms evolved, they are still the dominant plants over vast areas of the earth's surface. The coniferous forests of the Northern Hemisphere provide the 'soft woods' (e.g. deal, pitch pine, etc.) of commerce, which represent about seven-eighths of the world's timber.

Conifers differ from cycads in four important respects. (1) They are mostly tall trees; in fact the tallest trees in the world are conifers, one specimen of Californian redwood (*Sequoia sempervirens*) being nearly 120 m high. (2) The trunk and branches consist mainly of secondary xylem, which is dense and has very small xylem rays. The xylem consists mainly for long straight tracheids and is therefore easily workable; it is for this reason that coniferous woods are widely used as timber and for making paper. (3) Their leaves are typical needle-like or scale-like, in contrast to the large leaves of cycads.

(4) Fertilization is by means of non-motile male gametes which are conveyed direct to the egg cell by a pollen tube.

The genus *Pinus* (pine trees) has been extensively studied on account of its economic importance, and may be chosen as typical of conifers.

Vegetative features of *Pinus*

Pine trees form extensive forests in the Northern Hemisphere, but some of the eighty to ninety known species occur on tropical mountains where they are either native or introduced as plantation crops of fast-growing timber trees.

A pine tree typically has a single main trunk with whorls of spreading branches, which themselves branch in a similar fashion. In addition to the branches (the so-called long shoots) which contribute to the growth form of the tree, there are also numerous minute branches (the so-called short shoots) which live for a few years only and are then shed (Fig. 27.7). The leaves are also of two kinds, foliage leaves and scale leaves. The foliage leaves, popularly known as 'needles' on account of their long slender shape, are confined to the short shoots (except in young seedlings); according to the species there are two, three or five foliage leaves borne in a cluster at the apex of each short shoot. The foliage leaves are shed when the short shoot on which they grow is shed. The scale leaves occur on the long shoots and at the base of the short shoots. It is in the axils of scale leaves on the long shoots that both long and short shoots arise. Scale leaves are usually shed soon after they are formed, so that they are only present for a short distance behind the growing point of the branches.

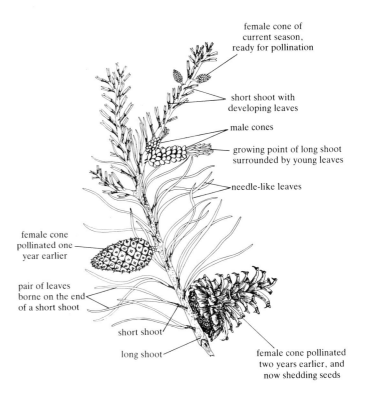

female cone of
current season,
ready for pollination

short shoot with
developing leaves

male cones

growing point of long shoot
surrounded by young leaves

needle-like leaves

female cone
pollinated one
year earlier

pair of leaves
borne on the end
of a short shoot

short shoot

long shoot

female cone pollinated
two years earlier, and
now shedding seeds

Fig. 27.7 Pinus. Composite sketch of a branch at time of pollination. (From Bold 1977, p. 216)

Reproduction of *Pinus*

The pine tree is the sporophyte and bears its microsporangia and ovules in two different kinds of cones. Both types of cone are borne on the same tree (i.e. *Pinus* is monoecious), but almost always on different branches. The female cones are produced on the end of short lateral branches (i.e. they replace long shoots) near the tips of some of the current year's twigs. Although they are formed in this subterminal position they take two years to complete their development and so come to occupy a lateral position remote from the terminal bud. The male cones are much smaller and far more numerous than the female cones. They occur as clusters of side shoots (i.e. they replace short shoots) just behind the terminal buds as these unfold at the beginning of the growing season. Shortly after the male cones have appeared the microspores are shed as pollen grains, whereupon the cones wither and fall from the tree. The clustered arrangement of the male cones suggests that the entire group on one twig may be the counterpart of a single female cone.

Male cone

The male cone consists of a central axis bearing spirally arranged microsporophylls, on the lower side of which are two microsporangia (Fig. 27.8). The formation of the microspores within the microsporangia is essentially the same as in cycads, the diploid microspore mother cells undergoing meiosis to produce haploid microspores.

The microspores, like those of cycads or of *Selaginella*, start to develop into the male gametophyte within the microsporangium. At the time the microspore is shed as a pollen grain its nucleus has divided three times, giving two slender prothallial cells (which soon degenerate), a generative cell and a tube nucleus. While the above divisions have been taking place inside the microspore the wall becomes two-layered and develops two lateral, air-filled bladders, or wings, where the outer wall grows away from the inner wall. The wings have the effect of lowering the density of the pollen grain and so facilitating its subsequent spread by wind.

When the pollen grains are ripe the axis of the cone elongates and each microsporangium opens by a longitudinal split. The pollen grains drop into the gaps between the microsporophylls and are dispersed from there by wind.

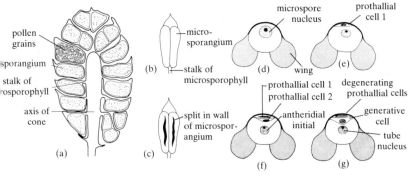

Fig. 27.8 Male cone and microspores of *Pinus*. (a) L.S. cone. (b) and (c) Ventral view of microsporophyll before and after the splitting of the microsporangia. (d)–(g) Stages in the development of the microspore.

Female cone

The female cone, like the male cone, has a central axis to which numerous lateral appendages, or cone scales, are attached spirally (Fig. 27.9). These

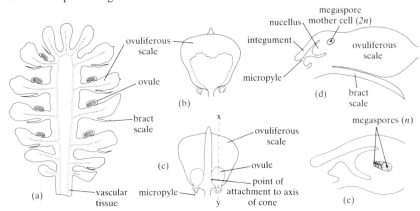

Fig. 27.9 Female cone and ovule of *Pinus*. (a) L.S. young cone at time of pollination. (b) Lower surface of individual cone scale. (c) Upper surface of individual cone scale. (d) L.S. cone scale, along line x – y in diagram (c), at time of pollination. (e) L.S. ovule after meiotic division of megaspore mother cell into four megaspores, three of which abort.

cone scales are paired structures, consisting of a lower, smaller *bract scale* and an upper, larger *ovuliferous scale* which bears two ovules on its upper surface. Although the cone scales resemble sporophylls (fertile leaves) in that they bear ovules, strictly speaking they should not be called megasporophylls because the ovule-bearing portion does not originate as a leaf. The most plausible interpretation of the cone scale is that the bract scale represents a modified leaf while the ovuliferous scale is a highly modified shoot in its axil.

When a female cone is several months old the ovules appear as two small swellings, side by side, on the upper side and towards the base of each ovuliferous scale. A longitudinal section (Fig. 27.9c) through such a swelling shows that the ovule consists at this stage of two parts, an oval mass of undifferentiated cells, the nucellus, and a single investing layer, the integument. The integument is fused to the nucellus except at the end facing the axis of the cone, where the integument is extended forwards to form a conspicuous aperture, the micropyle. At the opposite end of the ovule, the nucellus and integument are not distinct from the tissue of the ovuliferous scale.

At an early stage in the development of the ovule, a single megaspore mother cell in the centre of the nucellus becomes demarcated by its larger size and prominent nucleus. The megaspore mother cell divides by meiosis to give a linear tetrad of haploid megaspores but, of these, the three nearest the micropyle degenerate, leaving one functional megaspore. This enlarges but does not begin to produce the female gametophyte until after pollination has taken place.

Pollination

When the ovules have reached the stage at which the functional megaspores are differentiated, the female cone is ripe for pollination. The axis of the cone elongates and the ovuliferous scales become separated from one another. At this time the pollen grains, in the four-celled condition already described, are shed from the male cones in enormous quantities and are widely dispersed by the wind. Some of the airborne pollen grains penetrate between the cone scales of a female cone and are brought into direct contact with the apex of the nucellus by a pollination-droplet mechanism similar to that of cycads. Pollination complete, the cone scales enlarge so as to fit tightly together again, in which position they are further sealed by a secretion of resin.

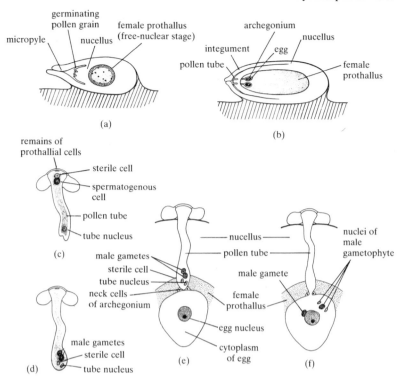

Fig. 27.10 Pollination and fertilization of *Pinus*. (a) Early stage in development of the female prothallus. (b) Mature female prothallus. (c) and (d) Stages in the development of the male gametophyte. (e) and (f) Egg cell just before and at time of fertilization.

Development of the male and female gametophytes

A peculiar feature in the reproduction of *Pinus* is that there is an interval of about a year between pollination and fertilization. During this period, while the male and female gametophytes are completing their development, the whole female cone gradually increases in size.

Within the now sealed-up cone the single functional megaspore continues to enlarge, and its nucleus divides many times (at first without wall formation, though walls are later laid down between the nuclei) to form a mass of haploid tissue (Fig. 27.10). This is the female prothallus and, though still enclosed within the megaspore wall, it now occupies a considerable space in the centre of the nucellus, at the expense of which it has grown. At the micropylar end of the female prothallus there arise two, three, or even more archegonia (the precise number depending partly on the species), which are very similar in appearance to those of cycads.

While the megaspores have been developing into female prothalli, the pollen grains, stranded on the surface of the nucellus, have also germinated. Shortly after pollination each pollen grain puts out a pollen tube, into which the tube nucleus passes. The pollen tube penetrates the tissue of the nucellus and grows extremely slowly towards the archegonia. During this time the generative cell divides into a sterile cell and a spermatogenous cell, which also pass into the pollen tube. A few days before fertilization the spermatogenous cell divides to form two male gametes; each has a large nucleus with a little cytoplasm but no flagella. The formation of these non-motile male gametes completes the development of the male gametophyte, which therefore consists of six cells (Fig. 27.11) and not five as in cycads.

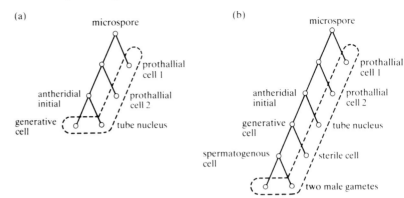

Fig. 27.11 Stages in development of the male gametophyte in *Pinus*. (a) Stage reached by pollen grain when shed from the male cone. (b) The six cells of the male gametophyte after completing its development shortly before fertilization.

Fertilization and formation of the seed

When the pollen tube eventually reaches the archegonium a year after pollination, the male gametes and egg cells are ready for fertilization. The tip of the pollen tube enters the archegonium and discharges the four cells it contains (i.e. the entire male gametophyte less the two prothallial cells) into the cytoplasm of the egg. One of the male gametes moves towards the egg nucleus and fuses with it to form a zygote, the other three cells degenerating within the egg cytoplasm.

Immediately after fertilization the zygote begins to develop into a pro-embryo, which is the name given to the earliest stages in the development of the embryo sporophyte. Since there are two or more archegonia within a single ovule (*simple polyembryony*), and since each pro-embryo develops in such a way (the details need not be considered here) that four potential embryos are formed from each zygote (*cleavage polyem-*

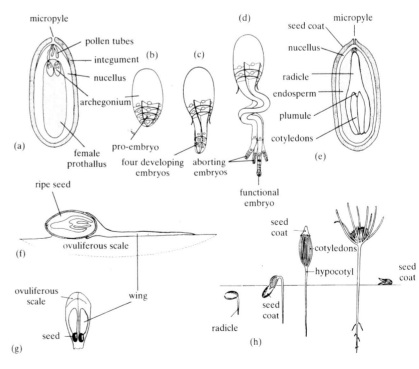

Fig. 27.12 Seed formation and germination of *Pinus*. (a) Young seed shortly after fertilization. (b) Archegonium containing pro-embryo with four tiers of four cells each. (c) Four potential embryos growing through base of archegonium into female prothallus. (d) One embryo outstrips the others to become the functional embryo. (e) Nearly ripe seed; note the relative reduction of the nucellus. (f) L.S. upper surface of ovuliferous scale showing attachment of wing to seed. (g) Upper surface of ovuliferous scale with two winged seeds. (h) Stages in the germination of a seed into a seedling. (b, c, d and h from Arnett and Braungart 1970, pp. 272–3)

bryony), more than one embryo may begin to develop (Fig. 27.12a–d). However, only one of the embryos reaches maturity in any one ovule. Each pro-embryo grows through the base of its archegonium into the sterile tissue of the female prothallus, which meanwhile has continued to grow at the expense of the surrounding nucellus. By the time the functional embryo has grown into a ripe seed the female prothallus has swollen to such an extent that the nucellus is represented only by a thin papery layer at the micropylar end of the ovule. The embryo, embedded in the female prothallus (now called the endosperm), rapidly grows into a young sporophyte plant with recognizable root, stem and cotyledons. During the development of the embryo the cells of the integument harden to form the testa or seed coat (Fig. 27.12e).

When the embryos are mature the cone scales separate from one another, so allowing the ripe seeds to escape. Attached to each ripe seed is a thin membranous wing which is not strictly part of the seed, because it is formed from a flap of tissue that splits off from the ovuliferous scale but remains attached to the seed (Fig. 27.12f, g.). This wing causes the seed to spin as it falls and thus, by retarding its rate of fall, assists its dispersal by the wind. Seeds which come to rest on damp ground germinate into pine seedlings. The germination of pine seeds does not differ significantly from that of angiosperm seeds discussed in Chapter 29, although the seedlings are unusual in appearance (Fig. 27.12h) on account of having several cotyledons instead of one or two as in angiosperms. The life cycle of *Pinus* is summarized in Fig. 27.13.

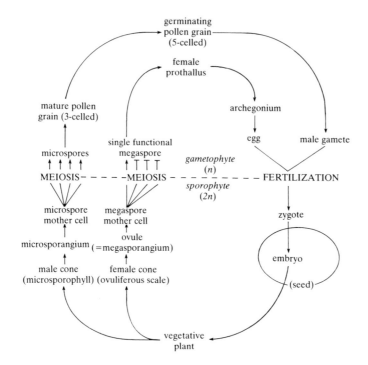

Fig. 27.13 The life cycle of *Pinus*.

Chapter 28

Angiosperms I. Flowers and pollination

The angiosperms, or flowering plants, and the gymnosperms, considered in the previous chapter, both produce seeds as dispersal units, but they differ in the extent to which the immature seeds (*ovules*) are exposed at the time of pollination. Whereas in gymnosperms the pollen grains have direct access to the ovules, in angiosperms the ovules are enclosed within modified megasporophylls called carpels and the pollen grains have to penetrate the carpellary tissue before they can reach the ovules to fertilize them.

Angiosperms are the most conspicuous and largest group of modern plants, both in number of species and in number of individuals. More than 250 000 species have been described, ranging in size from the small floating plant called *Wolffia* which is about 1 mm across, to gigantic trees like the Australian gums (*Eucalyptus* spp.) which are over 100 m tall. Although the angiosperms are the dominant land plants of the world today, there is no reliable evidence that they existed before Cretaceous times (130 million years ago). At the end of the Cretaceous period, however, they increased tremendously in frequency and abundance, and in the course of a few million years (a very short interval in geological time) became the dominant land plants which they still are today.

There is no satisfactory explanation for this 'biological explosion' of the angiosperms which, as the botanist Takhtajan remarks, had a decisive influence on the subsequent evolution of terrestrial animals and, in the final reckoning, made possible the appearance of man. Presumably the enclosure of the ovule has, or had in Cretaceous times, great biological value, but 'why' is one of the most puzzling features about angiosperms. It is difficult to see how enclosure of the ovules, which apparently makes fertilization more difficult, could have a selective advantage in evolution. It has been suggested that the angiospermous habit made possible the evolution of incompatibility mechanisms favouring outbreeding, but this is a teleological suggestion (i.e. it explains 'means' in terms of the 'ends' attained) and does not explain the initial enclosing of the ovules, the advantage of which would seem to be some form of protection. Protection from desiccation suggests itself as a possibility, but protection from potential seed predators would seem to be more likely. To whatever cause the success of the angiosperm may be attributed, the one thing which all angiosperms have in common is the possession of flowers, the structure of which will now be considered.

The flower

A flower is a shoot system (i.e. it consists of an axis and laterally borne leaves) specialized for sexual reproduction. It differs from a vegetative shoot system in that: (1) it has no buds in the axils of the leaves; (2) the internodes remain short, so that the vertical distance between successive leaves is very small; and (3) it shows limited growth, i.e. once the apical meristem has formed a flower, it ceases to grow any further. Although a flower is morphologically a shoot, there is nevertheless a terminology peculiar to flowers.

Stem terminology   When a plant has either a solitary flower or a single terminal cluster of flowers as in many 'lilies', the stem is called a *peduncle*. If, as is more usual, the plant bears several to many flowers along the length of a flower stalk, the stem of each individual flower is called a *pedicel*, and the word peduncle, if used at all, is applied to the floral axis to which the pedicels are attached. The extreme tip of the peduncle or pedicel on which the floral leaves are inserted is the *receptacle*.

Leaf terminology   Where either a pedicel or, in the case of a stalkless flower, the flower itself arises in the axil of a leaf, such a leaf is called a *bract*. Sometimes, as in bougainvillea and poinsettia (*Euphorbia pulcherrina*), the bracts are brightly coloured and more conspicuous than the flowers. If additional leaves are present on the pedicel itself, they are called *bracteoles*. The terminology of the leaves that make up a flower will be covered in the following section.

Flower structure

There are four possible types of floral leaves: sepals, petals, stamens and carpels. The last two (stamens and carpels) are described as the *essential* floral leaves because they contain the sexual organs, and the first two (sepals and petals) as the *accessory* floral leaves. All four types of leaves are, of course, essential in the sense that they are all parts of one working mechanism. The floral leaves are arranged in spirals or, more usually, in whorls (circles) on the receptacle, with the accessory leaves outside and/or below the essential leaves (Fig. 28.1).

The outermost, or lowest, whorl of accessory leaves consists of the sepals, which are usually green and resemble foliage leaves more closely than the other floral leaves. They serve to enclose and protect the flower when it is in the bud stage. Collectively the sepals are referred to as the *calyx*. Within, or above, the sepals is a whorl of petals, which are usually brightly coloured and form the conspicuous part of the flower. Collectively the petals form the *corolla*.

A collective name for the accessory leaves is the *perianth*, and where there is no marked distinction between the outer and inner whorls of accessory leaves the term *perianth members* is used, instead of sepals and petals (Fig. 28.1b). The same term is also used when there is only one whorl of accessory leaves, irrespective of whether they are green and sepal-like, or brightly coloured and petal-like.

The essential leaves are so called because they are the floral leaves directly concerned with sexual reproduction. The outermost whorl consists of the *stamens*, which collectively form the *androecium*. The stamens liberate pollen grains (microspores), which in turn produce male gametic nuclei. The way in which they do this will be considered in the next chapter, and here the only point to note is that each stamen consists of two parts, a sterile stalk

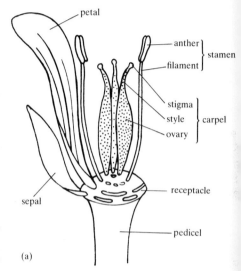

(a)

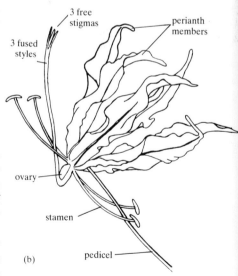

(b)

Fig. 28.1 Parts of a flower. (a) Diagram to show insertion of floral parts on the receptacle. (b) Flower with perianth members all alike.

portion, the *filament*, and a fertile terminal portion, the *anther*, in which the pollen grains are produced.

The apex of the receptacle is occupied by one or more *carpels* which, like the stamens, produce spores. In this case, however, the spores (megaspores) are not shed and give rise to female gametic nuclei inside the carpel. A single carpel consists of three parts: a box-like *ovary* in which the ovules (which, if fertilized, become the seeds) are formed, a stalk-like projection called the *style*, and a receptive surface on which pollen grains are deposited, the *stigma*. If there is more than one carpel on the receptacle, the carpels may be separate from one another, in which case there are as many stigmas, styles and ovaries as there are carpels. However, it is much more usual for two or more carpels to be fused together to form a single ovary. In this case the styles and stigmas may be fused or separate, but commonly there are as many stigmas as there are carpels forming such a fused ovary. Hence a clear distinction must be drawn between the terms carpel and ovary. Where two carpels are fused they form a *bicarpellary* ovary, where three a *tricarpellary* ovary, and so on. The collective name given to the carpels is the *gynoecium*, which is sometimes also called the pistil. The word pistil, however, can be confusing because it means either ovary or gynoecium according to the context, and its use should be discontinued. In any case, gynoecium is the counterpart of androecium, which is the collective term always used for the stamens.

Modifications of flower structure

Among the vast number of species of angiosperms there is considerable modification in floral organization, but the main departures from the 'typical' flower just described can be classified into four main categories as follows.

1. Loss of parts A flower which has all four whorls of floral members is said to be *complete*, but if one or more of the whorls is missing the flower is *incomplete*. Even though it may be incomplete, a flower that has both stamens and carpels is said to be *perfect*; if it lacks either or both of these, it is *imperfect*.

A perfect flower is *bisexual* (or hermaphrodite), but an imperfect one is either *unisexual* or *neuter*. If a plant has unisexual flowers, that is, separate staminate and carpellary flowers, both types of flower can be borne on the same plant, which is therefore called monoecious ('one house' accommodates both sexes). Examples of monoecious plants are *Zea mays* (Fig. 28.12) and most members of the Cucurbitaceae (see Fig. 32.5). By contrast, staminate flowers may occur on one plant and carpellary flowers on another, as in papaw (*Carica papaya*), and these plants are said to be dioecious ('two houses' required to accommodate the sexes).

2. Fusion of parts Members of any or all of the floral whorls may show fusion. In describing floral members as fused, it must be realized that the adjective fused does not imply that the members become fused during development. The members of a floral whorl are joined together because they originate from a ridge on the apical meristem and not from separate primordia. Sepals, petals, stamens or carpels are said to be fused only in the sense that these organs are assumed to have been separate in the primitive angiosperms.

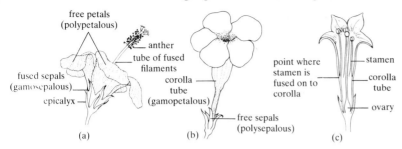

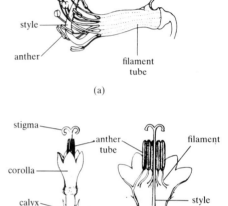

Fig. 28.2 Examples of flowers showing different types of fusion between floral members. (a) *Hibiscus*. (b) *Allamanda*. (c) *Nicotiana*. (a and b from Stone and Cozens 1969, p. 138)

Fig. 28.3 Fusion between members of the androecium. (a) Fusion of filaments. (b) Fusion of anthers: the corolla and anther tube of the flower on the right are split open to show the filaments.

Fusion of floral members may be between members of the same whorl or between members of different whorls (Fig. 28.2). Sepals may be fused by common basal growth to form a sepal tube (gamosepalous), as also may petals which grow as a petal tube (gamopetalous). The opposite conditions of separate, or free, sepals and petals are described as polysepalous and polypetalous. Stamens may be fused together by their filaments to form a staminal tube, as in many legumes of the subfamily Papilionoideae (Fig. 28.3a). Alternatively the anthers are united into a tube, as in all members of the family Compositae (Fig. 28.3b). It was mentioned in the previous section that most flowers have a single ovary composed of two or more fused carpels. In order to understand how this condition comes about it is necessary to consider the structural unit of the gynoecium, namely a single carpel. A single carpel, such as occurs in the flower of any member of the Leguminosae, can be compared to a leaf, with ovules attached at the margins, the two halves of the leaf being folded along the midrib and joined at the margins so that the ovules are enclosed (Fig. 28.4). The gynoecium of a flower which has one or more free carpels is said to be *apocarpous*. When, by contrast, a flower possesses two or more carpels which are fused together to form a single ovary, the gynoecium (and also the ovary itself) is described as being *syncarpous*. In a syncarpous gynoecium the carpels may be joined by their margins, in which case the ovary has a single cavity or loculus (unilocular) with its ovules attached in rows down the wall, the number of rows corresponding to the number of carpels forming the ovary. The regions of attachment of ovules are called placentas, and therefore a syncarpous ovary with a single cavity is said to have *parietal* placentation (literally, wall placentation). Figure 28.4e shows a tricarpellary unilocular ovary with parietal placentation. Alternatively, the carpels may be folded as in a single carpel and then united laterally so that there are as many cavities (loculi) in the single ovary as there are carpels forming it. In such a multilocular ovary the ovules are borne on a central column, and the placentation is *axile*. Figure 28.4f shows a tricarpellary trilocular ovary with axile placentation. In some plants the ovary is multicarpellary and unilocular, but the ovules are attached to a central column extending from the floor to the roof of the ovary. The placentation in this type of ovary is called *free-central* (Fig. 28.4g).

Fusion between floral members belonging to different whorls is more or less confined to fusion between stamens and petals. In this *epipetalous* condition (Fig. 28.2c) the stamens appear to arise from the corolla or perianth instead of from the receptacle. Epipetaly is a feature of almost all tubular flowers.

3. Methods of insertion of parts The way in which the floral members are inserted on to the receptacle may vary in either the horizontal or vertical plane.

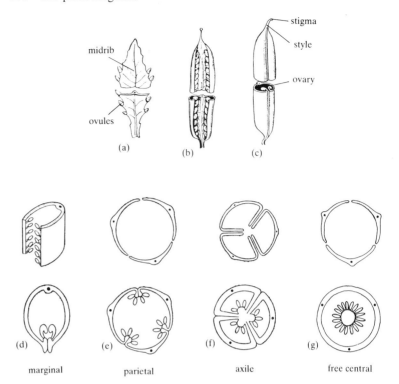

Fig. 28.4 Structure of the gynoecium. (a)–(c)
Hypothetical derivation of a single carpel from
a leaf with marginal ovules. (d)–(g) Types of
placentation.

The horizontal arrangement of the calyx and corolla in the flower bud
(aestivation) sometimes persists in the mature flower, where it may result in
a very striking pattern (Fig. 28.5). There are three main types of aestivation,
distinguished by the following names: (a) valvate, when the edges of the
sepals or petals meet without overlapping; (b) contorted or regular, when
each sepal or petal overlaps an adjoining one on one side, and is overlapped
by the other adjoining one on the other side; (c) imbricate or irregular, when
at least one sepal or petal is wholly internal and at least one other wholly
external. The aestivation of the corolla is one of the features which distinguish the Caesalpinioideae from the Papilionoideae, two subfamilies of the
Leguminosae.

Differences in the vertical level of insertion are due to different shapes of
the receptacle, which is the apical portion of the pedicel (or peduncle) on
which the floral whorls are borne (Fig. 28.5b). The receptacle may be a con-
vex structure, the gynoecium occupying its apex and the androecium, corolla
and calyx arising at successively lower levels. Such a flower is *hypogynous*,
because the outer whorls are inserted below the gynoecium (hypo–below,
gynous–female parts). The receptacle may also be concave to a greater or
lesser extent, the gynoecium occupying the centre of a shallow or deep cup,
and the outer whorls being inserted on the edge of the concavity (*perigynous*). Yet again, the ovary may be embedded in the receptacle, the outer
whorls being inserted above the ovary (*epigynous*). The ovary of an epigynous flower is termed inferior because it occurs below the level of insertion
of the other whorls, that of a hypogynous flower is superior. By convention,
the ovary of a perigynous flower is also classed as superior.

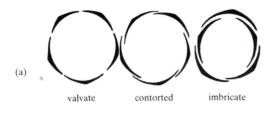

(a)

valvate   contorted   imbricate

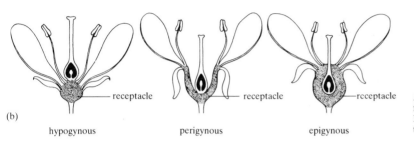

(b)

hypogynous   perigynous   epigynous

— receptacle   — receptacle   —receptacle

Fig. 28.5 Horizontal and vertical insertion of floral parts. (a) Types of aestivation. (b) Methods of insertion due to different shapes of the receptacle.

4. Unequal development of parts   In some flowers the floral members which form a single whorl are all alike in size and shape. Such flowers can be divided down the centre into two equal and similar halves by vertical cuts made in several different directions. The symmetry is radial (or, strictly speaking, multilateral) and the flower is therefore described as actinomorphic (literally, star-shaped). Flowers of *Hibiscus* and *Cucurbita* are actinomorphic (see Fig. 31.3c and Fig. 32.5c,f). On the other hand, if the members of any whorl are not all exactly alike, or if one or more of the members are missing, the flower cannot be equally halved by a cut in more than one direction. When such an irregular flower is bilaterally symmetrical, it is described as zygomorphic. Flowers belonging to the Bignoniaceae and Orchidaceae are zygomorphic (see Fig. 32.3c and Fig. 33.2g). A few flowers, e.g. *Canna* spp., cannot be divided symmetrically in any plane; they are said to be asymmetric.

Recording flower structure

In studying the structure of any flower the first points to be determined are the number of whorls present and the number of members in each of them. This basic information can be conveniently summarized by means of a *floral formula*. To write this, the types of floral leaves are each denoted by a capital letter (K for calyx, C for corolla, P for perianth, A for androecium, and G for gynoecium) and the number of parts in each of them is written as a subscript after the appropriate letter, e.g. C_5 for a corolla of five petals. If the members of any one floral category are arranged in more than one whorl, the fact is recorded by writing, for example, C_{5+5} and not C_{10} for a corolla made up of five inner and five outer petals. When there are more than about twenty members in a whorl the sign ∞ is used to denote numerous. Fusion between members of the same whorl is shown by enclosing the appropriate number in brackets, e.g. $K_{(5)}$ means a calyx of five fused sepals. Fusion between members belonging to different categories of floral whorl is represented by an overhead bracket linking the relevant letters together, e.g. $\overset{\frown}{C_{(5)}} A_5$ means that five stamens are joined to a corolla tube composed of five fused petals. The position of the ovary is recorded by drawing a horizontal line above or below the figure for the number of carpels. The line represents the level of insertion of the sepals, petals and stamens, so that a line below the number of carpels denotes a superior ovary (either a hypogynous or, by convention, a perigynous flower), whereas a line above the figure indicates an inferior ovary (an epigynous flower). Examples of floral formulae are:

$K_5 C_{(5)} A_4 G_{(2)}$ – most Bignoniaceae

$P_{3+3} A_{3+3} G_{(3)}$ – most Amaryllidaceae

The floral formula is a useful shorthand way of summarizing the basic data which must be accumulated before the structure of a flower can be illustrated. To supplement the data recorded in the floral formula, and to give them some spatial meaning, it is necessary to make two drawings, a so-called floral diagram (which is not a diagram in the sense of being a sketchy representation) and a half-flower drawing. These two drawings correspond to an architect's plan and elevation of a building.

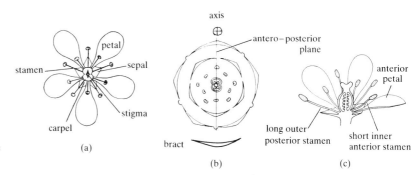

Fig. 28.6 Recording the structure of a hypothetical flower. (a) The flower seen from above looks something like this. (b) Floral diagram representing the plan of the flower. (c) The half-flower drawing representing the elevation of the flower. The surfaces cut in the antero–posterior plane are shown by double lines.

The *floral diagram* is a kind of ground-plan of the flower, in which each whorl is depicted in transverse section at the level giving the most information. It is a stylized, rather than an artistic, drawing in which the relative positions of the various floral parts are accurately represented on a series of concentric circles (Fig. 28.6b). The diagram is orientated by reference to an actual or 'hypothetical' bract subtending the flower. The half of the flower which faces the bract is said to be the anterior half, while the half of the flower which faces the main stalk, or axis on which the flower is borne, is the posterior half. On the floral diagram the position of the bract is indicated by a crescent moon underneath the ground-plan of the flower; the position of the axis is indicated at the top of the ground-plan by a small circle enclosing a cross. In practice it is often difficult to determine the antero–posterior plane of a flower, but, with a few exceptions, the following rule applies. In monocotyledons there is an odd outer perianth member in front, whereas in dicotyledons there is a sepal at the back or, in other words, a petal at the front (Fig. 28.7). The subfamily Papilionoideae of the Leguminosae is the most noteworthy exception to this rule because its members have a large petal (the standard) at the back of the flower. The various parts of the flower are shown on the diagram by the following widely accepted symbols: sepals, petals or perianth members by crescent moons, stamens by small kidneys, and the ovary by a transverse section showing the number of carpels and the type of placentation. Fusion between members of a whorl, or between mem-

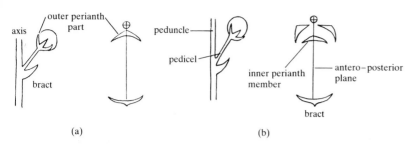

Fig. 28.7 Orientation of flowers. (a) Outer perianth part (sepal) in posterior position; this is the typical orientation of dicotyledonous flowers except for members of the subfamily Papilionoideae of the Leguminosae. (b) Inner perianth member in posterior position; this is the typical orientation of monocotyledonous flowers.

bers of adjacent whorls, is shown by lines which connect the fused members. When the sepals, petals or perianth members overlap one another, the method of their overlapping (aestivation) should also be depicted. The floral diagram, being a ground-plan of the flower, cannot record such features as position of the ovary or general shape of the flower, which can only be seen in vertical section. As a record of floral structure the floral diagram must therefore be supplemented by the half-flower drawing which should be made after the floral diagram.

The *half-flower drawing* (Fig. 28.6c) is precisely what it says it is, a drawing of half a flower, and requires more drawing ability than the floral diagram. The only rigid convention to be observed is that the cut surface of the half-flower drawn should coincide accurately with the antero–posterior plane. In practice it is virtually impossible to cut a flower into two exactly equal halves through this plane, and the floral diagram should be consulted while the half-flower drawing is being made. The latter will indicate which floral members will be bisected by a median antero–posterior cut. It is usual to draw the half of the flower which corresponds to the right-hand side of the floral diagram, i.e. the posterior side of the flower will be on the left of the half-flower drawing. Although the floral diagram is usually self-explanatory, it is a good idea to label the parts of a half-flower drawing, especially those which are characteristic of the flower under examination. Examples of floral diagrams and half-flowers can be seen in Chapters 31 to 33.

Inflorescences

A plant may produce flowers singly on the end of the main shoot or on the ends of branch shoots borne in the axils of ordinary foliage leaves. Such plants are rare, and in most plants the flowers are grouped together as a flowering shoot called an inflorescence. Just as vegetative stems may show monopodial or sympodial branching (see p. 5), so too may flowering stems. According to their pattern of branching, inflorescences are divided into two main types, racemose and cymose.

In a *racemose* inflorescence the main axis does not end in a flower but continues to grow, and the flowers are borne on side branches (i.e. monopodial branching). The flowers open in succession from below upwards, or, if the inflorescence axis is short and flattened, from outside inwards. The following types of racemose inflorescence are recognized (Fig. 28.8). If the flowers are separated by distinct internodes the inflorescence is either a *raceme* when the flowers are stalked, or a *spike* when they are sessile (i.e. have no pedicels). A *spadix* is a spike in which the inflorescence axis is thick and fleshy; it is well shown by aroids. Where the main axis of the inflorescence has not elongated, so that the stalks of the individual flowers arise from

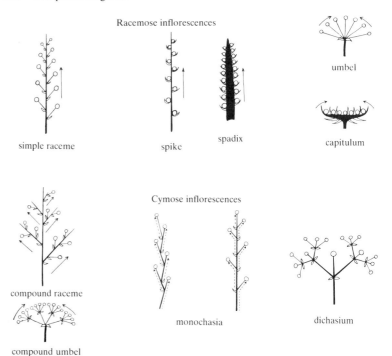

Fig. 28.8 Types of inflorescence. The arrows denote the order in which the flowers open. In the monochasia each dotted line denotes a single axis.

the same point at the top of the peduncle, the inflorescence is an *umbel*, with the oldest flowers of the cluster on the outside and the youngest in the centre. The *capitulum*, or head, has numerous, closely packed, sessile flowers which are borne on the flattened, laterally expanded top of the peduncle. As in the umbel, the oldest flowers are outermost and the youngest in the centre. The capitulum is the type of inflorescence found in all members of the large family Compositae, where a single capitulum often looks like, and is often erroneously called, a flower because the flowers around the edge of the inflorescence resemble petals. In some racemose inflorescences the main inflorescence axis does not bear flowers but lateral inflorescences and the resultant inflorescence is described as being compound racemose. Thus a compound umbel is an umbel of umbels and a compound raceme (or *panicle*) is a raceme of racemes.

In a *cymose* inflorescence the main axis ends in a flower, which is the first to open, and further growth takes place by the growth of one or more laterals, each of which behaves in the same way. The lateral branch overtops its parent axis, and bears a younger flower than the one borne on the parent axis. If each axis of the inflorescence bears only one branch, the cyme is a monochasial cyme or *monochasium*, but if each branch gives rise to two other branches (usually opposite) the cyme is a *dichasium*. In a dichasium the effect of continued overtopping of the parent axis by the branches is to give the inflorescence the form of an inverted cone, in contrast to the erect cone of a simple raceme.

Racemose and cymose inflorescences are not absolutely distinct types, and both types of branching may occur in the same inflorescence. Thus it is not uncommon for an inflorescence to be racemose in its first branching and then cymose in its subsequent branching. Such inflorescences are called mixed inflorescences.

Pollination

If an ovule is to develop into a seed, the female gametic nucleus must first be fertilized by a male gametic nucleus. Before fertilization can occur, however, pollen grains from a mature anther must be transferred to a receptive stigma. If this process of pollination does not take place the ovules die and the flower fails to set seed. The transfer of pollen from an anther of one flower to the stigma of either the same flower or another flower on the same plant is known as self-pollination, and the flower is said to be self-pollinated. When, on the other hand, the pollen is transferred to the stigma of a flower on a different plant of the same species, this is described as cross-pollination. Cross-pollination makes it possible for genes to be exchanged between different individuals of the same species (i.e. for outbreeding to occur) and, generally speaking, is much commoner than self-pollination. Nevertheless, many species are regularly self-pollinated and some are both cross- and self-pollinated.

Cross-pollination

In theory, most plants could be either self- or cross-pollinated but, in practice, the pollinating mechanism is usually such as to favour or ensure cross-pollination. The following are some of the methods by which this result is achieved.

1. Separation of the sexes in space   Some flowers have either stamens or carpels, but not both, and are therefore unisexual (see Fig. 32.5). In dioecious species, such as *Carica papaya* (papaw), where an individual plant is either male or female, cross-pollination is the only possible method. In monoecious species, such as *Cucurbita pepo* (pumpkin), *Ricinus communis* (castor oil plant) and *Zea mays* (maize), where male and female flowers occur on the same plant, the separation of the sexes into different flowers favours but does not ensure cross-pollination.

2. Separation of the sexes in time   In many flowers where self-pollination might be possible the stamens and stigmas mature at different times, with the result that the chances of self-pollination are decreased. This separation in time of maturity is known as dichogamy, of which there are two kinds: *protandry* where the stamens ripen first and *protogyny* where the stigmas ripen first. Both protandry and protogyny are widespread but protogyny is less common than protandry, which is characteristic of most Compositae and many Malvaceae.

3. Self-sterility   It has been shown that in some plants there is a mechanism under genetic control which prevents the ovules being fertilized by pollen from the same plant, and the same may be true for many other plants. In such self-sterile plants the pollen must often reach a stigma of its own flower but it either fails to germinate or fails to develop properly and fertilization is not effected. Certain species of *Passiflora* (passion flower) and *Abutilon* (mallow) have self-sterile flowers.

Since pollen grains have no power of independent movement, cross-pollination requires the aid of some external agent to transfer the pollen from

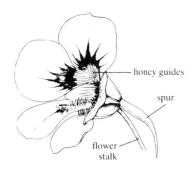

Fig. 28.9 Tropaeolum (usually, but erroneously, called *Nasturtium*) showing honey guides at front of flower and long spur containing nectar at back of flower. (From Scagel *et al.* 1965, p. 573)

one flower to another. Most flowers are pollinated by insects, birds or wind, but there are some which are pollinated by bats, and a few by water currents.

Insect pollination

Some insects visit flowers in search of all or part of their food requirements, and in the process accidently effect cross-pollination. Flowers offer two main kinds of food, pollen itself which is nutritious by virtue of the proteins and lipids inside the grains, and nectar which is a solution of various sugars (mainly sucrose and its breakdown products, fructose and glucose) with, in specific cases, small amounts of amino acids and other organic compounds. Nectar is secreted by specialized cells or hairs in localized areas, called nectaries, which vary in position from flower to flower. Nectaries usually occur at the base of the flower, most commonly as a complete or interrupted ring around the base of the ovary. In most flowers the nectar is retained in the cup formed by the bases of the petals. Generally speaking, therefore, gamopetalous flowers are better fitted for holding nectar than flowers with completely free petals. In a few flowers, of which *Tropaeolum*, the garden nasturtium (Fig. 28.9) and many terrestrial orchids are good examples, the nectar collects in long spurs formed as backward projections of the petals which contain the nectaries.

Types of insect visitors   The insects which visit flowers most frequently are bees. Bees have pollen-collecting hairs on their legs or abdomen, and mouthparts modified as a tube (proboscis) for sucking nectar. The length of the extended proboscis varies greatly in different kinds of bee and, since the nectar is placed at different depths in different species of flowers, this largely determines the types of flowers that are visited by a particular kind of bee. The weight and strength of large bees, such as bumble bees, enable them to open flowers with closed corollas, like *Crotalaria* (see Fig. 31.7) and *Antirrhinum*, and reach the nectar inside. Bees settle on the flower while collecting pollen and nectar, and bee-pollinated flowers usually have a convenient landing stage on which the bee can get a secure foothold before exploring for food (Fig. 28.10). This platform is commonly formed by the enlargement of one or more of the anterior petals, so that the flower becomes zygomorphic (e.g. many members of the Bignoniaceae, Acanthaceae, and Leguminosae subfamily Papilionoideae).

Butterflies and moths are another important group of insects which visit flowers. Unlike bees, they feed exclusively on nectar and never take pollen deliberately. Butterflies and moths have long sucking 'tongues' which are typically longer than those of bees; in tropical hawk-moths the tongue is sometimes as long as 25 cm. When not in use the tongue is coiled into a tight spiral. There are many flowers with long, narrow, tubular corollas, such as *Nicotiana* (tobacco), in which the nectar is so deeply placed that only butterflies and moths can reach it. Most butterflies and some moths settle on the flowers while feeding. The hawk-moths, most of which fly at night, do not alight on flowers but hover in front of them with the tongue inserted in the nectar. Butterflies are day-flyers and are attracted by brightly coloured flowers, such as *Lantana camara* (red sage), *Caesalpinia pulcherrima* (Barbados pride) and *Hibiscus* spp., which are not necessarily highly scented. Most moths fly at dusk or during the night, and are the chief visitors to flowers which are open only at night. Such flowers are usually pale yellow or white in colour (e.g. the night-blooming cacti) and are also strongly scented, especially at night.

Many flies also visit flowers and, with regard to pollination, may be divided into two classes: long-tongued and short-tongued. The long-tongued flies, of which the hover-flies (Syrphidae) are the most important, confine themselves to a floral diet and have mouthparts modified both for removing pollen and for sucking nectar. In the main they visit the same types of flowers as bees. The short-tongued flies have no particular specializations for feeding on flowers, and most of them derive their nourishment mainly from other sources, notably carrion, dung, humus, sap and blood. Flies of this type are usually attracted by flowers which are dull red in colour and have an offensive odour like rotten meat, e.g. those of stapeliads.

Beetles, which are important as pollinators only in the tropics, are as a rule not specially adapted for feeding on flowers, and their short tongues can only lick nectar which is freely exposed.

Members of other classes of insect, such as wasps, bugs and ants, are sometimes found in flowers and must occasionally effect pollination, but they have no special adaptation to a floral diet and derive most of their nourishment from other sources.

The colour and scent of flowers   Most insect-pollinated flowers appear to us to be brightly coloured, but it cannot be assumed that they appear like this to the bee, butterfly or fly. The question of whether bright colours make flowers conspicuous to insects depends on the colour sense of insects, and this can only be determined by experiment.

Research on colour vision in bees was undertaken first by Karl von Frisch in 1914. By training bees to associate colours with food, von Frisch and others have shown conclusively that bees can distinguish blue and yellow as distinct colours, but not red or green. These results agree with the observed fact that flowers visited by bees are usually blue or yellow, or some admixture of these colours, and are rarely red. When we say that such colours as blue and yellow attract the bee we mean simply that the bee can distinguish them and learns to associate them with food, so that it can pick out appropriate flowers very rapidly. Von Frisch's experiments have been repeated with other insects and give similar results for day-flying species but not for night-flying ones. The range of spectral sensitivity of the eyes of bees and most day-flying insects extends futher into the ultra-violet end but less far into the red end than our eyes do. As a result of this difference, two flowers which appear to be the same colour to us might appear differently coloured to bees. This could happen, for instance, if the pigments in the petals of the two flowers differ in that one of them reflects ultra-violet light whereas the other does not. It is also a fact that most red flowers have some blue pigment in them, pure-red flowers being relatively rare and usually pollinated by birds. Nevertheless, butterflies visit red flowers a great deal, and it is probable that their colour vision extends into the red end of the spectrum.

Night-flying insects tend to perceive brightness rather than colours as such. White reflects more light than other colours, so that white and pale yellow flowers are conspicuous against a less brightly reflecting background. If this applies during the day it applies still more at night, and this would help to explain why most flowers pollinated by night-flying moths are white or pale yellow, e.g. *Datura arborea* (moonflower or angel's trumpet), *Nicotiana tabacum* (tobacco) and *Yucca* (Spanish needle). Beetles and short-tongued flies have no evident colour preferences, even though most flowers pollinated by short-tongued flies happen to be dull red or reddish brown.

In addition to being brightly coloured, many flowers also have a very pronounced scent, and this suggests that insects are attracted to flowers by scent as well as by colour. There are also many small and inconspicuous flowers which are visited so regularly by insects that it is reasonable to suspect that such flowers appeal to senses other than sight. Experimental work on the role of scent is difficult, because there is no objective method which is readily available for classifying odours or for measuring their strength. Such exact knowledge as we possess is again due to von Frisch, who trained bees to certain scents by the same method which he used when studying their colour vision. He showed conclusively that bees have a sense of smell similar to our own, and they can rarely detect scent from a greater distance than we can. When bees were trained to colour and scent simultaneously and then offered the two attractions separately, colour alone was perceived from a distance, even when the scent was strong and a wind carried it to the bees. This does not mean that scent has no significance to bees. The conclusion drawn by von Frisch was that colour plays the major part in attraction from afar but that scent is useful in enabling a bee, flying among different flowers of similar colours, to confine its attention to one type of flower during a single flight.

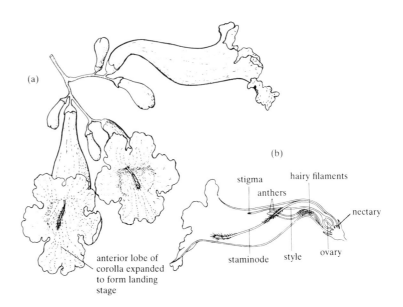

Fig. 28.10 Insect-pollinated flower of *Jacaranda*. (a) Small cluster of the tubular, zygomorphic flowers. (b) Half-flower, showing nectary around the base of the ovary. Pollination is effected either by bees which have to crawl up the tubular corolla to reach the nectar, or by long-tongued butterflies. (From Hall 1970, p. 84).

These results are not applicable to other insects and, in fact, there is very good evidence that the sense of smell is much more important in flies and night-flying moths than in bees. It has been proved that certain hawk-moths can detect the strong perfume of *Lonicera* (honeysuckle) from 100 m away, because when marked individuals were liberated at this distance from, and out of sight of, a plant of *Lonicera* they flew straight to it. Most flowers visited by night-flying moths have a strong scent and it seems that the moths are first attracted by this, and only when they approach more closely are they guided by seeing the white or pale yellow corollas.

Many insects are selective in the flowers they visit, collecting pollen or nectar from flowers of only one species at a time, thereby increasing the chances of cross-pollination. This phenomenon is known as *flower constancy*.

Bees show a high constancy, while butterflies and short-tongued flies show very little. It is interesting that flower constancy operates even when there is colour variation in the selected species. Thus bees in a garden of mixed flowers have been observed to visit those of a species which is represented by many different colour strains (e.g. *Cosmos* spp.) and ignore all the other species present. Although such behaviour could be due to scent alone, it is more likely that the bee is able to recognize at close quarters a combination of the characteristic scent, shape and colour pattern of the selected species. Experiments have shown that bees and day-flying hawk-moths prefer broken to solid patterns (e.g. a four-armed cross to a square), and can also distinguish between similar patterns in different colours, e.g. a yellow circle in a blue ring from a blue circle in a yellow ring. Some flowers have a central 'eye' of a different colour from the rest of the corolla, e.g. black-eyed Susan (*Thunbergia alata*), and others have lines on the petals usually converging towards the centre, e.g. *Tropaeolum* (Fig. 28.9). These eyes or lines are usually called *honey guides*, but this name is unfortunate because they are sometimes found in flowers which do not secrete nectar and 'offer' only pollen to visiting insects. If flowers are illuminated with ultra-violet light and then photographed with a camera having lenses which allow the passage of ultra-violet light, it is found that many flowers have ultra-violet patterns composed of reflecting and absorbing regions. Often the centres of ultra-violet-reflecting flowers have conspicuous ultra-violet-absorbing marks, producing an ultra-violet 'bull's-eye'. In some flowers these patterns correspond with, and often intensify, the honey guides that we can see, whereas in others they occur on petals which appear uniform in colour to us. The eye of a bee is very sensitive to ultra-violet light, and it will 'see' these patterns as honey guides. Since yellow, blue and ultra-violet, and mixtures of these, are the colours to which bees are most sensitive, it is perhaps not surprising that yellow-blue contrasts are particularly common in honey guides. The general conclusion would seem to be that, like scent, honey guides are important in aiding insects to select a particular species of flower from among others of a similar colour. This interpretation is supported by the absence of honey guides in night-blooming flowers, where it is unlikely that they could have any significance.

Bird pollination

It is a common notion that insects are the chief pollinators of flowers, but in some parts of the tropics birds may be even more important. Several groups of birds, particularly humming-birds and honey-suckers, regularly visit flowers to feed on flower-inhabiting insects or on nectar, which they are able to suck with their long beaks. These birds are very small, often little larger than moths, and most of them hover in front of the flowers on which they feed. They effect cross-pollination by touching the anthers and stigmas of the flowers with their breasts or heads.

Birds have a poorly developed sense of smell but powerful vision, and most bird-pollinated (ornithophilous) flowers are scentless but large and brightly coloured, usually red or scarlet. Brilliant colour contrasts are also frequent, as in *Strelitzia* (bird-of-paradise flower). Ornithophilous flowers usually have projecting stamens and stigmas, which are rigid and not easily damaged by birds as they hover in front of the flower to suck the nectar on the wing. The petals are typically fused into a tube which holds copious quantities of a thin watery nectar. The proportions of the tube frequently cor-

Fig. 28.11 Flower of a Jamaican species of *Columnea* (*C. argentea*) and head of the streamertail humming-bird (*Trochilus polytmus*) which pollinates it. Note how the size of the corolla appears to be correlated with the length of the beak and head of the hummingbird. (Drawn by D. Erasmus, from W. T. Stearn (1969), *Bulletin of the British Museum* (*Natural History*) *Botany*, **4**, 200)

respond to the length and curvature of the bird's beak (Fig. 28.11). Sometimes birds 'steal' the nectar by piercing the flower from behind with their beaks; this can often be seen in *Hibiscus* flowers which birds have visited.

Ornithophilous flowers include species of *Hibiscus, Erythrina* (coral tree) *Fuchsia, Passiflora* (passion flower), *Strelitzia* (bird-of-paradise flower), and *Bombax* (silk-cotton tree of tropical Asia and Africa).

Wind pollination

Flowers pollinated by wind (anemophilous) are generally inconspicuous, often green in colour and individually small. They are characterized mainly by the absence of the features associated with insect-and bird-pollinated flowers. Scent, nectar and bright colours are all lacking, and the calyx and corolla are either small, consisting of greenish scales, or even absent.

On the positive side there are peculiarities in the stamens, pollen grains and stigmas. The stamens have long filaments so that the anthers are freely exposed and easily swayed by the wind. The anthers themselves are often versatile (i.e. attached near the middle) and rock about the point of attachment; this arrangement is clearly seen in the grasses. The pollen is thus shaken into the air when conditions are such as to ensure its ready dispersal. The pollen grains of wind-pollinated flowers are very small in size, light in weight, smooth, dry and powdery (i.e. not sticking together in masses), and produced in enormous quantities. The pollen can sometimes be seen to scatter as a dense cloud of dust, especially when the plant is shaken. The stigmas, like the stamens, are fully exposed. They are comparatively large, and often lobed or feathery, so that wind-borne pollen is easily caught on their surfaces.

The success of wind-pollination depends on how far and how thoroughly the pollen is dispersed. In most herbaceous plants with wind-pollinated flowers, of which the grasses provide the most familiar examples, the flowers are borne on long stems well above the leaves, and airborne pollen therefore has free access to the flowers. Self-pollination is excluded in many wind-pollinated flowers by the fact that the plants are dioecious, the staminate and carpellary flowers being present on separate plants.

One of the best examples of wind-pollinated plants is *Zea mays* (Fig. 28.12). The male flowers occur at the top of the plant in a terminal panicle commonly called the *tassel*. When the anthers of the tassel burst a cloud of pollen dust is formed. It has been calculated that a single tassel may produce as many as 60 million pollen grains. The female flowers are produced lower down the plant in compound spikes, or *cobs*, which are closely enveloped in leaf sheaths called the *husk*. Projecting from the apex of the cob is a large number of long hairs, or *silks*, which are the stigmas. When

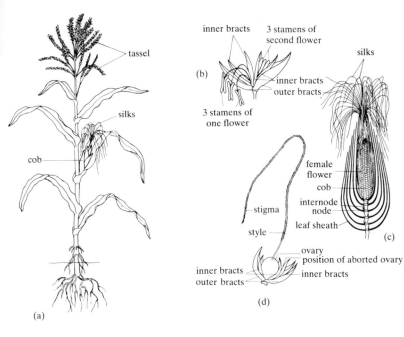

Fig. 28.12 Maize (*Zea mays*), a wind-pollinated plant. (a) Entire plant showing male and female inflorescences. (b) Single spikelet of tassel, opened to show the stamens. (c) L.S. cob showing ensheathing husk and projecting silks. (d) Single spikelet of cob to show attachment of silk to ovary. (After Hall 1970, p. 86)

receptive they are sticky so that any pollen grains which are blown on to them adhere. Each silk or stigma is connected by a long style to a single ovary which, following fertilization, becomes a grain or kernel. In maize, the male flowers generally shed their pollen before the female flowers of the same plant mature. When the stigmas of the female flower are receptive, therefore, pollen is blown to them from adjacent maize plants, so that cross-pollination takes place.

Maize is a member of the grass family, Gramineae, a family which covers more of the earth's surface than any other family of angiosperms, and also is one of the largest families as regards number of species. Since nearly all grasses are pollinated by wind, there can be no doubt that wind pollination in grasses is a very effective method of achieving cross-pollination.

Bat pollination

Many tropical trees are pollinated by species of bats which have long slender heads and extensible tongues, capable of being inserted into the flowers. Bats are nocturnal and are probably guided to the flowers by their acute sense of smell. They clamber on the flowers, hold on with the thumb-claws of their wings, and either extract nectar or small insects with their tongue or else chew the pollen or succulent petals. The claw holes in the corolla are so characteristic that they are considered to be good evidence of bat pollination, even in those plants where visits by bats have never been observed. Typical bat-pollinated flowers have a number of characteristics which are clearly related to visits by bats. The flowers produce an abundance of pollen and of sticky or mucilaginous nectar, and are frequently wide-mouthed so that the nectar is easily accessible; they open only at night when they emit a very peculiar fusty odour. Their position on the plant is such that bats can get at them easily. In *Kigelia*, the well-known 'sausage' tree (Fig. 28.13), the flowers are suspended by means of rope-like branches, whereas in *Crescentia* (calabash) they are attached to the tree trunk. The same result is achieved

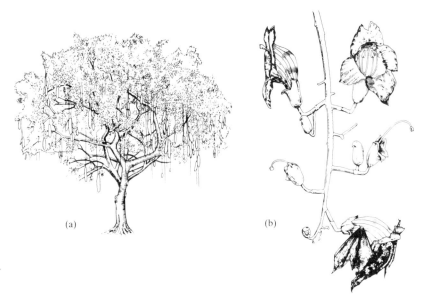

Fig. 28.13 Bat pollination. (a) Sausage tree
(*Kigelia*) to show fruits hanging on long stalks.
(b) Bat pollinating flower of sausage tree.
(From Scagel *et al.* 1965, pp. 579–80)

in *Ceiba* (silk-cotton tree of tropical America and Africa) which produces
its flowers during the dry season when the tree is leafless and bats can fly in
the canopy unhindered by leaves. Other bat-pollinated plants are *Adansonia*
(baobab), *Ochroma* (balsa), *Durio* (durian) and *Musa* (banana and plantain).

Self-pollination

Although most plants have devices which favour or ensure cross-pollination

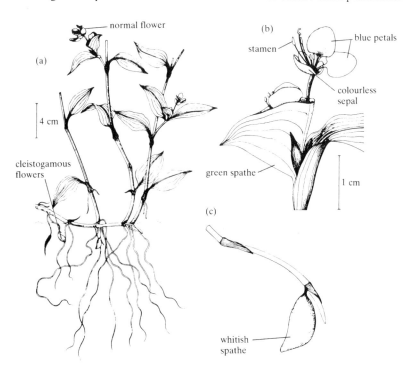

Fig. 28.14 Commelina benghalensis. (a) Habit.
(b) Normal flowers. (c) Cleistogamous
flowers, enclosed between two large whitish
bracts forming a spathe. (From Ewer and Hall
1973, p. 301)

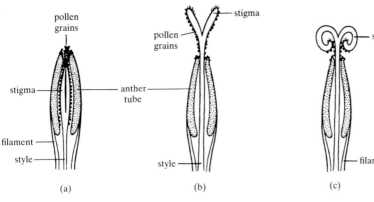

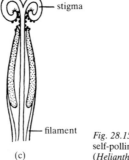

Fig. 28.15 Diagram to show how self-pollination takes place in sunflower (*Helianthus*) if cross-pollination fails to occur.

there are some which are regularly self-pollinated. Thus in the temperate-zone cereals, wheat (*Triticum*), barley (*Hordeum*) and oats (*Avena*), the anthers invariably shed their pollen on to the surface of the ripe stigmas before the flower opens. Other plants, such as *Commelina benghalensis* (Fig. 28.14), have both normal flowers and flowers which never open but yet produce abundant seed from their own pollen. These closed or *cleistogamous* flowers, which are produced at or below the surface of the soil, are small and budlike, and the petals are reduced or absent. Cleistogamous flowers are also found in *Arachis hypogaea* (groundnut), species of *Drosera* (sundew) and several other plants.

Apart from these special cases of obligatory self-pollination, many flowers which are apparently adapted for cross-pollination are so constructed that self-pollination can also take place if cross-pollination fails. This is usually brought about by movements of the filaments or stigmas, and is demonstrated by many members of the Compositae (Fig. 28.15). Here the inner receptive surfaces of the mature stigma are exposed above the surrounding anthers in a position favourable to receive pollen from another floret of either the same plant (self-pollination) or another plant (cross-pollination). If pollen from another floret does not reach the stigma, then the two branches of the stigma curve outwards and downwards bringing the stigmatic surfaces into contact with pollen from the same floret. In this way self-pollination can take place towards the end of flowering, whereas previously the receptive stigmas were in a position favourable for cross-pollination to occur.

Chapter 29

Angiosperms II. Life cycle and germination

Life cycle

The angiosperm plant body, with its roots, stems and leaves, belongs to the sporophyte generation. During flowering this plant produces microspores which develop into male gametophytes, and megaspores which develop into female gametophytes.

As in gymnosperms, both male and female gametophytes are reduced to microscopic proportions. The male gametophyte is ultimately shed as the pollen grain; the female gametophyte, which in angiosperms is represented by the *embryo sac*, is retained on the sporophyte as part of the ovule. If pollen grains are deposited on the receptive stigmatic surface of a compatible flower, they may germinate and one may fertilize the egg nucleus in the embryo sac. A zygote is formed, and the ovule develops into a seed which, if it germinates successfully, becomes the sporophyte of the next generation and the life cycle is complete. When represented by the type of life-cycle diagram standard in this book, the life cycle of angiosperms (Fig. 29.1) is seen to differ only in details from that of conifers (see Fig. 27.13).

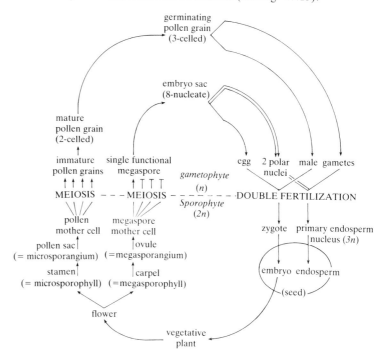

Fig. 29.1 The life cycle of an angiosperm.

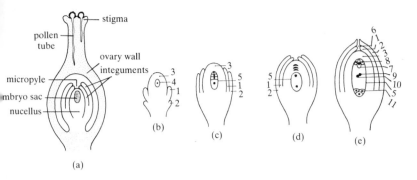

Fig. 29.2. Development of the ovule. (a) L.S. monocarpellary gynoecium containing a single erect ovule. (b) Very young ovule before meiosis. (c) Linear tetrad of megaspores. (d) First nuclear division within the embryo sac. (e) Ovule ready for fertilization. (b–e from Sporne 1974, p. 20) (1. inner integument; 2. outer integument; 3. nucellus; 4. megaspore mother cell; 5. functional megaspore, or embryo sac; 6. micropyle; 7. egg cell; 8. two synergidae; 9 and 10 polar nuclei, about to fuse to form the secondary endosperm nucleus; 11. three antipodal cells.)

The ovary and ovule

The structure of the ovary and ovule is described, for the sake of clarity, with reference to a hypothetical monocarpellary ovary containing a single erect ovule borne on a short stalk from the floor of the ovary (Fig. 29.2). By analogy with mammalian embryology, the portion of the ovary wall to which the ovule is attached is called the placenta. The ovule originates as an actively growing hump of tissue on the surface of the placenta. This hump develops into an ovoid mass of tissue, the nucellus (=megasporangium), and as it enlarges two collar-like rings of tissue, the integuments, grow up round it. Early stages in the development of the ovule are illustrated in Fig. 29.2. In Fig. 29.2b the inner integument partially envelopes the nucellus while the outer integument forms a ring around its base. Eventually (Fig. 29.2e) both integuments completely cover the nucellus except for a small hole, the micropyle, which is left at the apex.

The pattern of nuclear divisions described here is often called the 'normal' type because it occurs in about 70 per cent of angiosperms. At an early stage in the development of the ovule one hypodermal cell of the nucellus enlarges to become the spore mother cell, which divides by meiosis to form a linear tetrad of megaspores, of which only the one furthest from the micropylar end is functional (Fig. 29.2c). The other three megaspores degenerate. The functional megaspore, or embryo sac as it is called from now onwards, enlarges at the expense of the abortive megaspores and some of the surrounding nucellar tissue. Within the embryo sac the megaspore nucleus divides into two daughter nuclei which move to opposite ends of the now greatly enlarged embryo sac (Fig. 29.2d). Each then undergoes two further divisions. The group of four products nearest the micropyle consists of a three-celled *egg apparatus* and the *upper polar nucleus*, while the four at the opposite end develop into a group of three *antipodal cells* and the *lower polar nucleus*. The egg apparatus is made up of the female gamete or egg cell, and two other cells which are called the *synergidae* (literally, helper cells) because it was once presumed that they must assist fertilization in some way. Finally the two polar nuclei fuse to form a *secondary endosperm nucleus*, and the ovule is ready for fertilization. The female gametophyte is therefore represented by only seven cells (derived from eight nuclei), of which one is the egg cell, and any obvious traces of archegonia have been lost.

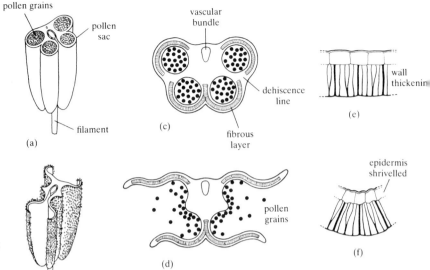

Fig. 29.3 The stamen. (a) Anther before dehiscence showing the four pollen sacs. (b) Anther after dehiscence. (c) and (d) Transverse sections of anther before (c) and after (d) dehiscence. (e) and (f) T.S. wall of anther showing epidermis and fibrous layer before (e) and after (f) dehiscence. (a and b from D.G. Mackean (1965), *Introduction to Biology*, p. 39, John Murray, London)

The stamen and pollen grain

A stamen consists essentially of two parts, a stalk or filament, and a head or anther. Although there is considerable variation in detail, there are very few stamens in which these two parts are not readily recognizable. In most stamens the anther is divided into two halves or *anther lobes* which are borne on either side of the *connective*, a continuation of the filament. In each anther lobe are two longitudinal compartments, the *pollen sacs*, and the position of the partition wall between them is marked externally by a groove running vertically down the side of the anther lobe (Fig. 29.3). As it approaches maturity, the pollen sac is full of pollen mother cells, each of which divides to give a tetrad of pollen grains (= microspores). Each pollen grain is covered with a two-layered wall, a thin inner wall of cellulose called the *intine*, and a thick outer cutinized wall called the *exine*. The cell wall usually has one or more places (apertures) where only a very thin layer of exine is deposited, and through one of these the pollen tube will later emerge. In wind-pollinated plants the exine is laid down evenly over the surface of the pollen grain except for any apertures, but in insect-pollinated plants it is sculptured or covered with spines in various, and often elaborate, patterns which are characteristic of the species. It is presumed that these modifications assist the pollen to cling to the bodies of insects or to receptive stigmas.

When the pollen grains are mature, the partition between the two pollen sacs in each anther lobe breaks down, so that the anther becomes bilocular, and the pollen grains are then released by longitudinal splits which develop along the lateral grooves marking the position of the original partitions (Fig. 29.3b). Like the forcible discharge of spores from the sporangium of a fern, the liberation of pollen grains depends upon tissue tensions caused by drying out of the tissues. The mechanism by which pollen grains are liberated can be understood by reference to a transverse section of a mature anther (Fig. 29.3c). Between the epidermis and the pollen sacs there is a layer of cells, called the *fibrous layer*, which have bands of lignified thickening on their radial and, particularly, on their inner tangential walls, but not on their outer tangential walls. The cells of the fibrous layer form a continuous sheet except under the lateral grooves on each side of the anther lobes. When the cells of the fibrous layer dry out, their radial walls are pulled together and the outer tangential wall, being unthickened, is thrown into folds. The surface area of the inner tangential wall remains more or less unchanged because the bands of lignified thickening on the surface of this wall form a meshwork which prevents it from contracting. The cohesion of the cell contents and their adhesion to the wall prevent the entry of air into the cells. With increasing desiccation, the strain on the radial walls builds up until the resultant tension causes the thin-walled cells along the groove of each anther lobe, which constitute a line of weakness, to tear apart and the anther walls to curl back (Fig. 29.3d and f). A long gaping slit on each side of the anther exposes the pollen grains, which are now free to be transported to a receptive stigma. Most anthers open by the formation of two longitudinal slits in this way, but there are other methods of dehiscence. For instance, some anthers open by an apical pore through which pollen is squeezed, and others by flaps which lift up as valves.

The formation of tetrads of pollen grains by meiosis marks the beginning of the male gametophyte generation. When first formed a pollen grain has dense cytoplasm and a centrally placed nucleus. Very soon, however, vacuolation takes place and the nucleus is consequently pushed to one side (Fig. 29.4). The nucleus then divides into a large *vegetative cell*, and a much smaller *generative cell* which later divides to form two male gametes. This last division may occur before the dispersal of the pollen grain or, more usually, during the germination of the pollen grain on a receptive stigma. The male gametophyte, consisting of only three cells, is thus even more reduced than the female gametophyte.

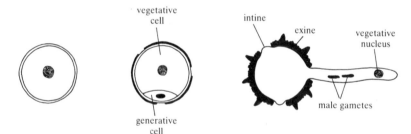

Fig. 29.4 Stages in the germination of a pollen grain.

Fertilization

By means of various agencies (such as wind, water and animals) pollen grains are transferred from the anther of a stamen to the stigma of a carpel. This transfer is called pollination, and has already been discussed. Once deposited on a receptive stigma, irrespective of how it got there, the pollen grain must germinate before it can reach the ovule, which is enclosed within the ovary. The pollen grain germinates by pushing out a pollen tube which grows through the tissue of the style and into the cavity of the ovary. The vegetative nucleus lies near the tip of the growing pollen tube, followed by the generative cell. While the tube is growing through the style, the generative cell (if it has not already divided before the pollen grain was shed) divides into two male gametes. In most angiosperms, the pollen tube enters the ovule through the micropyle, but in some species entry is effected through the stalk of the ovule. On reaching the embryo sac, the tip of the pollen tube forces an entry and discharges its contents. The nucleus of one of the male cells fuses with the egg cell to form the zygote, the first cell of the next sporophyte generation. The other male nucleus fuses with the secondary endosperm nucleus to form the *primary endosperm nucleus* which, in the type of embryo sac being described, is therefore triploid. This nucleus subsequently undergoes a series of divisions which result in the formation of a tissue known as the *endosperm*. This becomes rich in nutrients and surrounds the developing embryo.

The involvement of both male nuclei in fusion processes is called *double fertilization*. As far as can be judged, it is probably universal in angiosperms, and nothing like it occurs in any other group of plants. It is difficult to suggest any advantage which double fertilization could possibly confer on angiosperms, beyond the speculation that, because the endosperm contains both male and female genetic components, it might somehow provide a more favourable environment for the embryo (which itself contains the same genetic components) than one which, as in the female prothallus of gymnosperms, lacks the male component. Regardless of whether or not double fertilization has any biological advantage, its universal occurrence is very significant from a phylogenetic point of view, because it lends support to the belief that angiosperms are a natural group with a common ancestry.

Embryo and seed development

The development of the embryo plant from the zygote begins with the division of the zygote, or fertilized egg cell. Stages in the development of the embryo will be described with reference to the dicotyledonous plant called shepherd's purse (*Capsella bursa-pastoris*), the embryo of which was among the first to be studied in detail more than a century ago and has subsequently become the 'reference' type (Fig. 29.5). The first division of the zygote is at right angles to the axis of the embryo sac, and gives rise to a *basal* cell towards the micropyle, and a *terminal* cell towards the centre of the embryo

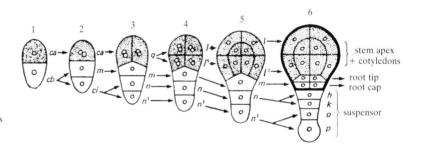

Fig. 29.5 Early stages in the development of the embryo of *Capsella*. (From Sporne 1974, p. 28) (Note: the products of the terminal cell are shaded, and the heavy outline in 6 encloses the 'embryo proper', as distinct from the suspensor.)

sac. The basal cell by transverse divisions becomes a filament (the *suspensor*) with a swollen end cell. This swollen cell seems to function both as an absorbing organ and as an anchor for the elongating filament, thus enabling the terminal cell of the pro-embryo (i.e. the complete structure derived from the zygote) to be thrust into the embryo sac. The cell of the suspensor adjacent to the terminal cell differs from the other suspensor cells in that it contributes directly to the development of the embryo, its derivatives subsequently forming the root cap and root tip of the embryo. The terminal cell divides first by anticlinal walls (i.e. at right angles to the surface) to give

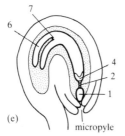

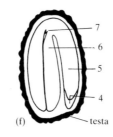

Fig. 29.6 Embryo formation in *Capsella*.
(a)–(d) Stages in the development of the
embryo. (e) L.S. ovule showing developing
embryo. (f) L.S. ripe seed. (a–d from Sporne
1974, p. 31) (1. end cell of suspensor; 2.
suspensor; 3. 'embryo proper'; 4. root cap; 5.
hypocotyl, with vascular system, 6. cotyledons;
7. stem apex.)

eight cells known as the 'octant stage'. Walls are then laid down periclinally, and a central region of cells is delimited from an outer series of surface cells which are the forerunners of the shoot epidermis. Subsequent divisions result in a mass of cells, which is at first spherical, then heart-shaped and finally bilobed (Fig. 29.6). At this stage the parts of the embryo destined to form the principal organs can be recognized. The two lobes form the cotyledons, while between them is the rudiment of the young shoot. The cylinder of tissue below the cotyledons is the root and *hypocotyl* (the portion of the axis of a young seedling connecting the root with the shoot). From the heart-shaped phase onwards the tissue in the centre of the embryonic axis is clearly differentiated from the tissues surrounding it. This tissue (shown by dotted lines in Fig. 29.6) is the forerunner of the vascular system of the seedling. Summarizing, one may say that the derivatives of the basal cell give rise to the suspensor and the root apex of the embryo, while the derivatives of the terminal cell provide the cotyledons and the stem apex.

While the zygote is developing into an embryo, changes are also going on elsewhere in the ovule. The primary endosperm nucleus divides repeatedly to form the endosperm, a triploid tissue which serves to nourish the developing embryo and often, later, the germinating seed. From an evolutionary point of view the endosperm is a 'new' structure which, like the double fertilization which precedes it, is characteristic of angiosperms only. As the embryo sac increases in size to accommodate the developing embryo, the nucellus gradually diminishes until in the mature seed it is obliterated. The ovule as a whole is also enlarging, and the two integuments differentiate to become the seed coat or testa.

The developing ovule is connected with the parent plant through the placenta, so that water and food materials can pass into the embryo sac. Part of this food is used at once for respiration and the formation of proto-

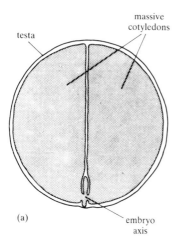

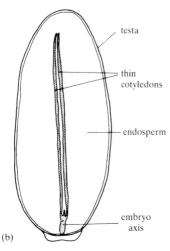

Fig. 29.7 Structure of (a) non-endospermic and (b) endospermic seeds. (After Villiers 1975, p. 16)

plasm and cell walls, but part of it is stored. In *endospermic* seeds the endosperm persists to become the storage tissue (e.g. castor-oil seed, cereal grains), whereas in *non-endospermic* seeds the endosperm dwindles and the cotyledons become the storage tissue, as in the seeds of legumes (Fig. 29.7).

Development of the embryo does not continue indefinitely. Cell divisions become less frequent and ultimately cease, and the seed dries out. Eventually it separates from the placenta, leaving a scar (the *hilum*). When this occurs the seed enters a phase of inactivity, in which the respiration rate may be so low as to be scarcely detectable. In this dry state seeds are extremely resistant to changes in the external environment, and some can even withstand immersion in boiling water or liquid air. Many biologists think that this property of delaying growth until conditions become favourable may be a major factor in the success of flowering plants over the rest of the plant kingdom.

The evolution of the seed habit

Plant life originated in water and then migrated on to the land. All land plants except gymnosperms and angiosperms bear witness to their aquatic ancestry in that external water is necessary for the motile male gametes to reach the female gametes. The seed plants have dispensed with the necessity of external water for fertilization, but the evolutionary steps which are thought to have been involved in the acquisition of the seed habit are intelligible only in terms of an aquatic ancestry. In their conquest of the land, plants had to solve many problems associated with the differences in the nature of the ambient medium, but the ones involved in the evolution of the seed habit were related mainly to the process of sexual reproduction. The seed is a reproductive body which is formed by a sexual process completely independent of the presence of external water.

Heterospory

The seed habit first occurred in plants during the Carboniferous period, and has undoubtedly arisen more than once in evolutionary history. It is thought that it evolved in a pteridophyte stock with megaphyllous leaves (i.e. fern-type leaves as opposed to *Selaginella*-type leaves, which are microphylls). In the life cycle of a 'typical' fern, such as that described in Chapter 26, the haploid gamete-bearing phase (the gametophyte) is bisexual, bearing both archegonia and antheridia. In seed plants, however, the haploid phase is unisexual, its sex being controlled by the preceding diploid phase (the sporophyte). As in all plants with a unisexual haploid phase, this is achieved by the production of two types of sporangia by the parent sporophyte. One type of sporangium contains a small number of large spores, called megaspores, which give rise to female haploid plants, while the other type contains a large number of small spores, called microspores, which give rise to male haploid plants. This condition is known as heterospory, and its essential features are clearly shown in the life cycle of *Selaginella* and other club mosses. The biological advantage of heterospory would seem to be that a megaspore with a large food reserve derived from the parental sporophyte enables the new embryo plant to have a better start than an embryo formed on a gametophyte which has to produce its own food supply after the megaspore has germinated. Conversely, haploid plants which develop antheridia need, and are provided with, only a small reserve of food. Heterospory thus achieves a division of labour between the sex cells. Although it may be looked upon as the first step in the evolution of the seed habit, heterospory as such is not an adaptation to life on land. Its importance in this context seems to lie in the fact that, once it had occurred, it made possible other changes which were stages directly involved in the evolution of the seed habit. These changes converted the megasporangium and female gametophyte into the ovule, and the microspore and male gametophyte into the pollen grain. Knowing the type of life cycle of the plants that must have been the ancestors of the seed plants on the one hand, and the life cycle of angiosperms on the other, it is possible to identify the changes which must have taken place.

Retention of the megaspore on the parent diploid plant

In the fern it is the haploid gamete-bearing phase which is ill-adapted to life on land, and a further advantage of heterospory over homospory would appear to be that heterospory, with its more adequate provision of food material for the next generation, allows the haploid phase to be passed through more rapidly. The haploid phase, relieved of the necessity to make an independent food supply, can largely dispense with its non-reproductive parts. A reduction in the number of functional megaspores to one per megasporangium would also be an advantage in this respect. A further step in the reduction of the female gametophyte would obviously be the retention of the megaspore within its megasporangium on the parent sporophyte until after fertilization had occurred. By this means the necessary food material could be supplied directly to the female gametophyte and to the new sporophyte developing from the fertilized egg. If fertilization is to occur, however, the male microspores would have to germinate in the vicinity of the female sporangia, instead of on the ground.

The step of retaining the megaspore is not just a hypothetical possibility, but was in fact taken by at least three separate lines of plants during the Carboniferous period. However, these lines did not replace the existing heterosporous lines which dispersed their megaspores. The reason may have been that, in acquiring the advantages of retaining the haploid phase within the megasporangium, they forfeited one of the primary functions of the ancestral megaspore, namely dispersal. It was not until further changes in the reproductive organs occurred to enable them to reassume a dispersal role that plants possessing attached megaspores became the dominant form of plant life on land. These further changes represent the final stages in the evolution of the seed habit.

The development of the seed as a dispersal unit

The development of an entirely new dispersal unit, called the seed, involved several major changes, and it is convenient to separate the changes which occurred on the female side from those which occurred on the male side, although the two are of course interdependent.

Changes on the female side   In the first plants to retain the megaspore it is probable that the female gametophyte was more or less freely exposed, and that the male gametes reached the archegonia from where the microspores germinated on the female sporangium by swimming in a film of rain water or dew. In the course of evolution not only does the female gametophyte become permanently retained within the megaspore, but the megaspore itself becomes permanently retained within the megasporangium. The megasporangium in its turn becomes surrounded by one or two protective coverings, called integuments, which are regarded morphologically as part of the megasporangium rather than as extensions of the megasporophyll. The composite structure so formed is 'an indehiscent, integumented megasporangium containing a single functional megaspore', and it is this structure which, after the egg cell it contains has been fertilized and started to develop into a new embryo plant, is shed as a dispersal unit called the seed. A mature seed is therefore composed of three generations: diploid tissues of the parent sporophyte (testa and nucellus); haploid, or in angiosperms triploid, tissue of the female gametophyte (endosperm); and the diploid tissue of the daughter sporophyte (embryo).

Changes on the male side   Of equal importance to the acquisition of the seed habit was the transformation of a free-living male gametophyte into a pollen grain. This change involved a reduction in the size of the microspore which allowed it to be produced in greater numbers and to be transported to the female gametophyte, which necessarily had to make provision for the reception of the microspores in a pollen chamber (gymnosperms) or on a receptive stigma (angiosperms). To bring about the fertilization of an egg cell produced by a female gametophyte retained in a megaspore retained in a closed megasporangium, the microspore or pollen grain forms a pollen tube to penetrate the intervening tissues and liberate motile male gametes (cycads) or non-motile gametic nuclei (conifers and angiosperms) in the immediate vicinity of the egg cell. Internal fertilization by means of a pollen tube thus completely eliminates the necessity of a supply of external water for the process, which is an obvious limitation in the life cycle of a land plant.

The success of the seed plants as a group is somehow tied up with the adoption of the seed habit. It is difficult to decide just what advantage the seed plants have over other plants, but it should be realized that the seed is a great deal more than a mere dispersal unit. Being a multicellular structure it has been able to evolve all kinds of structural and physiological devices not available to a single-celled megaspore, and it has become the thoroughly efficient dispersal and perennating unit that it is today.

Fruits

After a flower has been pollinated it withers, and some or all of the floral parts other than the ovary (or ovaries in the case of flowers having an apocarpous gynoecium) are shed. The ovary persists and develops into the fruit, which is the modified gynoecium that encloses the seeds until they are ripe. There are two separate processes involved in fruit formation, the development of the ovules into seeds and the modification of the ovary wall into the fruit wall (*pericarp*). In a few flowers, of which cashew (*Anacardium occidentale*) is an example, the floral receptacle also undergoes changes which are involved in the formation of the mature fruit.

The structure of fruits largely determines the method by which the seeds inside are dispersed, and also provides a very useful character in the classification of flowering plants. To understand the structure of any given fruit there are five questions which have to be answered. (1) Is the ripe fruit dry or succulent? (2) Does the ripe fruit split at maturity to release the seeds (dehiscent), or does it remain closed and itself become the dispersal unit (indehiscent)? (3) Does a single fruit contain one or more seeds? (4) Is a single fruit the product of one, or more than one, carpel, and if the latter how many carpels were involved in the formation of the fruit? Exceptionally, as in the fig (*Ficus spp.*) and pineapple (*Ananas comosus*), the fruit is formed from a whole inflorescence. (5) Is the fruit the product of a superior or an inferior ovary?

Types of fruits

In describing the different types of fruits it is convenient to divide them into two major types, dry and succulent, according to the condition of the fruit wall in the ripe fruit.

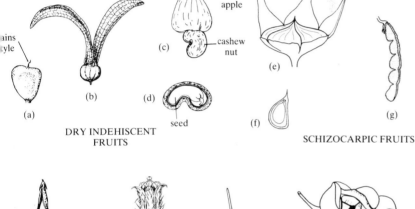

DRY INDEHISCENT FRUITS

SCHIZOCARPIC FRUITS

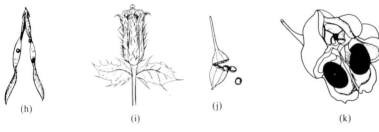

Fig. 29.8 Dry fruits. (a) 'Achene' of maize. (b) Samara of *Dipterocarpus*. (c) Cashew nut on swollen receptacle and pedicel. (d) Section of cashew nut showing seed inside hard pericarp. (e) Schizocarp of *Sida carpinifolia* (family Malvaceae) which splits into five fruitlets (f) Single one-seeded fruitlet of *Sida carpinifolia*. (g) Lomentum of *Desmodium*. (h) Legume. (i) Capsule of Mexican poppy (*Argemone*). (j) Capsule of *Celosia argentea* with 'lid', (k) Succulent capsule of akee apple (*Blighia sapida*). (k from Purseglove 1968, vol. 2, p. 645)

Dry fruits may be further subdivided according to whether the fruit is indehiscent or dehiscent (Fig. 29.8). Dry indehiscent fruits usually contain a single seed, in which case they are called achenes. The *achene* is typically the product of a single carpel, but the fruits of Compositae and Gramineae, which develop from bicarpellary ovaries, are also regarded as achenes because only one of the carpels, containing a single seed, comes to maturity. In the achene of maize (Fig. 29.8a), *Sorghum* or any other grass the pericarp is exceedingly thin and fused with the testa of the enclosed seed. The *nut* differs from the achene only in having a very hard pericarp. The cashew nut is a true nut (Fig. 29.8c), but the 'cashew apple' to which it is attached is the swollen, pulpy pedicel and receptacle. An achene in which the pericarp is extended to form a wing is called a *samara* (Fig. 29.8b).

Dry dehiscent fruits contain many seeds, and are of two basic types; *capsular*, in which the fruit wall splits along predetermined lines as a result of tensions set up in the tissues as they dry out, and *schizocarpic*, in which

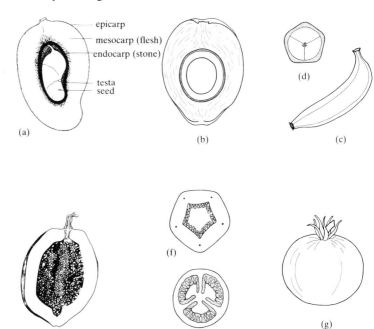

Fig. 29.9 Succulent fruits. (a) Drupe of mango. (b) 'Drupe' of coconut. (c) Inferior berry of banana. (d) T.S. banana to show derivation from a three-carpellary ovary with axile placentation. (e) Superior berry of papaw. (f) T.S. papaw to show parietal placentation. (g) Berry of tomato (*Lycopersicon esculentum*). (h) T.S. tomato to show axile placentation. (e) from Purseglove 1968, vol. 1, p. 47)

the ripe fruit separates into closed one-seeded portions. The simplest type of capsular fruit is the *legume*, or pod, which is the distinguishing feature of the Leguminosae. The legume is formed from a superior monocarpellary ovary, and, when ripe, splits open along both margins from the tip towards the point of attachment (Fig. 29.8h). The *follicle* is similar but differs in that it splits along one margin only. Dry dehiscent fruits derived from two or more carpels form *capsules*. A three-carpellary capsule is common among monocotyledons, while many dicotyledons have a five-carpellary capsule. Most capsules dehisce by longitudinal slits (Fig. 29.8k) but in a few, for example *Portulaca* and *Celosia* (Fig. 29.8j), there is a horizontal line of weakness so that the top of the capsule comes off like the lid of a box. Schizocarpic fruits, like capsular fruits, are dehiscent but, instead of exposing the seeds, the fruit splits into as many one-seeded fruitlets as there are carpels. The seeds remain enveloped in carpellary tissue so that each fruitlet is comparable to an achene. *Schizocarps* are typical of the Malvaceae (Fig. 29.8e). A type of fruit resembling the schizocarp is the *lomentum*, which is a legume which does not open by longitudinal slits in the normal way but splits transversely into one-seeded segments when the fruit dries out (Fig. 29.8g).

Succulent fruits are of two main types, the drupe and the berry, both indehiscent (Fig. 29.9). On account of their fleshy nature many fruits in this category are edible and grown as crops. The *drupe* is a 'stone fruit', of which the mango (*Mangifera indica*) is a good example. In the formation of a drupe the ovary wall gives rise to a pericarp consisting of three distinct layers. The outer one, the *epicarp*, forms the outside thin skin; the middle layer, the *mesocarp*, gives rise to the succulent portion of the fruit; while the innermost layer, the *endocarp*, forms the hard stone which is the characteristic feature of this fruit type. Within the stone is a single seed or 'kernel'. Most drupes are formed from superior monocarpellary ovaries. Even though it is not a succulent fruit, the coconut (*Cocos nucifera*) is usually classed as a drupe, because the thick, fibrous layer (from which coir is prepared) and the stony 'nut' are comparable with the fleshy and stony layer of the mango.

The *berry* differs from the drupe mainly in not having a hard stone, though it may have many pips. In a berry the whole of the pericarp is soft and juicy, and during development it obliterates the loculi of the ovary and envelops the seeds. Each seed may have a hard coat, but this is derived from its own testa and not from the pericarp. A date (*Phoenix dactylifera*) is a one-seeded berry, because the pip is the seed and does not contain a seed as the stone of a mango does. A berry may be formed from either a superior or an inferior ovary, which usually consists of two or more carpels. In the banana (*Musa paradisiaca*) it is inferior, for the withered remains of the perianth can usually be seen at the top of the fruit. In this case the berry is formed from a tricarpellary ovary with axile placentation because, although the ovules of a banana do not normally develop into seeds, the flesh is divided into three longitudinal strips and the aborted ovules are axile in position. In the papaw (*Carica papaya*) the berry is formed from a superior ovary because the remains, if any, of the calyx are to be seen at the base of the fruit. In contrast to the banana, the papaw is formed from a five-carpellary ovary with parietal placentation.

All the above types of fruit are '*simple*' in the sense that they are formed from one ovary contained in a single flower, but there are '*compound*' fruits which are formed from a collection of separate ovaries. If these represent the apocarpous gynoecium of a single flower, as in the sweetsop (*Annona squamosa*) (Fig. 29.10), the fruit is said to be an *aggregate* fruit. If, on the other hand the ovaries represent the flowers of an inflorescence, as in the breadfruit (*Artocarpus communis*), fig (*Ficus*), and pineapple (*Ananas comosus*), the fruit is classified as a *multiple* fruit.

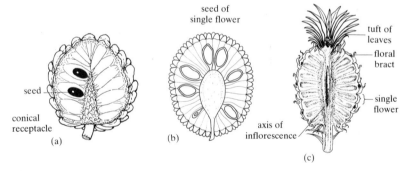

Fig. 29.10 Fruits derived from more than one ovary. (a) Aggregate fruit of sweetsop (*Annona squamosa*) formed from one flower. (b) Multiple fruit of breadfruit (*Artocarpus communis*) formed from an inflorescence. (c) Multiple fruit of pineapple (*Ananas comosus*). (c after Stone and Cozens 1969, p. 153)

The classification of fruits described here is empirical but useful for the classification and identification of angiosperms. Like most biological classifications, there are exceptions. Thus some legumes are indehiscent, for example the groundnut (*Arachis hypogaea*), and some capsules are succulent, for example the akee apple (*Blighia sapida*). It should be emphasized, however, that fruits can only be understood in terms of their function, which is the dispersal of seeds.

Dispersal of fruits and seeds

If seedlings grow directly beneath or very near the parent plant, there is intense competition and only a few seedlings survive. When the seeds are dispersed, however, more seedlings are likely to survive, and the area colonized by the plant may also be extended. Another reason for dispersal is that it lessens the effects of seed predators. An individual plant producing a seed crop provides a concentration of food for seed predators, but if the seeds are dispersed there is a greater chance that more of them will escape being eaten. Seeds may be distributed as seeds or still within the fruit, but from an ecological viewpoint the mechanism of dispersal is more important than the morphological nature of the dispersal unit. Seed-bearing plants show many and varied mechanisms of dispersing their fruits and seeds but these may be divided into four categories – dispersal by wind, dispersal by water, dispersal by animals, and explosive dispersal.

Dispersal by wind

Wind dispersal mechanisms are of five general kinds:

1. Dust seeds   A few families of plants, notably the Orchidaceae, produce seeds which are dispersed by wind merely by virtue of their minute size and light weight. Some idea of the size of orchid seeds may be gained from the fact that a single plant may produce as many as several thousand million seeds.

2. Winged fruits and seeds (Fig. 29.11)   Many fruits and seeds have a large surface/mass ratio as a result of developing one or more wings. The disseminule (the word used for any structure which functions as a dispersal unit) either glides or spins as it falls, and the longer time thus taken for it to reach the ground increases the distance that it can travel.

In winged fruits the wings are commonly outgrowths of the ovary wall (e.g. *Combretum*, *Piscidia*, *Securidaca* and *Terminalia*) but they may represent sepals which become enlarged and membranous in fruit (e.g. *Dipterocarpus*, *Petrea* and *Triplaris*). Winged seeds, in which the testa is the only structure that can become winged, are less common than winged fruits. Common examples are found in the Bignoniaceae (e.g. *Spathodea* and *Tecoma*), although the most spectacular is provided by *Macrozanonia* which has a wingspan of up to 15 cm.

3. Plumed fruits and seeds   Many fruits and seeds have a 'parachute' mechanism of dispersal as a result of possessing plumes or coverings of hairs. Long silky plumes may develop on any part of a fruit and its related structures, but are developed most profusely on the sepals. In the Compositae, for example, the calyx lobes are frequently plumed to form the pappus, which is one of the most quoted examples of a dispersal mechanism. Plumed seeds are always borne in capsules or follicles and are well illustrated by members of the Asclepiadaceae (e.g. *Asclepias* and *Calotropis*) and Apocynaceae (e.g. *Strophanthus*). Seeds with a dense covering of hairs are sometimes called woolly seeds to distinguish them from plumed seeds. This condition attains its greatest development in the Malvaceae and Bombacaceae, whose seeds are the commercial sources of cotton and kapok.

4. Censer mechanism   This is the method whereby seeds are scattered by being shaken out of a capsular fruit when the rigid stalk is swayed to and fro by a strong wind. In such capsular fruits the ripe capsule develops holes or

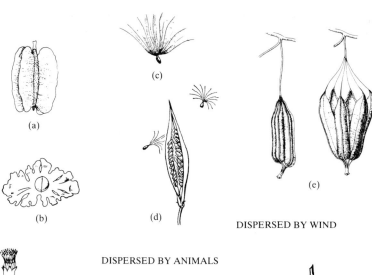

(a)

(b)

(c)

(d)

(e)

DISPERSED BY WIND

DISPERSED BY ANIMALS

(f)

(g)

(h)

(i)

(j)

Fig. 29.11 Dispersal of fruits and seeds by wind and animals. (a) Winged fruit of *Combretum*. (b) Winged seed of *Bignonia*. (c) Plumed fruit of Compositae. (d) Plumed seed of *Asclepias*. (e) Capsule of Dutchman's pipe (*Aristolochia*), before and after opening, to show censer mechanism. (f) Flowering and fruiting heads of *Bidens pilosa*. (g) Single fruit of *Bidens pilosa* showing barbed appendages. (h) Spiny fruit of *Tribulus*. (i) Split segment of fruit of *Tribulus*. (j) Succulent fruit of *Sterculia*, brilliant red in colour with black seeds. (e from Hall 1970, p. 80; j from Holttum 1954, p. 92)

slits at the top through which the seeds escape a few at a time; hence the name 'censer' mechanism. Since the seeds can escape only when the capsule is shaken, they are dispersed precisely at a time when they have a reasonable chance of being carried some distance by the wind. As a means of dispersal the efficiency of the censer mechanism will be enhanced where, as in *Aristolochia*, the seeds themselves are flattened and therefore adapted to wind dispersal. The censer mechanism is found in a number of familiar plants, including *Argemone mexicana*, *Datura stramonium* and *Aristolochia* spp.

5. Tumble weeds Usually ripe seeds are dispersed either separately or contained in fruits, but occasionally the whole inflorescence comes loose and is light enough to be blown along the ground, scattering the seeds as it does so. Such plants are called tumble weeds, and occur in open habitats like deserts and grasslands. The South African *Cybistes longiflora* is a typical example. In this case the spherical umbellate head breaks off as a unit from the flowering stalk and is then rolled bodily along the ground by the wind. As it rolls along, the papery walls of the indehiscent capsules are rubbed away and the seeds are scattered along the ground.

Dispersal by water
Plants which grow in or near water or by the sea often have fruits or seeds which float in water and can be dispersed by currents or blown by surface

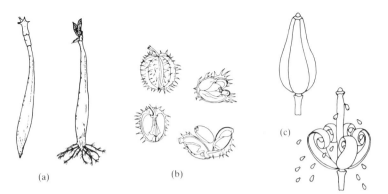

Fig. 29.12 Dispersal of fruits and seeds. (a) Dispersal of fruits of red mangrove (*Rhizophora*) by water. Seedling developing on parent tree (left), and germinating in mud (right). (b) Explosive dispersal of seeds of castor-oil plant. (c) Explosive dispersal of seeds from wet capsule of balsam (*Impatiens*). (a from Mitchelmore 1967, p. 9; c from Stone and Cozens 1969, p. 154)

wind movements even if there is no current. Water dispersal is an important dispersal mechanism for plants which grow along tropical coastlines. The buoyant fruit of the coconut palm is an obvious example of a fruit adapted for long-distance dispersal by floating in the sea. The widespread distribution of the coconut on tropical islands may be due partly to this natural dispersal mechanism, although intentional human introduction certainly started a very long time ago.

Mangroves are an ecological group of small tree species which are not related systematically but which form dense woodlands on muddy tidal estuaries in many areas of the tropics. The seeds of some of the species germinate inside the fruit while still on the parent plant (Fig. 29.12). The root emerges and grows vertically downwards, becoming considerably swollen at its lower end. When eventually the fruit falls from the tree it either drops straight into the mud or, if the tide is in, it may float for a considerable distance before it finally lodges in some obstruction and continues growth.

Water in the form of rain-wash is also probably more important as an agent of dispersal than is generally realized because of the part it may play in carrying away the seeds that fall beneath a parent plant. Although it is difficult to visualize any extensive carriage by this means, rain-wash may be significant in the colonization of bare ground.

Dispersal by animals

Animal dispersal is of two kinds – outside or inside the animal. Where transport is external, fruits (but only rarely seeds) possess hooks and so form 'burrs', or use other structures such as glandular hairs for easy attachment to the fur of mammals or the feathers of birds. Hooks may develop on the pericarp of the fruit itself (e.g. *Triumfetta* spp.) or on the accessory floral members such as the calyx (e.g. *Bidens* spp.) or involucral bracts (e.g. *Achyranthes indica*). Most hooked fruits are small and may be carried unnoticed for long distances before they are rubbed off. Others (e.g. *Cenchrus echinatus* and *Tribulus* spp.) are relatively large and the 'hooks' are developed into hard spines. The fruit lies upon the soil until it is carried away sticking in the foot of a passing animal. Animals, particularly water birds, are also responsible for transporting seeds in mud adhering to their feet, and for this no special adaptation other than a relatively small size is needed.

Where distribution is through the gut of an animal the fruit is succulent and the seed protected by a hard testa or endocarp. In some instances only the fleshy part of the fruit is eaten and the seeds are rejected, but in others the whole fruit is swallowed and the seeds pass unharmed through the gut. Most fleshy fruits are brightly coloured and develop on shrubs or trees. Birds are the chief dispersing agents, at least in the tropics, and the frequent occurrence of red fruits may be related to the fact that birds can distinguish this colour as a sharp contrast to green. In the tropics there are many groups of fruit-eating birds, some of which, like the fruit pigeons and hornbills, are large and capable of dealing with fruit of substantial size. Fruit pigeons fly for long distances from forest to forest in search of fruit, and the widespread distribution of the nutmeg (*Myristica fragrans*) tree in Malaysia is attributed to them. In South America the toucans play a similar role in seed dispersal.

Explosive dispersal

Here the seeds are scattered by the explosive opening of the fruit. This mechanism is common among members of the Leguminosae, the pods of which dry out in the sun and shrivel. The tough diagonal fibres in the pericarp shrink and set up internal tensions, until finally the pod splits into two halves. In splitting down the lines of weakness along the two margins, the two halves of the pod become spirally twisted and flick out the seeds for distances up to 3 or 4 m. The 'clicks' that accompany the splitting of many legume pods (e.g. *Clitoria ternatea*) can often be heard on hot days. Other common examples of dry fruits with explosive and audible dehiscence are those of castor-oil bean, sandbox (*Hura crepitans*) and rubber (*Hevea*). The seeds of the sandbox tree are regularly dispersed for distances of up to 10 m. Most self-dispersing fruits explode when dry, but the juicy fruits of the balsam (*Impatiens* spp.) explode only when wet (Fig. 29.12c). When the capsule is ripe, the inner part of the fruit wall is so tense that when the fruit is moistened or touched it explodes, scattering the many small seeds.

Morphological aspects of germination

A mature seed is enclosed by a protective coat, the testa, on the surface of which are a small hole, the micropyle, and a scar, the *hilum*, which marks the place where it was attached to the placenta on the fruit wall. Within the testa is an embryo plant, and a supply of food which is housed either outside the embryo as endosperm or, more usually, within the embryo itself in the cotyledons. When the seed is shed from the parent plant the embryo is a miniature plant with root and shoot ends. The embryo of a dicotyledonous plant consists of a young root (radicle), a prospective shoot (plumule), two seed leaves (cotyledons), and the embryonic axis which, because it occurs below the level at which the cotyledons are attached, is called the hypocotyl (*hypo*, below). It is continuous with the radicle at its lower end, and with the plumule at its upper end.

Under suitable growing conditions, a viable seed germinates to produce a young plant, or seedling. The first external sign of germination is the bursting of the testa in the region of the micropyle, through which the radicle grows and becomes anchored in the soil by the production of root hairs (Fig. 29.13). At the same time as the radicle is emerging active growth is starting in another region of the embryo. If this region is the hypocotyl, the plumule and the cotyledons, still covered by the testa, will be carried above the ground where the cotyledons soon expand as the first foliar structures. Often the cotyledons turn green, and function as photosynthetic leaves; the

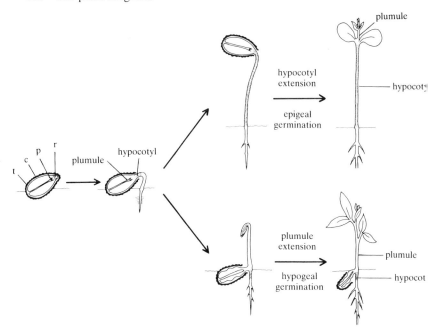

Fig. 29.13 Types of seed germination. (t= testa, c = cotyledons, p = plumule, r = radicle)

cotyledons of the castor oil seedling increase tremendously in size and look like foliage leaves. In a few non-endospermic seeds, for example the common bean (*Phaseolus vulgaris*), the cotyledons are so full of stored food reserves that by the time they are exhausted the seedling has grown sufficiently to be independent, and the cotyledons wither and drop off. This type of germination in which the hypocotyl elongates, is described as *epigeal* (literally, above the earth) because the cotyledons are carried above ground. Common examples of seeds showing epigeal germination are cowpea (*Vigna unguiculata*), castor oil and gourds.

If, however, the region that grows actively while the radicle is emerging is the *epicotyl* (i.e. the stem axis of the plumule above the cotyledons but below the node of the first true leaves), as in the pigeon pea (*Cajanus cajan*) and *Citrus* spp., the cotyledons remain in the soil. Until the plumule has reached the surface of the soil it is bent back on itself like a hook, so that the delicate shoot meristem is directed downwards and thus protected from abrasion by soil particles. When the plumule reaches the surface it straightens out and grows into the shoot. The cotyledons shrivel up as their food content is depleted. Germination of this type, in which the epicotyl elongates, is said to be *hypogeal* (literally, below the earth). Hypogeal germination is probably less common than epigeal germination.

Physiological aspects of germination

Provided the temperature is suitable, all that is needed to start most seeds growing is the presence of water. The seed takes up water by imbibition, as a result of which the desiccated colloidal contents of the seed are rehydrated. This is essentially similar to what happens when water is added to dehydrated vegetables in packet soups. Imbibition is accompanied by swelling of the seed, and occurs regardless of whether or not the seed is alive. Great forces, up to 2000 bar, can develop within dry seeds as a result of their imbibing water. This is why wooden grain ships have been known to break apart and sink after a leak has developed in the hull. As more water enters the seed by imbibition, the imbibitional forces for the intake of water decrease, but the hydrated cells develop osmotic forces which facilitate further intake. Eventually the tissues regain the size and shape they had before the seed dried out during ripening, and the structural organization of the cells is restored. Active metabolism now begins, and this is manifest by a sudden rise in the rates of respiration and protein synthesis, both of which are associated with the hydration of existing enzymes.

When the embryo plant within the seed resumes growth, it is necessary for it to establish an absorbing root system and a photosynthetic shoot system so that it can become self-supporting. To achieve this object the embryonic axis requires food material to carry out active respiration and to synthesize new protoplasm and cell walls. During the very early stages, before any visible signs of germination are apparent, the embryonic axis has sufficient material within itself to satisfy its immediate requirements. In cereal grains, for example, the embryo contains sucrose, reserve protein and lipids. Before long, however, the embryo becomes dependent upon reserves in the storage region (cotyledons or endosperm) of the seed. These reserves, which are insoluble, must be converted into soluble products before they can be transported to the growing regions of the very young seedling. This is accomplished by the synthesis and subsequent action of hydrolytic enzymes such as amylases which convert starch to sugars, proteases which break down proteins into amino acids, nucleases which hydrolyse nucleic acids into nucleotides, and lipases which split fats into glycerol and fatty acids. The activity of such hydrolases leads to a rapid breakdown of the insoluble reserve materials.

The type of controlling mechanisms which bring about the regulated mobilization of seed reserves will be illustrated by reference to barley (*Hordeum distichon*) grain the germination of which, because it underlies the malting process in the brewing of beer in temperate regions, has been more extensively studied than the germination of any other seed. The basic structure of the barley grain is shown in Fig. 29.14. The embryo lies at one end of the grain, most of which is occupied by endosperm. This is a dead tissue whose cells are packed with starch grains and some protein. Shortly after the embryo has begun to germinate the endosperm starts to 'liquefy'. The cell walls are degraded, the proteins are hydrolysed to amino acids, and starch is hydrolysed to reducing sugars which are later transformed into sucrose for transport to the embryo. All this happens because of the secretion of enzymes by a special layer of cells called the *aleurone layer*, which is found at the periphery of the endosperm in seeds belonging to the grass family.

During normal germination the production of hydrolases by the aleurone cells is induced by the embryo. If the embryo is removed, the aleurone cells fail to produce hydrolases. The experiment shown in Fig. 29.15 demonstrates this phenomenon for amylases. It is now known that the signal from the embryo which instructs the aleurone cells to produce active enzymes is the plant hormone *gibberellic acid*. From the results of a number of experiments it would appear that the effect of gibberellic acid is to induce the synthesis of hydrolases rather than to activate already existing ones. When, for example, barley grains are treated with inhibitors such as actinomycin C and cycloheximide, which inhibit the synthesis of all proteins including en-

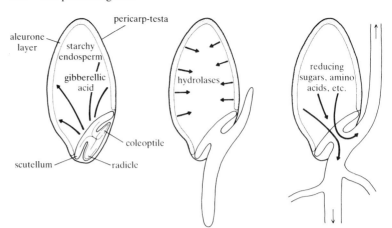

Fig. 29.14 Stages in the germination of a barley grain, showing the activities of the various regions.

zymes, the endosperm fails to liquefy because the gibberellic acid is prevented from acting in the normal way. Also, when an aleurone layer is treated with a mixture of gibberellic acid and a radioactive amino acid, the hydrolases which are poured into the endosperm are also labelled.

The above account of the hormonal control of the mobilization of food reserves is an oversimplification. It is known that the embryonic axis and the scutellum (a shield-like outgrowth of the axis of grass embryos, which is homologous with the cotyledon) start to synthesize gibberellic acid at the onset of germination, and that the gibberellic acid moves to the aleurone layer through the newly differentiating vascular tissues of the embryo and scutellum. This route has been proved by experiments in which the embryonic axis is treated with labelled gibberellic acid. Autoradiographs reveal radioactivity only along the vascular bundles of the embryo and scutellum. Vascularization has itself been shown to depend on the auxin indole-acetic acid, which is released by the coleoptile of the growing embryo.

Two hormones are therefore involved in the control processes for the mobilization of reserves in the endosperm of barley, gibberellic acid to induce the synthesis of hydrolase enzymes, and indole-acetic acid to assist transport of the gibberellic acid to the aleurone layer which is its target tissue.

Fig. 29.15 Experiment to demonstrate that barley embryos secrete a substance which induces aleurone cells to release an α-amylase enzyme. The dialysis membranes prevent movement of any α-amylase which might come from the embryo itself (right). After 3 days, iodine is added to the starch-agar medium and the resulting staining indicates where the starch remains. Starch disappears only around the embryo–aleurone combination (centre).

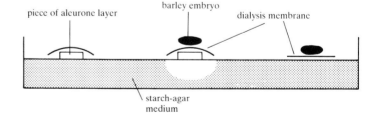

Seed dormancy

During their development on the parent plant the embryos of all seeds pass from a state in which they continue to grow and develop to one in which further growth and development ceases. In many plants this period of inactivity need not last for more than a few days or weeks, and shortly after the seeds of such plants have been shed they can be induced to germinate by supplying them with water and suitable growing conditions. In some plants, however, the period of inactivity may last for a relatively long time, and germination cannot be induced, no matter how favourable the external conditions, until the seeds have passed through this state. Such seeds are said to be dormant.

Dormancy is a special kind of rest which can only be terminated by certain, often very precise, environmental cues. The ability to rest in this way enables plants to survive periods of water scarcity or, in the case of temperate-zone plants, cold temperatures. The seeds of most wild plants have a dormancy requirement, which has evolved in the course of time by natural selection. By contrast, most commercial seeds are artificially selected for their readiness to germinate promptly when exposed to favourable conditions, so that all the plants in a crop will be at the same stage of maturity at harvest time. There is no one cause of seed dormancy, and it is convenient to recognize at least five different types.

Mechanical dormancy

A hard and impervious testa may prevent germination for no other reason than that it is too strong for the embryo to be able to split it open before growing through it. More often, however, the testa prevents water and /or oxygen from entering the seed, and the essential physiological processes of germination cannot take place. Resistance of the testa to water uptake is most widespread in the bean family (Leguminosae), the seed coats of which, in addition to being hard, may possess a waxy covering. Germination of such seeds becomes possible only after complete removal or weakening of the seed coat. In field conditions, hard testas may be softened by the natural processes of decay resulting from the action of fungi and bacteria, or by the action of the digestive juices and bacteria in the gut of fruit-eating animals, especially birds and ruminants. Far from damaging the embryo, the passage of hard seeds through the digestive tracts of animals increases their chances of germination, and furthermore deposits them in a moist, manured environment. In horticulture and agriculture hard testas are deliberately damaged or weakened by man through a process called *scarification*. In chemical scarification, seeds may be dipped into strong sulphuric acid, organic solvents such as acetone, or even boiling water. In mechanical scarification, seeds may be shaken with some abrasive material such as sand or be scratched with a knife.

After-ripening

Another type of dormancy is that in which the embryos, although apparently fully formed and viable, remain dormant even when the seed coats are removed and conditions are suitable for growth. Germination in these cases takes place only after a series of little-understood changes, collectively called after-ripening, has taken place in the embryo as it lies dormant in the soil or is kept in storage. In some species, 1 year is sufficient for after-ripening to occur, but in others the process is drawn out over several years, with some germination occurring each year. This can be regarded as an insurance by the species against sudded catastrophes that might completely wipe out all the seedlings of any one age group.

The period of after-ripening is usually shortened by exposure to high temperatures, and this fact is used in establishing plantations of the oil palm (*Elaeis guineensis*). Moistened seeds are heated at 42 °C for 60 days, after which the temperature is reduced to 27 °C. The dormancy of the seeds is broken by this treatment, and it is thus possible to establish a plantation of uniform plants instead of using self-sown seeds which have been lying in the soil for a variable number of years.

In the above two types of dormancy, the effect of the dormant period is to stagger the onset of germination, and so ensure that all seeds do not germinate simultaneously. In the next three types, dormancy has a more specific function in ensuring that germination occurs only when conditions for establishment are favourable.

Temperature related dormancy

Many dry seeds are remarkably resistant to extremes of temperature, and some can even withstand short periods of immersion in boiling water or liquid air (–140 °C) without losing viability. Ecologically, such heat resistance is important in vegetation-types periodically ravaged by fire, such as the savannas of Africa where the germination of the seeds of the valuable forage grass *Themeda triandra* and of other species is stimulated by fire. Also important ecologically is a germination requirement for a moderate daily alternation between a maximum and a minimum temperature. Especially in deserts, extreme temperature fluctuations are an unavoidable feature of the surface of the ground, but with increasing depth these fluctuations are gradually evened out. A requirement for a modest fluctuation, e.g. from 20 °C at night to 30 °C in the daytime (the fluctuation required by the grass *Oryzopsis miliacea*), virtually ensures germination within a layer between two depths from the surface. This is advantageous because a seed germinating in soil has to strike a balance between two conflicting demands, both depending on depth. It is an advantage to germinate deep in the soil because the water supply is more dependable than near the surface, but it also is an advantage to germinate near the surface because this enables the seedling to reach air and light rapidly and become self-supporting before the food reserves in the seed are exhausted.

Light-related dormancy

The seeds of maize (*Zea mays*) and many legumes such as cowpea (*Vigna sinensis*) germinate equally well in light and in darkness. The germination of some seeds, e.g. four-o'clock (*Mirabilis jalapa*) and onion (*Allium cepa*), is inhibited by exposure to light, whereas the seed germination of lettuce (*Lactuca sativa*), many weed species such as *Tridax procumbens*, and many grasses is stimulated by light.

The controlling influence of light on seed dormancy has important ecological consequences. When an area of vegetation is cleared, it is quickly colonized by 'weeds' which are the opportunists of the plant world. If, for example, a road is cut through a West African forest the verges are often quickly covered by a dense growth of the fast-growing umbrella tree (*Musanga cecropioides*), although this tree is almost absent from the mature forest. The question immediately arises as to where the seeds have come from. Soil collected from mature forests, and then watered in the greenhouse, yielded abundant *Musanga* seedlings, showing that *Musanga* seed is widely present in soil, even sometimes when it is far from parent trees. The tiny one-seeded fruits are borne on a fleshy receptacle, and are distributed widely in the droppings of birds and bats which feed upon them. Although not confirmed by laboratory evidence, it is probable that the stimulus for the germination of *Musanga* seed after the forest cover has been removed is the increase in intensity and, particularly, the change in quality of the light reaching the ground. The germination of the seeds of lettuce and many other plants has been shown to be stimulated by red light (with a peak at 660 nm) and inhibited by far-red light (with a peak at 740 nm). This property depends on the presence of the bluish pigment called phytochrome (see p. 265). Green leaves absorb red light strongly but transmit far-red light more readily, with the result that the light beneath a forest canopy is relatively rich in the far-red wavelengths. The removal of a forest canopy alters the balance between the stimulatory and inhibitory wavelengths of light, and allows germination of *Musanga* seeds to take place at precisely the time when competition from other plants is minimal. The same phenomenon is probably involved in the flush of germination which often follows the burning or cutting of grassland.

Rainfall-related dormancy

The seeds of certain plants which grow in arid or semi-arid regions have been shown to have built-in 'chemical rain gauges'. These are inhibitory substances which are water-soluble, and therefore leached out by rain. The amount of rain necessary to leach out these inhibitors is directly related to the supply of water the plant needs to complete its short life cycle from seed through flower to seed again. The possession of chemical rain gauges by the seeds of plants living in extremely seasonal habitats ensures that the species is not wiped out as would happen if simultaneous germination of all the seeds were to be followed by a severe drought which killed off the seedlings.

Although five types of dormancy have been described, it should be pointed out that more than one type is commonly found in a single seed. The seeds of the sagebush (*Artemisia*), a very common shrub of the cold deserts of the USA, need light, but at an extremely low intensity, to germinate, but their ability to respond to light of the right intensity is increased as water is imbibed by the testa. This means that, in the field, the seeds germinate more readily the longer the sand remains moist.

During recent years botanists have become increasingly interested in the physiological mechanisms underlying seed dormancy. Most current theories envisage most examples of dormancy as resulting from an excess of natural growth inhibitors over natural growth promoters. Abscisic acid is widely favoured as the main inhibitor in many species, but there are probably a large number of endogenous inhibitors. Gibberellins and cytokinins are often credited with being the active promoters, although treatment with other substances will also break dormancy. Part of the stimulus responsible for current work on seed dormancy is the world-wide interest in developing seed banks, with the aim of preserving the genetic characteristics of the original wild-type varieties of various crop plants for use in future breeding programmes. The need for such seed banks arises because the genes of the older varieties are being rapidly lost as new varieties progressively replace the older ones.

Chapter 30

The classification of angiosperms

Principles of classification

Classifying objects is an everyday activity which is used as a means of putting and then keeping things in order. Any one who uses a library knows that there are usually two catalogues of the books, one in which the index cards are arranged alphabetically according to the surname of the author, and another in which the cards are grouped according to the subject matter of the books. Both classifications are necessary for the efficient running of a large library, but the most appropriate one to use will depend on whether you want to know the titles by a given author which the library has in stock, or what books are available in the library on a given subject. The classification of plants follows exactly the same principles as those which govern the classification of any objects, and this means that plants, like books, can be classified in different ways according to the purpose in view.

It is obvious that different plants show varying degrees of resemblance to one another, and close resemblances are often reflected in the common names of plants. For example, there are many different plants which are called *beans*, all of which have long pods containing kidney-shaped seeds. According to the concept of evolution, the possession by different plants of similarities which are constant (i.e. not subject to modification by the environment) implies that the plants concerned are related to one another by descent from common ancestors. The botanist tries to classify plants so as to show what he believes to be their evolutionary relationships as deduced from their resemblances, i.e. to devise as 'natural' a classification as possible. It is found in practice that a classification which is based on agreement in only one or a few characters rarely indicates the true relationship between plants; such a classification is known as an *artificial classification*. Plants may be artificially classified in many ways. For instance, one might classify them according to their colour. On this basis a red-flowered plant of the common garden hibiscus (*Hibiscus rosa-sinensis*) would be grouped with flamboyant (*Delonix regia*) with which it has little in common, but separated from a yellow-flowered form of hibiscus from which it differs markedly only in flower colour. By contrast, a *natural classification* is based on the correlation of a wide variety of characters, so that two plants which agree in a large number of characters are considered to be more closely related to each other than two which agree in only a few characters. The aim of systematic botany, therefore, is to devise a filing or indexing system which expresses, as accurately as possible, how closely existing plants are related to one another.

Systems of classification

Although no two plants are ever exactly alike, the different individuals belonging to one species usually have so many points of resemblance that it is impossible to doubt their genetic relationship. For most purposes the species is therefore accepted as the basic unit of classification. Those species which have most characters in common are placed together into larger, more inclusive groups called genera, and on the basis of overall similarities between the species of a genus it is difficult to doubt that most genera are also natural groups. The different genera assembled in one family share a certain number of very basic characters which distinguish them from genera belonging to other families. Most families are probably natural groups, but because the points of resemblance between the genera of a family are fewer than those between the species of a genus, it may be doubted whether or not certain families are assemblages of closely related genera. The arrangement of families into larger units so as to maintain a natural grouping is difficult or impossible, because there are so few characters which different families have in common. It is not surprising therefore, that no satisfactory phylogenetic system has yet been devised. Many botanists think it is pointless even to try and formulate one unless more evidence becomes available from fossils. There are, however, two systems of classification in common use; these are the systems of Bentham and Hooker and of Engler. Both these systems are artificial in so far as they use single alternative floral characters, such as the freedom or fusion of petals, to separate large groups of families. In this connection, it cannot be overemphasized that vegetative characters should also be taken into account if anything approaching the true relationships between families is to be established. Vegetative characters which are not liable to modification by the environment, such as the arrangement of leaves on the stem or the venation of the leaf, may be indicative of genetic relationship just as much as floral characters.

The considerable element of artificial grouping in both the well-established systems of classification does not matter much in practice. Students should become familiar with one system, preferably that adopted in the standard Flora of their local region, and then adhere to it.

The system of the British botanists G. Bentham (1800–84) and J. D. Hooker (1817–1911) was expounded in their *Genera Plantarum* (Genera of Plants), written entirely in Latin and published in three volumes between 1862 and 1883. This system has been widely used throughout the countries of the Commonwealth and is still used in the herbaria of the Royal Botanic Gardens, Kew, England and the British Museum (Natural History), which contain extensive collections of plants representative of the whole world. Bentham and Hooker based their system on the pre-Darwinian ones of A. P. de Candolle (published between 1824 and 1873) and B. de Jussieu (published in 1789), and had no phylogenetic aim. Consequently this system is primarily one of convenience, supplying an easy means for determining the generic names of plants. Some attempt is made to classify the dicotyledonous families on a logical basis, but the monocotyledonous families are arranged arbitrarily into series. The dicotyledonous families are divided into three primary groups on the basis of the presence or absence of petals, and on whether the petals are free or united. Each group is further subdivided into series mainly according to ovary position (hypogynous, perigynous, or epigynous). As it happens most of the families fit quite well into these arti-

Table 30.1 Summary of the two main systems of classification of flowering plants

Bentham and Hooker's System	
Dicotyledons	
Primary groups	*Important families dealt with in this book*
I. Polypetalae – Flowers with two distinct perianth whorls, the inner one with free petals	Capparaceae Malvaceae Leguminosae Myrtaceae Cucurbitaceae
II. Gamopetalae – Flowers with two distinct perianth whorls, the inner one with petals united	Rubiaceae Compositae Apocynaceae Convolvulaceae Bignoniaceae
III. Incompletae – Flowers usually with only one perianth whorl which is green, or with none	None
Monocotyledons	
There are no primary groups. The Monocotyledons are divided arbitrarily into seven series	Liliaceae Gramineae Orchidaceae

Engler's System	
Monocotyledons	
The families of the Monocotyledons are arranged in order of increasing floral complexity	Gramineae Liliaceae Orchidaceae
Dicotyledons	
Series	
1. Archichlamydeae – Flowers with perianth either (a) absent, (b) in one petaloid or sepaloid whorl, or (c) in two whorls, the inner one with free petals	Capparaceae Leguminosae Malvaceae Myrtaceae
2. Metachlamydeae – Flowers with perianth in two whorls the inner one with petals united	Apocynaceae Convolvulaceae Bignoniaceae Rubiaceae Cucurbitaceae* Compositae

* The relationships of this family have been much disputed, and Bentham and Hooker include it in the Polypetalae.

ficial divisions but, as one would expect, there are several notable exceptions. Although Bentham and Hooker's system as a whole is now regarded as out of date, their *Genera Plantarum* continues to be a standard work (and was reprinted in 1965) because the descriptions in it are based on meticulous first-hand observation. A simplified form of Bentham and Hooker's system is shown in Table 30.1.

The second important classification is that of the German Adolf Engler (1844–1930), who based it on an earlier classification by Eichler (1839–87). Engler's classification is used in the enormous twenty-volume *Die natürlichen Pflanzenfamilien* (published between 1887 and 1909) which contains descriptions in German of all the families and genera of plants known at that time. Because Engler's system is so comprehensive it has been adopted in many national herbaria and in most modern regional Floras. Engler's system is based on the assumption that flowers with no petals are the most primitive, those with separate sepals and petals are more advanced, and those with sepals and petals showing varying degrees of fusion are the most highly advanced of all. In his classification Engler arranges the families of both the Monocotyledons and the Dicotyledons in order of increasing floral complexity. It is not surprising that it is impossible to make a linear sequence in this way to cover either all the monocotyledonous or all the dicotyledonous families. In the Dicotyledons the succession is broken, for convenience, into two series, the Archichlamydeae and the Metachlamydeae, according to whether the petals (if present) are free or fused together.

It must be emphasized that Engler's system, although it arranges families in linear sequences, is not truly phylogenetic. Its great merit is that it provides a logical basis for classifying the families of flowering plants, whereas the Bentham and Hooker system does not follow such well-defined principles. From the summary of Engler's system (Table 30.1) it will be seen that the Metachlamydeae are equivalent to the Gamopetalae of Bentham and Hooker, and that the Archichlamydeae contain the families placed in both the Polypetalae and Incompletae by Bentham and Hooker.

During the last few decades several attempts have been made to replace the systems of Bentham and Hooker and of Engler by one that is more in accord with modern ideas on phylogeny. Any phylogenetic system, however, must be built on a theory of floral evolution, and so far no agreement has been reached as to the nature of the primitive flower. Further, with the realization that morphological evidence alone is rarely sufficient to indicate the genetic relationships between families, much work has been done to investigate these relationships by means of cytological, genetical, biochemical, and other techniques. The chemical characteristics of plants, for example, are now extensively used in the investigation of taxonomic problems. The main reason for this has been the development of quick and efficient screening techniques, such as chromatography and electrophoresis, which make possible the rapid identification of large numbers of organic compounds. The standard approach is to make a systematic survey of plant groups for the presence of secondary compounds of low molecular weight, which are by-products of the major metabolic pathways. Unlike glucose or the twenty protein amino acids which are present in practically all plants, these substances (such as non-protein amino acids, alkaloids, various groups of pigments, and terpenoids) are irregularly distributed in the plant kingdom, and therefore their presence or absence can be used as criteria for classification.

Betanin, a betacyanin pigment   Cyanin, an anthocyanin pigment

A very striking example of the use of biochemical systematics, as these studies are now called, is the taxonomic distribution of the red and yellow pigments, the betacyanins and betaxanthins, known collectively as the betalains. These are confined to ten Angiosperm families (Aizoaceae, Amaranthaceae, Cactaceae, Nyctaginaceae, Phytolaccaceae, etc.) which make up the bulk of a group of families known as the Centrospermae. This group was recognized on morphological evidence, in particular the structure of the embryo which is usually strongly curved. The betalains do not occur in plants containing anthocyanins, which are the pigments normally found in most other Angiosperm families. The betalains differ both chemically and biosynthetically from the anthocyanins. The discovery that most of the families included in the Centrospermae contain betalains has led to the suggestion that the other families containing the more usual anthocyanins should be removed from the group. However, the combined evidence from morphology, anatomy, fine structure, pollen and embryology emphasizes the similarities between all the families traditionally included in the Centrospermae, even though the biochemical evidence implies that several of the families may not be as closely related to the rest as was once thought.

As a result of the application of modern techniques, such as those used in biochemical systematics, much has been learnt about aspects of classification, and some botanists consider that a change to a more realistic classification is long overdue. Computers are now being used extensively in plant classification, and it is possible that a more generally acceptable arrangement than any existing one will be produced as a result of this new approach. Part of the reason for the continued adherence to the earlier systems is that in large herbaria it is quite impossible to make drastic changes of arrangement today to suit a modern system that may be rejected tomorrow.

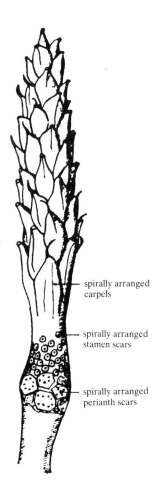

Fig. 30.1 Axis of *Magnolia* flower showing the spiral arrangement of the floral parts. The perianth and stamens have been shed but scars indicate their previous position. (From Sporne 1974, p. 14)

spirally arranged carpels

spirally arranged stamen scars

spirally arranged perianth scars

Floral complexity

According to one view of floral evolution the primitive flower of the angiosperms resembled a cone, or *strobilus*, which consisted of an elongated central axis bearing a large and indefinite number of spirally arranged and separate floral parts (sepals, petals, stamens and carpels). From such a strobilus-like flower evolution is thought to have taken place to more complex floral types. Flowers of *Nymphaea* (water lily) and *Magnolia* (Fig. 30.1) bear some resemblance to a strobilus and are therefore called 'primitive', whereas more complex flowers are called 'advanced'.

Plants with flowers having both sepals and petals are arranged in Engler's system on the assumption that they have developed along certain lines of evolution called progressions, of which the following are the most important:

1. From spiral to cyclic arrangement of floral parts   The primitive condition of a spiral arrangement of floral parts persists in only a few existing families, and even in these it never extends to all four sets of floral members. Thus in *Michelia* (Fig. 30.2), which is representative of one of the most primitive families known, the stamens and carpels are spirally arranged but the sepals and petals are cyclic. The arrangement in most modern flowers is cyclic throughout.

2. From a large and indefinite to a small and definite number of floral parts In the more advanced members of the Dicotyledons the number of floral parts is commonly two or five or multiples of these, while in the Monocotyledons the number is usually three or multiples of three. Accompanying progressions 1 and 2 the floral axis becomes greatly shortened.

3. From free petals to fused petals It is on the basis of this progression that Engler separates the Dicotyledons into two main groups, the *Archichlamydeae* in which the individual petals are entirely separate from each other, and the *Metachlamydeae* in which the petals are fused into a gamopetalous corolla.

4. From actinomorphy to zygomorphy In some flowers all the members of either the perianth or corolla are alike and symmetrically arranged around the central axis, so that the flower is radially symmetrical (actinomorphic). In other flowers these members differ among themselves in size and shape, so that there is only one plane in which the flower can be divided into two equal halves (bilaterally symmetrical, or zygomorphic). Bilateral symmetry is found in advanced flowers of both the Monocotyledons (e.g. orchids) and the Dicotyledons (e.g. *Jacaranda* and many other Bignoniaceae), where it is often associated with the provision of a landing stage for insect pollinators.

5. From apocarpy to syncarpy The possession of separate carpels (apocarpy) is considered more primitive than the more frequent condition where two or more carpels are fused to form a single ovary (syncarpy).

6. From hypogyny to epigyny With regard to the insertion of the stamens and the perianth or corolla, hypogyny is generally considered the most primitive and epigyny the most advanced condition, with perigyny intermediate between the two. It is a curious fact that epigyny and zygomorphy (progression 4 above) rarely occur together among dicotyledonous families, and in the Compositae, where the combination is found, only one type of floret is zygomorphic.

7. From conspicuous individual flowers to conspicuous inflorescences of small massed flowers This tendency manifests itself along several divergent lines, and can be regarded as a general feature of floral evolution. The progression reaches its highest development in the Compositae where the so-called 'flower' is actually an inflorescence (capitulum) of flowers, each 'petal' being a single greatly modified flower. Such an aggregation doubtless encourages the visits of insect and other animal pollinators.

The above progressions have occurred independently and at different rates in separate lines of descent, so that the flowers of any particular family may be advanced in some respects and primitive in others. Thus the Bignoniaceae are advanced in progressions 1–5 but primitive in progressions 6 and 7. Many other combinations of advanced and primitive characters occur in other families (see Fig. 31.1). The Compositae are usually considered to represent the most advanced angiosperm family, their flowers being advanced in six and sometimes all seven of the progressions mentioned.

The concept of progressions is useful when comparing any two flowers. If, for example, the flowers of *Michelia* and *Coffea* (Fig. 30.2) are compared on this basis, it will be seen that the flowers of *Michelia* are primitive in all seven progressions (although only partially so for the first two), whereas those of *Coffea* are advanced in all except progression 4. *Michelia* may therefore be considered to have a very primitive type of flower, and *Coffea* to have a relatively advanced type.

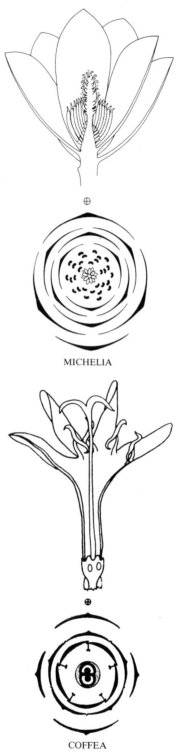

MICHELIA

COFFEA

Fig. 30.2 Comparison of a primitive flower (*Michelia* sp.) with a relatively advanced flower (*Coffea arabica*). (*Coffea arabica* from Purseglove 1968, vol. 2, p. 491).

Botanical keys

Before the information which is recorded about a particular plant can be obtained, the scientific name of the plant must be known. In order to find out this name, it is necessary to exclude the names of all the plants which the particular plant cannot be, until there is only one name left which must be that of the plant under consideration. This identification is done by means of a botanical key which consists of 'a series of contrasting statements of outstanding and presumably constant features arranged so as to lead by a process of elimination to the name of the plant concerned and thus to distinguish it from related or more or less similar plants' (W. T. Stearn). At each stage in working through a key, the user is offered two or more alternative statements, called *leads*, of which only one should apply to the plant being identified. A given lead either directs the user to another set of leads, or ends in the name of a plant. The user chooses successively the leads fitting the specimen until its name is reached, and he can go no further. When using keys it is not sufficient merely to accept this identification as inevitably correct. The specimen should always be checked by reference to a published description of the plant, in order to confirm that no mistake has been made somewhere along the route chosen. In most modern Floras the keys are constructed on a strictly dichotomous plan, which means that at every stage in the process of identification there are only two leads, which are jointly referred to as a *couplet*. Sometimes the leads are given numbers to indicate the contrasts, and this procedure is particularly useful for long keys. The key (Table 30.2) to the various families described in this book is an example of a dichotomous key with numbers.

To illustrate the use of keys, let us assume that we wish to determine, by means of the key in Table 30.2, the family to which the well-known flowering tree jacaranda (*Jacaranda mimosifolia*) belongs. Jacaranda is obviously a dicotyledonous plant because it has a thick trunk formed by secondary growth, its leaflets have reticulate venation, and its tubular flowers are based on a five-rayed symmetry. It is therefore legitimate to start with the first couplet of the portion of the key headed DICOTYLEDONS. Jacaranda cannot belong to the Leguminosae (the alternative offered by the first lead), because its flowers have a syncarpous bicarpellary ovary and therefore fit the second lead. The colon at the end of this lead directs us to the next couplet which is numbered 2 on the left-hand side. It will be noted that in the key being used the contrasting statements are indented by the same amount from the left margin, so that the first word of the second lead lies directly beneath the first word of the first lead. In couplet 2 there are alternative statements about whether the petals are free or fused. Since the plant in question has a gamopetalous corolla it fits the second lead which takes us down to couplet 5, skipping couplets 3 and 4 which apply only to plants with polypetalous flowers. Because the plant has hypogynous flowers, it agrees with the first lead of couplet 5 which directs us to couplet 6. In having zygomorphic flowers with five petal lobes and four functional stamens, it agrees with the second lead of couplet 6, skipping couplets 7 and 8. This lead identifies jacaranda as a member of the family Bignoniaceae.

In a good dichotomous key there should be no ambiguity between the two alternatives, and wherever possible each couplet should consist of several contrasting characters with the most obvious ones first. It ought to be an unbreakable rule that both the leads of a couplet should be read in their entirety before deciding which is applicable to the specimen. The con-

Table 30.2 An example of an indented dichotomous key with numbers. This short key aims to show how a botanical key works, but it cannot be used for determining the family to which any unknown flower belongs because it covers only the thirteen families of flowering plants described in this book.

---

1. Embryo with 2 cotyledons; vascular bundles of the stem usually arranged in a circle; leaves typically net-veined; parts of the flower usually in fives or fours, seldom in threes DICOTYLEDONS
1. Embryo with only 1 cotyledon; vascular bundles of the stem closed and scattered; leaves typically parallel-veined; parts of the flower nearly always in threes MONOCOTYLEDONS

DICOTYLEDONS
1. Gynoecium composed of 1 carpel .... *Leguminosae*
1. Gynoecium composed of 2 or more united carpels:
  2. Petals free from one another:
    3. Ovary superior:
      4. Sepals and petals 4; stamens free; ovary with a distinct gynophore .... *Capparaceae*
      4. Sepals and petals 5; filaments of stamens ± united into a tube; ovary without a gynophore .... *Malvaceae*
    3. Ovary inferior; stamens numerous .... *Myrtaceae*
  2. Petals ± united:
    5. Ovary superior:
      6. Flowers actinomorphic; stamens as many as the corolla lobes:
        7. Leaves present, well developed:
          8. Leaves opposite or verticillate; corolla lobes contorted; ovules more than 2 in each loculus .... *Apocynaceae*
          8. Leaves alternate; corolla lobes not contorted; ovules 1–2 in each loculus .... *Convolvulaceae*
        7. Leaves absent or reduced to scales; slender twining parasitic plants (dodders) .... *Convolvulaceae*
      6. Flowers ± zygomorphic; stamens 4, fewer than the corolla lobes .... *Bignoniaceae*
    5. Ovary inferior:
      9. Flowers in capitula; ovary 1-locular .... *Compositae*
      9. Flowers not in capitula; ovary 2-locular or more:
        10. Trailing or climbing herbs or shrubs with tendrils; leaves alternate, often palmately lobed or deeply divided; flowers unisexual; anthers often curved or flexuous or folded .... *Cucurbitaceae*
        10. Trees or shrubs, rarely herbs, sometimes climbing but without tendrils; leaves opposite or verticillate, entire, with interpetiolar or intrapetiolar (sometimes leaf-like) stipules; flowers bisexual; stamens with straight anthers .... *Rubiaceae*

MONOCOTYLEDONS
1. Perianth present:
  2. Ovary superior; stamens 6; flowers actinomorphic .... *Liliaceae*
  2. Ovary inferior, often spirally twisted; stamens 1 or 2; flowers zygomorphic .... *Orchidaceae*
1. Perianth absent or reduced to lodicules; flowers arranged in spikelets and in the axils of scaly bracts (glumes) .... *Gramineae*

---

trasting alternatives should refer to characters which are readily observable not only in freshly picked plants but also in dried plants preserved as herbarium specimens. The colour of the petals, for instance, is an unsatisfactory character because colour is usually lost in the process of drying. If measurements are given they should be expressed in exact terms, such as leaves 6–10 cm long, 2–3 cm broad; adjectives such as 'large', 'small' and 'very large' are practically useless for plant identification purposes.

The maker of a botanical key uses characters from any part of the plant. The user of the key, however, often has incomplete material at his disposal and, for this reason, cannot decide which of the leads offered applies to his specimen. In this event, both leads should be followed through the key. Often it soon becomes apparent that one of the leads is following a route where the statements and the plant are incompatible, and when this happens the alternative route can be assumed to be correct. If both leads reach a name, then the decision as to which is the correct one has to be made by comparing the specimen with detailed descriptions of the two species.

In some dichotomous keys the leads of all the couplets are printed flush with the left margin, but there can be little doubt that, space permitting, the indented dichotomous key is the best type. It has the advantage over other keys of showing more clearly which plants most resemble one another, and what characters they have in common to distinguish them from other groups of plants. For this reason, the indented dichotomous key is used in most modern tropical Floras (e.g. *Flora of Barbados*, *Flora of West Tropical Africa*, *Flora of Tropical East Africa*, *Flora Zambesiaca* and *Flora Malesiana*) but it is not used as widely as it should be because it takes more space, and is more costly to set up in print.

In most Floras there are separate keys to the families, to the genera and to the species. Keys for separating the species in a genus can often be constructed in several ways because the sequence of subdividing the relatively small differences between species is often a matter of individual preference. Keys for separating families, however, tend to be less flexible because they reflect the fundamental characters of families. Although the purpose of using keys is to identify a plant in the least possible number of steps, the experience gained from their use (particularly keys to families) is valuable in itself because it develops the realization that plants conform to a number of recognizable patterns.

Understanding plant families

There are about 250 000 different species of flowering plants which can be separated into about 450 more or less naturally defined groups or families. The recognition of natural grouping is more familiar among higher animals than flowering plants. If, for instance, you are told that an animal has hair, then you can be almost certain that it will also have four limbs, be warm-blooded, and its young will be born alive and nourished by milk from the mother. It is because characters like hairiness and suckling tend to be connected that distinct groups of animals are readily recognized and given such common names as mammals. Natural groups also exist among flowering plants, but they are often less obvious from general impressions.

The separation of animals and plants into discontinuous groups is not a state of affairs that could have been predicted on logical grounds, but is a consequence of the fact of nature that characters are not distributed at random but tend to be correlated. Many combinations of characters which could theoretically exist are in fact missing. For example, an animal with feathers never suckles its young, although there is no apparent reason why feathers and suckling should not occur together. Most modern biologists consider that the correlation of characters can best be explained in terms of the occurrence of evolution. If two plants have many features in common, then the most likely reason would seem to be that they inherited those characters from a common ancestor.

There are three main difficulties facing the beginner in his effort to learn the characters of plant families. The first difficulty arises from the large number of families of flowering plants. In the tropics many families are widespread either because their natural distribution is pantropical or because they have been accidentally or deliberately introduced by man into many tropical countries. Obviously it would be difficult to learn something about all the families present in any one country, but, if not more than about fifteen are studied, the task assumes manageable proportions. The families treated in the next three chapters have been selected mainly to give an idea of the range of floral structure found in tropical plants which are both relatively easy to obtain and not too difficult to examine. However, it should be realized from the outset that vegetative characters are also important in distinguishing families, and that even in the absence of flowers many families can be identified on vegetative characters alone. The student must learn how to consider a plant as a whole, and not regard the union of petals and the number of stamens as more important characters than whether the plant under consideration is an annual herb or a lofty tree. It is also an interesting fact that the characters which make certain plants important to man are often important in classification. For example, the Apocynaceae are characterized by a white milky latex which in many species is highly poisonous, and clearly this fact should be known as well as their possession of a gamopetalous corolla and superior ovary. The next three chapters will point out features, both vegetative and floral, which are important in the recognition of some of the commonest tropical families, but it should be emphasized that the information contained in these chapters is no substitute for the actual experience of handling specimens.

The second difficulty that frightens some students is the technical language which is used in Floras and other reference books. It goes without saying that botany, like any other science, must have its technical terms but, in reality, the number with which a student (as opposed to an expert) must be familiar is surprisingly small. Provided one is prepared to make the effort to learn the meaning of such common botanical words as apocarpous, syncarpous, hypogynous, epigynous, actinomorphic and zygomorphic, then one is well on the way to overcoming the language difficulty. It is true that, in the past, many more botanical terms were used than were absolutely necessary, but the modern trend is to reduce their number to the minimum. It is now appreciated that there is no scientific value in using a Latinized word when the English equivalent is equally convenient and just as accurate. Thus the use of the word 'alae' instead of 'wings' for the two lateral petals of a pea or bean flower has nothing to commend it. Similarly many old-fashioned words like 'pistil' for ovary are best forgotten. When such words are encountered in working with Floras they can always be looked up in a glossary. The point is that there is no need for a student to become acquainted with more than a relatively small number of technical terms, the meanings of which can be easily and quickly mastered.

The third difficulty is admittedly a very real problem, and concerns the abundance of exceptions. The description of any large family, as given in a Flora, is apt to read like a catalogue of every possible eventuality. Consider the following description of the Leguminosae: 'Trees, shrubs or herbs, often climbing. *Leaves* alternate or opposite or whorled, simple or (usually) compound, sometimes ending in a tendril, usually stipulate. *Inflorescences* in panicles, racemes, spikes or heads. *Flowers* regular or irregular, generally hermaphrodite.... *Stamens* usually twice as many, sometimes as many, as the petals, occasionally fewer or numerous . . .' and so on. Such precise descriptions are necessary but their proper place is in reference books, where they can be consulted when required. In view of the complexity of family descriptions, it is reasonable to ask what is the best way to approach the systematics of flowering plants. To answer this question one must first consider the reason for the exceptions. The variation between the plants of any family represents the facts of nature, and there is no reason why this variation should be simple or convenient to record. Indeed, for anyone who accepts the occurrence of evolution as a fact, strict conformity to standard patterns of organization is unlikely because the essence of evolution is the creation of new types. To think of families as the natural result of evolution therefore gives a clue to the proper understanding of exceptions. Although new kinds of plants arise by descent from existing forms, they never change so completely that *all* the ancestral characteristics are lost. A plant which diverges from the 'typical' pattern of a family usually does so in some special direction. For example, flowers may change from being radially symmetrical (actinomorphic) to bilaterally symmetrical (zygomorphic) as a result of the adoption of some new method of pollination, but the plant remains 'true to type' in other respects. Of course different plants within a family may diverge in different directions, and when this occurs (as it has in the case of the Leguminosae) a precise definition of the family affected becomes difficult because it has to cover all the variations. Fortunately, there is absolutely no point in trying to memorize precise family descriptions, because the exceptions serve only to obscure the basic pattern. The correct way to understand a given family is to build up a mental picture of its general pattern of organization, but at the same time to realize that almost every plant within the family (especially if it is a large one) may be exceptional in some respect. Once the general pattern has been appreciated, any exceptional feature will be seen to be of secondary importance. Provided that the plant as a whole and not just one feature of its flowers has been examined, any exceptional feature should not cause real difficulty in deciding the family to which it belongs. Consideration of the other features of the plant provide alternative ways of recognizing the family. In fact it is never safe to assign a plant to its family from the observation of a single character because the limits of families depend on the correlation of several characters. The guiding rule, therefore, is to examine the whole of a plant thoroughly and so learn something of the family to which it belongs, not to look up a description of the family in some reference book and then check that the plant agrees with the description. When this rule has been followed for several plants within the same family, a picture of the basic pattern will be formed in the mind without undue effort. Even when the number of families to be studied is small, there can be no substitute for first-hand observation of specimens.

Chapter 31

Selected dicotyledonous families: Archichlamydeae

The Dicotyledons include all flowering plants with two cotyledons. One might imagine from the name of the group that its recognition depends entirely on this feature, but such is not the case. Although the possession of two cotyledons is the most constant feature, there are several associated characters which are more obvious but, individually, less reliable than the number of cotyledons. These characters include:
1. Stems with vascular bundles arranged in a circle (with a few exceptions).
2. Leaves typically with reticulate venation.
3. Flower parts usually in fives or fours, seldom in threes.
The Dicotyledons are thus distinguished not by a single character but by the above combination of characters.

The arrangement of the dicotyledonous families in both the well-established systems of classification is artificial in that large groups of families are distinguished on the basis of the single character, freedom or fusion of petals. However, since the families selected here are used to illustrate the concept of floral progressions (Fig. 31.1), they can be studied in a more logical order if the system of Engler is followed rather than that of Bentham and Hooker. Engler's system will therefore be adopted for the Dicotyledons, although it has been found convenient to depart from this system as regards the position of the Leguminosae. Some important families belonging to the series Archichlamydeae will be considered in this chapter, and important families belonging to the series Metachlamydeae in the next.

Series 1 – Archichlamydeae

The Archichlamydeae (i.e. those dicotyledonous families with either no petals or separate petals) are placed before the Metachlamydeae in Engler's system because, generally speaking, their flowers are less elaborate than those of the Metachlamydeae. Only families of Archichlamydeae which have flowers with both a calyx and a corolla will be described in this book, but the series includes about forty families with flowers having either no perianth (i.e. naked flowers) or a perianth consisting of a single whorl of members, usually inconspicuous and bract-like. It would serve no useful purpose here to discuss whether such flowers are simple because they are primitive or because they are reduced or degenerate. It is nevertheless an interesting fact that, with the exception of the family Rosaceae and the sub-family Papilionoideae of the Leguminosae, the vast majority of native trees and shrubs in the north-temperate region have inconspicuous flowers with perianth absent or simple. In the tropics, on the other hand, trees and

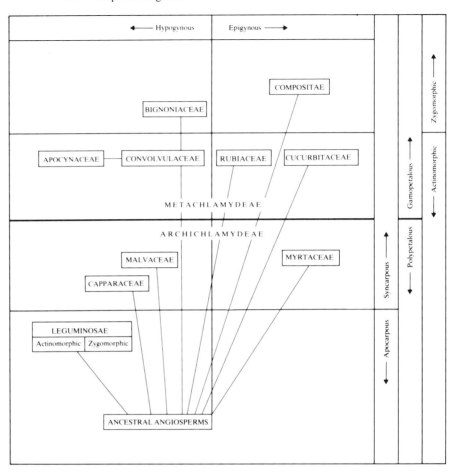

Fig. 31.1 Some important tropical families of Dicotyledons arranged according to the degree of advancement of their flowers.

shrubs belonging to many families, which are fairly evenly scattered throughout the angiosperms, bear conspicuous flowers: the variety of so-called flowering trees and shrubs in tropical botanic gardens is well known. This difference in distribution between trees and shrubs with inconspicuous flowers and those with conspicuous flowers may be significant in view of the widespread belief that flowering plants originated in the tropics.

Capparaceae

The Capparaceae (Fig. 31.2) are dealt with first because they have flowers which may be regarded as more primitive than those of the families considered subsequently. Apart from being typically advanced in progressions 1 (i.e. cyclic arrangement) and 2 (i.e. definite number of floral parts), states which have been reached by almost all living families, the most obvious advance they show on the primitive arrangement of floral parts is their syncarpous gynoecium. This condition results from the fusion of two or more carpels to form a single, superior ovary with parietal placentation.

The Capparaceae also illustrate the fact that, as a consequence of the various floral progressions having occurred at different rates, related species and genera within a single family sometimes show very different degrees of

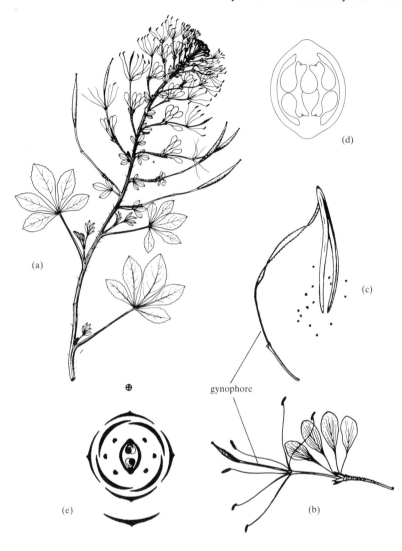

gynophore

Fig. 31.2 Capparaceae. *Gynandropsis gynandra.* (a). Stem with flowers and fruits (b) Flower. (c) Fruit of *Cleome spinosa.* (d) Section of ovary showing the two parietal placentas. (e) Floral diagram. (a and b from *Flora Zambesiaca,* **1**, 206, 1960).

advancement in floral structure. Thus, whereas some species of the Capparaceae have a definite number of stamens, others have an indefinite number. Also, whereas the corolla of most members of the Capparaceae is actinomorphic, in some members there is a distinct tendency to zygomorphy. This variable degree of advancement in floral structure between different species and genera within a single family is most marked in the Rosaceae, a very important family in temperate regions but not described in this book because it is not well represented in the tropics.

Family description

Herbs, shrubs, trees, or sometimes woody climbers. *Leaves* alternate, simple or digitately compound; stipules absent, minute or spiny. *Flowers* bisexual, actinomorphic or zygomorphic. *Sepals* 4, usually free. *Petals* 4, free, often clawed. *Stamens* 4 to numerous, usually free but sometimes united to the stalk of the ovary and thus appearing to be borne from it.

Ovary superior, borne on a long stalk (gynophore), syncarpous, usually 1-locular with 2 parietal placentas, but sometimes 2-locular by intrusion of the placentas, or multilocular. *Fruit* an elongated capsule, or a berry.

The family can be roughly divided into two groups: (1) herbs or undershrubs with elongated, dehiscent capsules; and (2) trees, shrubs and woody climbers with subglobose or broadly oblong, indehiscent berries. Although some of the characters of this family (e.g. the number of stamens) are variable, its members can easily be recognized by the presence of the gynophore which often elongates in the fruit, and by the four sepals and four petals. The leaves are always alternate, and either simple or digitately compound.

The Capparaceae are mostly tropical and subtropical. A few annual species are cultivated as garden plants, including *Gynandropsis speciosa*, which is similar to the widespread weed *G. gynandra* (Fig. 31.2a) but differs in having larger flowers and in being almost glabrous. The most widely known member of the family is *Capparis spinosa*, whose unexpanded flower buds are pickled as capers.

Malvaceae

The Malvaceae (Fig. 31.3) may be regarded as a family that has evolved parallel to the Capparaceae. Both families illustrate the progression towards syncarpy, but, apart from characters peculiar to each family (e.g. the gynophore in the Capparaceae, and the staminal column in the Malvaceae), they differ from each other in the basic number of members in the floral whorls. Thus, whereas the Capparaceae have typically four sepals and petals and two carpels, the Malvaceae have typically five sepals and petals and two to many carpels (i.e. the number of carpels is not constant throughout the family as in Capparaceae). This pattern can be conveniently summarized by the floral formula $K_5 C_5 A_\infty G_{(2-\infty)}$.

Family description

Herbs and shrubs, sometimes trees, often with stellate hairs. *Leaves* alternate, simple, often palmately divided, palmately veined at least at the base, stipulate. *Flowers* bisexual, actinomorphic, often with bracteoles forming an epicalyx outside the true calyx. *Sepals* 5, ± united, lobes valvate (i.e. with the edges meeting but not overlapping). *Petals* 5, free but fused to the base of the staminal column and thus appearing to be slightly gamopetalous, twisted in bud. *Stamens* numerous, the filaments united below into a characteristic tubular column with the style protruding through the top of it; anthers 1-lobed. *Ovary* superior, (2–) 5–many-locular; style simple at the base, often branched towards the apex; ovules 1 to numerous in each loculus; placentation axile. *Fruit* a schizocarp or capsule.

It is usually possible to recognize members of the Malvaceae by their obvious relationship to the cultivated *Hibiscus*. On a more precise basis, they can readily be distinguished by the staminal column, which gives the false impression that the stamens arise from the side of the style.

The family Malvaceae is of world-wide distribution. In the tropics *Hibiscus* spp., especially *H. rosa-sinensis* (common hibiscus) and *H. mutabilis* (changeable rose or changeable hibiscus), are among the most conspicuous shrubs in gardens. The young fruits of okra (*H. esculentus*) are used as

(a)

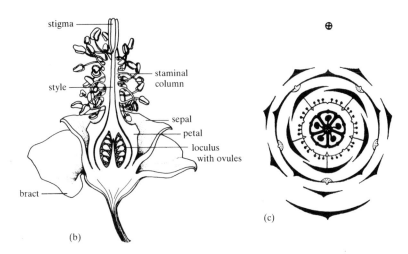

stigma

style

staminal
column

sepal

petal

loculus
with ovules

bract

(b)

(c)

Fig. 31.3 Malvaceae. (a) Flowering twig of
Hibiscus rosa-sinensis. (b) Half flower of
Gossypium showing the staminal column
characteristic of the family. (c) Floral diagram
of *Hibiscus rosa-sinensis*. (b from Raven,
Evert and Curtis 1976, p. 368)

vegetables throughout the tropics. Great economic importance is attached
to the family because of *Gossypium* spp. (cotton). Commercial cotton is
obtained from long fibrous hairs attached to the seeds, and the crop is har-
vested shortly after the capsule has burst open. After removal of the fibrous
hairs the seeds yield the valuable cotton-seed oil. Several *Hibiscus* spp.,
notably *H. cannabinus* (Deccan hemp), are sources of strong fibres, but
here the fibre is derived from the bark. Many tropical weeds, belonging to
such genera as *Malvastrum* and *Sida*, also belong to the Malvaceae.

Myrtaceae

The Myrtaceae (Fig. 31.4) resemble the Capparaceae and Malvaceae in
being syncarpous, but they show an advance on these two families in their
epigyny. Although their flowers are cyclic throughout and have a definite
number of sepals, petals and carpels, the staminal whorl remains primitive
in retaining an indefinite number of members.

Family description

Evergreen trees or shrubs. *Leaves* usually opposite, simple, entire,

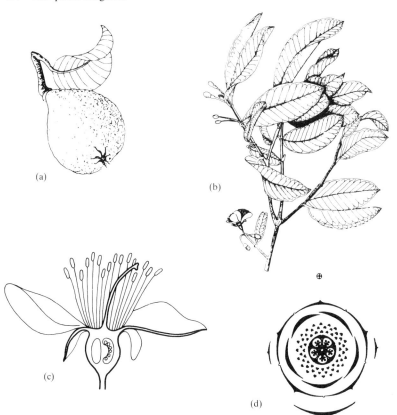

Fig. 31.4 Myrtaceae. *Psidium guajava* (a) Fruit. (b) Flowering twig. (c) Half flower. (d) Floral diagram. (b from Little, E. L., Jr. and Wadsworth, F. H. (1964), *Common trees of Puerto Rico and the Virgin Islands*, p. 417, published by the United States Department of Agriculture, Washington)

gland-dotted (i.e. translucent dots are seen when the leaf is held up to the light) and aromatic when crushed. *Flowers* bisexual and actinomorphic. *Sepals* 4–5, leathery and persistent on top of the fruit. *Petals* 4–5, imbricate (i.e. overlapping one another irregularly) and deciduous. *Stamens* numerous and bent inwards in the bud. *Ovary* inferior, crowned by a small disc; style single, due to the fusion of the carpels extending to the tip of the style. *Fruit* a berry, sometimes a capsule opening cross-wise at the top as in *Eucalyptus*, usually with a persistent calyx.

Trees and shrubs belonging to the Myrtaceae are easily recognized by their evergreen, gland-dotted leaves which are opposite, simple and entire. The numerous stamens and inferior ovary with a single style are also characteristic.

The Myrtaceae are a large tropical family, of which the following are well known members:

Psidium guajava (guava) – cultivated throughout the tropics for its edible fruit.

Eucalyptus spp. (Australian gums) – widely planted because of their hardiness and adaptability. Many yield useful timber, while eucalyptus oil is obtained from their leaves. The name *Eucalyptus* alludes to the fusion of the petals into a cap-like structure (calyptra) which, before the stamens expand, covers the young flower but soon falls off in the form of a lid.

Syzygium spp. – many have edible fruits, e.g. *S. jambos* (rose-apple), and *S. malaccense* (Malay or otaheite apple). *Syzygium* is included by some authorities in the large genus *Eugenia*.

Eugenia caryophyllus – cloves are the flower buds of this tree.

Leguminosae

The Leguminosae represent a stock that has evolved from the primitive angiosperms along its own separate line. Whereas the three previous families all illustrate the progression towards syncarpy, the Leguminosae have reduced the gynoecium to a single carpel which is therefore free as in the primitive apocarpous condition. On the other hand, whereas the Capparaceae exhibit only a tendency towards zygomorphy, the Leguminosae have advanced so far in this respect that some of its members (i.e. those belonging to the subfamily Papilionoideae) are the most zygomorphic of all dicotyledonous flowers. The Leguminosae are a very large family (with about 13 000 species) whose members, while exhibiting a considerable range of floral structure, are united by the possession of a flower with a single elongate carpel which develops into a typical or modified *legume*, the characteristic fruit from which the family derives its name.

Family description

Trees, shrubs or herbs, sometimes climbing, most species possessing root nodules colonized by strains of the nitrogen-fixing bacterium *Rhizobium leguminosarum*. *Leaves* alternate, usually compound and generally stipulate, often exhibiting sleep movements. *Flowers* bisexual, actinomorphic or zygomorphic, usually somewhat perigynous. *Sepals* more or less united, 5 or sometimes 4. *Petals* as many as the sepals. *Stamens* usually twice as many as the petals or numerous, free or united. *Ovary* superior, often shortly stalked, elongate, composed of 1 carpel and therefore 1-celled, with numerous ovules (rarely 1) on a parietal placenta. *Fruit a legume*, or some modification of it (e.g. a *lomentum* in which the pod splits transversely into 1-seeded segments, as in *Desmodium*).

The family divides naturally into three categories which correspond with three easily recognizable types of flower. In some systems of classification these categories are raised to the rank of separate families, but here they are treated as subfamilies which are 'keyed out' as follows:

1. Flowers actinomorphic, usually small; petals valvate; stamens 10 to many; leaves bipinnate, rarely once pinnate . . . subfamily I. Mimosoideae
1. Flowers zygomorphic:
    2. Flowers with the uppermost (posterior) petal inside and enclosed by the others in bud; stamens typically 10, their filaments free or slightly united at the base; leaves pinnate or bipinnate . . . subfamily II. Caesalpinioideae
    2. Flowers with the uppermost (posterior) petal outside and enclosing the others in bud; stamens 10, usually diadelphous but sometimes monadelphous or free; leaves pinnate or palmate . . . subfamily III. Papilionoideae

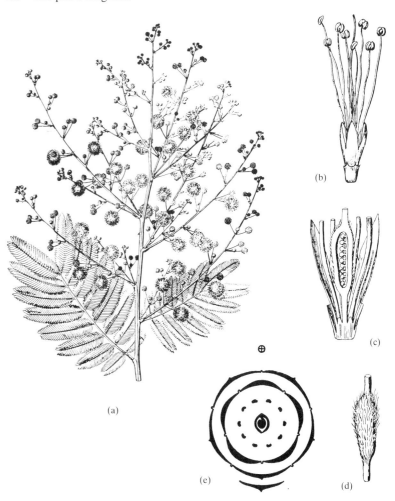

Fig. 31.5 Leguminosae *subfamily* Mimosoideae. *Mimosa sepiaria* (a) Flowering twig. (b) Flower. (c) Half flower. (d) Ovary. (e) Floral diagram of member of Mimosoideae with ten stamens. (a–d redrawn from Martius, K. F. Ph. von, *Flora Brasiliensis,* **15** (2), t.91, 1876)

Subfamily I. Mimosoideae

Trees or shrubs, less frequently herbs. *Leaves* bipinnate or rarely pinnate; stipular spines common. *Flowers* actinomorphic, individually small but usually aggregated in dense heads or spikes. *Sepals* united into a tube, 4–5-lobed or toothed, the lobes valvate. *Petals* 4–5, free or united below, valvate. *Stamens* often numerous, or as many as or twice as many as the petals; filaments free or united below, long and coloured. *Ovary* and *fruit* as for the family.

The Mimosoideae (Fig. 31.5) are mostly tropical and include the following well-known examples:*Acacia* spp. (which often form the characteristic feature of *Thorn woodland* in the dry tropics), *Mimosa pudica* (sensitive plant) and *Samanea saman* (rain tree).

Subfamily II. Caesalpinioideae

Trees or shrubs, less frequently herbs. *Leaves* pinnate (2 pinnae partly fused in *Bauhinia*) or bipinnate. *Flowers* zygomorphic, rarely regular. *Sepals* 5, imbricate. *Petals* 5, free, imbricate, the uppermost (posterior) petal overlapped by the others in bud (cf. reverse situation in

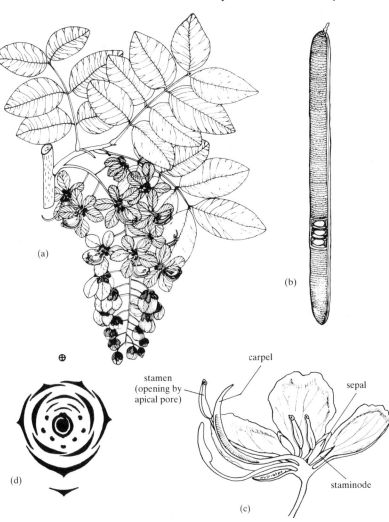

carpel

stamen
(opening by
apical pore)

sepal

staminode

(a)

(b)

(c)

(d)

Papilionoideae). *Stamens* 10, often more, rarely fewer, filaments free or sometimes slightly united at the base. *Ovary* and *fruit* as for the family.

The Caesalpinioideae (Fig. 31.6), like the Mimosoideae, are mostly tropical and are represented by many well-known flowering trees and shrubs, including species belonging to the following genera: *Amherstia* (the species *A. nobilis*, pride of India, is often regarded as the most attractive of all tropical flowering trees), *Bauhinia*, *Caesalpinia* (notably *C. pulcherrima*, Barbados pride or peacock flower), *Cassia* and *Delonix* (including *D.regia*, flamboyant). The pods of *Tamarindus indica*, tamarind, contain an acid pulp which is either eaten or made into a refreshing drink.

Subfamily III Papillionoideae (= Faboideae)

Herbs, less frequently shrubs or trees. *Leaves* palmately or pinnately compound, (occasionally simple by reduction), sometimes ending in a tendril. *Flowers* zygomorphic, somewhat perigynous, relatively large and with a very characteristic appearance remotely resembling a butterfly (hence the name of the subfamily, from the Latin *papilio*–butterfly).

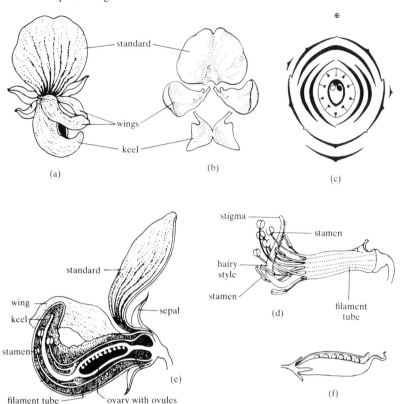

Fig. 31.7 Leguminosae *subfamily*
Papilionoideae. *Crotalaria* (a) Typical
'butterfly' flower characteristic of the
Papilionoideae. (b) Petals detached from
flower. (c) Floral diagram. (d) Filament tube.
(e) Half flower. (f) Legume or pod. (a, d and
e from Stone and Cozens 1969, p.143)

Sepals usually 5, partly united into a tube. *Petals* 5, one large posterior
petal (*standard*), 2 lateral petals (*wings*) and 2 anterior petals more or less
united by their lower margins to form a *keel*; the standard outside and
enclosing the other petals in bud (cf. reverse situation in
Caesalpinioideae). *Stamens* 10, usually (9)+1 but sometimes (10), (5) +
(5), or 10 free as in *Sophora*. *Ovary* and *fruit* as for the family.

The Papilionoideae (Fig. 31.7), in contrast to the two previous subfamilies,
are mainly herbaceous and temperate in distribution. They are economically
important as food and forage plants because their seeds are large and rich in
nitrogen. Groundnuts (*Arachis hypogaea*) and many varieties and species of
beans and peas are grown for human consumption, while *Cajanus cajan* (pigeon pea, Congo pea, etc.) and *Canavalia ensiformis* (horse bean, overlook
bean, sword bean, etc.) are important forage crops in the tropics.

The three subfamilies of Leguminosae are easily distinguishable by the
general appearance of their flowers. In the Mimosoideae the flowers are
actinomorphic and have long stamens giving a 'pin-cushion' appearance. In
the Caesalpinioideae the flowers are somewhat zygomorphic but the petals
and stamens remain distinct. In the Papilionoideae the flowers are markedly
zygomorphic and the stamens are more or less united.

These three subfamilies show a derivation of an advanced, predominantly
herbaceous, temperate group (i.e. Papilionoideae) from more primitive,
woody, tropical groups (i.e Mimosoideae and Caesalpinioideae). The
Mimosoideae, with their typically numerous stamens and actinomorphy, are
considered more primitive than the Caesalpinioideae. The Leguminosae,
whether we regard them as one family or as three separate families, thus
represent the type of sequence which is probably typical of the evolution of
many present-day flowering plants.

Chapter 32

Selected dicotyledonous families: Metachlamydeae

Series 2–Metachlamydeae

The chief characteristic of the Metachlamydeae is that their petals are fused to form a corolla tube. The entire group has reached a condition of definite numbers for all the floral whorls, and therefore is constantly cyclic. The stamens are typically attached to the corolla tube (i.e. epipetalous), and this character is so constant that in the early nineteenth century the botanist de Candolle called the group the 'Corolliflorae'. Since freedom of the floral parts is doubtless a primitive condition, epipetaly may be looked upon as a progression. Woody plants are less common in the Metachlamydeae than in the Archichlamydeae, and this may also be regarded as an advanced character of the series.

In the most primitive Metachlamydeae the flowers are usually pentacyclic, having five whorls of floral parts, one each of sepals, petals and carpels, and two of stamens. Their typical floral formula is $K_5 \widehat{C_{(5)}} A_{5+5} G_{(5)}$. In the more advanced families, which includes all those described here, the flowers are tetracyclic, and the number of carpels is less than the number of members in the other three whorls. The floral formula is typically $K_5 \widehat{C_{(5)}} A_5 G_{(2)}$. Three hypogynous and three epigynous families with this basic ground-plan will be considered.

Hypogynous families

Apocynaceae

The family Apocynaceae (Fig. 32.1) is representative of several families whose flowers conform to the tetracyclic ground plan mentioned above. Apocarpy, however, tends to linger in this family because the carpels are rarely completely united.

Family description

Trees, shrubs, woody climbers or herbs, with white latex. *Leaves* usually opposite or whorled, rarely alternate, simple, pinnately veined, entire, usually without stipules. *Flowers* bisexual and actinomorphic. *Sepals* 5. *Corolla* lobes 5, contorted (i.e. overlapping each other all in the same direction) in the bud. *Stamens* 5, included within the corolla tube; filaments short. *Carpels* superior, 2, more or less free below but fused above to form a single style and stigma. *Fruit* sometimes a berry or

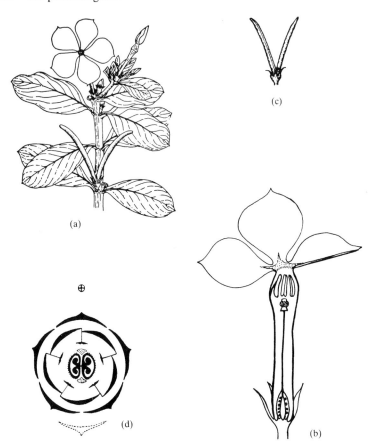

Fig. 32.1 Apocynaceae. *Catharanthus roseus*
(= *Vinca rosea*) (a) Flowering and fruiting
shoot. (b) Half flower. (c) Fruit. (d) Floral
diagram. (a and c from Dutta 1964, p. 611)

prickly capsule, but usually a pair of long narrow follicles containing
many seeds, each seed bearing a tuft of hairs.

Members of the Apocynaceae can for the most part easily be recognized by
their milky sap, opposite or whorled entire leaves, the contorted aestivation
of the corolla lobes, and the two superior carpels which may be more or less
free below but are united above by the single style.

The Apocynaceae are mostly tropical, but a few species occur in temperate regions. The flowers are often large and showy, and frequently pleasant-
ly scented, with the result that several species are commonly cultivated as
ornamentals, for example allamanda (*Allamanda cathartica*), oleander
(*Nerium oleander*), frangipani or temple tree (*Plumeria* spp.) and periwin-
kle (*Catharanthus roseus*)

The Apocynaceae are rich in alkaloids and glucosides; several genera are
sources of drugs, notably *Rauvolfia* from which the sedative drug rauwolfia
is obtained. Many are exceedingly poisonous, including the common garden
ornamental oleander (*Nerium oleander*).

Convolvulaceae

The Convolvulaceae (Fig. 32.2) have reached about the same evolutionary
level as the Apocynaceae, and the floral formula for both families is typically the same. The Convolvulaceae may perhaps be regarded as more ad-

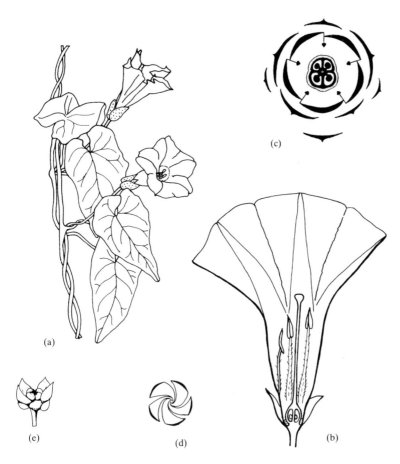

Fig. 32.2 Convolvulaceae (a) Typical habit of member of the Convolvulaceae. (b) Half flower of *Ipomoea tiliacea*. (c) Floral diagram of *Ipomoea tiliacea*. (d) Folding of corolla of Convolvulaceae while in the flower bud. (e) Fruit.

vanced than the Apocynaceae in that the gynoecium is completely syncarpous.

The Convolvulaceae can always be distinguished from the Apocynaceae by their alternate leaves, but in any case nearly all members can be recognized as belonging to this family by their showy funnel-shaped flowers, obviously related to the cultivated morning glory.

Family description

Herbs or shrubs, usually twining or trailing. *Leaves* alternate, simple, often with cordate base and long petiole, exstipulate. *Flowers* actinomorphic and bisexual, often large and showy. *Sepals* 5, equal or unequal, usually free, often very strongly imbricate, persistent in fruit. *Petals* 5, united into a funnel-shaped corolla with 5 longitudinal 'rays', the corolla strongly twisted in the flower bud. *Stamens* 5, alternating with the rays of the corolla. *Ovary* superior, with a nectar-secreting disc at the base, bicarpellary, 2- or 4-celled, the 2-celled forms with 2 ovules in each cell, and the 4-celled forms with 1 ovule in each cell. *Fruit* usually a capsule, sometimes a berry.

The name Convolvulaceae derives from the fact that the family is composed mostly of twiners, which usually twin in a clockwise direction. The funnel-shaped flowers with five conspicuous rays are also easy to recognize.

The Convolvulaceae are world-wide in distribution but chiefly tropical. On account of their large and showy flowers, many species are cultivated as ornamental climbers: these include *Argyreia nervosa* (elephant climber), and various species of *Ipomoea* including *I. purpurea* (morning glory) and *I. alba* (*I. bona-nox*; moonflower). The roots of some of the herbaceous perennial species develop into large tubers, those of *Ipomoea batatas* (sweet potato) being a common vegetable. The cosmopolitan weeds called dodders (*Cuscuta* spp.) are twining parasites which attach themselves to their host by means of numerous haustoria. Their parasitism is complete and this is reflected both in their stems which are never green, and in their leaves which are reduced to scales or absent.

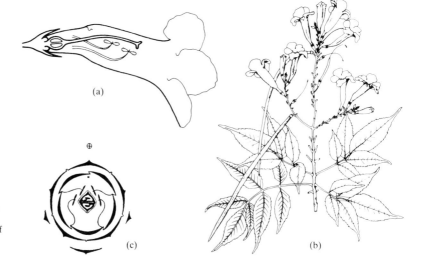

Fig. 32.3 Bignoniaceae. *Tecoma stans* (a) Half flower. (b) Flowering and fruiting twig. (c) Floral diagram (a from Little, E. L., Jr. and Wadsworth, F. H. (1964), *Common trees of Puerto Rico and the Virgin Islands*, p. 503, published by the United States Department of Agriculture, Washington)

Bignoniaceae

The flowers of the Bignoniaceae (Fig. 32.3) are built on the same tetracyclic ground-plan as those of the two previous families but they differ in two respects: (1) the corolla is more advanced in showing the progression towards zygomorphy; and (2) the posterior stamen is reduced to a staminode, or is absent, so that the staminal whorl is reduced to only four functional members. The typical floral formula therefore becomes $K_{(5)}\widehat{C_{(5)}A_4}G_{(2)}$.

Family description

Trees, shrubs or woody climbers, rarely herbs. *Leaves* usually opposite, often pinnate, in the climbers often with twin leaflets and a terminal tendril, exstipulate. *Flowers* large, more or less zygomorphic, and bisexual. *Sepals* 5, gamosepalous. *Corolla* gamopetalous, usually bell- or funnel-shaped, 5-lobed, more or less zygomorphic and 2-lipped. *Stamens* 4, didynamous (i.e. in two pairs of unequal length). *Ovary* superior, seated on a nectar-secreting disc, bicarpellary, usually 2-locular; ovules numerous, attached to 2 placentas borne on the partition wall of the ovary. *Fruit* a long, narrow, 2-valved capsule; seeds usually flattened and with a large membranous wing.

The Bignoniaceae can usually be recognized by their woody habit, their opposite pinnate leaves, their zygomorhic two-lipped flowers, and their long narrow capsules with winged seeds.

Members of the Bignoniaceae are commonly found throughout the tropics, although most of them have their origin in tropical America with a few in Africa and Asia. The reason for their widespread distribution is that many species are cultivated because of their large, brightly coloured flowers. Common examples include the following:

(a) *Shrubs or trees–Jacaranda mimosifolia* (jacaranda), *Spathodea campanulata* (African tulip tree), *Tecoma stans* (yellow bignonia).

(b) *Climbers – Pandorea jasminoides* (pandorea), *Podranea brycei* (Zimbabwe creeper) and several species of *Bignonia* including *B. magnifica* (purple bignonia), *B. venusta* (golden shower) and *B. unguis-cati* (cat's claw).

Epigynous families

Rubiaceae

The Rubiaceae (Fig. 32.4) occupy a position among the epigynous Metachlamydeae comparable to that of the Apocynaceae among the hypogynous Metachlamydeae. Both families have the same number of floral parts in each whorl and therefore conform to the same basic ground-plan, namely $K_5C_{(5)}A_5G_{(2)}$. The only major floral difference is that the Rubiaceae are epigynous.

Family description

Herbs, shrubs or trees. *Leaves* opposite, rarely whorled, simple, with usually interpetiolar (i.e. between the petioles of opposite leaves) or sometimes intrapetiolar (i.e. between the petiole and the stem) stipules. *Flowers* actinomorphic, usually bisexual. *Calyx* of 4–6 free or fused sepals, often rudimentary, persistent on top of the ovary. *Corolla* tubular, of 4–6 petals. *Stamens* as many as the corolla lobes. *Ovary* inferior, usually 2-celled, rarely many-celled; ovules 1 to many in each cell, with axile placentation; style long and slender, often forked at the end (i.e. bifid). *Fruit* a capsule, berry or drupe.

The Rubiaceae is one of the largest tropical families, with between 5000 and 6000 species. The most distinctive characteristics of the family are opposite leaves with inter- or intra-petiolar stipules, and actinomorphic flowers with an inferior ovary. The phenomenon of *floral dimorphism* in which the flowers of a given species occur in two forms, each on a separate plant, is very common in this family; the two forms of flowers are *heterostylous* (i.e. the ratio between the lengths of the style and the stamens varies in different flowers).

Various species of *Gardenia* and *Ixora* are common garden shrubs, while coffee and quinine are obtained from the fruit of *Coffea arabica* and the bark of *Cinchona* spp. respectively.

Cucurbitaceae

The relationships of the Curcurbitaceae (Fig. 32.5) have been much dis-

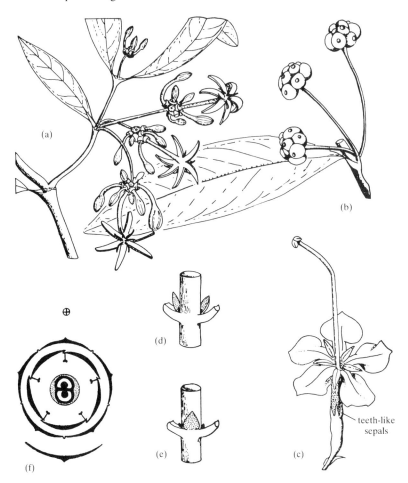

Fig. 32.4 Rubiaceae (a) Flowering shoot of *Morinda lucida.* Although the flowers drawn have six corolla lobes, flowers with five or four lobes are also common. (b) Fruits of *Morinda lucida.* (c) Flower of *Macrosphyra longistyla.* The five teeth-like sepals mark the position of the ovary within the receptacle. (d) Intrapetiolar stipules. (e) Interpetiolar stipules. (f) Floral diagram typical of the family, although variations occur in the gynoecium. (After Nielsen 1965, p. 172)

puted, but Engler's arrangement of placing them between Rubiaceae and Compositae is adopted here. The Cucurbitaceae are on much the same level of evolutionary advancement as the Rubiaceae, but the two families have different numbers of stamens and carpels.

Family description

Trailing or climbing herbs, with lateral tendrils, simple or branched. *Leaves* alternate, often palmately lobed or deeply divided. *Flowers* actinomorphic, unisexual, monoecious or dioecious, usually bright yellow or white in colour. *Calyx* of 5 sepals, fused below into a bell-shaped tube. *Petals* 5, usually gamopetalous or sometimes free. *Male flowers*: stamens 3 (or 5), variable in arrangement, anthers often twisted; rudiments of the gynoecium sometimes present. *Female flowers*: ovary inferior, 3-carpellary, at first one-celled with parietal placentation, but becoming 3-celled by ingrowth of the placentas until they meet in the centre and the placentation appears axile; fruit usually a large berry.

The Cucurbitaceae are better represented in tropical than in temperate countries. Among the families useful to man, this family is prominent because a large number of its members provide vegetables or fruits, including

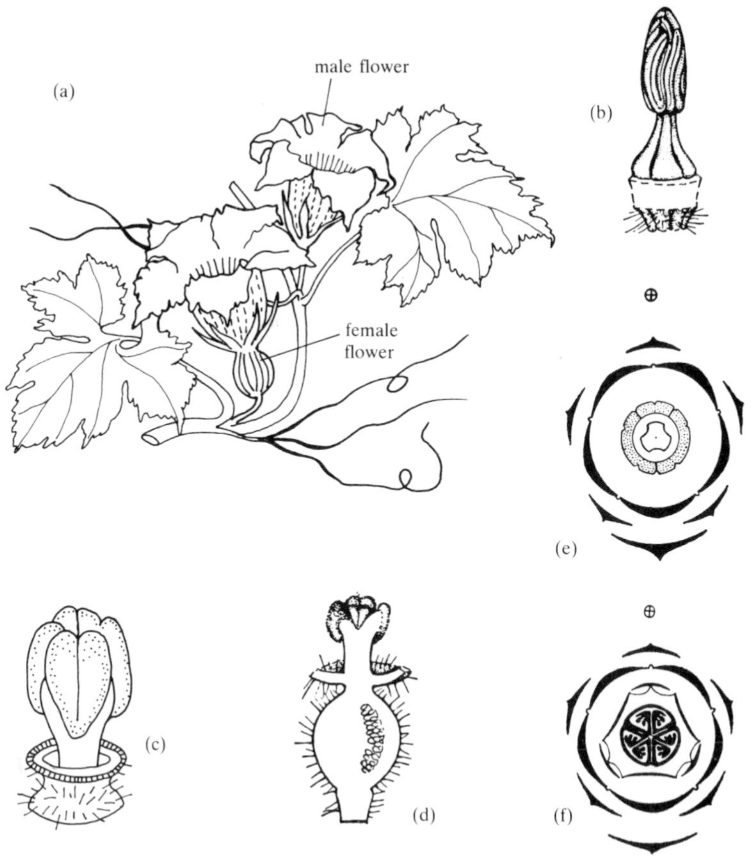

Fig. 32.5 Cucurbitaceae. *Cucurbita pepo* (a) Branch with male and female flowers. (b) Male flower, with calyx and corolla removed. (c) Floral diagram, male flower. (d) Female flower less calyx, corolla and the inferior ovary. (e) Vertical section of female flower less calyx and corolla. (f) Floral diagram, female flower. (b and e from Lawrence 1955, p. 144)

Cucurbita spp. (e.g. pumpkin, squash and vegetable marrow), *Cucumis* spp. (e.g. cucumber and melon), *Citrullus lunatus* (watermelon) and several others. *Luffa cylindrica* is the source of the well-known loofah or bath sponge, which consists of the vascular framework of the pericarp.

Compositae

The Compositae reach the extreme limit in the progression from conspicuous individual flowers to conspicuous inflorescences of small massed flowers. What is usually called a flower in this family is in fact a *capitulum* (i.e. an inflorescence of small sessile flowers, called *florets*, borne on a common receptacle) which is surrounded by a calyx-like involucre of bracts (Fig. 32.6). The name Compositae refers to the composite nature of the structures commonly, but erroneously, called 'flowers'. Another constant feature of the family is the fusion of the anthers into a tube. There are a few other families with a similar aggregation of small flowers, for example the Proteaceae, but these have superior ovaries, free anthers, and other differences from the Compositae.

Family description

Mostly herbs, sometimes shrubs or trees. *Leaves* opposite or alternate, simple, variously dissected, exstipulate. *Flowers* in heads (capitula), each

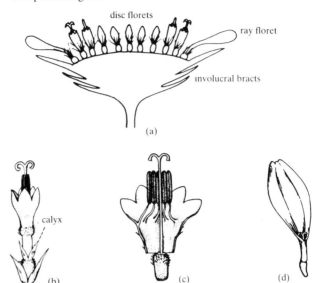

Fig. 32.6 Compositae I. (a) Diagrammatic vertical section of a capitulum. (b) Disc floret with calyx represented in this case by scales (*Helianthus*). (c) Disc floret (with corolla and anther tubes split open) to show fused anthers and free filaments. (d) Ray floret (*Helianthus*). (b, c and d from Lawrence 1955, pp. 65 and 145)

surrounded by a calyx-like involucre of bracts. *Florets* sessile on the dilated head of the receptacle (peduncle), each usually in the axil of a receptacular bracteole, epigynous. The florets in a head may be (a) all tubular (discoid heads, Fig. 32.7), (b) all ligulate, i.e. strap-shaped (ligulate heads) or (c) the inner or disc florets tubular and the outer or ray florets ligulate (radiate heads, Fig. 32.8).

Tubular florets: actinomorphic and bisexual. *Calyx* wanting or represented by scales (e.g. *Helianthus*), by bristles (e.g. *Bidens*) or, most commonly, by a ring of hairs persisting in the fruit as a pappus. *Corolla* gamopetalous, tubular, with 5 (rarely 3 or 4) lobes. *Stamens* 5 (or as many as the corolla lobes), usually with free filaments but always with the anthers united into a tube. *Ovary* inferior, 2-carpellary, one-celled with one basally attached ovule; style bifid, the 2 arms stigmatic on their inner surfaces. *Fruit* an achene, often crowned by the persistent pappus which acts as a parachute in the dispersal of the fruits by wind.

Ligulate florets: zygomorphic, bisexual, female or neuter. *Calyx* as for tubular florets. *Corolla* gamopetalous, strap-shaped. *Stamens* (if present), *ovary* and *fruit* as for tubular florets.

The family Compositae is divided into two subfamilies: (1) the *Asteroideae* which possess tubular florets with or without ray florets, and (2) the *Lactucoideae* which have only ligulate florets. The fact that this division is a natural one is supported by the physiological property that all members of the Lactucoideae exude a white latex when cut, whereas no member of the Asteroideae does so.

The Compositae are numerically the largest family of flowering plants with about 1100 genera and 25 000 species of world-wide distribution. This family is often regarded as the culmination of evolution in the Dicotyledons, because its flowers have departed from the primitive condition along more progressions than any other dicotyledonous family. The prevalence of herbs in the Compositae may also be regarded as an advanced character. The only respect in which the Compositae tend to remain primitive is in the occurrence of actinomorphy in their tubular florets, but it is an observed fact that

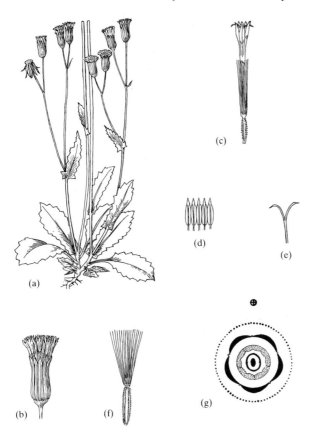

Fig. 32.7 Compositae II. *Emilia sonchifolia*
(with discoid heads) (a) Portion of plant. (b)
Discoid head. (c) Tubular floret, with calyx
represented by ring of hairs. (d) Anther tube,
split open. (e) Forked style. (f) Fruit with
pappus. (g) Floral diagram of a tubular floret.
(From Fawcett, W. and Rendle, A. B. (1936),
Flora of Jamaica, vol 7, p. 274, British
Museum (Natural History), London)

the combination of epigyny and zygomorphy rarely occurs together in the
Dicotyledons. Even though zygomorphy is not constant in the Compositae,
it does occur in the ligulate florets which are therefore advanced in all the
seven progressions mentioned on p. 392.

The Compositae probably include more cultivated plants than any other
single family. Those commonly cultivated in the tropics include *Lactuca sativa* (lettuce), and species of *Chrysanthemum* (including *C. cinerariifolium*
which yields the insecticide pyrethrum), *Cosmos*, *Dahlia*, *Gerbera*, *Helianthus* (including *H. annuus*, the seeds of which are a poultry feed and also
yield sunflower oil), *Tagetes* (marigold), *Tithonia* and *Zinnia*. The family
also includes numerous weeds, many of them introductions which have become naturalized, like *Acanthospermum* spp. (star-burrs), *Bidens* spp.
(blackjacks), *Sonchus* spp. (sowthistles), *Tagetes minuta* (Mexican marigold) and *Tridax procumbens*.

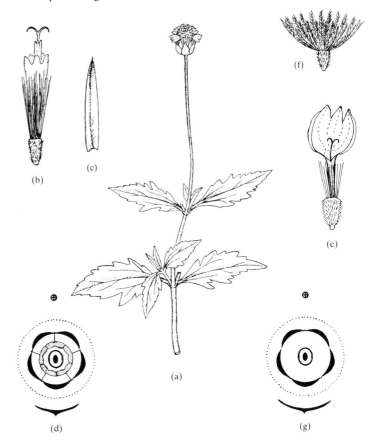

Fig. 32.8 Compositae III. *Tridax procumbens* (with radiate heads) (a) Flowering stem. (b) Disc floret, actinomorphic and hermaphrodite. (c) Bracteole. (d) Floral diagram, disc floret. (e) Ray floret, zygomorphic and female. (f) Fruit, with calyx persisting as the pappus. (g) Floral diagram, ray floret. (b, c, e and f from Dutta 1964, p. 609).

Chapter 33

Selected monocotyledonous families

Just as the Dicotyledons are not distinguished by a single character, recognition of the Monocotyledons does not depend entirely on the possession of a single cotyledon. This feature is associated with several other characters, including:

1. Stems with closed, scattered vascular bundles.
2. Leaves alternate, typically with parallel veins.
3. Flower parts nearly always in threes.

Since the Monocotyledons are characterized by a combination of vegetative and floral characters, there can be no doubt that they represent a natural group. There are in fact no Monocotyledons which have more than one cotyledon, although there are a few Dicotyledons which have only one cotyledon.

The Monocotyledons have long been recognized as a distinct group, but the arrangement of the families within the group varies widely in different systems of classification. For example, in Engler's system the grass family (Gramineae) is placed near the beginning of the Monocotyledons, whereas in the Bentham and Hooker system it is placed at the extreme end. In an attempt to overcome this difficulty John Hutchinson of the Royal Botanic Gardens, Kew, published in 1934 a new classification of the Monocotyledons and this has become widely accepted as the best treatment of the group. In this book, therefore, Hutchinson's arrangement will be followed for the Monocotyledons, even though Engler's system is followed for the Dicotyledons.

The underlying concept of Hutchinson's system is that the most primitive Monocotyledons are those having apocarpous, hypogynous flowers with a distinct calyx and corolla; such types seem to form a link with the primitive Dicotyledons which are also assumed to be apocarpous. Starting with families showing the above characters, the Monocotyledons are arranged in a sequence of families ending with the Gramineae, which is regarded as the climax of monocotyledonous evolution. Three main groups of Monocotyledons are recognized:

1. *Calyciferae* ('calyx-bearers') with a distinct calyx, usually green, and a corolla.
2. *Corolliferae* ('corolla-bearers') in which the two whorls of perianth leaves are more or less similar and petaloid (these are sometimes called the 'petaloid Monocotyledons').
3. *Glumiflorae* in which the perianth is much reduced, or represented merely by scales (lodicules) as in the grasses.

The group Calyciferae contains no families of major importance and will not be dealt with here. To illustrate the range of floral structure within the Monocotyledons, three families, the Liliaceae, Orchidaceae and Gramineae, will be considered. The Liliaceae, at the beginning of the Corolliferae, is chosen to represent a generalized type of monocotyledonous flower, the Orchidaceae and Gramineae as the peak or climax of floral evolution in the Corolliferae and Glumiflorae respectively.

Group – Corolliferae

The most primitive families of Monocotyledons have a distinct calyx and corolla, but in the course of evolution these two whorls became similar in appearance and petaloid. Families with such a petaloid perianth are included in the group Corolliferae, which begins with the Liliaceae and ends with the Orchidaceae.

Liliaceae

The Liliaceae (Fig. 33.1) are of fundamental importance among the Monocotyledons because they represent a central stock from which many other families can be traced either directly or indirectly.

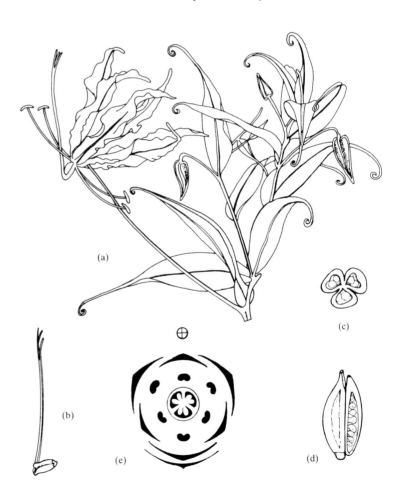

Fig. 33.1 Liliaceae. *Gloriosa superba.* (a) Portion of plant with flowers. (b) Gynoecium. (c) T.S. ovary. (d) L.S. ovary. (e) Floral diagram. (a–d redrawn from C. Letty in Pole-Evans, I. B. (Editor) (1931), *Flowering plants of South Africa*, vol. 11 published by L. Reeve and Co. Ltd)

Family description

Herbs, rarely shrubby, sometimes climbing, with rhizomes, bulbs or corms. *Leaves* simple, variable in position. *Flowers* actinomorphic, bisexual, usually arranged in racemes (never in umbels). *Perianth* of 6 segments in two similar whorls. *Stamens* 6 in two whorls, opposite the perianth segments, inserted on them or free. *Ovary* superior, 3-locular, with axile placentation; ovules numerous, in 2 rows on each placenta. *Fruit* a capsule or berry.

The Liliaceae are very uniform in floral structure but very diverse in habit and appearance. They are the third largest family of Monocotyledons, and may be divided into two main series: (1) a generally more primitive group having an underground rootstock (rhizome) with fibrous or tuberous roots; and (2) a more advanced group which has developed a different form of root system, the bulb or corm. Liliaceae with a corm or bulb are well adapted for life in drier regions, and are abundant in the flora of the Mediterranean and of certain parts of South Africa. Species belonging to many genera, including the type-genus *Lilium*, are commonly cultivated in gardens, but otherwise the family is of little economic importance.

All members of the Liliaceae have a superior ovary, and formerly this was the character that distinguished them from the closely related family, the Amaryllidaceae. Hutchinson considers the type of inflorescence to be a more important and natural character, and so regards the possession of an umbellate inflorescence (rather than an inferior ovary) as the feature separating the Amaryllidaceae from the Liliaceae. The adoption of this view involves the transfer from the Liliaceae to the Amaryllidaceae of a number of genera with a superior ovary but possessing an umbellate inflorescence; these genera include the well-known genus *Allium* (containing the onion and shallot) that is used, and illustrated, in some books as a typical example of the Liliaceae. In both the Liliaceae and Amaryllidaceae there are six perianth segments and six stamens, but whereas in the Liliaceae (as delimited by Hutchinson) the flowers are never in umbels and the ovary is always superior, in the Amaryllidaceae (as delimited by Hutchinson) the inflorescence is always an umbel subtended by one or more membranous bracts, the root stock is always a bulb or corm, and the ovary is typically but not invariably inferior. Hutchinson's concept of the Liliaceae is supported by several independent lines of evidence based on morphological, anatomical and pollen studies. The difference between the criteria formerly and currently used for distinguishing between the Liliaceae and the Amaryllidaceae is illustrated by the two ways in which the genera, *Lilium, Allium* and *Crinum* are classified.

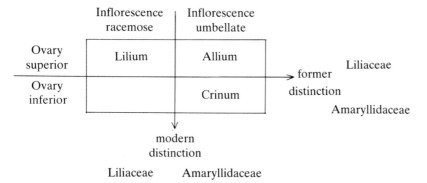

Orchidaceae

The Orchidaceae (Fig. 33.2) represent the peak or climax of floral structure in the Corolliferae. Their flowers are built on the same ground-plan as those of the Liliaceae, but differ in the following respects:

1. The perianth, although consisting of six segments in two whorls, is very much modified; the middle member of the inner whorl ('lip' or *labellum*) is much larger than the other perianth segments, and this results in the formation of a zygomorphic flower.
2. The number of stamens is reduced from six to one (or rarely two.)
3. The ovary is inferior, and has parietal placentation.

Family description

Perennial herbs, growing either in the ground (terrestrial) or perched on trees (epiphytic); almost invariably mycorrhiza (i.e. their roots or rhizomes are associated with fungi) and sometimes saprophytic, deriving their organic matter from humus. *Leaves* simple, often fleshy; in saprophytic species reduced to scales lacking chlorophyll. *Flowers* bisexual, zygomorphic, often very spectacular in form and colour. *Perianth* of 6 segments in 2 whorls of 3 each, usually all petaloid though often green; the median or morphologically posterior segment (i.e. that

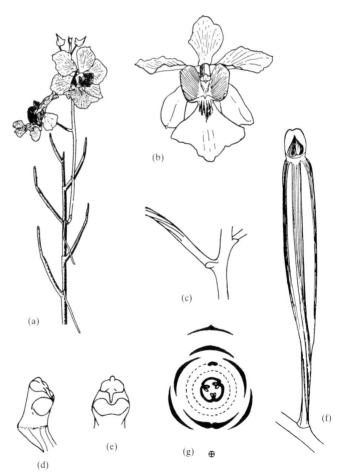

Fig. 33.2 Orchidaceae. *Vanda teres ×
hookeriana* (= *Vanda Miss Joaquim*). (a)
Climbing stem bearing leaves, two roots and
an inflorescence. (b) Flower. (c) Base of ovary
showing twist and bract. (d) Column and top
of ovary (perianth removed). (e) Front of
column. (f) Fully grown fruit. (g) Floral
diagram. (a, b and f from Holttum 1954,
pp. 41 and 68).

nearest to the supporting stem) of the inner whorl forming the 'lip' or labellum, which is commonly larger than the other segments and often spurred or otherwise modified; the labellum is usually directed downwards owing to the inversion of the flower in a pendulous inflorescence, or to the twisting of the ovary or its stalk through 180° in an erect inflorescence. *Stamens* 1 or 2, borne together with the stigmas on an extension of the floral axis (corresponding in position to what would conventionally be called a style) known as the 'column' or *gynostemium*, a structure peculiar to orchids; pollen usually aggregated into waxy masses (*pollinia*) instead of consisting of separate grains. *Ovary* inferior, tricarpellary but unilocular, at flowering time resembling a flower stalk (pedicel) and frequently twisted so that the labellum assumes an anterior position; ovules numerous, attached to 3 parietal placentas. *Fruit* a 3- or 6-valved capsule, containing numerous powdery seeds.

Within the above limits, orchid flowers assume a great variety of shapes and forms, due mainly to modifications of the labellum. These modifications are often associated with the structure and habits of pollinating insects, to whose bodies the pollinia become attached in such a way that they are placed on the stigmas of other flowers. Orchid flowers with their much modified perianth, column, peculiar pollen and inferior ovary are highly specialized in comparison with the relatively simple flowers of the Liliaceae. There are, however, several families which bridge the gap and make the orchid flower intelligible as a modification of the basic liliaceous type.

The Orchidaceae are widely distributed throughout the world, and are found in the most diverse habitats. They are, however, most abundant as epiphytes in humid tropical forests, where they form a conspicuous feature of the vegetation. The Orchidaceae are numerically the second largest family of flowering plants (Compositae are the largest) with about 750 genera and 18 000 species; there are also many natural and artificial hybrids obtained by crossing different species and even different genera. Except for the genus *Vanilla* which is used for flavouring, orchids have no direct economic importance other than as the basis for a vast floricultural industry. Immense sums of money are spent on the importation, cultivation, breeding and sale of choice orchids. The industry is centred in the United States of America, but it is also a major source of income in such countries as Singapore, Hawaii and Thailand. As a result of the extensive collecting that accompanied the growth of the orchid industry, thousands of wild species are in danger of becoming extinct. Fortunately, legislation prohibiting the removal and export of wild orchids has been introduced in many countries, so that orchids are now among the best-protected plants in the world.

Group – Glumiflorae

The Glumiflorae comprise a few families in which the perianth is much reduced or even absent. The group starts with the Juncaceae ('rush' family), which were probably derived directly from the liliaceous stock because they still retain a small perianth of six parts even though it is dry and calyx-like. It ends with the Gramineae, which is the most highly evolved and successful of all angiosperm families.

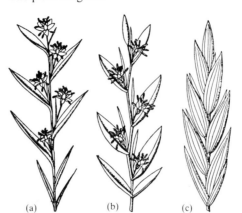

Fig. 33.3 Theoretical stages in the evolution of the grass spikelet. (a) Inflorescence of presumed ancestral type. (b) Intermediate stage in the development of the spikelet. (c) Typical spikelet. (From J. Hutchinson (1948), *British flowering plants*, p. 286, published by P. R. Gawthorn Ltd.)

Gramineae

When the typical flower of the Gramineae is compared with that of the generalized monocotyledonous type, as represented by Liliaceae, it is found that the perianth is reduced to two small scales (lodicules), the stamens are reduced to three, one stigma is no longer present and only one carpel of the ovary is functional. In the flowers of some bamboos (which are regarded as being the most primitive tribe of grasses) there are three lodicules, six stamens and three stigmas. Such flowers are considered to resemble most closely the ancestral type. Although it seems a far cry from the grass spikelet to a more conventional type of inflorescence, the way in which the spikelet may have been evolved is represented in Fig. 33.3. The flowers of grasses are so reduced and exhibit such uniformity in their structure that the unit of the grass inflorescence is not the single flower, as in most other plants, but the spikelet which is a miniature inflorescence. The spikelets show great variability, and their modifications and arrangement are very useful in identification and in suggesting relationships.

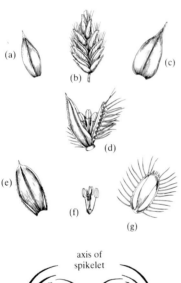

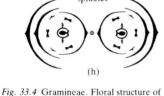

Fig. 33.4 Gramineae. Floral structure of *Eragrostis ciliaris*, a grass with 8 fertile florets in each spikelet. (a) Lower glume. (b) Spikelet. (c) Upper glume. (d) Floret. (e) Lemma. (f) Floret with lemma and palea removed, showing stamens and immature ovary with lodicules at base. (g) Palea. (h) Floral diagram. All much enlarged. (a–g drawn by Priscilla Fawcett, from Gooding, E. G. B., Loveless, A. R. and Proctor, G. R. (1965), *Flora of Barbados*, p. 31, published by Her Majesty's Stationery Office, London)

Family description

Annual or perennial herbs, rarely woody (as in bamboos). *Stems (culms)* jointed, usually hollow in the internodes (solid in maize and sugar cane) and solid at the nodes. *Leaves* alternate, arranged in two opposite rows, consisting of a lamina, sheath and ligule; lamina usually long and narrow; sheath encircling the stem, but split down to point of insertion on the side opposite the lamina; ligule a small membranous appendage, or reduced to a fringe of hairs, at the junction of lamina and sheath. *Flowers* usually bisexual, sometimes unisexual, consisting of 3 stamens (occasionally 1, 2 or 6), a one-celled superior ovary with 2 plumose stigmas, and usually 2 small delicate scales (lodicules) which are thought to represent all that remains of the perianth. *Fruit* a *caryopsis*, a dry one-seeded indehiscent fruit in which the ovary wall (pericarp) becomes fused with the testa of the single seed; this type of fruit is characteristic of the family.

Flowers (Fig. 33.4) enclosed between 2 bracts, a lower (which is called the *lemma*) and an upper (called the *palea*), the whole forming a floret. Florets one to many, more or less sessile on a short slender axis (*rachilla*) usually with 2 bracts at its base (*glumes*), the whole forming a *spikelet*. One or more florets in a spikelet sometimes sterile. (Figure 33.5c shows

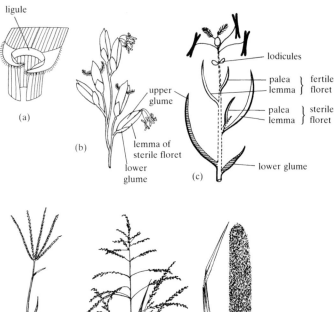

Fig. 33.5 Gramineae. (a) Grass leaf at junction of blade and sheath, showing ligule.

Panicum maximum, a grass with two florets in each spikelet. (b) Portion of inflorescence, showing six spikelets (each containing two florets, the upper fertile and the lower sterile or male). (c) Diagram of a single spikelet expanded to show position of parts, the dotted lines indicating the axis as if expanded.

Types of inflorescence. (d) Spike of *Cynodon dactylon* (Bahama grass, Bermuda grass, and many other common names). (e) Panicle of *Panicum maximum* (Guinea grass). (f) Spike-like panicle of *Pennisetum typhoides* (pearl or bulrush millet).

the relative position of these parts in *Panicum maximum* in an expanded diagram, the dotted lines showing the axis as if elongated; actually the axis is very short and the spikelet is much condensed as in Fig. 33.5b.)

The spikelets are arranged in various ways to form an inflorescence (Fig. 33.5). If the spikelets are sessile along the main axis the inflorescence is spicate, as in *Cynodon dactylon* (Bermuda grass) or *Eleusine coracana* (finger millet); if, as is more common, the spikelets are stalked on a branching axis, the inflorescence is paniculate. The panicle may be open as in *Panicum maximum* (Guinea grass) and *Oryza sativa* (rice), or much contracted and spike-like as in *Pennisetum typhoides* (pearl or bulrush millet) and *Setaria* spp. (hair grasses).

Although the Gramineae are only the fourth largest family as regards number of species (Leguminosae, Compositae and Orchidaceae are all larger), they certainly cover more area of the earth's surface than any other family. They are extremely important as sources of food and fodder. The most important food and fodder grasses in the tropics are: *Eleusine coracana* (finger millet), *Oryza sativa* (rice), *Panicum maximum* (Guinea grass), *Panicum miliaceum* (broom corn), *Pennisetum clandestinum* (Kikuyu grass), *Pennisetum purpureum* (elephant or Napier grass), *Pennisetum typhoides* (pearl or bulrush millet), *Saccharum officinarum* (sugar cane), *Sorghum* spp. (millets) and *Zea mays* (maize). Grasses are also useful in other ways, for instance as soil binders to prevent erosion. Some grasses can be very troublesome weeds, e.g. *Cynodon dactylon* (Bermuda grass).

Suggestions for further reading

General surveys of the plant kingdom

Bell, P. R. and Woodcock, C. L. F. (1971) *The diversity of green plants*, 2nd edn. Edward Arnold.

A concise, modern account of the groups of green plants from algae to angiosperms. Understandably a popular undergraduate textbook.

Bold, H. C., Alexopoulos, C. J. and Delevoryas, T. (1980) *Morphology of plants and fungi*, 4th edn. Harper & Row.

This classic text on the morphology of the entire plant kingdom is probably the most comprehensive in the field. The last two co-authors have collaborated for the first time in the writing of this latest edition.

Corner, E. J. H. (1964) *The life of plants*. Weidenfeld & Nicolson.

A renowned botanist who has studied in the tropics for many years surveys the evolution of plant life, describing how the structure and function of plants were modified to meet the challenge of a new environment as they invaded the shore and spread across the land. Controversial in places, but always stimulating.

Coulter, M. C. and Dittmer, H. J. (1964) *The story of the plant kingdom*, 3rd edn. University of Chicago Press.

A very readable, introductory account of the plant kingdom. Background reading.

Delevoryas, T. (1977) *Plant diversification*, 2nd edn. Holt, Rinehart & Winston.

This short, readable book emphasizes the evolutionary aspects of the plant kingdom.

Jamieson, B. G. M. and Reynolds, J. F. (1967) *Tropical plant types*. Pergamon.

A book of plant types, using tropical examples wherever suitable, which has been written to cover the plant diversity and angiosperm morphology required for A-level or equivalent examinations.

Scagel, R. F. and others. (1965) *An evolutionary survey of the plant kingdom*. Blackie & Son.

A comprehensive, thoroughly illustrated review of the plant kingdom, in which the editor has supplemented his own contribution with chapters submitted by a panel of botanists who are each specialists on a particular group of plants.

Books on particular groups

Alexopoulos, C. J. and Bold, H. C. (1967) *Algae and fungi*. Macmillan.

The algal section of this book combines brief descriptions of illustrative types with a discussion of general principles.

Doyle, W. T. (1970) *The biology of higher cryptogams*. Macmillan.
A concise account of the biology of the bryophytes and pteridophytes, with an emphasis on developmental and evolutionary relationships.

Everard, B. and Morley, B. D. (1974) *Wild flowers of the world*. Octopus Books.
A collection of paintings of over 1000 of the world's exotic and colourful plants. This book complements Heywood's *Flowering plants of the world*, especially because the plants are here grouped on a geographical rather than a taxonomic basis.

Foster, A. S. and Gifford, E. M. (1974) *Comparative morphology of vascular plants*, 2nd edn. W. H. Freeman.
A well-written general account of the vascular plants which, by its emphasis on basic principles, synthesizes its subject matter into a satisfying whole.

Heywood, V. H. (Ed.) (1978) *Flowering platns of the world*. Oxford University Press.
The best book on angiosperm systematics currently available. The text has been written by a panel of recognized authorities, and is copiously illustrated with numerous well-chosen plates, most of them in colour. Very highly recommended.

Holttum, R. E. (1954) *Plant life in Malaya*. Longman.
A delightful book which, although written for residents in Malaya, should be essential reading for anyone who wishes to learn something about the biology of tropical plants. Much of the book is based on the first-hand observations of the author, who is one of the most distinguished tropical botanists of this century.

Kochhar, S. L. (1980) *A textbook of economic botany in the tropics*. Macmillan.
Combines in one volume a wealth of factual information about economic plants, ranging from their history and chemical composition to the methods used in their cultivation.

Kumar, H. D. and Singh, H. N. (1979) *A textbook on algae*, 2nd edn. Macmillan.
A comprehensive and up-to-date account of the more important genera of algae together with information on physiological and other aspects of the group. Intended for use by students in tropical and subtropical countries.

Lawrence, G. H. M. (1955) *An introduction to plant taxonomy*. Macmillan.
Although intended specifically for use in North America, this little book gives a clear and concise account of the theory and practice of plant taxonomy which is applicable everywhere.

Nielsen, M. S. (1965) *Introduction to the flowering plants of West Africa*. University of London Press.
Although the aim of this book is to provide West African students with a lucid account of the principal families of flowering plants in Nigeria and Ghana, it will be useful in all tropical countries.

Proctor, M. C. F. and Yeo, P. (1973) *The pollination of flowers*. Collins.
A modern and beautifully illustrated account of all aspects of pollination biology. Although most of the examples refer to European plants, the authors have not hesitated to use tropical examples where appropriate.

Purseglove, J. W. (1974–75) *Tropical crops: Dicotyledons* (1974); *Monocotyledons* (1975). Longman.
These two volumes cover the major crop plants which are grown anywhere

within the tropics. With their authoritative text, copiously illustrated with excellent line drawings, these books are deservedly the standard works in English on the subject.

Round, F. E. (1973) *The biology of the algae*, 2nd edn. Edward Arnold.
Assuming a previous knowledge of the standard algal types, this book deals with selected topics, particularly those of an ecological and physiological nature.

Smith, G. M. (1955) *Cryptogamic botany*, 2nd edn. Vol. 1 *Algae and fungi*. Vol. 2 *Bryophytes and pteridophytes*. McGraw-Hill.
Old-fashioned in its format, this classic text is still useful as a reference book. The quality of many of its illustrations has never been improved upon.

Sporne, K. R. (1974) *The morphology of angiosperms*. Hutchinson.
A short, well-written account of the range of form and structure in flowering plants. This definitive book summarizes the author's research over 30 years on angiosperm evolution.

Vickerman, K. and Cox, F. E. G. (1967) *The protozoa*. John Murray.
A short, clearly-written text is complemented by an excellent set of drawings, photographs and electron micrographs of representative types which include the unicellular 'plants' *Euglena* and *Chlamydomonas*.

Part IV

Ecology and genetics

Chapter 34

The plant community

The concept of plant communities

Previous chapters have described plants in terms of their structure and the processes which enable them to grow and reproduce. A survey of plant life, however, would be incomplete if these aspects alone were considered because, under natural conditions, plants (and animals) live together in communities. If all the plants on a high oceanic island were removed and you were given the task of replanting them in their original places, you would very soon realize that the plants to be put in any particular place were not a random selection of the island's flora but a more or less definite assemblage. Some plants would be widespread in their distribution, others confined to the coastal regions, some to the mountain tops and others even to gardens. When your task was finished you would in fact have reconstructed a number of characteristic and recognizable communities (Fig. 34.1).

Although the precise reasons for the presence or absence of plants in any given community are difficult to determine because of the interaction of many controlling factors, the general reason is straightforward. Just as some people must be very particular about their diet while others can be more catholic in their tastes, so different plants vary greatly in their physiological requirements. Any plant, if it is to grow successfully and reproduce, must have these requirements satisfied, and it will exist only where the surroundings supply them. Consequently every plant can be said to have a certain *range of tolerance* to external conditions. Some plants can grow successfully

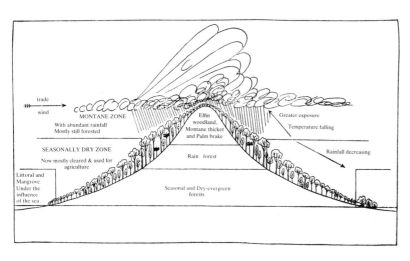

Fig. 34.1 Diagrammatic cross-section of a high oceanic island showing typical arrangement of the vegetation into communities which, in this case, are related to altitude. (From Beard, J.S. (1949), *Oxford Forestry Memoirs* **21**, 55)

in a wide variety of environmental conditions, and therefore tend to be widely distributed. Others with a narrower range of tolerance are restricted to those places which meet their more specialized requirements. Lastly, a few species have such exacting requirements that these can be satisfied only in one particular type of locality; striking examples are the mangroves which grow only under the influence of salt water, and the sundews (*Drosera* spp.) which are found only in acid bogs. Plants belonging to this last group are called *indicator plants* because their presence gives a sure indication of the prevailing conditions. Some of them are used by agriculturists and foresters for deciding the most appropriate use to which land should be put. Others are used in prospecting because their distribution is closely tied to soils exceptionally rich in certain mineral ores; for example, in Zaire, *Crotalaria cobalticola* and *Silene cobalticola* are excellent indicators of cobalt deposits.

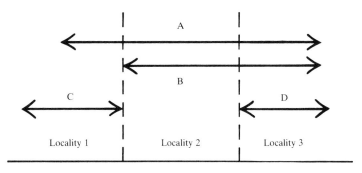

Fig. 34.2 The concept of tolerances illustrated by two pairs of species, A and B with different but overlapping requirements, and C and D with completely different requirements. Species A and C could occur together in locality 1, A and B in locality 2 and A, B and D in locality 3.

Index of environmental conditions

The reason why different plants occur together as a community is that their ranges of tolerance fall within, or extend into, the environmental limits of a particular place (Fig. 34.2). The influence of the environment on the floristic composition of a community is thus chiefly a selective one, i.e. the environmental conditions at any one place are such that they permit the growth of only a proportion of the plants available in the area. The general similarity in the species composition of communities occurring at similar sites within the same geographical area is readily recognizable. For example, a strand community along any tropical sandy coastline is always characterized by the presence of certain creeping plants, such as *Ipomoea pes-caprae* (beach morning-glory, or seaside yam), which are restricted to sandy soils near the sea. It should be pointed out, however, that a particular species in a community at one place may be absent from a similar community at another place simply because seed parents do not occur within dispersal distance of the second locality.

The various environmental conditions which influence the growth of plants and determine the species composition of plant communities are called habitat factors. The word *habitat* (Latin, it inhabits) was originally used to denote the place or type of locality (e.g. a lagoon) in which a given plant grew but, as an ecological term, it has come to mean the sum total of the factors under which a plant or plant community exists. These factors include not only the physical and chemical conditions (e.g. climate and soil) which prevail, but also the effects which organisms have on one another. The effect of animals, especially man, on plant communities can be very far-reaching and animals must therefore be regarded as one of the factors in the environment of plants. The study of the relations between organisms and their environment is the subject matter of ecology. Plant ecology is a thoroughly scientific discipline which investigates such problems as why certain species regularly colonize certain habitats, or why the vegetation of an area may change with time. In short, its aim is to understand the life of plants in their natural habitats.

The structure of plant communities

Plant communities are the natural units of vegetation and their reality has long been recognized, as is evident from such words of common speech as forest, veld and swamp. The first task of plant ecology is to describe plant communities scientifically so that they can be compared, classified, or studied in terms of the energy transformations which occur within them. However, before any community can be described adequately the identity of the component species must be known. In all serious ecological work it is essential to build up a reference collection of plants, carefully pressed, dried, and mounted on herbarium sheets. A label giving the name of the plant, the precise locality where it was collected, by whom and when, should be stuck on each herbarium sheet. Other relevant information not evident from the dried specimen, such as the colour and scent of flowers, should also be added. (Full details of techniques for collecting and pressing plants are given in a booklet *Instructions for Collectors No. 10: Plants* published by the British Museum (Natural History), London).

It cannot be overemphasized, however, that when the plants in a community have been listed the task has only just begun because the structure of the community must then be described. In every stable community the plants are of different sizes, and they are distributed in such a way as to form a three-dimensional pattern which is not apparent from a mere list of names. A description of the structure of a plant community involves the assessment of the relative abundance and the vertical arrangement of the component species.

Assessment of abundance

The simplest and most rapid way of expressing abundance is to attach to each species in the community a visual estimate of its abundance, denoted by conventional symbols such as a = abundant, f = frequent, o = occasional, and r = rare. Such subjective assessment inherently contains a large degree of error because factors other than abundance unconsciously influence the observer's judgement. These include the size of the species under consideration (e.g. large plants tend to be rated higher than small ones), the pattern of distribution (e.g. a species forming obvious clumps is more conspicuous than one which is evenly distributed) and the state of flowering (e.g a species in full flower at the time of assessment receives a higher rating than one which is in the vegetative state). Furthermore, different workers employ different standards so that a species which one person describes as 'frequent' may be classed as 'abundant' or 'occasional' by others.

The absence of a common standard for 'abundance' or 'rarity' is an inevitable source of error in the use of conventional symbols. However, symbols which define measures of abundance in exact terms are less subject to errors of interpretation because, in theory at least, the only discrepancy will be the accuracy with which different workers estimate the measure of abundance as defined by the symbol. Various methods of describing vegetation by the use of clearly defined symbols have been devised and, as an example, the Braun–Blanquet scale of cover-abundance, which has been employed extensively by European (continental) ecologists, will be mentioned. According to the Braun–Blanquet system the species of a plant community are each scored on the basis of both their abundance and the area of ground which they occupy, i.e. their *cover*. The Braun–Blanquet scale is as follows:

x = sparsely or very sparsely present, cover very small
1 = plentiful but of small cover value
2 = very numerous, or covering at least $\frac{1}{20}$ of the area
3 = any number of individuals covering $\frac{1}{4}$ to $\frac{1}{2}$ of the area
4 = any number of individuals covering $\frac{1}{2}$ to $\frac{3}{4}$ of the area
5 = covering more than $\frac{3}{4}$ of the area

It will be noted that the higher numbers in this scale are defined by percentage cover, and the lower ones by abundance. The advantage of this compromise is that the value of cover for indicating the relative importance of the various species within a community is stressed, but at the same time the practical difficulty of trying to assess the cover of very numerous small individuals is avoided. Despite its shortcomings the Braun–Blanquet scale has much to recommend it, especially when large areas of vegetation have to be surveyed in a relatively short time. Botanists who use the scale claim that, with practice, different workers obtain the same results from the same site.

Quantitative assessment of abundance

Realization of the limitations in the visual estimation of abundance has led to the development of quantitative methods which aim to exclude the personal judgement of the observer. Quantitative methods for assessing abundance depend on recording the actual abundance in a number of sample areas and then using the results to estimate the abundance in the community as a whole. Since the sample areas are assumed to be representative of the whole community it is necessary that they should be taken 'at random'. This method of random sampling ensures that the sample areas are distributed in such a way as to eliminate the conscious or unconscious tendency to try and include at least one individual of a large and conspicuous species which it is felt ought to be included in the sampling. Sample areas are called *quadrats* (literally, squares) because traditionally they have been square, but they are still called quadrats when their shape is circular, rectangular or even a straight line. Convenient sizes of quadrat for analysing herbaceous vegetation are squares of either 25 cm or 1 m side length, but the type of vegetation must determine the size of the quadrat chosen – a small quadrat if the plants to be sampled are small and numerous (e.g. mosses), a large one if the plants are large and widely scattered (e.g. trees). Adequate sampling of most plant communities is a time-consuming process, and usually as many quadrats should be taken as time will allow, although there are several empirical calculations which indicate when further sampling becomes pointless.

There are three common methods of expressing abundance quantitatively, and the one which should be used depends on the purpose for which the information is required and on the time available for sampling the vegetation. The first of the following quantitative methods expresses abundance in absolute terms, the last two in relative terms.

Density The density of a species is the average number of individuals per unit area. It is estimated by counting the number of individuals of each species in a quadrat of known size, and then repeating the operation at other randomly distributed places. The results from all the quadrats are then added up and an average density calculated for each species. The method is not subject to personal bias but, apart from the practical difficulty of deciding for certain plants (e.g. rhizomatous grasses) where one individual ends and the next begins, it suffers from the defect that absolute numbers are not necessarily indicative of relative importance in a community. For example, five large trees are more significant in the structure of a forest than five moss plants. To some extent this limitation of the density method is overcome by using percentage cover as a measure of abundance.

Percentage cover This is defined as the percentage of ground covered by the aerial parts of a given species. The nature of cover is perhaps more easily understood by pointing out that, if a community composed of only one species growing on level ground were illuminated directly from above, the proportion of ground in shadow would represent the cover of the species. Approximate cover values can be obtained by visually estimating the percentage of the total area of a quadrat that is covered by a given species, but more exact values can be measured by the *point quadrat* method. This depends on recording which species, if any, is covering the surface of the ground at a number of points in the community being described. The procedure is to use long thin pins or knitting needles suspended in a frame which can be adjusted to the height of the vegetation. The pins are lowered one at a time and the species touched by each pin in turn recorded. For each species, the total number of 'hits' from a succession of sample frames is then expressed as a percentage of the total number of pin 'shots'. The main error of the point quadrat method is that it overestimates the true cover value of a species because a stout pin, like a knitting needle, will touch plants that would not make contact with the axis of the pin. For comparative purposes this exaggeration of the percentage cover can be ignored provided the pin diameter is kept constant from site to site.

Percentage cover is probably the most informative measure of plant abundance, and is widely used for describing grassland vegetation where it is impossible to define where one individual starts and another ends. The main disadvantage of the method is that, owing to the large number of observations which must be made before any reliable estimates of cover are obtained, the sampling is tedious and slow. The use of the method is also more or less restricted to low herbaceous communities.

Frequency The frequency of a species measures the chance of its occurrence in any one quadrat, of a specified size, taken at random in a given community. Frequency is determined by recording merely the presence or absence (not the number of individuals) of a species in a series of randomly distributed quadrats. For example, if fifty quadrats are taken and a given species occurs in ten of them, then its frequency is 20 per cent. The main advantage of frequency as a measure of abundance lies in the speed and ease with which an area can be sampled. In using the frequency method,

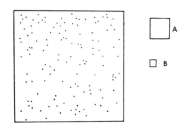

Fig. 34.3 Diagram to show how values of percentage frequency are affected by the size of quadrat used for taking the samples. If the diagrammatic community (containing only one species) were sampled by the two sizes of quadrat, A and B, widely different frequency values would be obtained.

however, it is most important to realize that frequency is dependent on quadrat size, and completely different frequency figures can be obtained from sampling the same community with different sizes of quadrat. In the diagrammatic community shown in Fig. 34.3 it is clear that quadrat A would give a frequency value of 100 per cent whereas quadrat B would give a frequency value of about 60 per cent. Hence it is impossible to use frequency data for comparing the floristic composition of two similar sites unless the same quadrat size has been used for both. For this reason it is essential in recording frequency values to state the size of quadrat on which they are based. Because frequency values depend on quadrat size, careful consideration must be given to deciding what size of quadrat should be used. If the quadrats are much too big for sampling a particular community, similar frequency values will be obtained for two species which clearly differ in abundance. However, properly used, frequency provides a convenient means of expressing the *relative* abundance of the component species of a community, but frequency data can be most misleading unless the limitations of the method are fully appreciated.

Stratification, the vertical arrangement of species

The above-mentioned measures of abundance depend on the horizontal distribution of the plants, but every stable plant community has also a vertical arrangement or stratification of the species. This may be illustrated by considering the general structure of tropical Rain forest. At first sight a tropical Rain forest might appear to be a bewildering chaos of vegetation, in which all the available space is filled with stems and leaves. This impression is certainly true of the forest edge where the increased illumination results in a very dense 'fringe vegetation', but if the interior of a forest can be viewed from the side (as, for example, where it comes to an abrupt edge in a *recent* clearing) it will be seen that it can be divided into five more or less distinct layers or strata, conventionally labelled A to E.

1. A layer of very tall trees over 30 m (100 ft) high (stratum A). This tree layer may form a more or less continuous canopy, as in the profile diagram shown (Fig. 34.4), or it may be represented by isolated 'emergent' trees which rise above the general canopy.
2. A layer of tall trees from about 15–27 m (50–90 ft) tall (stratum B), which is densest in forests with a discontinuous A layer. Strata A and B together thus form a very dense canopy.
3. A layer of smaller trees about 4–12 m (12–40 ft) tall (stratum C). Many trees in this layer are young individuals of species which reach strata A and B when mature, the remainder are small trees peculiar to this stratum.

   Below the tree strata there are two other strata not represented on the profile diagram.
4. A discontinuous shrub layer (stratum D) consisting not only of shrubs but also of young trees, tall herbs and large ferns. Their average height is about 1–2 m (3–6 ft).
5. A ground layer (stratum E), generally a sparse covering of herbaceous plants and undershrubs, less than 1 m (3 ft) high.

   Layering or stratification of vegetation is not confined to the aerial parts of plants, and can also occur in the soil. Thus the root systems of different species of plants may occupy and absorb water and mineral salts from differ-

Stratum A

Stratum B

Stratum C

Fig. 34.4 Profile diagram of tropical Rain forest in Trinidad showing the three tree strata. As shown by the letters on the trees, which refer to their local names (e.g. M=Mora), stratum A of this profile consists of only one tree species; this is unusual in most tropical Rain forests. The diagram represents a strip of forest 61 m (200 ft) long by 7.6 m (25 ft) wide. (From J. S. Beard (1946), *Journal of Ecology*, **33**, 176)

ent layers of the soil. The net result of the phenomenon of layering is that a greater volume of both soil and air can be occupied by plant organs than would be possible if all the species belonged to one layer.

In tropical Rain forest many different species, perhaps several hundred, live together as members of one community. All the members of the community interact in one way or another. For instance, some plants compete directly with one another for living space, water and light, so that certain individuals may eventually be killed. Competition of this type is especially severe between plants occupying the same layer, and one result of stratification is to minimize the competition between members of different layers. In addition to competition, however, some members of the community create conditions favourable for the growth of other plants. Thus climbers can hang like long ropes from tree branches, and epiphytes (mosses, liverworts and ferns as well as flowering plants) can 'perch' on trunks and branches. Similarly, many of the herbs would not be present except for the shade provided by the foliage of the canopy. As a result of such interactions of competition and dependence between the different members, the community as a whole functions as a single system. Consequently a marked change in the abundance of one member of the community may have far-reaching repercussions on the structure of the whole community.

Although community structure is shown most strikingly in tropical Rain forest, all plant communities have a recognizable structure dependent on the diversity in size and form of the species present, their relative abundance and their arrangement in space. Stratification is most obvious in complex communities, but it can be detected on a smaller scale even in such apparently uniform communities as grassland.

Life forms of flowering plants

Since the different plants that make up a complex community vary greatly in size and form, it is necessary to have some means of characterizing them. The simple classification of plants into herbs, shrubs and trees does not go far enough, and a more detailed classification into what are called *life forms* is often used. This method of classifying plants recognizes that, besides differing in size and form, plants differ also in their seasonal behaviour and in their life cycle. Thus in some plants the vegetative parts remain dormant during the unfavourable season whereas in others they die back to a variable extent. Some plants live several to many years (perennials) whereas others die after only one or two growing seasons (annuals and biennials respectively). In its widest sense, the term life form denotes the sum total of all the features of a plant by which it is adapted to its environment. Because we do not know all the adaptations of any plant and we are often unable to measure the value of those we do understand, it is convenient to consider only those which are concerned with surviving the unfavourable season. These adaptations are found particularly at the shoot apices, on which the continued existence of plants depends.

There is good reason to suppose that flowering plants first evolved in moist tropical climates with little seasonal variation in temperature or humidity. Taking this for granted, it is thought that the primitive flowering plants were evergreen trees, the life form best suited to a continuously favourable climate. Modifications which allowed colonization of less favourable habitats would have had considerable survival value, because plants possessing them could exploit such habitats with little or no competition. The most widespread adaptation to habitats subject to seasonal drought or cold is the deciduous habit, i.e. shedding the leaves during the unfavourable season.

If it is to be effective, the deciduous habit involves at least two requirements. First, the leaves must be shed before they become a liability to the plant as a result of excessive water loss during transpiration. Second, the dormant (perennating) buds which remain after leaf fall must themselves be protected against excessive drying, otherwise they also become a liability. The presence of bud scales (especially when supplemented by sticky resinous substances) and a reduction in water content of dormant buds undoubtedly give some protection against desiccation. Besides this natural resistance, however, probably the most important factor in the protection of dormant buds is their position in relation to ground level. In trees the buds are exposed to the extremes of atmospheric temperature and to the full force of drying winds, whereas in herbs the buds may lie beneath the soil where they will experience a more equable microclimate. The advantage gained from increased protection of the buds is often considered an important factor in the evolution of herbaceous plants.

At the beginning of this century the Danish botanist Raunkiaer (1860–1938) devised a system of classifying life forms which is based on the distance between ground level and the position of the highest buds which carry the plant through the unfavourable season of the year. Adaptation to increasingly severe seasons of drought or cold is achieved by having the highest dormant buds borne closer and closer to the ground until they are actually buried in it. The extreme method for protecting buds is shown by those plants (annuals) which complete their life cycle in a single growing season and then die, surviving the adverse season solely in the highly resistant form of dormant seeds. In these plants the bud for starting the next season's growth is inside the seed in the form of the shoot meristem of the embryo plant. Raunkiaer's classification (Fig. 34.5) may be summarized as follows:

1. Phanerophytes (literally, conspicuous plants) These are comprised of trees and shrubs which have their highest perennating buds on shoots projecting into the air. Because this class is so large, it is subdivided according to the following arbitrary limits of size:

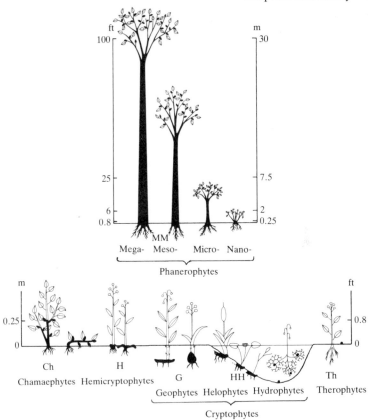

Fig. 34.5 Raunkiaer's life forms showing the position of the perennating parts relative to ground level. The persistent axes and surviving buds (seeds in therophytes) are shown in black.

Megaphanerophytes – over 30 m (100 ft) high
Mesophanerophytes – 7.5 – 30 m (25–100 ft) high } often grouped together and represented jointly by the symbol MM

Microphanerophytes (M) – 2–7.5 m (6–25 ft) high
Nanophanerophytes (N) – 0.25–2 m (10 in–6 ft) high

2. Chamaephytes (literally, ground plants) These have their perennating buds close to the ground (0–0.25 m or 0–10 in) where they are less exposed than those of the previous group. If the shoots grow taller than about 0.3 m (1 ft) during the growing season, they die back at the end of the season and are replaced the following year by new shoots arising from low down on the old stems. This class (abbreviated as Ch) includes small undershrubs, and many plants whose stems either trail along the ground or form more or less compact 'cushions' (the so-called cushion plants).

3. Hemicryptophytes (literally, half-hidden plants) Hemicryptophytes have their perennating buds at soil level where they are protected by the surrounding soil and by the withered shoot system of the previous season. Plants of this group (abbreviated as H) often have a *rootstock*, i.e. a large swollen root capped at soil level by a condensed stem from which new leaves and branch shoots are produced annually. This class of life form is

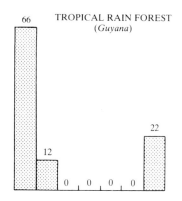

TROPICAL RAIN FOREST
(*Guyana*)

66

22

12

0 0 0 0

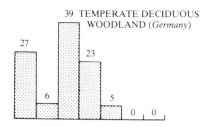

39 TEMPERATE DECIDUOUS
WOODLAND (*Germany*)

27

23

6

5

0 0

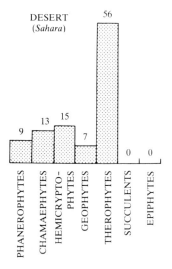

DESERT
(*Sahara*)

56

9 13 15

7

0 0

PHANEROPHYTES
CHAMAEPHYTES
HEMICRYPTO-
PHYTES
GEOPHYTES
THEROPHYTES
SUCCULENTS
EPIPHYTES

Fig. 34.6 Biological spectrum for tropical Rain forest compared with those for temperate Deciduous woodland and Desert. The columns represent the number of species of each life form as a percentage of the total flora. (Spectra for tropical Rain forest and temperate Deciduous woodland from Richards (1952), p. 9)

typified by the 'rosette plants', so-called because most of their leaves arise in the form of a rosette just above the soil surface.

4. Cryptophytes (literally, hidden plants) These achieve a greater degree of protection than any of the preceding types by having their perennating buds buried *below* the soil or, if swamp plants, beneath the water surface. The group is subdivided into:

Geophytes (literally, earth plants) – plants with perennating buds buried below the soil surface, e.g. plants with bulbs, corms, rhizomes, etc. Geophytes are represented by the symbol G.

Helophytes (literally, marsh plants) – marsh (seasonal swamp) plants with perennating buds in waterlogged mud.

Hydrophytes (literally, water plants) – aquatic plants with perennating buds beneath the water surface.

} symbolized as HH

5. Therophytes (literally, summer plants) Therophytes grow rapidly and complete their life cycle during the favourable season (i.e. they are annuals). The vegetative parts of plants belonging to this group (abbreviated as Th) have no special adaptations for the adverse season, which is passed in the form of seeds.

6. Stem succulents (S) and *7. Epiphytes* (E)   Both are highly specialized life forms which do not fit into the above series but are nevertheless characteristic of certain habitats. Cacti are good examples of stem succulents, while the bromeliads and orchids growing on the branches of Rain forest trees are typical epiphytes.

Raunkiaer's classification of life forms has been described in some detail because it has proved very useful for describing and comparing the structure of plant communities. The life forms are used to construct a *biological spectrum*, which indicates the relative importance of different life forms within a given community. The species in a community are grouped according to their life forms, and the number of species within each life-form group is then expressed as a percentage of the total number of species. An example of the technique is shown in Fig. 34.6 which gives the biological spectra for tropical Rain forest in Guyana, for temperate Deciduous woodland in Germany and for Desert in the Sahara. The spectra clearly demonstrate important differences between the three communities. The spectrum for tropical Rain forest is notable for the enormous preponderance of phanerophytes, the abundance of epiphytes (which probably reflects the constantly humid atmosphere within the forest), and the absence of life forms showing adaptation to seasonal drought or cold (i.e. hemicrytophytes, geophytes and therophytes). By contrast, the spectra for temperate Decidous woodland and for Desert are characterized by a preponderance of hemicryptophytes and therophytes respectively.

For an individual stand of a community whose floristic composition has been determined by quantitative methods, the structure of the community can be defined with greater precision by *weighting* the biological spectrum in such a way that more weight is attached to the commoner members of the community than to the rare and incidental members. This is done by giving every species a numerical rating (e.g. its frequency) which measures its abundance. The sum of the numbers given to the species within each life-form group is then calculated as a percentage of the sum total of the numbers for all the species in the stand. Often there is little differenc between the two types off spectrum, but a discrepancy is liable to occur where a small number of species of the same life form are much more abundant than the other species in the community. If, for example, the community under consideration is rich in species, a small number of very common species will contribute a relatively low percentage in the biological spectrum based solely on number of species, but a much higher percentage in the weighted biological spectrum.

Biological spectra provide a useful basis for comparing the structure of communities that occur in different parts of the world. A list of species, even with some indication of their relative abundance, will convey no information about the nature of a community to a botanist who is not familiar with the flora of that region. Presentation of the same data in the form of a biological spectrum, however, will enable botanists in other countries to form a mental picture of the community.

Chapter 35

Factors of the habitat

If a plant is to grow successfully in any particular environment, the environment must be able to supply its requirements for growth and the completion of its life-cycle. Since the properties of an environment depend not only on the non-living physical and chemical conditions but also on the presence of other organisms, it is often convenient to divide the contributory factors into three main groups:
1. Climatic
2. Edaphic (= soil factors)
3. Biotic (= influence of other plants and animals).
This subdivision is based on the observation that any one group of these factors may be more important than the others in determining the nature of a particular community. However, environmental factors never act independently but always interact to some degree. In the tropics climatic and edaphic factors are indeed so inextricably interrelated because of their effects on moisture relations that they will here be considered together. This is not to deny the overriding influence of edaphic factors on certain tropical communities, but these rather special cases will be mentioned in passing.

Climatic and edaphic factors

An outstanding feature of lowland tropical climates is that the temperature remains more or less constantly high throughout the year, the mean daily variation being usually greater than the mean seasonal variation. For example at Kingston, Jamaica, with a mean annual temperature of 26.1 °C (79 °F) the mean monthly temperatures show an annual range of only 3.2 °C (5.8 °F) whereas the mean daily range is about 9.4 °C (16.9 °F). Temperature therefore is not a limiting factor for the growth of plants in the lowland tropics. Modern ecological work has shown that the main factor controlling the distribution and structure of most natural plant communities in the tropics is the 'moisture factor'. Used in this sense the moisture factor must not be equated simply with the total rainfall, because not all rainfall is effective in promoting plant growth. Some is lost by surface run-off and by drainage, and there are also losses due to evaporation from the surface of the soil. The moisture available for plant growth thus depends on many contributory factors. There have been several attempts to devise satisfactory formulae which measure the 'effective rainfall' and, although these vary somewhat, they all take account of the factors mentioned above. Seasonal variation in the moisture factor is of tremendous importance to vegetation, and occasional extremes may be of more significance in determining the presence or absence of certain species than the average values.

The most luxuriant vegetation will be found only on sites where soil moisture is always readily available, but where drainage is also sufficient to prevent waterlogging. Such conditions occur in large areas of the great tropical river systems of the Amazon, Congo, Irrawaddy and others, and in certain islands such as those in the East Indies. Here, under conditions of steady high temperature (about 27 °C) and continuously 'high' rainfall (from over 500 cm per year to as low as 125 cm provided that the rainfall is distributed regularly throughout the year) with correspondingly high atmospheric humidity, tropical Rain forest develops. This type of forest has the most complex structure of any plant community in the world. Although the species composing it are largely different in each of the areas mentioned, tropical Rain forest has a similar biological spectrum wherever it is found because its basic structure is always the same. Since the climate shows little seasonal change, there is little evidence of seasonality in the vegetation. Most of the trees have 'medium-sized' evergreen leaves (those placed in the leaf-size class 20–45 cm^2 are usually predominant), and flowering and fruiting are not confined to any particular time of the year.

Simplification in community structure, and the development of modifications such as the seasonal shedding of leaves or the possession of hard leathery (i.e. sclerophyllous) leaves, begin to appear in the vegetation where, for one reason or another, there is not enough moisture to support optimum growth throughout the year.

The part played by climate and soil, either jointly or singly, in controlling vegetation can be illustrated by a brief survey of the various natural plant communities that occur in tropical America. The American tropics have provided exceptional opportunities for the study of relatively undisturbed communities, because until recently human settlement had not modified the vegetation by fire and grazing to the same extent as in the Old World tropics. In consequence many ecological problems have been easier to define in tropical America because they have not been unduly complicated by additional biotic factors.

The natural plant communities of tropical America have been classified by Dr J. S. Beard* according to a system which stresses the importance of the moisture factor in controlling tropical vegetation. According to Beard the habitats where conditions for plant growth are adverse to some extent at some period of the year can be divided into five categories:
1. Well-drained sites where there is a marked seasonal distribution of rainfall, and a consequent excess of evaporation over precipitation at certain times of the year.
2. Well-drained sites where there is a constant lack of available moisture, evaporation exceeding moisture supply all the year round.
3. Badly drained sites subject to flooding.
4. Badly drained sites subject alternately to flooding and desiccation.
5. Mountainous sites subject to the climatic changes associated with increasing altitude. These changes include falling temperature, exposure to violent winds, cloudiness with resultant high humidity, and lack of sunshine.

* The author wishes to acknowledge his gratitude to Dr Beard for allowing him to draw so freely on his publications in this chapter.

When conditions in any of these five categories of habitat become extremely adverse, terrestrial plant life ceases altogether. Thus on well-drained sites (habitat categories 1 and 2 above) one finds a barren desert where lack of moisture is so extreme that no plants of any kind can survive; or in badly drained sites (habitat categories 3 and 4 above) one finds open water where the ground is permanently flooded and the water is too deep and too mobile to support even the growth of free-floating plants. Similarly on mountain slopes (habitat category 5 above) the vegetation does not extend beyond the level of permanent snow. Throughout the range between conditions of extreme adversity and those optimal for plant growth throughout the year there is a distinct series of vegetation types, or plant formations, for each type of habitat. Plant formations can accordingly be divided into five *formation-series* corresponding to the five categories of suboptimal habitat conditions. Within each formation-series there are plant formations with structures and life forms expressing every degree of transition from extremely favourable to extremely adverse. The five formation-series are named as follows:

1. Seasonal formations.
2. Dry Evergreen formations.
3. Swamp formations.
4. Seasonal-swamp formations.
5. Montane formations.

All five series converge on Rain forest at the optimum end. Thus they can be regarded as radiating outwards from Rain forest like the spokes of a wheel (Fig. 35.1) or, more accurately, like lines drawn from the centre of a sphere. Since different habitat factors may sometimes produce similar responses by plants, it cannot be assumed that the structures of the various formation-series will necessarily diverge in straight lines. It is possible for the structure of a formation belonging to one series to approach that of a formation belonging to another series. This situation is most likely to occur when different series are either converging on Rain forest or perhaps reconverging to end in the complete absence of vegetation.

1. Seasonal formation-series

Seasonal formations occur on well-drained sites with a marked seasonal distribution of rainfall, which causes desiccation of the soil and lowered atmospheric humidity at certain times of the year. In assessing the influence of rainfall on vegetation the annual-rainfall figure by itself can be very misleading because, during rainy months, excess rainfall merely drains away and is lost to plants. The important factors to be considered are the length and severity of the dry season. It has been estimated that throughout the American tropics, under normal drainage conditions, a monthly rainfall of approximately 10 cm (4 in) is necessary if evaporation is not to exceed precipitation, otherwise plant growth will be subject to drought conditions. The length of the dry season may therefore be measured by the number of consecutive months with less than 10 cm of rainfall. Rainfall in excess of 10 cm per month will probably not have any appreciable effect on the vegetation because the excess drains away, leaving the soil constantly moist.

The length of the seasonal drought determines the degree to which the vegetation diverges from Rain forest, so that when the seasonal drought is very short the vegetation will be very similar to Rain forest. As the sea-

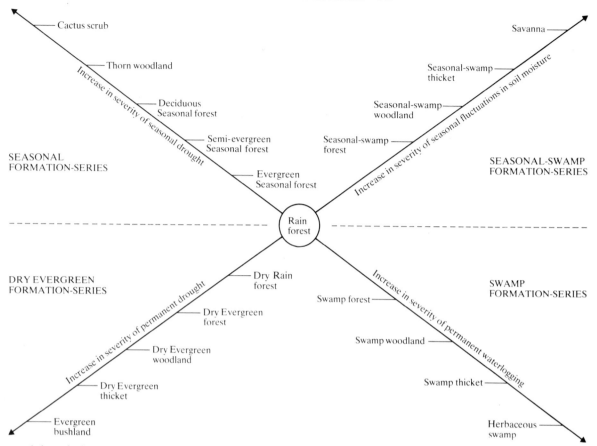

Cactus scrub

Thorn woodland

Deciduous
Seasonal forest

Increase in severity of seasonal drought

Semi-evergreen
Seasonal forest

SEASONAL
FORMATION-SERIES

Evergreen
Seasonal forest

Savanna

Seasonal-swamp
thicket

Seasonal-swamp
woodland

Increase in severity of seasonal fluctuations in soil moisture

Seasonal-swamp
forest

SEASONAL-SWAMP
FORMATION-SERIES

Rain
forest

DRY EVERGREEN
FORMATION-SERIES

Dry Rain
forest

Dry Evergreen
forest

Increase in severity of permanent drought

Dry Evergreen
woodland

Dry Evergreen
thicket

Evergreen
bushland

Swamp forest

Increase in severity of permanent waterlogging

Swamp woodland

SWAMP
FORMATION-SERIES

Swamp thicket

Herbaceous
swamp

sonal drought becomes more severe, the stature of the vegetation decreases and its structure is simplified due to a decrease in the number of strata. The deciduous habit becomes more common, appearing first in the upper storeys and then lower down. It is not surprising that the lower storeys should remain evergreen when the top storey has become mainly deciduous because their microclimate will be more moist. There is also a tendency towards a general reduction in leaf size; plants with comparatively small leaves, which are rare in Rain forest, become increasingly common towards the dry end of the series. Thorniness also becomes an increasingly common feature.

The Seasonal formation-series is envisaged as one continuous series starting with Rain forest and ending in barren desert. Between these extremes the series may be divided into five stages. Though the dividing lines are more or less arbitrary, the divisions proposed by Beard are convenient in practice. The five formations so delimited are illustrated in Fig. 35.2.

Each formation shows a regular stepping-down of structure. Rain forest, the starting point of all tropical formation-series, has three tree storeys, the uppermost one continuous. Evergreen Seasonal forest, the highest member of the Seasonal series, also has three tree storeys but the uppermost is discontinuous and the canopy is formed by the second layer. In Semi-evergreen Seasonal forest only two storeys remain, of which the upper is closed, and in Deciduous Seasonal forest this upper layer becomes discontinuous. Thorn woodland has only one tree layer left, and in Cactus scrub we have only the bushes and succulents.

Fig. 35.1 Diagram showing how the different formation-series diverge from Rain forest as the habitat conditions become progressively more adverse. (N. B. The montane formation-series is omitted because the sequence of formations is too complicated to be represented on a diagram of this type.)

Fig. 35.2 The five formations recognized in the Seasonal formation-series. (From J. S. Beard (1955), *Ecology*, **36**, p. 91)

Fig. 35.3 The Dry Evergreen formation-series. (From J. S. Beard (1955), *Ecology*, **36**, p. 91)

It is clear that the Seasonal formations are predominantly controlled by climate. The climatic limits for the three forest members of the series may be summarized as follows:

	Evergreen Seasonal forest	Semi-evergreen Seasonal forest	Deciduous Seasonal forest
Total annual rainfall	Over 180 cm	130–180 cm	90–130 cm
Duration of dry season	Three consecutive months each with under 10 cm, but over 5 cm rainfall	Five consecutive months each with under 10 cm, but over 5 cm rainfall	Five consective months each with under 10 cm, two under 5 cm, rainfall

2. The Dry Evergreen formation-series

Dry Evergreen formations occur in areas which are subject to a more or less permanent lack of available moisture. This may be due to excessive drainage (e.g. on shallow soils overlying porous limestone rock, or on deep sandy soils) or to excessive evaporation as a result of strong winds and low relative humidity (e.g. in exposed coastal areas). Plants which can permanently withstand such conditions must be drought resistant, and it is therefore not surprising that the primary characteristic of Dry Evergreen plants is the possession of evergreen, sclerophyllous (i.e. hard and leathery) leaves. A further characteristic is that a high proportion of the plants contain latex or essential oils.

The intensity of the drought conditions prevailing during most of the year varies from very slight to extremely severe. Such variation is reflected in a progressive reduction in stature. Unlike the Seasonal formations, where the successive strata become first discontinuous and then absent, the formations of the Dry Evergreen series always have a canopy layer with emergent trees above it and little or no understorey, but the height of these two layers is reduced as the lower end of the series is approached. The Dry Evergreen series is envisaged as one long unbroken series, in which the recognition of distinct formations (Fig. 35.3) represents an attempt to divide it into a regular sequence of types.

Dry Evergreen formations are often found under regional climatic conditions similar to those supporting Seasonal formations. The crucial factor deciding which of the two formation-series occurs in a given area is undoubtedly the water-supplying capacity of the soil. Thus, under a rainfall regime of five consecutive dry months (each with less than 10 cm of rain) and on a soil that is not rapidly depleted of soil moisture, it is almost certain that Deciduous Seasonal forest will develop because the incidence of drought would be seasonal. Under the same rainfall regime but on a shallow soil overlying porous limestone, Dry Evergreen thicket could be expected to develop because under these conditions drought would be felt by tree growth for most of the year. The dominating influence of climate in controlling the distribution of Seasonal formations is thus modified on soils where drainage is excessive. This provides an example of the importance of soil factors in controlling vegetation. However, climate is still important in Dry Evergreen vegetation because, on any one soil type, a given Dry Evergreen formation is replaced by the next member higher up the series as the climate becomes more favourable to plant growth. Dry Evergreen formations therefore reflect the combined influences of climate and soil, which cannot easily be separated because they interact with one another.

One of the basic assumptions underlying Beard's system of classification is that the properties of a plant community express the habitat; in other words, the possession of a similar structure by two floristically different communities implies that they are growing in similar habitat conditions. It is on this basis that the well-known littoral communities of windy sea shores are regarded as Dry Evergreen vegetation under the influence of the sea. Thus, Littoral woodland and Littoral thicket are equated, on the basis of their structure, with Dry Evergreen woodland and Dry Evergreen thicket respectively. It is not suggested that the well-established names for these littoral communities should be discarded but merely that their equivalence with Dry Evergreen vegetation of inland sites should be recognized.

3. The Swamp formation-series

Whereas Seasonal and Dry Evergreen formations diverge from Rain forest because of a scarcity of water, Swamp and Seasonal-swamp formations diverge from it because of a permanent or seasonal excess of water due to waterlogging of the soil. Owing to the low solubility of oxygen in water, waterlogged soils are deficient in oxygen and only specialized plants can grow in them.

Swamp formations grow in permanently or more or less permanently waterlogged ground where the soil never dries out. The ground may be inundated all the year round or for varying periods but it always has a high water table. The shorter the period of inundation the more nearly optimal are the conditions and the nearer the approach to Rain forest. The longer the period of inundation the more reduced is the community structure until at the extreme limit Herbaceous swamp gives way to open water. This sequence of conditions is operative along the banks of slow-flowing tidal rivers where it results in a lateral zonation of swamp formations. At low tide, only the extreme edge of Herbaceous swamp is covered by free water. Away from the river the ground is inundated for progressively shorter periods until a level is reached at which flooding occurs only at exceptionally high tides or during the rainy season. Although the four members (listed in Fig. 35.1) of the Swamp-formation series show a progressive reduction in complexity of structure, there is a discontinuity between Swamp forest (the highest member recognized in this series) and Rain forest. This is probably due to the marked specialization of plants that are able to grow in permanently waterlogged ground.

4. The Seasonal-swamp formation-series

Seasonal-swamp formations grow where there is a marked seasonal fluctuation in soil moisture, varying from waterlogging to desiccation. Such conditions are produced when seasonally heavy rainfall and impeded drainage jointly cause stagnation of the water in the soil and consequent shortage of oxygen. This will occur on flat land having a 'pan' of impervious subsoil, which may lie near the surface or be more than 1 m below it. In the wet season the ground is waterlogged for varying periods. In the dry season the ground dries out, allowing aeration of at least the topsoil. The sequence of formations in the Seasonal-swamp series is determined by the severity of the alternations of waterlogging and desiccation. Generally speaking a level topography, a shallow pan and a strongly seasonal rainfall regime produce the most adverse conditions. Alternation between very wet and very dry soil conditions is too severe for tree growth and is the cause of most natural savannas in tropical America. Savannas in Africa are usually due to other causes, notably man-made fires.

The forest-like Seasonal-swamp formations are characterized by a highly irregular canopy. It may be possible to define strata, but they are erratic and the height of the forest canopy may vary considerably, even within a small area. Nevertheless, the formations show a progressive reduction in stature as conditions deteriorate. As in the Swamp formation-series, there is an obvious discontinuity between the head of the series (Seasonal-swamp forest) and Rain forest, and no intermediates are known to bridge the gap. In the past the presence or absence of palms in Seasonal-swamp formations was considered an important diagnostic feature. Since there are palm-dominated variants of Rain forest, Seasonal forest and Swamp communities, it seems that the Seasonal-swamp communities with palms should also be regarded merely as variants.

From what has been said about the habitat conditions of both Swamp and Seasonal-swamp formations, it will be clear that both are predominantly controlled by soil factors. Being under the influence of ground water, these formations will appear in substantially the same form in any tropical climate where the appropriate soil conditions occur.

As previously mentioned, moisture is the 'master' factor controlling the distribution and structure of tropical vegetation. Where soils are important as habitat factors, it is the physical properties (e.g. porosity) of these soils which are usually the important consideration because of their effect on soil moisture. There are cases, however, where the chemical nature of the soil seems to be the primary influencing factor. Thus Mangrove forest, which is the brackish-water equivalent of Swamp forest, will not survive either in pure sea water or fresh river water. It is, therefore, more or less confined to tidal mud flats covered with brackish water at high tide because only in such situations are its specialized chemical requirements met. Because Mangrove forest requires certain conditions of salinity in the soil, it may be regarded as a more or less pure expression of soil conditions. Even this community, however, is dependent also on climate to the extent that it is confined to tropical latitudes.

5. Montane formation-series

On climbing a high mountain in the tropics, one passes through a series of vegetational zones which are not found in the lowland tropics. Many lowland tropical vegetation-types do ascend into the mountain ranges, often very far up the valleys, but only where their habitat requirements are met. The vegetational changes which accompany increasing altitude are often compared with the general sequence of vegetation-types that occur in belts parallel with the equator. Such comparisons should not be accepted without reservation because the alternation of warm summers and cold winters, so characteristic of higher latitudes, does not exist in the tropics. As with the other formation-series, the Montane formations progressively decrease in stature as the habitat conditions deteriorate towards the mountain summits. If the mountain is high enough, the formation nearest the top (tundra) gives way to perpetual snow. The sequence of formations in montane vegetation is too complicated to discuss here because altitudinal position is not the only factor concerned. For instance, the sequence on mountains experiencing seasonal drought is different from that occurring on mountains with continuously cloudy, humid conditions.

Montane formations reflect essentially the climatic zonation associated with altitude. Although temperature falls steadily with altitude, there are many climatic factors additional to temperature which are important in montane habitats. These factors include the amount and distribution of rainfall, atmospheric humidity, cloudiness, light intensity and 'exposure'. Exposure as an ecological factor is compounded of numerous effects, of which probably the most important is the mechanical effect of violent winds. On steep mountain slopes this effect will be enhanced by shallow soil and resultant lack of room for root growth, so that in such situations trees above a certain size are bound to be broken or uprooted by wind. It is impossible to overstress the importance of exposure in determining the nature of montane vegetation. This is particularly true in the mist belt, where the distributional pattern of the various formations is closely correlated with the degree of exposure.

The above sumary of Beard's classification of tropical American vegetation illustrates how climate and soil, through their effects on the moisture relations of the habitat, influence the composition and structure of plant communities. Although only tropical American vegetation-types have been mentioned, the underlying principles apply throughout the tropics. We now turn our attention to the third broad category of habitat factors, the so-called biotic factors which have their origin in the activities of living organisms.

Biotic factors

This heading covers the influence which plants exert upon one another as well as the effects of animals, including man, upon the community.

Competition

Competition is a feature of all closed plant communities (i. e. those in which the ground is completely covered), and root competition may occur even in some open communities. Among the resources of the habitat for which plants compete are mineral salts, nitrogenous compounds, water and sunlight, of which the last two are usually most important. Although, in general, competition is greatest between plants of similar life-form, i.e. those belonging to the same layer of the community, it occurs also between plants of diverse life-form. For example, the roots of trees may enter into competition with those of members of the field layer. Work in Zimbabwe on veld management has shown that cattle do not deteriorate in condition when the bushes and trees on which they browse during the dry season are removed, because the removal of the 'bush' results in a great increase (sometimes up to 60 per cent) in the amount of grass. This increase cannot be accounted for by the removal of shade (which in any case is minimal) or the slightly increased amount of ground made available for the growth of grass, but can be explained by reduced root competition.

Just as all plants have their characteristic ranges of tolerance to climatic and soil conditions, so some plants tolerate competition less than others. For this reason the vegetation of an area seldom includes all the species which could, according to their climatic and edaphic tolerance, flourish there. A species will fail to establish itself if it lacks the ability to compete successfully with surrounding individuals, first as a seedling or later as a mature plant.

The controlling influence of competition on plants is convincingly shown by the dramatic and 'explosive' growth that sometimes occurs when competition does not operate. An example is the explosive growth of the free-floating water fern *Salvinia molesta* (originally thought to be *S. auriculata*, and referred to as such in the literature) which accompanied the formation of Lake Kariba in Central Africa. The closure of the Kariba Dam in December 1958 resulted in the formation of one of the largest man-made lakes in the world. The lake flooded the entire middle section of the Zambesi valley and took $4\frac{1}{2}$ years to fill, by which time it was 280 km (174 miles) long with a surface area of 5546 km² (2141 square miles). The plant *Salvinia molesta* is of unknown origin but was first recorded in 1948 in the Upper Zambesi. Although *S. molesta* had thus been present in the Upper Zambesi for at least 10 years before the closure of the Kariba dam, its growth exploded only when the river downstream began to flood into a lake. From the floating mats first observed on the developing lake in May 1959, the weed multiplied so rapidly that a year later it was estimated by aerial survey to cover about 195 km² (75 square miles). The infested area continued to increase until it reached a peak in 1962 of about 1000 km² (386 square miles), or approximately one fifth of the lake surface at that time. After this the infested area began to decline as violent storms on the full lake battered the mats of weed, and now infestations are largely confined to the river mouths and sheltered bays. The massive invasion of Lake Kariba by *S. molesta* was not only of theoretical interest but constituted a threat to navigation, to the development of commercial fishing, and even to the operation of the turbines in the power station at the dam.

There were probably two main reasons for the explosive growth of *Salvinia* – the absence of competition for space and the absence of competition for nutrients. In the early stages of the formation of Lake Kariba a very large area of shallow calm water became available and *S. molesta*, which happened to be present at the time, exploited it. It is possible that some other plant, such as the Nile cabbage (*Eichhornia crassipes*) might have exploded if *Salvinia* had not first done so. As the lake filled, increasing wave action progressively limited the availability of suitable habitats because *Salvinia* cannot withstand storms. While the new lake was being formed, the water was almost certainly enriched by nutrients released from the newly flooded soils. The area finally occupied by Lake Kariba was flooded in stages during the rainy seasons of 1959–63, and it is probably significant that each period of flooding was followed by a 'flush' in the growth of *Salvinia*. These seasonal flushes indicated a growth response by the plants when an abundance of nutrients became available. A similar response was observed in the phytoplankton along the shoreline.

The creation of a large area of shallow nutrient-rich water, in which *Salvinia* could increase without competition, was a temporary phase in the formation of Lake Kariba. As the lake deepened the extent of the favourable habitat diminished and *Salvinia* correspondingly declined. When the potential danger of the infestation to navigation, fishing, and the hydroelectric power installation was first appreciated, extensive herbicide trials were undertaken and recommendations made. As it turned out, no control measures proved necessary because the threat disappeared before it became a reality.

Dominance

In addition to direct competition, plants also affect one another in more indirect ways, in particular by modifying the microclimate and thus influencing the environment occupied by their associates. In general the larger the plant the greater is its influence on habitat conditions and thus the greater the control it has over the resulting community. In any community we may therefore speak of a *dominant* life form, meaning the life form which, by virtue of its size or numbers or both, has the greatest influence upon the habitat and dominates the whole community. Normally this is the largest life form present in abundance; for example, in a forest the dominant life form is that of the trees, whereas in a savanna, even if scattered trees are present, the grasses are dominant.

The controlling influence or dominance of the trees upon forest structure may be illustrated by comparing a stand of Rain forest with a plantation of *Eucalyptus* trees. In a Rain forest there are shrub and ground layers present, whereas in a *Eucalyptus* plantation no shrub layer is able to develop and, at most, only a very sparse ground layer is present. In addition to bringing about conditions which prevent the development of certain plants, forest trees may also influence the species composition of the community by providing conditions which encourage the growth of other plants. For example, the tall trees of a Rain forest characteristically support not only a number of woody climbers (lianes) which climb up the trunks and flower on the forest roof, but also provide a 'niche' for a number of epiphytes which grow on the trunks and branches.

The detailed study of any plant community always leads to questions about the role and behaviour of particular species. Under what soil conditions does a certain species occur? Does it spread mainly by seed or vegetatively? How is it affected by shade and by fire? What numbers of its seeds germinate? How many reach maturity? Such questions can be solved only by observation and experiment. Investigation of the ecology of individual species (autecology as it is called), especially of the dominant members of the community, is an essential part of ecological work. Until the more important facts are learned about the dominant species, it is impossible to penetrate at all deeply into the ecology of any community. In the tropics the study of the ecology of single species, which is the most time-consuming part of ecological work, is only just beginning.

Animals as biotic factors

Animals are intimately concerned in the life of most plant communities; for example, insects pollinate many flowering plants, and birds and mammals often disperse fruits and seeds. Of the animals which influence plant communities man is undoubtedly the most important. The effects of human activity are evident in all but the uninhabited or the most sparsely populated regions of the world. In the tropics man must be considered a key factor in the creation and maintenance of many types of the so-called savanna grasslands which are prevalent throughout this region.

Man-made fires

If an area of grassland in practically any part of Africa with an average yearly rainfall of more than about 50 cm (20 in) is completely protected from fire (and grazing) and observed closely, a number of changes soon become apparent. First the grass cover begins to grow coarser because of the spread of perennial tussock grasses. The spread of these more vigorous grasses causes the elimination of weaker competitors, including most of the annual grasses, and after a few years significant changes in the species composition of the grass cover can be recorded. At the same time, provided there are seed parents in the neighbourhood, seedlings of trees and shrubs make their appearance and in 20 or 30 years the area will develop into woodland. Replacement of grassland by woodland has been demonstrated at several centres in Africa, including the Grasslands Research Station at Marandellas in Zimbabwe. Here a plot of land that had previously been grassland for decades has been protected from fire and grazing since the 1930s, and is now covered by dense woodland.

More direct evidence for the ecolgical importance of fire is provided by a long-term experiment on the influence of grass fires on *Brachystegia–Julbernardia* woodland in Zambia (Fig. 35.4). Under normal conditions fires are liable to occur in this type of woodland at any time during the dry season, but they are particularly frequent at the end of the dry season when the dead litter is highly inflammable. In 1933 plots of indigenous woodland (Woodland plots) and plots of indigenous woodland stripped of all woody growth down to ground level (Coppice plots) were established. Every year since 1933 both sets of plots have received the following three treatments: (1) late-burning, i.e. a severe burn at the end of the dry season; (2) early-burning, i. e. a very light burn at the beginning of the dry season; and (3) complete protection from fire. The ecological trends resulting from these annual treatments were clearly evident by 1944 when a complete enumeration of all the plots was made, and subsequent observations have only served to confirm these trends. Broadly speaking, the results show that late burning progressively destroys the indigenous woodland (as shown by the destruction of the shrub layer and the thinning out of the tree layer in the Woodland plots) and effectively prevents any coppice regeneration (as shown by the dense grassland that has developed on the Coppice plots, and the heavy grass growth on the floor of the Woodland plots). This contrasts with the much lesser effects of the early-

(a)   (b)

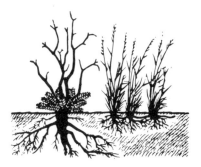

Fig. 35.4 The woodland burning experiment in Zambia. (a) Development of grassland on a late-burnt coppice plot at the woodland burning experiment in Zambia. The photo was taken in 1944 after eleven annual treatments. The indigenous woodland is shown in the background. (b) Natural regeneration of woodland on an early-burnt coppice plot at the woodland burning experiment in Zambia. The photo was taken in 1944, eleven years after all woody growth in the plot area had been cut down to ground level. The forest guard stands under a regenerating *Julbernardia paniculata*. (Photographs taken by G. Foyster)

burning treatment which allows existing woodland (Woodland plots) to survive as a closed woodland and does not interfere appreciably with natural regeneration (as seen by the woodland that has regenerated on the Coppice plots). The early-burnt plots are in fact very similar to those under complete protection.

The devastating effects of late-burning on trees (Fig. 35.5) suggest that many of the world's grasslands are maintained in their characteristic condition because of the frequent bush fires that occur during the dry season. It is certainly true that the greater part of the land surface of Africa south of the Sahara is burnt about once a year. Usually fire does not spread continuously over extensive areas but over small patches, one at a time, so that a mosaic pattern is produced. Under these conditions some areas may be burnt twice in one year, while others may miss being burnt in any one year.

Fire has a number of direct and indirect effects on vegetation. A late bush fire in savanna country can cause the temperature at ground level to rise to about 700 °C. The heat is fatal to all leaves and to cambium which is not insulated by a thick periderm. Fortunately, heat is rapidly lost from the soil surface by convection, and at depths of more than 1–2 cm the temperature of the soil scarely rises. Where fire is a regular feature of the environment, plants must have adaptations to render them fire-resistant if they are to survive. Because the exposed position of their perennating buds renders trees extremely susceptible to the damaging effects of fire, fire-resistant features are more highly developed in trees than in herbs, although the less obvious adaptations of the latter may be just as critical in enabling them to survive. Most savanna trees and shrubs have a very thick bark which undoubtedly insulates them against the intense, but short-lived, heat of bush fires. Even so, their growth is checked and, instead of producing straight erect stems, they have twisted and gnarled trunks. Many fire-tolerant trees are also able to regenerate themselves from their root system, which typically consists of a main tap root with most of its lateral roots growing more or less horizontally at between 10 and 20 cm below ground level and often extending outwards for considerable distances. Thus a savanna soil, at about this depth, is occupied by a network of long intertwining roots of the tree and shrub

Fig. 35.5 Diagrams showing the devastating effect of grass fires on trees. In the absence of fire (above) trees restrict or prevent the growth of grasses. After a severe fire at the end of the dry season (below) the grasses recover quickly and grow with increased vigour under conditions of reduced competition. Persistent burning will eliminate the trees and maintain grassland. (From Odum 1975, p. 116)

species. The shoot systems of such plants are often very small in comparison with the size of their roots. The shoots have been repeatedly burnt back to ground level by the dry-season fires, but the roots have produced new shoots known as 'suckers'. This sucker regeneration is very common and accounts for most of the woody plants below 2 m tall. In addition to their thick bark and their ability to regenerate, some savanna trees are deciduous. In savanna most leaves are killed by the annual fires, a new flush of leaves being produced shortly after the old ones are burnt, and so it is not always easy to say whether a tree is naturally deciduous or not.

The herbaceous vegetation of savanna consists of two main groups of plants, the grasses (most of which are hemicryptophytes) and the geophytes. Most savanna grasses are perennial and grow in tussocks, the outer portion of which appears to protect the resting meristems in the centre from damage by fire. After burning, the grasses regenerate very rapidly and produce new green shoots. In fact, there is evidence that the growth of many grasses is stimulated by the burning of the old parts. The geophytes, which have an underground storage organ from which a leafy aerial shoot arises during the growing season, are even better protected from fire than the grasses because their perennating buds are well below ground level where the heat of surface fires has little or no effect. For the same reason therophytes, which perennate as seeds, may also be well represented in the vegetation of certain types of savanna.

The frequency of fires in most of the world's grasslands raises the question of the natural status of savanna. In most savanna areas the fires are usually started deliberately by hunters to drive out animals, or by farmers either to clear the land for planting or to remove the accumulation of dry litter and thereby encourage a lush growth of young grass in the following season. For many sites there is good evidence that, if fire is excluded, the existing grassland would develop into woodland, but this must not be taken to mean that all savannas are man-made. Some fires occur naturally as a result of lightning, and it is quite possible that such fires may have maintained savanna long before the appearance of man. In most savanna regions there seems to be a very delicate ecological balance between trees and grasses, and the frequency and severity of fires may tip the balance in favour of one or the other. The fact remains, however, that man-induced fires have been used as hunting and agricultural techniques for many thousands of years (e.g. archaeological evidence from near the Kalambo Falls, Zambia has proved that man has been using fire in Africa for at least 57 000 years), and will continue to be used as the most convenient and cheapest method of managing 'natural' pastures in the tropics and subtropics. In the past 50 years or so, as a result of increasing numbers of hunters and farmers, savannas have often been burnt too frequently and there is now a great risk that the nutrient reserves in the soil and the consequent yields of grass will decline drastically unless this trend is checked. Phillips, one of the leading authorities on tropical agriculture, has summed up the ecological effects of fire on vegetation by pointing out that fire is a good servant but a bad master.

Grazing by domestic animals

Modification of vegetation which results from grazing by domestic animals is an effect due indirectly to man's activities. On grazed land certain plants flourish because there is less competition from others which are weakened or killed by grazing. According to its intensity and frequency, grazing can cause changes varying in degree from a temporary replacement of one dominant grass by another, to a complete and more or less permanent alteration in the whole character of the vegetation. A very striking example of change in a plant community caused by grazing is provided by *bush encroachment*, which is an agricultural problem in the more arid parts of all tropical countries. An immediate result of severe overgrazing is the spread of species unpalatable to cattle. This is the reason why thorny trees and shrubs such as *Acacia* spp. are so numerous in overgrazed areas, even where the wet season appears to be sufficiently pronounced and prolonged to support the growth of much less xerophytic trees. If a pasture is continuously overgrazed, the grasses are progressively exterminated with the result that the soil water hitherto absorbed by grasses is made available for a more luxuriant growth of thorny shrubs and for the establishment of their seedlings. The cattle assist in the propagation of the thorny shrubs because, on overgrazed areas, they supplement their diet by eating the pods of the thorny *Acacia* spp. The seeds pass through their intestines undamaged and are deposited with the dung, where they germinate and grow rapidly. Therefore after several years the grassland turns into a dense thorn bush, which is useless for farming. That bush encroachment is not a climatic phenomenon, but is caused by overgrazing, is shown by the fact that it does not occur when pastures are properly managed on a rotational grazing system which avoids overgrazing.

The effects of burning and grazing are merely two of the effects that man and his domestic animals have on the vegetation, but they provide striking examples of how community structure can be greatly influenced by biotic factors.

Chapter 36

Change and succession

After reading the two previous chapters one might be left with the impression that a plant community is a relatively stable and unchanging unit, representing the equilibrium point of the prevailing climatic, edaphic and biotic factors. Such a concept is completely erroneous because all communities are, to a greater or lesser degree, dynamic systems.

Cyclical change

Even in communities which appear to be relatively stable and permanent, such as forests, there is a continual element of change. For instance, from time to time in a forest an old tree dies leaving a gap in the canopy. The increased intensity of light reaching the interior stimulates a tremendous growth of the vegetation below the canopy until one or more of the tree species outstrips the rest and the original state is progressively re-established. Cyclical changes of this type are continually going on throughout a forest, even though the overall appearance of the community remains the same.

More obvious cyclical change is typical of communities which occur in exacting or unstable habitats, such as high mountain slopes. Here it sometimes happens that an area is stripped of vegetation by high winds or heavy storms. A pioneer phase follows in which the area may be colonized by a completely different set of species, then a building stage during which the original species reappear and gradually take over until finally the original vegetational cover is restored. The ground covered by such an unstable community consists of a mosaic of patches, each at a different stage of the same cycle. Although the appearance of the vegetation at any one place is constantly changing, the community as a whole remains patchy and no general blending takes place.

Succession

Whereas cyclical change of the type described above does not alter the overall appearance of a plant community, in some places it can be shown by observations extending over several years that the whole vegetation is undergoing progressive change. Such change is obvious when an area of bare ground becomes available for colonization by plants. When, for example, the trees and soil of an area of forest are removed to quarry the underlying rock and the site is then abandoned, the forest returns only after a series of temporary communities have prepared the way. The successive communities may be entirely different in structure from the forest that eventually develops on the site. This type of change is called plant succession. The sequence of distinct vegetation-types that can be recognized in a particular succession is called a *sere*. The final stage of a sere is called the *climax*, which may be defined as the vegetation-type in equilibrium with the climatic, edaphic and biotic factors of the area. There are two main types of causes giving rise to succession: gradual changes in the habitat factors independent of the plant community, and changes in the habitat caused by the influence of the vegetation itself. Both types of causes are normally operative, but one or other may be of greater importance.

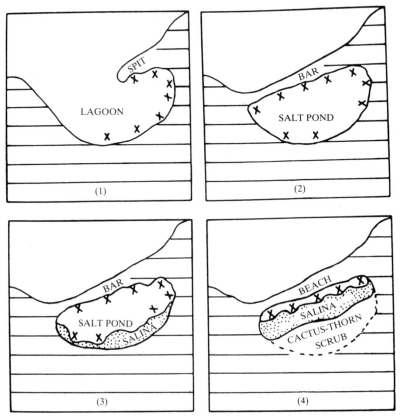

Fig. 36.1 Formation of a salina from an open bay on the south coast of Jamaica. Diagrams illustrate the formation of a sand spit (1), and the subsequent development of a sand bar and salt pond (2), followed by a salina (3). In the final stage (4) a beach, with the usual strand flora, is backed by mangrove woodland (x) and a salina that is being invaded by cactus-thorn scrub. (From Asprey, G. F. and Robbins, R. G. (1953), *Ecological Monographs*, **23**, 374)

Succession caused by a changing habitat occurs, for example, when an open bay is cut off from the sea by the development of a sand bar (Fig. 36.1). Starting with an open bay the first stage is the formation of a narrow sand spit stretching from one shore towards the opposite one (stage 1). A sheltered lagoon is produced in which silt is deposited and mangroves appear round the edge. Next the spit is thrown completely across the bay to form a sand bar and an inner salt pond (stage 2). Later the salt pond progressively dries up by evaporation until the open water is replaced by a salt flat, or salina, with a soil of heavy marine clay (stage 3). This soil may at first be so saline that the area is bare of vegetation and salt crystals can be seen on the surface. More frequently, however, the salinity, while high, is not sufficient to prevent the growth of certain succulent, salt-tolerant plants such as *Batis*, *Salicornia* and *Sesuvium*. Salinas formed from coastal lagoons are inevitably low-lying and it is therefore common to find that they are fed, at least during

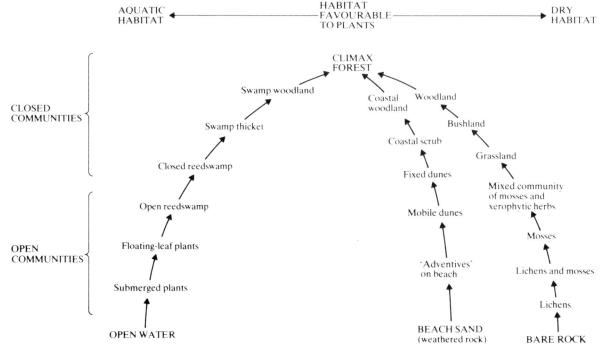

AQUATIC HABITAT ←——————— HABITAT FAVOURABLE TO PLANTS ——————→ DRY HABITAT

CLIMAX FOREST

Swamp woodland

Coastal woodland   Woodland

Swamp thicket

Coastal scrub   Bushland

Closed reedswamp

Fixed dunes   Grassland

Open reedswamp

Mobile dunes   Mixed community of mosses and xerophytic herbs

Floating-leaf plants

'Adventives' on beach   Mosses

Submerged plants

Lichens and mosses

Lichens

OPEN WATER   BEACH SAND (weathered rock)   BARE ROCK

CLOSED COMMUNITIES

OPEN COMMUNITIES

Fig. 36.2 Generalized diagram showing the pattern of successions starting from open water (hydrosere) and from beach sand and bare rock (xeroseres).

the rainy season, by fresh-water streams. As a result the salt is progressively leached from the soil and, when the salinity falls sufficiently, the salina vegetation is followed by less specialized plants. Still further leaching results in a series of communities following each other until the land vegetation typical of the area develops (stage 4).

Successions caused chiefly by the influence of plants themselves upon the habitat may be divided into those following the colonization of open water and those following the colonization of bare rock or weathered rock (e.g. scree, sand, etc.). Examples are given in Fig. 36.2. Until a plant community is in harmony with its habitat (i.e. it has reached the climax stage) the plants give rise to new habitat conditions which are less favourable for their own survival but more favourable for other plants which follow in due course. In successions starting in water (*hydroseres*) the plants cause changes in the habitat first by increasing the rate of silting and later by depositing their remains. The succession round the edge of a fresh-water pond or lake, for instance, begins with submerged plants whose stems and leaves increase the rate of silting until the water is sufficiently shallow for the growth of floating-leaf plants such as water lilies, which are the next stage in the succession. The floating-leaf plants, by continued silting and by depositing their remains to form peat, raise the level of the bottom still further and prepare the habitat for reed swamp vegetation, consisting of such genera as *Phragmites* (common reed), *Typha* (reedmace, bulrush) and *Cyperus* (sedges). As the peat level approaches the surface the reed swamp is colonized by shrubs and trees which progressively dry out the soil until it is too dry for reed swamp. In its turn, therefore, the reed swamp is succeeded by thicket or woodland.

Successions from bare or weathered rock are called *xeroseres*; an example is the colonization of wind-blown sand on windward coasts where the beach slopes very gently towards the sea. Here the changes caused by the plants first take the form of binding the mobile sand into dunes (Fig. 36.3), followed by the addition of humus and the development of a soil retentive of water. Some distance above the high tide mark scattered 'adventives' (i.e. plants established temporarily) may appear in the sand of the seashore. They do not modify the habitat and thus play no part in the succession which begins with such plants as *Ipomoea pes-caprae* (beach morning-glory, or seaside yam) and certain grasses, e.g. *Sporobolus*, *Spinifex*, etc. These plants have either rhizomes or long trailing prostrate stems, capable of growing through accumulations of sand and of stabilizing its surface sufficiently for seedlings of other plants to gain a foothold. In the next stage mat-forming plants, mostly grasses and sedges, further stabilize the sand and, by the addition of humus, gradually develop a soil in which water is retained. With the development of a soil the way is prepared for shrubby plants (e.g. *Calotropis*, *Mallotonia*, *Scaevola*, *Sophora*) which form a zone behind the fixed dunes. In a typical succession the shrubby zone is replaced inland by coastal woodland (e.g. *Thespesia populnea* and *Hibiscus tiliaceus*).

In both hydroseres and xeroseres there is a definite trend for an environment of extreme conditions, either wet or dry, to be modified in the direction of moderation. The successions lead from communities of hydrophytes (aquatic plants) or xerophytes (plants living in dry places) to a community of mesophytes (plants whose water requirements are intermediate between hydrophytes and xerophytes). Hydroseres and xeroseres thus converge until they ultimately meet in the climax of the area

Parallel with the development of the vegetation there is a development of the habitat. In the succession on wind-blown sand, for instance, the various vegetational zones follow one another as the soil becomes richer in humus, less saline and more retentive of water. By contrast, in the colonization of bare soil, such as where a woodland has been felled, the vegetational changes take place in a habitat which has already matured in equilibrium with a climax community. This is an example of what is called a secondary succession to distinguish it from a primary succession starting in a habitat (open water, bare or weathered rock) completely devoid of soil. Generally, secondary successions resemble telescoped versions of primary successions, with the pioneer stages omitted and the species of the climax beginning to establish themselves from the outset together with those of the intervening stages. In areas affected by human activity a primary succession may be deflected to form a community different from any in the natural succession. Grassland maintained by man-made fires in a region where it would, if left alone, revert to forest is an example of deflected succession. Such grassland forms a distinct community with a floristic composition different from any community in the primary succession.

Successions are often very slow processes so that it may be impossible, at least within the span of a lifetime, to demonstrate the changes between consecutive stages by the direct method of marking out permanent sample areas (quadrats) and recording their composition at regular intervals. If, however, an area has been progressively invaded by vegetation over a long period, the stages in the succession can often be reconstructed from observing the distributional pattern of the vegetation. For instance, in a succession starting in the water of a pond which is being silted up, the first stage is assumed to be represented by the zone of vegetation nearest the open water, with the later stages following one another until the shore is reached. In such ways the zonation in space sometimes reflects the succession which takes place in time at any given spot.

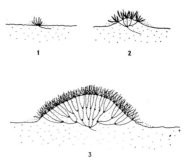

Fig. 36.3 Dune formation by the gradual deposition of wind-blown sand particles around the aerial shoots of plants capable of growing through accumulations of sand (e.g. *Ipomoea pes-caprae* and *Spinifex littoreus*). (From K. A. Kershaw (1973), p. 40)

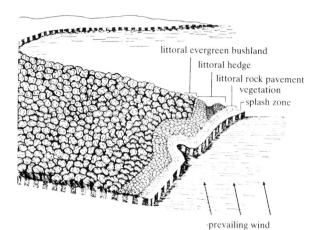

littoral evergreen bushland

littoral hedge

littoral rock pavement
vegetation

splash zone

prevailing wind

Fig. 36.4 A diagrammatic representation of the zonation of the littoral communities on a raised coralline limestone terrace in Jamaica. (From Asprey, G. F. and Loveless, A. R. (1958), *Journal of Ecology*, **46**, 549).

Zonation

Although zonation of vegetation often provides valuable information about the sequence of communities in a succession, in many instances it is merely indicative of a gradient in the habitat conditions and not of successional change. A good example of this is the zonation of flowering plants on coastal rock terraces formed by the elevation of coralline limestone (Fig. 36.4). Where such a terrace is under the direct influence of the sea there is a gradient in such factors as exposure to wind and quantity of salt arriving in splash and spray. At the extreme edge of the terrace, where the waves break over to produce a splash zone, there is no vegetation. Further inland there is an open community, called Littoral Rock Pavement vegetation, composed of a few dwarf shrubby species of extreme halophytes (i.e. plants tolerant of high salt concentrations in the habitat). On proceeding inland this community gradually merges into a dense growth of windswept shrubs, up to 2 m tall, which is distinguished as Littoral hedge. On the landward side Littoral hedge increases rapidly in stature and variety until it culminates in the more sheltered and taller growth of either Littoral Evergreen bushland or Palm thicket.

There is a similar zonation of seaweeds on rocky shores subject to tidal influence. Here the zonation depends on such factors as average depth of water and hence the intensity and quality of the light, frequency and duration of tidal exposure, salinity and temperature. In this zonation of seaweeds, as in all zonations which reflect a gradient in habitat conditions, the separate seaweed communities stand in a spatial relationship to one another and, so long as there is no subsidence or elevation of the shore line, there is no tendency for one community to be replaced by another.

Chapter 37

Mendelian inheritance

The handing on of a piece of land or other form of property to somebody else upon the death of its owner is a legal practice which has been carried out for at least 10 000 years. In contrast to the legal type of inheritance is the transmission from parents to offspring of biological characters, and this type of inheritance is distinguished as *heredity*. Because the biological units of heredity were given the name of genes, the study of heredity is called *genetics*.

All living organisms have hereditary characters because every individual has arisen from other individuals of a similar kind. It is a matter of everyday experience that any dog resembles its father and mother more closely than it resembles any cat, and any cat resembles its father and mother more closely than it resembles any dog. Dogs and cats thus have certain hereditary characters which are possessed by all members of the species, but not all dogs or all cats are alike and even members of the same litter may show marked differences. Thus, superimposed on the basic similarities due to heredity there is variation, the occurrence of differences between individual organisms. Some of these differences are also inherited, but others are not. Sometimes a difference can be traced to a difference of upbringing. For example, a puppy of a large breed of dog may grow up small and weakly because it has not been fed adequately, whereas another puppy of the same breed which receives a healthy diet grows into a large and strong adult. At other times, however, the difference cannot be traced to any difference of upbringing, and must be due to some inborn property. In a litter resulting from a cross between a small dog and a large dog, the puppies may grow into adults which vary greatly in size even though they are all adequately fed.

It can be seen, therefore, that differences between individual organisms are of two kinds: (1) those imposed from the outside, i.e. traceable to the environment; and (2) those which are inborn, i.e. traceable to heredity. This means that every adult organism must be regarded as the outcome both of the natural inheritance it received from its parents and of the conditions in which it has lived. In the language of genetics an individual's *phenotype* (its appearance and mode of functioning) is the result of the interaction between its *genotype* (its natural inheritance) and its *environment*. From this first general principle of genetics it is clear that no character of an individual can be attributed solely to its heredity or solely to its environment. Both are always involved because any character must have an environment in which to express itself. Even though the characteristics of any given phenotype cannot be traced solely to genotype or environment, it is nevertheless possible to attribute a difference in phenotype between separate individuals to a difference in genotype or environment, or both. Thus, if two individuals of like genotype are brought up in different environments, any differences which may develop in their phenotypes must be environmental. Similarly if any two individuals are brought up in a constant environment, any phenotypic differences which they may show must be genotypic. It is therefore by the study of differences that we can untangle the part that heredity plays and the way in which it works. There is, in fact, only one sure way of deciding that a difference in phenotype is due to heredity, and that is by showing that the difference can be passed on from one generation to the next. The rules governing the transmission of characters from parent to offspring are associated with the name of Mendel, to whose outstanding work consideration will now be given.

Mendel's rules

Man has been interested in heredity since the beginning of recorded history, and it is well established that as early as 6000 years ago he kept records of pedigrees of horses. Until about 120 years ago, however, all attempts to explain the results of recorded pedigrees failed, because the investigators concentrated on too many characters at a time and consequently did not 'see the wood for the trees'. It was an Austrian monk, Gregor Mendel (1822–84), who first worked out the basic rules of heredity and recorded his findings in a scientific paper published in 1866. Mendel's experiments were done entirely with plants, mainly *Pisum sativum* (a type of garden pea), but his conclusions are now known to apply to most forms of life. Unlike previous investigators, Mendel restricted his attention to the transmission of one or a few characters in a planned breeding experiment, i.e. an experiment in which parents exhibiting contrasting expressions of the same character or characters are cross-bred, or crossed, and careful records of the results kept through several generations. His success was mainly due to the logical way in which he approached the problem. Not only by his method of approach but also by his suggestion of particulate 'elements' (which correspond to what are now called genes) to explain his results, Mendel made a tremendous advance towards a real understanding of heredity. The greatness of Mendel's contribution lay not in his observations but in his deductions. His theoretical analysis of heredity was far in advance of his time and, to his great disappointment, his work was ignored during his life. It was not until 1900 that Mendel's results were read by scientists who were in a favourable position to appreciate their significance and, as a result, he became world famous in the course of a few months. Since the rediscovery of Mendel's work genetics has become one of the major branches of biology, impinging on virtually every aspect of the subject from the biochemistry of cell processes and the behaviour of viruses at one end to the behaviour of populations and the course of evolution at the other. The concept of genes as the agents of heredity is undoubtedly one of the foundations of modern biology, comparable in importance to the concepts of the cell and of evolution.

From the way in which Mendel tackled his breeding programme it is clear that he realized two practical requirements for success in planning a critical experiment on the heredity of any sexually reproducing organism. First, crosses must be made between parents showing sharply contrasting expressions of the same character. If, for instance, we wished to learn something about the inheritance of shortness of stem in a given plant, we would cross an individual having a stem of normal size with an individual having a short stem, all other characters being ignored. The second requirement is that the parents must each come from a pure-breeding line (i.e. a strain of individuals which breeds true for the characters being considered) because, unless this requirement is met, any differences visible in the progeny cannot be attributed to hereditary differences between the parents. Such 'pure lines' can be obtained with certainty only in organisms with regular self-fertilization (which, in the case of flowering plants, means that each flower is regularly pollinated by its own pollen).

Monohybrid inheritance

Variegated coleus (*Coleus blumei*), a common garden plant, meets the above requirements of an organism suitable for genetic experiments, and may be used to demonstrate the features characteristic of a cross in which only one character is considered. Such a cross is called a *monohybrid cross*.

The character of *C. blumei* to be considered is the indentation of the leaf margins. Some plants have shallowly crenate margins, others have rather deeply incised leaves (Fig. 37.1). Plants displaying one or other of these characters will, in future, be referred to as *shallow* and *deep* respectively. If a true-breeding *shallow* plant is crossed with a true-breeding *deep* plant (by cutting off the immature anthers of one plant and then dusting the ripe stigmas with pollen from flowers of the other plant), all the hybrid offspring have deeply incised leaves and thus resemble one of the parental types only. Such a characteristic which expresses itself in all the offspring of a cross between two true-breeding parents (as *deep* does in this case) is termed *dominant*, and its alternative which fails to be expressed (here *shallow*) is termed *recessive*. The result is the same no matter which way round the cross is made, i.e. irrespective of whether the *shallow* or *deep* plant is used as the seed parent (female parent) or as the pollen parent (male parent).

If these *deep* hybrids, which may be called the F₁ because they are the first filial generation, are allowed to self-pollinate, a remarkable fact emerges in the F₂ or second filial generation. Instead of these F₁ *deep* plants giving rise, as the parental (or the P generation) *deeps* would have done, exclusively to *deep* offspring, they produce a mixture of *deep* and *shallow* offspring resembling the two original parental phenotypes with no individuals intermediate in appearance between them. Furthermore *deep* outnumber *shallow* plants in the approximate ratio of 3 : 1. (For a reason which will shortly become clear, quite large samples may have to be taken for this ratio to be approached closely.) If these F₂ individuals are now allowed to self-pollinate, it is found that the plants in the F₃ families derived from the F₂ *shallows* are all *shallow*, and further selfing confirms that they are true-breeding for this character. The F₂ *deeps*, however, do not all behave alike and reveal themselves on selfing to be of two kinds. About one-third of them give *deep* only and can be shown by further selfing to breed true for this character, but the remaining two-thirds give *deep* and *shallow* types in the proportion of 3 : 1 as in the F₂ itself. Thus the F₂ consists of three genotypes: true-breeding *deep* (like one parent), hybrid *deep* (like the F₁) and true-breeding *shallow* (like the other parent). These occur, respectively, in the ratio of 1 : 2 : 1, or ¼ : ½ : ¼ when expressed as fractions of the whole.

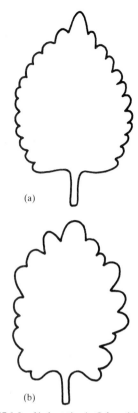

(a)

(b)

Fig. 37.1 Leaf indentation in *Coleus*. (a) Shallow-lobed leaf caused by a recessive allele. (b) Deep-lobed leaf caused by a dominant allele.

Explanation of the monohybrid cross

From the fact that *shallow* plants are absent in the F_1 but reappear, together with *deep* plants and without any transitional types, in the F_2, it is clear that 'something' from the parental plants with *shallow* leaf margins must have passed unchanged through the hybrid F_1 and separated out in the F_2. It is equally clear that this something is not the *shallow* character itself (which did not appear in the F_1) but some 'factor' which determines the development of that character at the appropriate stage in the leaf development of certain plants. The determinant of an hereditary character such as leaf indentation is called a *gene*, which is now known (although the evidence will not be given until Chapter 39) to be a particular segment of a DNA molecule located at a fixed point along the length of a chromosome in the nucleus. When a gene has more than one state, each of which produces a difference in the phenotype, these alternative states are known as *alleles*. In the cross being considered, the gene for leaf indentation has two states or alleles, *deep* and *shallow*.

Since the sex cells, or gametes, form the only link in sexually reproducing organisms between parent and offspring, it is also clear that the gametes must be involved in transmitting the genes from one generation to the next. In the monohybrid cross under consideration the *deep* parent plant was produced by the fusion of two gametes from a true-breeding *deep* line, the *shallow* parent plant by the fusion of two gametes from a true-breeding *shallow* line, and the F_1 by the fusion of two gametes one from each line, it being immaterial which way round the cross was made. The results of the cross can be explained by assuming that: (1) every cell (except the gametes) of a *Coleus* plant has two alleles of the gene controlling leaf indentation, one derived from the female parent and the other from the male parent; (2) in the formation of the gametes (ovules and pollen) the two alleles separate from one another, or *segregate*, with the result that each gamete has only one allele; and (3) at fertilization the gametes fuse in pairs at random so that the zygotes, and the new individuals which develop from them, again have two alleles. The cycle is repeated when gametes are produced by the new generation. How do these assumptions work out when applied to the monohybrid cross? If the dominant allele for *deep* indentation is designated as *A* and the recessive for *shallow* indentation as *a*, then the two pure-breeding P individuals can be represented as *AA* and *aa* because both possess two identical alleles in each body cell. Similarly the F_1 individuals can be represented as *Aa* because they have in each of their body cells one *A* allele and one *a* allele. On the basis that the alleles segregate when gametes are formed, each P individual gives gametes, both ovules and pollen, of one kind only, which carry either the *A* allele (if the individual is *AA*) or the *a* allele (if the individual is *aa*). In the same way, an F_1 individual gives two kinds of female gametes (ovules), *A* and *a*, in equal numbers, and two kinds of male gametes (pollen), *A* and *a*, also in equal numbers. If an *A* ovule is fertilized by an *A* pollen grain the result is an *AA* plant which can only give in its turn *A* gametes and so must breed true. If an *a* ovule is fertilized by an *a* pollen grain we get an *aa* plant which also breeds true. If, however, an *A* ovule is fertilized by an *a* pollen grain, or vice versa, we get an *Aa* plant which, like an F_1 individual, will produce both *A* and *a* gametes and so repeat the F_2 pattern of offspring when self-pollinated.

Chance decides which of the two alleles present in the body cells of an individual enters a given gamete. The probablity that any one F_1 gamete,

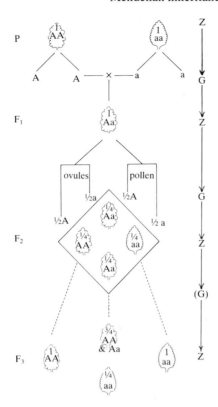

Fig. 37.2 Monohydrid inheritance of leaf-margin shape in *Coleus*. The alternation of zygote (Z), which is double in respect of the hereditary factor, and gamete (G), which is single in respect of it, is shown on the right. The gametes arising from F₂ and giving rise to F₃ are not shown in the diagram.

ovule or pollen grain, will carry the dominant allele *A* is one in two, or $\frac{1}{2}$, just as a coin tossed into the air has an equal chance (i.e. a probability of $\frac{1}{2}$) of landing 'heads' or 'tails'. The probability that any given gamete, ovule or pollen grain, will carry the recessive allele *a* is similarly $\frac{1}{2}$. Chance also decides whether an *A* or *a* ovule will be fertilized by either an *A* or *a* pollen grain because fertilization is a random event. The rules governing gene transmission thus depend on the operation of the laws of chance, which apply to all chance or random events.

An F₂ individual of genotype *AA* must have resulted from an *A* ovule being fertilized by an *A* pollen grain. Since the probability of an F₁ ovule being *A* is $\frac{1}{2}$ and the probability of an F₁ pollen grain being *A* is also $\frac{1}{2}$, it follows that the probability of obtaining an F₂ *AA* seed by random fertilization is $\frac{1}{2} \times \frac{1}{2}$ = $\frac{1}{4}$. This conclusion is an example of the application of the 'product law of probability' which states, in essence, that the probability of the simultaneous occurrence of two or more independent events is equal to the product of the probabilities of their separate occurrences. Following the same line of reasoning for the other F₂ genotypes, it is possible to represent the selfing of the F₁ in the following way:

F₁	$Aa \times Aa$
F₁ ovules	$\frac{1}{2}A + \frac{1}{2}a$
F₁ pollen	$\frac{1}{2}A + \frac{1}{2}a$

F₂ genotypes		$\frac{1}{4}AA + \frac{1}{4}Aa + \frac{1}{4}aA + \frac{1}{4}aa$
	=	$\frac{1}{4}AA + \frac{1}{2}Aa + \frac{1}{4}aa$

Genotypes of these constitutions in these proportions give a 1 : 2 : 1 genotypic ratio or, since *AA* and *Aa* individuals look alike, a 3 : 1 phenotypic ratio. Our assumptions have thus led to an explanation in complete agreement with the observed results. The monohybrid cross from the parental to the F_3 generation, as interpreted by the probability method of calculating gametic and zygotic ratios, is summarized in Fig. 37.2.

Some useful genetical terms

At this point it is convenient to introduce some terms which will help to simplify the description and explanation of crosses which are more complicated than the monohybrid one just described.

One of the assumptions made to explain the monohybrid ratio is that every body cell of a *Coleus* plant (or any other diploid organism) carries two alleles of any given gene, one derived from the female parent and the other from the male parent. An individual may thus have either two identical alleles, *AA* or *aa*, or two dissimilar alleles, *Aa*. The former true-breeding individual is called *homozygous*, or a homozygote, for the given character, whereas the latter hybrid individual is called *heterozygous*, or a heterozygote. The word *phenotype* to describe the appearance of an individual with regard to the character or characters under consideration, and the word *genotype* to indicate the genetic make-up of an individual with regard to the same character or characters, have already been introduced. It is worth mentioning, however, that the phenotype of an individual is usually expressed by a descriptive word or phrase (e.g. *deep, shallow, wild-type*), whereas its genotype is customarily given by letters of the alphabet (e.g. *AA, Aa, aa*).

The monohybrid testcross

As has already been indicated, both *AA* and *Aa* individuals of *Coleus blumei* have deeply indented margins because allele *A* is completely dominant over allele *a*. This raises the question of how homozygous and heterozygous individuals of a dominant phenotype can be distinguished from each other. One useful method of obtaining the answer is to cross the dominant phenotype with the recessive phenotype which is, of course, homozygous. Such a cross is called a testcross and, according to whether the dominant phenotype is homozygous or heterozygous, the results conform to one or other of two contrasting patterns.

Testcross of homozygous dominant	Testcross of heterozygous dominant
AA × *aa*	*Aa* × *aa*
(*deep*) × (*shallow*)	(*deep*) × (*shallow*)
ovules all *A*	ovules $\frac{1}{2}A + \frac{1}{2}a$
pollen all *a*	pollen all *a*
progeny all *Aa*	progeny $\frac{1}{2}Aa + \frac{1}{2}aa$
(*deep*)	(*deep*) (*shallow*)

It will be seen that when the individual under test is homozygous for the dominant allele, all the progeny of the testcross show the dominant character. When the individual under test is heterozygous, however, the progeny consists of 50 per cent showing the dominant character and 50 per cent showing the recessive. The number of offspring from a testcross must be high enough (at least seven or eight) to ensure that any possible absence of recessive phenotypes among the progeny is not due to chance deviation from a 1 : 1 ratio.

Dihybrid inheritance

Having seen what happens in a cross involving one gene, what happens when two genes, each showing typical monohybrid inheritance, are involved in the same cross?

In addition to the leaf indentation character of *Coleus blumei*, which was used to demonstrate monohybrid inheritance, another hereditary character of the leaves of this plant is the venation pattern. There are two alternative phenotypic expressions of this character, which will be described as *regular* and *irregular* respectively (Fig. 37.3). By making a monohybrid cross it can be shown that *irregular* is completely dominant to *regular*.

Any cross between two true-breeding *C. blumei* plants differing in both leaf indentation and leaf venation gives *deep irregular* F_1 individuals. The F_1 is the same irrespective of whether both dominant alleles, *deep* (*A*) and *irregular* (designated as *B*), come from one parent and both recessives, *shallow* (*a*) and *regular* (*b*), come from the other, or whether one dominant (*A* or *B*) and one recessive (*a* or *b*) come from each. Furthermore, just as in the monohybrid cross, it is immaterial from which parent, male or female, any particular allele comes. Expressed in symbols, the F_1 is the same whichever of the four possible ways the cross is made:

Female parent		Male parent
AA BB	×	aa bb
aa bb	×	AA BB
AA bb	×	aa BB
aa BB	×	AA bb

From the fact that the appearance of the F_1 is the same (i.e. *deep irregular*) no matter whether the two dominant alleles come from the same or from different parents, it may be concluded that the genes for leaf indentation (*A/a*) and leaf venation (*B/b*) do not interfere with each other in producing their effects on the phenotype, i.e. they are *independent in action*.

If the F_1 individuals are now selfed to give F_2 families, all the four possible phenotypes (*deep irregular, deep regular, shallow irregular* and *shallow regular*) appear. Furthermore, no matter from which of the four types of cross the F_1 individual being selfed has arisen, the numerical proportions of the F_2 phenotypes approach an ideal ratio of

9 A– B–	: 3A – bb	: 3 aa B–	: 1 aa bb
deep irregular (showing both dominants)	*deep regular* (showing the first dominant and the second recessive)	*shallow irregular* (showing the second dominant and the first recessive)	*shallow regular* (showing both recessives)

(*Note*: This method of labelling phenotypes by the combined use of letters and dashes, such as *A– B–* to indicate the phenotypically double dominant, is useful in recording phenotypic ratios of crosses where the dominance of the character or characters is complete. An individual labelled *A– B–* can be any one of the four possible genotypes, *AA BB, Aa BB, AA Bb, Aa Bb*).

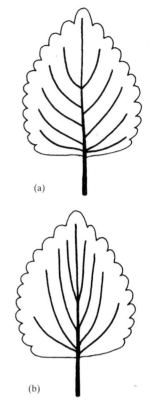

(a)

(b)

Fig. 37.3 Venation pattern of *Coleus* leaves. (a) Regular pattern caused by a recessive allele. (b) Irregular pattern caused by a dominant allele.

When the F_2 plants are selfed to give F_3 families, as was done in the monohybrid cross, it is possible to classify the F_2 individuals into their genotypes on the basis of whether they do or do not breed true as regards leaf indentation or leaf venation or both. The genotypes of the F_2 plants are given in the following table where the columns show constitution for leaf indentation (A/a gene), the rows show constitution for leaf venation (B/b gene), and the figures represent the ratios of plants obtained for each of the nine classes.

Leaf venation	Leaf indentation		
	AA	Aa	aa
BB	1	2	1
Bb	2	4	2
bb	1	2	1

From the above table it is clear that the F_2 genotypic ratio for the A/a gene is not merely $1 : 2 : 1$ for all the plants taken together, but is also $1 : 2 : 1$ in each of the three classes for the B/b gene. Similarly the B/b gene gives a $1 : 2 : 1$ ratio in each of the three classes for the A/a gene. This result must mean that the A/a and B/b genes segregate independently of each other and recombine at random. The proportions of the nine F_2 genotypic classes (shown by the figures in the table) therefore represent the square of the $1 : 2 : 1$ ratio characteristic of the F_2 of a monohybrid cross.

Explanation of the dihybrid cross

We have already shown how the F_2 genotypic ratio of $1AA : 2Aa : 1aa$ for a monohybrid cross can be explained by the random combination of male and female gametes, both of which are of two types (A and a) in equal numbers. In order to explain the results of the dihybrid cross it is necessary to make the additional assumption that the genes for the two characters are segregating and recombining independently of one another, so that during gamete formation there is an equal chance that A or a will be associated with either B or b. The doubly heterozygous F_1, of constitution $Aa\ Bb$, would thus be expected to give equal numbers of four kinds of gametes AB, Ab, aB, ab. Applying the probability method used for the monohybrid case, these four kinds of gametes may be expected to combine randomly as follows:

F_1 ovules	$\frac{1}{4} AB$	$\frac{1}{4} Ab$	$\frac{1}{4} aB$	$\frac{1}{4} ab$
F_1 pollen	$\frac{1}{4} AB$	$\frac{1}{4} Ab$	$\frac{1}{4} aB$	$\frac{1}{4} ab$

F_2 genotypes

$\frac{1}{16} AA\ BB$	$\frac{1}{16} AA\ Bb$	$\frac{1}{16} Aa\ BB$	$\frac{1}{16} Aa\ Bb$	(Entries in top row $\times \frac{1}{4} AB$)
$\frac{1}{16} AA\ Bb$	$\frac{1}{16} AA\ bb$	$\frac{1}{16} Aa\ Bb$	$\frac{1}{16} Aa\ bb$	(Entries in top row $\times \frac{1}{4} Ab$)
$\frac{1}{16} Aa\ BB$	$\frac{1}{16} Aa\ Bb$	$\frac{1}{16} aa\ BB$	$\frac{1}{16} aa\ Bb$	(Entries in top row $\times \frac{1}{4} aB$)
$\frac{1}{16} Aa\ Bb$	$\frac{1}{16} Aa\ bb$	$\frac{1}{16} aa\ Bb$	$\frac{1}{16} aa\ bb$	(Entries in top row $\times \frac{1}{4} ab$)

These sixteen zygote combinations can be resolved into nine genotypes which produce a $9 : 3 : 3 : 1$ phenotypic ratio:

$\frac{1}{16} AA\ BB$ $\frac{1}{16} AA\ bb$ $\frac{1}{16} aa\ BB$ $\frac{1}{16} aa\ bb$

$\frac{2}{16} Aa\ BB$ $\frac{2}{16} Aa\ bb$ $\frac{2}{16} aa\ Bb$

$\frac{2}{16} AA\ Bb$

$\frac{4}{16} Aa\ Bb$

$\frac{9}{16} A-B-$ $\frac{3}{16} A-bb$ $\frac{3}{16} aa\ B-$ $\frac{1}{16} aa\ bb$

(*deep*   (*deep*   (*shallow*   (*shallow*

irregular)   *regular*)   *irregular*)   *regular*)

Of the four phenotypes produced in the F_2 of a dihybrid cross, two are always the same as those of the original parents but the other two are new, or *recombinant*, types in which the characters of the parents are recombined in a different way. For example, if the two phenotypes of the parents are *deep irregular* and *shallow regular*, then the two recombinant phenotypes are *deep regular* and *shallow irregular*. This feature of a dihybrid cross is readily explained in terms of the assumptions being made. An F_1 plant from the cross *deep irregular* × *shallow regular* (i.e. *AA BB* × *aa bb*) must result from the fusion of parental gametes of types *AB* and *ab*. Such an F_1 in its turn produces gametes not only of these two parental types, but also of the two recombinant types *Ab* and *aB*, and, what is more, it produces all four types in equal numbers. If, on the other hand, the F_1 plant results from the cross *deep regular* × *shallow irregular* (i.e. *AA bb* × *aa BB*) the parental gametic types are *Ab* and *aB*, and the recombinant types are *AB* and *ab*, but again the frequency of the recombinant gametes equals that of the parental types. The production of parental and recombinant types of gametes in equal numbers is, in fact, the criterion of independent segregation of two genes.

The dihybrid testcross

Just as the monohybrid testcross provides a useful way of determining homozygosity or heterozygosity in respect of a single gene (A/a), so the dihybrid testcross (crossing an individual phenotypically dominant for two characters with the doubly recessive phenotype) is equally valuable for revealing the genotype of an individual in respect of two genes (A/a and B/b). In a cross involving two independently segregating characters, a doubly dominant phenotype ($A-B-$, *deep irregular*) could be any one of the four following genotypes:

AA BB (homozygous for both dominants)

AA Bb (homozygous for the first dominant, heterozygous for the second dominant)

Aa BB (heterozygous for the first dominant, homozygous for the second dominant)

Aa Bb (heterozygous for both dominants)

There are therefore four possible testcrosses, which can be summarized as follows:

AA BB × *aa bb* → 100 per cent *Aa Bb*;   1 phenotype

AA Bb × *aa bb* →   50 per cent *Aa Bb* + 50 per cent *Aa bb*; 2 phenotypes

Aa BB × *aa bb* →   50 per cent *Aa Bb* + 50 per cent *aa Bb*; 2 phenotypes

Aa Bb × *aa bb* →   25 per cent *Aa Bb* + 25 per cent *Aa bb* +

25 per cent *aa Bb* + 25 per cent *aa bb*; 4 phenotypes.

Thus, from the number of phenotypes in the progeny and whether they show one or both the dominant characters, it is possible to determine the genotype of the individual being tested.

Modifications of the 9 : 3 : 3 : 1 ratio

In the *Coleus* cross used to demonstrate the pattern of dihybrid inheritance, the genes responsible for leaf indentation and leaf venation were independent of each other not only in their transmission but also in their action. Such dihybrid crosses give the standard phenotypic ratio of 9 : 3 : 3: 1 in the F_2 generation. Many characters, however, are controlled not by one gene but by two or more genes interacting with one another. Depending upon the type of interaction, the standard dihybrid or polyhybrid ratios of the F_2 phenotypes (not the genotypes) are found to be modified in various ways, even though the genes are transmitted independently of one another in the usual way.

Confining our attention to examples of interaction between two genes, each of which has dominant and recessive alleles, the following types of interaction represent most of the ways in which the four expected phenotypic classes can be reduced in number as a result of two or more of the genotypic classes becoming indistinguishable (Fig. 37.4).

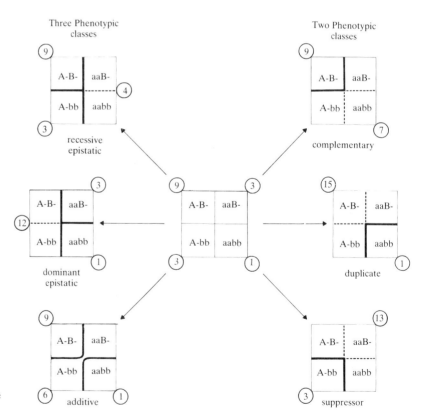

Fig. 37.4 Summary of the modifications of the F_2 dihybrid ratio caused by various types of gene interaction. (Modified from Darlington, C. D. and Mather, K. (1949), *The elements of genetics*, p. 156, published by George Allen & Unwin Ltd., London).

Epistasis Where two different genes both affect the same character, the effect of one gene may be nullified by that of the other gene which is said to be *epistatic* to it. This masking phenomenon is known as epistasis and must not be confused with dominance which occurs between two alleles of the same gene and not (as in epistasis) between two alleles belonging to different genes. Epistasis is well illustrated by coat colour in mice. Wild mice normally have hair with alternating black and yellow bands, known as 'agouti' type, which is due to a dominant allele *A*. The recessive allele *a* suppresses the yellow banding to give entirely black hair. However, whether or not any colour develops at all depends on the alleles of a second gene, *C/c*, pigmentation (*C*-) being dominant to albinism (*cc*). When the albino allele is present in the homozygous condition *cc* it masks the action of either *A*- (agouti) or *aa* (black) with the result that albinos can be of two types: those that would have been agouti if they had had any colour at all (*A*- *cc*) and those that would have been black (*aa cc*). If a true-breeding albino of the 'potential agouti' type (*AA cc*) is crossed with a true-breeding black (*aa CC*), the F$_1$ (*Aa Cc*) will be pigmented because it has *C*, and agouti because it has *A*. Inbreeding (i.e. brother/sister mating) of the F$_1$ gives a 9 : 3 : (3 + 1) or 9 : 3 : 4 ratio as follows:

9 *A* – *C* – :	3 *aa C* – :	3 *A* – *cc*	:	1 *aa cc*
agouti	black	albino		albino
		(unable to show agouti)		(unable to show black)

This particular example of epistasis is one of recessive epistasis because the recessive allele *c* is epistatic to *A* and *a*. The converse of this situation, where absence of colour is due to a dominant allele which is epistatic to another gene controlling colour, is shown by fruit colour in squashes (*Cucurbita pepo*). Putting *W* for dominant white and *w* for colour potential, *Y* for yellow fruit-colour, and *y* for green, then *W*– *Y*– and *W*– *yy* would be white, *ww Y*– yellow and *ww yy* green. The doubly heterozygous F$_1$ (*Ww Yy*) would clearly be white and the F$_2$ of 9 *W* – *Y* – : 3 *W* – *yy* : 3 *ww Y* – : 1 *ww yy* would show as (9 + 3) white : 3 yellow : 1 green or a 12 : 3 : 1 ratio which is characteristic of dominant epistasis.

Gene interactions which are reciprocal There are three kinds of interaction involving two genes where the relationship between them is reciprocal and not one-sided as in epistasis. In such interactions only two phenotypic classes are produced in the F$_2$ as opposed to three with epistasis. The three kinds of interaction depend on whether both the dominants, either of the dominants, or one dominant and one recessive are needed for the expression of the jointly controlled character. The classical example of reciprocal gene interaction is that of flower colour in the sweet pea (*Lathyrus odoratus*). When two pure-breeding white strains were crossed, they gave a coloured F$_1$ which, on selfing, produced an F$_2$ containing coloured and white in the proportion of 9 : 7. On further testing, three different types of breeding behaviour were recognized among the F$_2$ white flowers in the proportions of 3 : 3 : 1. This suggests that, although only one character is involved in the cross, the 9 : 7 is really a 9 : (3 + 3 + 1) where the last three classes are indistinguishable. Such a modification can readily be explained on the assumption that colour is produced only when two different dominant alleles (*C* and *R*) are present simultaneously in one individual. On this assumption the cross may be represented as follows:

P $CC\,rr$ × $cc\,RR$
    white   white

F$_1$ $Cc\,Rr$
           coloured

F$_2$ 9 $C-R-$:   3 $C-rr$:   3 $cc\,R-$:   1 $cc\,rr$
    coloured   white   white   white
              all white because they lack C or R or both

Genes which interact in this way to give a 9 : 7 ratio are called *com-plementary* genes because jointly they produce an effect which is not produced by either of them acting separately.

The second type of reciprocal gene interaction is that in which the character is produced when either of the dominant alleles is present, so that only the doubly recessive $aa\,bb$ is phenotypically distinct. Because the phenotypes $A-B-$, $A-bb$ and $aa\,B-$ are all alike, the F$_2$ ratio is (9 + 3 + 3) : 1 = 15 : 1. In the cross $AA\,bb \times aa\,BB$ both the parents and the F$_1$ are alike but the F$_2$, like that of all crosses involving this type of gene interaction, gives a 15 : 1 ratio. The two genes are described as *duplicate* genes because both produce the same effect, i.e. they duplicate each other. The third type is when the dominant allele of one gene and the recessive alleles of another gene, i.e. $A-bb$, are necessary for the expression of the character. The $A-B-$, $aa\,B-$ and $aa\,bb$ genotypes are all alike so that the F$_2$ ratio is (9 + 3 + 1) : 3 = 13 : 3. One interpretation of this ratio is to regard the dominant allele B as a *suppressor* gene which prevents the expressions of a character normally produced by the dominant allele A, so that the genotypes of the F$_2$ express themselves as follows:

9 $A-B-$ character not shown (because of inhibitor B)
3 $A-bb$ character shown (because of presence of A and absence of B)
3 $aa\,B-$ character not shown (because of both B and aa)
1 $aa\,bb$ character not shown (because of aa)

Additive genes   Distinct from the types of interaction discussed so far is that where the $A-bb$ and $aa\,B-$ combinations are indistinguishable from each other but distinguishable from $A-B-$, so giving a 9 : 6 : 1 ratio. Genes A/a and B/b are said to be *additive* because both influence the same process in the same way but in a cumulative fashion. Additive gene interaction between two alleles accounts for three fruit shapes found in squashes (*Cucurbita pepo*). A cross between two pure-breeding strains with spherical fruits produced an F$_1$ with disc-shaped fruits. In the F$_2$ disc-shaped, spherical and elongate (a new shape) fruits appeared in the proportions of 9 : 6 : 1, suggesting the following interpretation:

    9 $A-B-$:   3 $A-bb$:   3 $aa\,B-$:   1 $aa\,bb$
      disc   spherical   elongate

On this interpretation one parent strain had spherical fruits because it lacked B, being of constitution $AA\,bb$, the other because it lacked A, being $aa\,BB$. The F$_1$, $Aa\,Bb$, had both dominant alleles and so had disc-shaped fruits, a shape not shown by either parent.

The chi-squared test for testing the validity of Mendelian ratios

Because monohybrid, dihybrid and testcross ratios depend on the random union of gametes at fertilization, the results observed in any given cross are unlikely to agree exactly with the theoretical (ideal) ratio which one would expect to get on the assumption that a particular mechanism is operating in the cross under consideration. This immediately raises the question of how much deviation between observed and expected values is acceptable as being due solely to chance, rather than due to the possibility that the postulated mechanism (i.e. the hypothesis) may not be the correct one. A mathematical way of indicating the 'goodness of fit' between the observed and expected values of any cross is provided by the chi-squared (χ^2) test, which gets its unusual name from the Greek letter chi, χ.

The χ^2 test is carried out in two stages. Firstly, a formula is used to calculate the value of a mathematical quantity, called χ^2, which measures how widely the observed results deviate from those expected on the basis of the hypothesis. Then, having calculated a value for χ^2, the second stage is to ask the question: 'How often can a deviation of this magnitude be attributed to chance alone when, on theoretical grounds, the results would be expected to fit a particular ratio?'

Stage 1

The formula for calculating chi-squared is

$$\chi^2 = \Sigma \left[\frac{(o - e)^2}{e} \right]$$

where o = the observed number in a given class
e = the expected number in the same class
Σ = the 'sum of' for all the classes

Both o and e must be actual numbers and not percentages.

It will be noted that each of the numerators in the χ^2 formula (i.e. the squares of the differences between the observed and the expected numbers) measures the magnitude of the deviation in each class, and that the denominators (i.e. the expected numbers) take into account the size of the experiment because a deviation of a given magnitude is less striking *relative* to a large total than to a small total. In general, however, it is not safe to use the χ^2 test where the expected number in any one class is less than 5. The device of squaring the numerator, because it eliminates the sign (+ or −) of the deviations, means that numbers less than expected carry as much weight in the calculation as those more than expected.

Despite its rather frightening name, χ^2 is in fact easy to calculate. For each class of a Mendelian ratio the difference between the observed number and the expected number is obtained. This difference is squared. The square of the difference is then divided by the expected number. This is repeated for each class, and the figures obtained are totalled. The sum is the value of χ^2 for the ratio being tested. The formula for χ^2, expressed in words, is therefore

$$\chi^2 = \text{the sum of} \left[\frac{(\text{observed number} - \text{expected number})^2}{\text{expected number}} \right] \text{for all the classes}$$

The χ^2 test never states directly that the results of an experiment do or do not fit the hypothesis, but gives an indirect answer by indicating how probable it is that the calculated value of χ^2 could have been produced solely by chance. To use the χ^2 test, therefore, it is necessary to understand the concept of probability.

Probability   Probability (abbreviated as P) is the numerical expression of the likelihood that an event will occur. In problems in probability we are dealing with a trial, yet to be made, that can have a number of different outcomes. The simplest problems can be solved by writing down all the different possible outcomes of a trial, and recognizing that these are equally likely. Thus, in throwing a 6-sided die, the probability of obtaining a four is 1 in 6 because there are 6 outcomes. In general terms this result may be stated as follows: If a trial has x *equally likely* outcomes, of which one and only one will happen, the probability of any individual outcome is $\frac{1}{x}$ (Rule 1).

In some problems the event in which we are interested will happen if any one of a specific group of outcomes turns up when the trial is made. Suppose that the seven letters a, b, c, d, e, f and g are written on identical balls which are placed in a bag and mixed, and that one ball is then drawn out blindly. What is the probability of drawing a vowel? The event is now 'a vowel is drawn'. This will happen if either an a or an e is the outcome. Since the probability of drawing an a is $\frac{1}{7}$, and similarly the probability of drawing an e is also $\frac{1}{7}$, the probability of drawing a vowel (i.e. either an a or an e) is $\frac{2}{7}$. This result is an example of a second rule of probability, which states: If an event is satisfied by any one of a group of *mutually exclusive* outcomes, the probability of the event is the sum of the probabilities of the outcomes in the group (Rule 2, or the Addition Rule). This rule contains the condition that the outcomes in the group must be *mutually exclusive*, i.e. if any one of the outcomes happens, all the others fail to happen.

In most numerical problems in genetics we have to consider the results of repeated trials from a population. The successive trials are assumed to be *independent* of one another, i.e. the outcome of a trial does not depend in any way on what happens in the other trials. With a series of trials, it follows from Rules 1 and 2 that, if the probability that an event may occur is P, we would expect that, when a large number (N) of trials is made, the number of times that the event would occur is $P \times N$. In other words, probability can be defined mathematically by the equation

$$\text{Probability} = \frac{\text{No. of times an event occurs}}{\text{Total no. of trials}}$$

Probability is expressed on the scale 0 to 1, a value of 0 indicating that an event will certainly not occur (or that a hypothesis is not true) and a value of 1 indicating that the event is certain to occur (or that the hypothesis is definitely true). On this scale a value of 0.5 implies that the event is just as likely to occur as not (or that the hypothesis is just as likely to be true as not). This scale of fractions can be expressed in an alternative way by stating how often in every 100 cases the event may be expected to happen by chance alone, e.g. a result with a P of 0.5 is the same as saying that in 50 times out of every 100 it could have happened by chance.

Stage 2

In order to convert the calculated value of χ^2 from stage 1 into a probability figure, a table of χ^2 values (Table 37.1) is consulted. To use a χ^2 table it is necessary to know how many 'degrees of freedom' there are in any particular example, because the table relates probability to values of χ^2 with a specified number of degrees of freedom. In practice, when testing Mendelian ratios, this number always equals the number of classes less one; thus if

Table 37.1 Simplified table of chi-squared

Degrees of freedom	$P =$	0.99	0.94	0.90	0.70	0.50	0.30	0.20	0.10	0.05	0.02	0.01
1		0.000	0.004	0.02	0.15	0.46	1.07	1.64	2.71	3.84	5.41	6.64
2		0.020	0.103	0.21	0.71	1.39	2.41	3.22	4.61	5.99	7.82	9.21
3		0.115	0.352	0.58	1.42	2.37	3.67	4.64	6.25	7.82	9.84	11.34
4		0.297	0.711	1.06	2.20	3.36	4.88	5.99	7.78	9.49	11.67	13.28
5		0.554	1.145	1.61	3.00	4.35	6.06	7.29	9.24	11.07	13.39	15.09
10		2.558	3.940	4.87	7.27	9.34	11.78	13.44	15.99	18.31	21.16	23.21

there are four classes as in a $9 : 3 : 3 : 1$ ratio there are three degrees of freedom. The basic idea behind the concept of degrees of freedom is that any measure, such as χ^2, which summarizes the deviation from expectation of a number of classes is based mathematically on the number of independent deviations that must be specified in order to define the total deviation. For example in a Mendelian ratio with n classes, when the deviations of $n-1$ classes have been fixed, the value of the deviation of the remaining class is automatically fixed. There are therefore only $n-1$ independent deviations possible, which is the reason why the number of degrees of freedom for this ratio is $n-1$.

The values of P across the top of the table indicate the probability of obtaining by chance a χ^2 value equal to or greater than a specified value, for a given number of degrees of freedom. For any particular number of degrees of freedom, the value of χ^2 increases with increased discrepancy between the observed and expected values; in other words, the higher the value of χ^2 the smaller is the probability that the hypothesis on which the expected values are based is correct. It is standard practice to accept a probability of 0.05 (i.e. one in twenty) or lower as sufficient evidence of a significant discrepancy between the hypothesis and the observed values; similarly it may be assumed that, for probabilities greater than 0.05, there is no reason to doubt the truth of the hypothesis. Acceptance of a probability level of 0.05 in this way is arbitrary, and ratios whose χ^2 values give a probability of about 0.05 should be re-investigated if possible. Values of χ^2 which give probabilities much higher or much lower than this accepted standard of 0.05 can generally, but not infallibly, be interpreted as supporting or disproving the hypothesis.

An example of the use of the χ^2 test

To show how the χ^2 test is used, the following example of its application to a modified dihybrid ratio is given. Fruits of squash (*Cucurbita pepo*) may be disc-shaped, spherical or elongate. Selfing an F_1, all of which had disc-shaped fruits, produced an F_2 of which 188 plants had disc-shaped fruits, 135 spherical fruits, and 29 elongate fruits. On the assumption that fruit shape in squash can be explained in terms of the action of two additive genes, the expected result for an F_2 having three phenotypes would be a $9 : 6 : 1$ ratio. Do the observed numbers agree with the expected ratio?

To calculate χ^2 the data may conveniently be set out as follows:

Class	o	e (i.e. $P \times N$)	$o - e$	$(o - e)^2$	$\dfrac{(o - e)^2}{e}$
disc-shaped	188	$\frac{9}{16} \times 352 = 198$	-10	100	0.505
spherical	135	$\frac{6}{16} \times 352 = 132$	$+3$	9	0.068
elongate	29	$\frac{1}{16} \times 352 = 22$	$+7$	49	2.227
					$\chi^2 = 2.800$

Having calculated the value of χ^2 for this result as 2.80, we consult a χ^2 table and read across the row of figures for two degrees of freedom (because there are three classes of phenotypes in the F_2) until we find this value of χ^2 or the two values which lie on either side of it. In this case a value of 2.80 does not appear in the table but there are two values, 2.41 and 3.22, between which it lies. Reading up the columns to the values of P at the top of the table, we see that a χ^2 value of 2.80 corresponds to a probability value of between 0.30 and 0.20. This means that we can expect a deviation by chance as large as that observed in this experiment in about 25 trials out of every 100 of this size (i.e. involving 352 individuals), or once in about every four.

Accepting the standard that a discrepancy is not significant unless P is 0.05 (one in twenty) or less, the results of this experiment can be assumed to agree with an ideal ratio of 9 : 6 : 1 and therefore with the hypothesis on which it is based.

Linkage

Mendel's two rules have been tested in nearly all groups of sexually reproducing organisms, both plant and animal, and they have been found to be universally applicable with one modification, which itself is also universal. The modification is that genes do not always segregate independently in the way discussed under dihybrid inheritance.

The characteristic behaviour of genes showing non-independent segregation may be illustrated by two crosses, carried out in the second decade of this century by the famous American geneticist T. H. Morgan, with the fruit fly *Drosophila melanogaster*. Wild fruit flies are grey in colour and have wings which extend beyond the tip of the abdomen. Two alleles, one for black body colour (*b*) and another for vestigial wing (*v*) are both recessive to their corresponding wild-type alleles (*B* and *V*). In the first experiment Morgan crossed a true-breeding wild-type fly, grey with normal wings (*BB VV*), with a true-breeding black fly with vestigial wings (*bb vv*) and obtained an F_1 which was wild-type in appearance (*Bb Vv*); this is to be expected because the two characters in which the wild-type differed from its mate are both dominant. Females of this F_1 were then testcrossed with males of the black vestigial parental stock, i.e. *Bb Vv* × *bb vv*. The black vestigial male can only contribute *bv* to his offspring since he is doubly homozygous. On the other hand the F_1 female, being doubly heterozygous, could reasonably be expected to produce four types of egg: *BV*, *Bv*, *bV* and *bv*. On being fertilized by a sperm of constitution *bv* these different types of egg will give offspring as follows:

Egg		Sperm	Offspring	Phenotype
BV	×	*bv*	*Bb Vv*	normal body, normal wing (wild-type)
Bv	×	*bv*	*Bb vv*	normal body, vestigial wing
bV	×	*bv*	*bb Vv*	black body, normal wing
bv	×	*bv*	*bb vv*	black body, vestigial wing

The relative frequency of the different types of offspring thus reflects the relative frequency of the different types of egg produced by the doubly heterozygous mother.

Morgan found all four types of offspring in the testcross, the numbers being:

Bb Vv	*Bb vv*	*bb Vv*	*bb vv*
586	106	111	465

After due allowance has been made for the fact that fertilization is a random process, it is clear from these figures that *BV* and *bv* eggs had been produced in equal numbers as also had *Bv* and *bV* eggs. However, the former were about five times as common as the latter ($586 + 106 + 111 + 465 \times 100 = 17.1$ per cent), and this explains the excess of the parental types and the corresponding shortage of recombinant types. Since *BV* and *bv* represent the constitution of the gametes of the original parental types, it is the genes which go into the F_1 together which tend to come out of the testcross together.

Morgan undertook a second experiment in which the F_1 was obtained by crossing pure-breeding normal body, vestigial wing (*BB vv*) with pure-breeding black body, normal wing (*bb VV*). Females from this F_1 gave, on testcrossing to *bb vv* as in the first experiment, offspring as follows:

Bb Vv	*Bb vv*	*bb Vv*	*bb vv*
338	1552	1315	294

Here the types of offspring, and hence the types of egg, which were deficient in the first experiment are now in excess. Nevertheless it is again the original parental combinations of alleles which are produced in excess of the recombinant types, and it is also significant that the recombinants again form the same proportion (something over 17 per cent) of the total offspring. Comparison of the results of this experiment with those of the first experiment demonstrates that the ratios between the classes of offspring in the testcross depend on the way in which the genes are fed into the F_1 from the parental stocks. The F_1 from the *BB VV* × *bb vv* cross produces *BV* and *bv* gametes in excess, whereas the F_1 from the *BB vv* × *bb VV* cross produces an equal excess of *Bv* and *bV* gametes.

This type of hereditary transmission in which certain genes tend to remain together in passing from one generation to the next is known as *linkage*. The strength of linkage between two genes varies inversely with the amount of recombination between them, which is expressed quantitatively by the *recombination percentage* (here just over 17 per cent). The recombination percentage has a characteristic value, within narrow limits, for any given pair of genes (such as between *B/b* and *V/v*), but varies widely, from 0 per cent to nearly 50 per cent, between different pairs of genes. A recombination percentage of 50 per cent for two particular genes would of course indicate independent segregation, i.e. the absence of any linkage. The explanation of linkage will be given in the next chapter.

Chapter 38

The chromosomal basis of heredity

The nucleus as the bearer of heredity

Each individual of a sexually reproducing species develops from a single cell, the fertilized egg or zygote. This cell in its turn arises from the fusion of two single cells, the gametes (egg and sperm) which were produced by the parents, and anything that the zygote gets from its parents, including its heredity, must therefore have been supplied by these gametes. It is thus clear that the materials of heredity are to be found inside the single cell of the zygote.

All cells consist essentially of two living parts, the cytoplasm and the nucleus. The nuclei of the two gametes which fuse at fertilization are approximately equal in size, but the female gamete (egg) usually contributes most of the cytoplasm to the zygote. In fact, the male gamete of many species is almost devoid of cytoplasm. One of the basic assumptions of Mendelian heredity is that both parents contribute equally to the heredity of their offspring and, since predictions based on this assumption fit the observed facts, it would seem likely that the hereditary material is located in the nucleus rather than the cytoplasm. That this is the case has now been confirmed by nuclear transplant experiments and by other even more sophisticated methods, and it is now accepted as a proven fact that the nucleus is the bearer of heredity.

Examination of a non-dividing nucleus under the light microscope offers little chance of finding out how the hereditary material is organized within the nucleus, because all that can be seen is a nearly homogeneous material containing one or two rather denser bodies, the nucleoli. Attempts to gain more information about nuclear structure by various staining techniques do not give any significant improvement. If, however, a nucleus can be studied when the cell containing it is dividing into two, then the material of which it is composed can be seen. Clues as to how the hereditary material is organized can therefore be sought in a detailed consideration of the processes involved in nuclear division.

Mitosis

The terms mitosis and cell division are widely but wrongly used as though they were synonymous. Strictly speaking, mitosis refers only to the division of the nucleus, whereas cell division includes the division of the cytoplasm which usually follows mitosis. To avoid the ambiguity inherent in the phrase 'cell division', the division of the cytoplasm is sometimes distinguished as *cytokinesis*.

Mitosis is similar in both plants and animals, but in plants the stages can readily be seen in thin, suitably stained, longitudinal sections through an apical meristem, usually the root tip of a flowering plant. The sequence of events can then be reconstructed from a study of cells which have been killed and stained in different stages of division. At one time this was the only method by which the process could be studied, but using cells in culture solutions it is now possible to follow under the microscope the changes going on in a living cell from the beginning to the end of mitosis. The time taken for the entire process of mitosis is usually somewhere between 30 minutes and 3 hours, but in different organisms it can vary from 10 minutes to 24 hours.

Although mitosis is a continuous process, it is divided for convenience of description into four stages, prophase, metaphase, anaphase and telophase.

Prophase

The first indication that mitosis is about to take place is the appearance within the nucleus of a tangled mass of very long threads (Figs. 38.1a and 38.2b), which can be stained by basic dyes such as Feulgen's stain. These threads are called *chromosomes* and from the time they first appear they can be seen to be double, consisting of two single threads called *chromatids*, except at one point, the *centromere*. Whereas the rest of the chromosome takes up basic dyes readily, the centromere remains unstained. The centromere may occur anywhere along the length of a chromosome except at the extreme ends, but its position is constant for any given chromosome. As prophase continues the two chromatids become shorter and thicker (Figs. 38.1b and 38.2c) as a result of becoming coiled into a tight spiral, a process known as *spiralization*. Some chromosomes have bladder-like structures, called *nucleoli*, attached to them, which may be very conspicuous in the non-dividing nucleus and in the early stages of prophase, but shrink in size and usually become completely absorbed into the substance of the chromosome by the end of prophase. The end of prophase is marked by the disappearance of the nuclear envelope, and the chromosomes, which have hitherto been scattered around the inner surface of the nucleus, are dispersed into the cytoplasm.

At the same time as the nuclear envelope begins to break down, a structure shaped like two cones joined at their bases is formed in the cytoplasm just outside the nucleus. This structure is called the *spindle*, and consists of longitudinally arranged contractile protein microtubules which run along its length from pole to pole. (Microtubules are submicroscopic unbranched tubes which have been shown by the EM to occur in groups in the cytoplasm of virtually all eukaryotic cells.)

Metaphase

With the formation of the spindle, the chromosomes move into the plane across the middle of the spindle to form the equatorial or metaphase plate. The chromosomes are attached to the metaphase plate by their centromeres, the chromosome arms lying passively in the cytoplasm as space allows (Figs. 38.1c and 38.2d). The two chromatids of each chromosome are still in close contact with each other. Since all the chromosomes come to lie on the equator of the spindle at the same time, the metaphase position affords the most favourable opportunity in the nuclear cycle for observing the chromosomes. From numerous surface or polar views of metaphase plates (i.e.

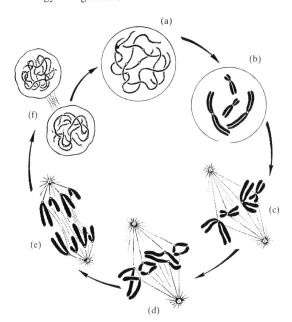

Fig. 38.1 Diagram of mitosis, illustrated by a cell nucleus containing two pairs of chromosomes. (a) Early prophase. (b) Late prophase. (c) Metaphase. (d) Early anaphase. (e) Late anaphase. (f) Daughter nuclei entering interphase. (From R. J. Berry (1965), *Teach yourself genetics*, p. 10, published by The English Universities Press Ltd, London)

looking along the long axis of the spindle), the following important characters of the chromosomes can be made out.

1. The number of chromosomes in the body cells of all individuals of a given species is the same, ranging from a minimum of two to a maximum of over two hundred.
2. The chromosomes differ among themselves in length, shape, and position of the centromere. Some chromosomes have the centromere near the midpoint so that they have approximately equal arms (metacentric type), while others have it close to one end so that one arm is very much longer than the other (acrocentric type). Each chromosome type in a set is, however, represented in every nucleus of an individual, thus demonstrating that each chromosome has its own individuality.
3. Each specific type of chromosome occurs twice in the nucleus of a body cell, one member of each pair being derived from the male gamete and the other from the female gamete that fused at fertilization. The maternal and paternal chromosomes which are similar in length and shape are referred to as *homologous* pairs, because they are mates or *homologues* of each other.

Anaphase

Anaphase begins with the simultaneous division of the centromeres of all the metaphase chromosomes. The two resulting half centromeres then move to the opposite poles of the spindle, pulling one chromatid after them (Figs. 38.1d,e and 38.2e, f). The spindle changes shape at anaphase, possibly due to the contraction of the protein microtubules, and this may be associated with the movement of the chromosomes to the poles.

Telophase

Telophase begins when the diverging chromatids reach the poles. During this phase the groups of chromatids at each pole undergo the changes of prophase in the reverse order (Figs. 38.1f, and 38.2g). The chromatids become longer

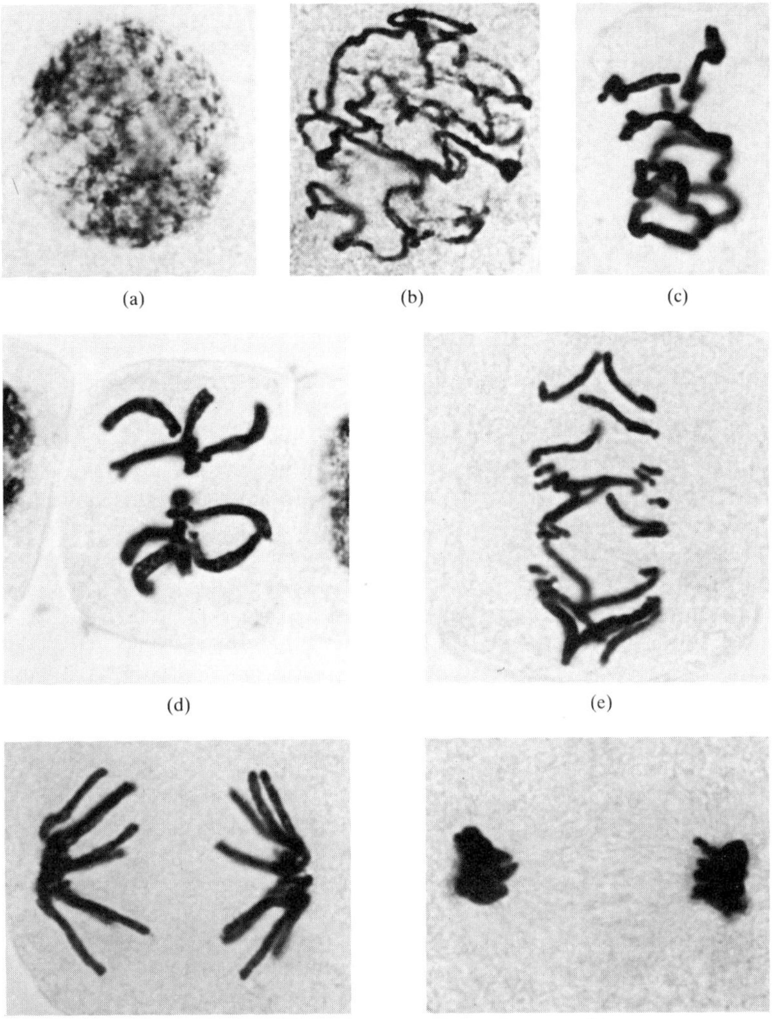

(a)

(b)

(c)

(d)

(e)

(f)

(g)

Fig. 38.2 Stages of mitosis in the root tip of *Aloe*. (a) Interphase. (b) Early prophase. (c) Late prophase. (d) Metaphase; in this and the following three photographs the spindle is orientated horizontally. (e) Early anaphase, the sister chromatids moving sideways to opposite poles of the spindle. (f) Late anaphase. (g) Telophase, the individual chromatids being no longer identifiable. (Crown copyright, and reproduced by permission of the Controller of Her Majesty's Stationery Office and the Director of the Royal Botanic Gardens, Kew. Microscopic preparations made by Dr P. E. Brandham)

and less distinct as they uncoil, the nuclear envelope re-forms, and the nucleolus reappears. After telophase the cytoplasm usually also divides, and two new cells are formed.

Cytokinesis and wall formation

In plant cells the division of the cytoplasm into two (cytokinesis) begins in early telophase when a system of fibres (microtubules), called the *phragmoplast*, develops in the region of cytoplasm between the two telophase nuclei. At first the phragmoplast consists entirely of spindle fibres, but soon small membrane-bound vesicles appear in the plane of the equator of the disappearing spindle. These vesicles gradually fuse with one another, and this gives the first physical sign of the new cell wall. At this stage the spindle-fibre/vesicle complex is termed the *cell plate*, although it should be understood that this name refers to a stage in the formation of the wall rather than a specific part of the wall itself. The cell plate expands laterally as more vesicles are added to its rim. The vesicles presumably contain pectic substances because their contents fuse to form the middle lamella and the plasmalemma on each side of it. Lengths of endoplasmic reticulum 'trapped' between the fusing vesicles become plasmodesmata. The expanding cell plate eventually unites with the plasmalemma and the wall of the mother cell, thus completing the separation of the two new daughter cells. Following the formation of the new middle lamella each daughter cell deposits a primary wall on the surface of the middle lamella facing it. Each daughter cell also deposits around the entire protoplast a layer of new wall which is continuous with the wall material deposited on the cell plate. Subsequent enlargement of the daughter cells stretches and breaks the original wall of the parent cell.

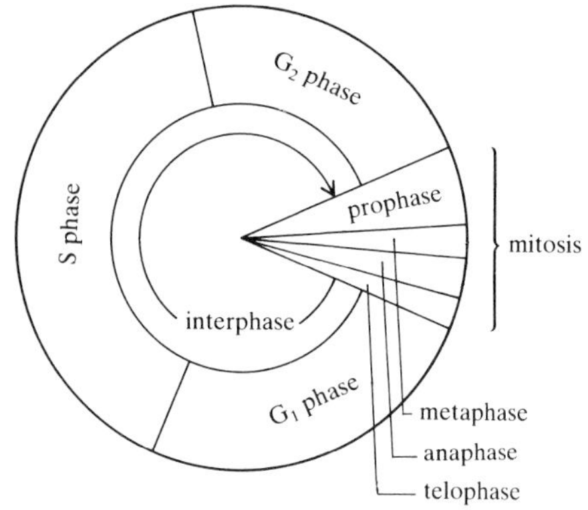

The cell cycle

With cytokinesis as well as mitosis completed, the two daughter nuclei enter *interphase*, i.e. the stage at which a nucleus is not dividing by mitosis or meiosis. A nucleus at this stage is sometimes described as a *resting nucleus*, but this description is a misnomer because during interphase the nucleus is active in carrying out such functions as protein synthesis and the duplication of the chromatids preparatory to undergoing another division. The complete series of events that occurs in a dividing cell from one mitosis to the next is known as the *cell cycle*. It is divided into four main phases (Fig. 38.3). When a nucleus enters interphase, only one half of each chromosome (i.e. one chromatid), and hence only half the usual amount of the genetic material (DNA) is present. The phase of the cell cycle in which each chromatid duplicates itself is called the S (synthesis) phase. This is preceded and followed by a so-called G (gap) phase. The G_1 phase follows mitosis, and is essentially the period in which the cytoplasm, including all its various organelles, and the cell wall are actively growing. It is also thought that during the G_1 phase substances are synthesized that either inhibit or stimulate the onset of the S phase and the rest of the cell cycle, thus determining whether or not cell division will be repeated. The S phase, if it occurs, is the period during which the genetic material in the nucleus is duplicated. It is followed by the G_2 phase during which the cytoplasmic structures specifically involved in mitosis, such as the spindle fibres, are synthesized. Mitosis itself, the M phase, follows the G_2 phase of one cell cycle and precedes the G_1 phase of the next cell cycle. The G and S phases together are called the interphase.

Some cells, like unicellular organisms or the initial cells which remain permanently within a meristem, pass through successive cell cycles one after another, whereas other cells become highly specialized and lose the ability to divide once they are mature. Such differentiated cells remain permanently in the G_1 phase, or G_0 phase as it is sometimes called in these circumstances. A third group of cells, of which those forming the tissue collenchyma are an example, retain the capacity to pass through the S and G_2 phases but do so only under special circumstances.

Meiosis

In all sexually reproducing organisms two gametes fuse at fertilization to form a zygote, which in diploid organisms is the first cell of the new individual. Since fertilization involves the fusion of two nuclei, it would result in a progressive doubling of the chromosome number unless, at some stage in the life cycle, the number was halved. This halving of the chromosome number occurs in animals during the formation of the gametes, but in *most*

Fig. 38.3 The cell cycle.

plants during the production of spores (sporogenesis) which have the haploid number of chromosomes. The spores undergo mitosis to form a haploid body, the *gametophyte*, which in due course gives rise to the gametes. Gametes and spores therefore have the haploid or half the full number of chromosomes, whereas the zygote and all the cells derived by mitosis from it (i.e. the *sporophyte* of plants) have the full or diploid number. However, in addition to merely reducing the chromosome number from diploid to haploid, meiosis also has the function of producing genetic diversity as a result of an exchange of genetic material between homologous chromosomes which accompanies the reduction in chromosome number. The pattern of meiosis is extremely similar in all organisms in which it occurs.

In essence meiosis consists of two successive divisions of a diploid nucleus accompanied by only one division of its chromosomes. The two divisions are known as division I and division II. Both nuclear divisions resemble mitosis, and similarly each may be divided into prophase, metaphase, anaphase and telophase.

At the beginning of prophase I the chromosomes assume a form in which they can be identified when stained with basic dyes. So much happens in prophase I, however, that it is usual to subdivide it into the five stages *leptotene, zygotene, pachytene, diplotene* and *diakinesis*. In leptotene (lepto- means slender) the chromosomes, which are present in the diploid number, appear as long thin threads (Figs. 38.4a and 38.5a) which superficially resemble those appearing at the beginning of mitosis, but with the important difference that they are single not merely at their centromeres but throughout their entire length, i.e. they are *not* longitudinally divided into chromatids. Meiosis therefore differs from mitosis right from the start. In zygotene (zygo- means paired) the two members of each homologous pair of chromosomes come together and associate side by side along their entire length, starting at one end and zipping up like a zip fastener (Figs. 38.4b and 38.5b). This process is called pairing or *synapsis*, and the double chromosome unit so formed is a *bivalent*. The number of bivalents present in the nucleus of a given cell corresponds, of course, to the haploid number of chromosomes. In pachytene (pachy- means thick) the two chromosomes of each bivalent become individually coiled so that the bivalent as a whole shortens and thickens (Fig. 38.4c). The homologous chromosomes also coil around each other. Pachytene is followed by diplotene (diplo- means double) during which the chromosomes do what they would have done during the interphase of a mitosis, if the nucleus in which they are present had not been destined to embark on meiosis. In other words the homologues of each bivalent, which in the preceding pachytene were still single, divide longitudinally except at their centromeres to give two chromatids (Fig. 38.4d and Fig. 38.5c). Each bivalent, therefore, now consists of four chromatids. During diplotene the two chromosomes of each homologous pair tend to separate from each other, but are prevented from achieving complete separation because they are held together at one or more points along their length. These points are called *chiasmata* because they present a characteristic cross-shaped appearance suggestive of the Greek letter *chi*. The significance of chiasmata will be considered after the events of meiosis have been described. In diakinesis (Fig. 38.4e) the bivalents are fully contracted and arranged around the periphery of the nucleus. At the end of this stage the nuclear envelope breaks down, the nucleolus (or nucleoli) disappears, and a nuclear spindle forms as in mitosis.

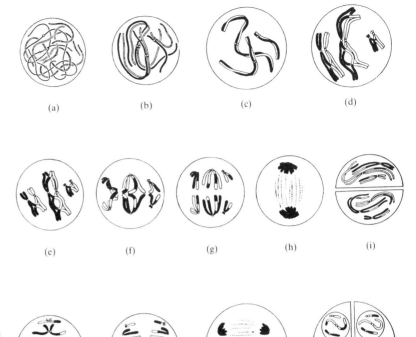

Fig. 38.4 Diagram of meiosis, illustrated by a nucleus containing three pairs of homologous chromosomes. Chromosomes of paternal origin are shown in black, those of maternal origin in white. (a)–(e) Prophase I. (a) Leptotene. (b) Zygotene. (c) Pachytene. (d) Diplotene. (e) Diakinesis. (f) Metaphase I. (g) Anaphase I. (h) Telophase I. (i) Prophase II. (j) Metaphase II. (k) Anaphase II. (l) Telophase II. (m) Four daughter cells, each with the haploid chromosome number of three. (From R. J. Berry (1965), *Teach yourself genetics*, p. 12, published by The English Universities Press Ltd, London)

Metaphase I consists in the bivalents arranging themselves on the equator of the spindle (Figs. 38.4f and 38.5d). Each bivalent has two centromeres, one paternal and one maternal in origin, which orientate themselves on either side of the metaphase plate and equidistant from it. It is a matter of chance, quite independent of the behaviour of the other bivalents, which centromere faces which pole of the spindle.

At anaphase I the two centromeres of each bivalent start to move to opposite poles of the spindle, each dragging two sister chromatids behind it (Figs. 38.4g and 38.5e, f). In contrast to mitosis, anaphase movement is not preceded by the division of the centromeres.

When the centromeres reach the opposite poles of the spindle telophase I begins, during which the spindle disperses and a nuclear envelope forms around each group of half-bivalents. At telophase I (Fig. 38.4h) it will be noticed that each chromosome (or half-bivalent) is double, not single as in mitosis, so that interkinesis, i.e. the resting stage between the first and second divisions of meiosis, starts with the chromosomes already in the state which is characteristic of the end of a mitotic interphase (Fig. 38.6a). It is therefore not surprising that interkinesis is usually short, the nucleus passing quickly into the second division of meiosis. Telophase I is, however, always followed by the division of the cytoplasm to form two cells.

The second division of meiosis is mechanically very similar to a mitosis. By the end of prophase II (Figs. 38.4i and 38.6b, c) the chromosomes have again become short and thick, although they may have only partially uncoiled

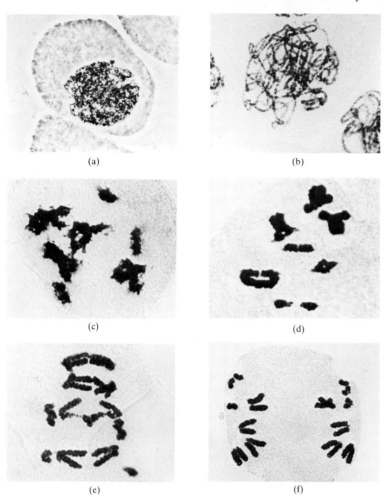

(a)

(b)

(c)

(d)

(e)

(f)

Fig. 38.5 Meiosis in the pollen mother cells of *Aloe* ($n=7$). Meiosis I. (a) Early leptotene, the chromosome threads just becoming visible. (b) Zygotene. (c) Diplotene. (d) Early metaphase I, the bivalents not yet orientated on the equator of the spindle: in this and the following photographs the first-division spindle is orientated left to right. (e) Early anaphase I, showing three smaller bivalents separating earlier than four larger ones. (f) Late anaphase I. (Crown copyright, and reproduced by permission of the Controller of Her Majesty's Stationery Office and the Director of the Royal Botanic Gardens, Kew. Microscopic preparations made by Dr P. E. Brandham)

during the preceding telophase I. In metaphase II the chromosomes become orientated on the equatorial plate, the chromatids diverging more or less widely except at the centromere (Fig. 38.4j). In anaphase II the centromeres divide and each daughter centromere takes one of the chromatids to a pole of the spindle (Figs. 38.4k and 38.6d). As in metaphase I the distribution of the chromatids is a matter of chance, except that the two from each chromosome must go to opposite poles. In telophase II each set of chromatids, or chromosomes as they may now be called, re-forms a nucleus, and the four nuclei go into interphase (Figs. 38.4l, m and 38.6e, f).

The significance of chiasmata

In describing meiosis mention was made of the fact that during the diplotene stage of prophase I the two homologous chromosomes of each bivalent, which in the preceding zygotene stage had come together in pairs, now tend to repel each other, but are held together at one to several places along their length by cross-shaped linkages called chiasmata. A chiasma is formed when, due to the strain caused by the twisting of the two members of a bivalent around each other, two homologous chromatids break at cor-

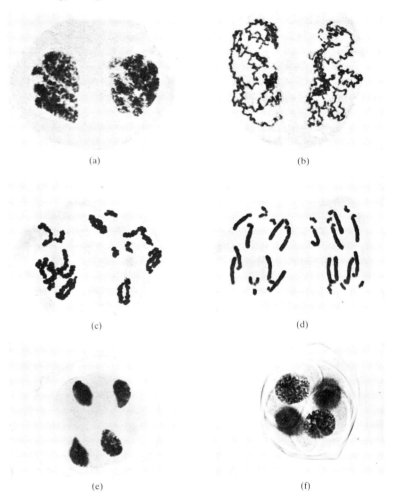

(a)   (b)

(c)   (d)

(e)   (f)

Fig. 38.6 Meiosis in the pollen mother cells of *Aloe* (*n* = 7). Meiosis II. (a) Interkinesis. (b) Prophase II. (c) Late prophase II, the haploid set of seven chromosomes being clearly visible in the right-hand group. (d) Anaphase II, the second-division spindles orientated vertically. (e) Telophase II. (f) Tetrad of pollen grains. (Crown copyright, and reproduced by permission of the Controller of Her Majesty's Stationery Office and the Director of the Royal Botanic Gardens, Kew. Microscopic preparations made by Dr P. E. Brandham)

responding points along their length, and the broken ends of one chromatid join up with the broken ends of the other. (Fig. 38.7). Each bivalent consists of two pairs of sister chromatids but, in the formation of a chiasma, breakage and reunion occur only between non-sister chromatids, that is, chromatids belonging to different parental chromosomes. Where there is more than one chiasma in a single bivalent they may involve different pairs of chromatids, but always one from each of the two homologous chromosomes. Thus at each chiasma two of the four chromatids remain unchanged, but the other two form new chromatids, each containing part of both the parental chromosomes. It follows that, even if there were only one chiasma in a bivalent, the four chromatids would all be different. Assuming the formation of only one chiasma for each bivalent, the four haploid nuclei which result from meiosis are all different in respect of every chromosome. When it is realized that there are usually several chiasmata in each bivalent, and that where they form varies in different nuclei, some inkling of the wide variety of new combinations of genes created by meiosis will be appreciated. In fact, it is most unlikely that any two gametes of any one individual are ever completely alike, or that any of them is like one or other of the gametes that fused to form the parent zygote. The formation of chiasmata must therefore be

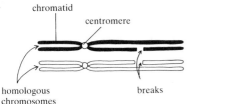

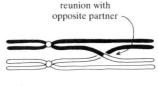

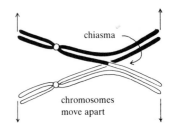

Fig. 38.7 Chiasma formation during prophase I of meiosis.

regarded not as an unfortunate breakdown in the machinery of meiosis, but rather as a 'deliberate' means of increasing the variation between characters which would otherwise be unable to recombine. This aspect of chiasma formation emphasizes that meiosis has a more fundamental function than merely halving the number of chromosomes in the gametes prior to fertilization.

Linkage groups and chromosome mapping

The phenomenon of linkage in *Drosophila* was discussed at the end of Chapter 37, but no explanation was given for it. Now that meiosis and the formation of chiasmata have been described, it is possible to give an explanation for its occurrence. In the first cross illustrating linkage, the F_1 hybrids (*Bb Vv*) were testcrossed with the double recessive flies which had black bodies and vestigial wings (*bb vv*). Seventeen per cent of the offspring were recombinants, both possible types being equally frequent (106 and 111). Since the double recessive parent can produce only one type of gamete, it follows that it is the F_1 hybrid parent which produces four different types of gametes, with an excess of the parental-type gametes (83 per cent) and a shortage of recombinant-type gametes (17 per cent). In this particular cross the F_1 partner in the testcross was a female, and therefore 17 per cent of the eggs must have been recombinant-type gametes. In the knowledge that chiasmata are a normal feature of meiosis, only two assumptions are necessary to explain this result. The first is that the genes for body colour (*B/b*) and wing-type (*V/v*) are present on the same chromosome. The second is that in 8.5 per cent of the meiotic divisions giving rise to the eggs of the F_1 females, a chiasma occured on this chromosome between the points, or loci, at which the genes for body-colour and wing-type occur. This means that 83 per cent of the resultant eggs had the same combination of these two genes as had the parents, and 17 per cent were the two recombinant types.

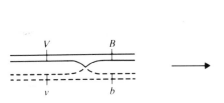

In the earlier discussion of linkage it was stated that the strength of the linkage between two genes is measured by the recombination percentage, i.e. the percentage of recombinants in the total number of offspring from a given cross. It will now be appreciated that it also indicates the frequency of chiasma formation between the loci of two linked genes. If chiasmata are as likely to form at one point as at any other point along the length of a bivalent, it follows that the further apart any two genes lie on the chromosome the greater will be the chance of a chiasma forming between them. If chiasma formation is at random, then the value of the recombination percentage is also a measure of the distance between two linked genes. On the assumption that chiasmata are formed at random along the length of a chromosome, it is possible to construct a map of each chromosome showing the relative position of the different genes resident on it. To illustrate how this is done, assume that linkage has been detected between two genes A/a and B/b, and that the recombination between them is 17.5 per cent. Linkage has also been detected between gene A/a and gene C/c, and the recombination between them has been shown to be 5.5 per cent. It is possible to construct two chromosome maps for the location of genes A/a, B/b and C/c, as follows.

$$
\begin{array}{cccc}
 & \longleftarrow5.5\longrightarrow\!\!\longleftarrow\!\!-17.5\longrightarrow & & \longleftarrow\!-\!17.5\longrightarrow \\
 & & & \longleftarrow5.5\longrightarrow \\
\text{Possibility} & \overline{\qquad\qquad\qquad\qquad} & \text{Possibility} & \overline{\qquad\qquad\qquad\qquad} \\
1 & C \qquad A \qquad\qquad B & 2 & A \qquad C \qquad\qquad B
\end{array}
$$

To decide which of the two possibilities is correct, it is necessary to determine the recombination percentage between genes B/b and C/c which must be linked because they are both linked to gene A/a. The value obtained from making the appropriate cross (e.g. $BC/bc \times bc/bc$) must approximate to either 23 per cent $(17.5 + 5.5)$ or 12 per cent $(17.5 - 5.5)$. If it is found that the recombination between these two genes is, say, 22.5 per cent, then possibility 1 is the correct map. In drawing chromosome maps from recombination data, a recombination of 1 is taken as the unit for expressing linkage relationships.

The recombination percentages of some hundreds of genes have been tested in *Drosophila* among animals, and in maize among plants. A recombination of 50 per cent, it will be recalled, indicates a complete absence of linkage, and a value of 0 indicates complete linkage. In *Drosophila*, which has four pairs of homologous chromosomes, the hundreds of genes which have been tested fall into four linkage groups. Genes from different groups segregate independently and those within a group maintain a constant linear order. These four linkage groups are the genetical equivalents of the four pairs of homologous chromosomes. The conclusions reached for *Drosophila* can be applied to other organisms and it is always found that the number of linkage groups is equal to the number of pairs of homologous chromosomes.

The chromosomal theory of heredity

Having described separately the basic rules of genetics and the behaviour of the chromosomes during mitosis and meiosis, it now remains to correlate the two sets of information. If the chromosomes, i.e. the structures in which the nuclear material is organized, are to be regarded as the carriers of heredity, then it is necessary to show that they behave in a way which parallels the observed facts of heredity.

Mendel's conclusions were deduced from observed results, but the current state of knowledge about cell structure was insufficient for him to identify his 'factors' with any structural units in the cell. When the details of mitosis were discovered it was noticed that each species had a definite number of chromosomes, constant in shape and relative size, and that there were two chromosomes of each kind. The question then arose as to why there was not a progressive doubling of the chromosome number at each fertilization. The answer was provided when it was shown that the eggs and sperm of the horse threadworm (*Ascaris*), which has four chromosomes in its somatic cells, contain only two chromosomes, or half the number which is characteristic of the somatic cells. This observation was quickly seen to be one example of a general property of the chromosomes of diploid organisms, namely that there are twice as many in the somatic cells as there are in the gametes. Attention was then focused on the nuclear changes leading up to the formation of gametes, and it was found that, during two special divisions which jointly constitute the process of meiosis, the chromosomes come together in pairs, become linked by cross-shaped bridges, and their number is halved. It then only remained to demonstrate a clear parallelism between the rules of Mendelian inheritance and the behaviour of the chromosomes during mitosis and meiosis. These parallelisms include the following.

1. Each chromosome, and hence each of its parts, is represented twice in the zygote but only once in the gamete, just as is the Mendelian factor or gene. As will be explained later in this chapter, this generalization does not apply to the sex chromosomes of dioecious organisms.

2. The members of each pair of homologous chromosomes separate at meiosis just as each pair of Mendelian factors or alleles segregate from each other.

3. The members of each pair of homologous chromosomes segregate independently of one another, just as do the two alleles of each separate gene in a Mendelian dihybrid cross. Direct evidence for the independent segregation of chromosomes, and therefore of genes, is shown, for instance, by the orthopteran *Circotettix* in crosses between a race with an unequal (heteromorphic) pair of homologous chromosomes and a race with an equal pair. The unlike chromosomes segregate relative to the other chromosomes exactly like an independent Mendelian factor.

4. One of the modifications that had to be made to Mendel's original rules is that genes do not always segregate independently of one another. On the chromosome theory of heredity the occurrence of non-independent as well as independent segregation is to be expected, because different genes which are located on the same chromosome will inevitably be transmitted together, i.e. linked in their inheritance, unless they become separated as a result of chiasma formation. In a given organism the number of linkage groups that can be identified by breeding experiments is invariably equal to or less than the total number of homologous pairs of chromosomes of that organism. In maize, for example, several hundred characters can be arranged into ten linkage groups, which correspond in number and size with ten pairs of homologous chromosomes.

Every feature of gene transmission has, therefore, a physical counterpart in chromosomal behaviour. The reason is, of course, that the same hereditary materials are being observed in two different ways, genetically in the breeding experiments and cytologically under the microscope. The linkage map worked out from a long series of breeding experiments is the genetical counterpart of the chromosome.

To summarize, therefore, the strict parallelism between Mendel's conclusions and chromosome behaviour in meiosis, plus the evidence from linkage groups, would seem to provide irrefutable evidence for the view that the chromosomes are the bearers of hereditary characters.

Sex-linked inheritance

Although there is no longer any doubt that the chromosomes are the bearers of hereditary characters, there are certain traits whose inheritance does not appear to fit into any of the patterns of Mendelian inheritance described in the preceding chapter. It is found, however, that whenever the genes behave unusually, the behaviour of the chromosomes is also found to be correspondingly modified. In the circumstances, therefore, the apparent exceptions to the chromosomal theory of heredity provide even more cogent evidence for its validity. One of these unusual patterns of inheritance is *sex linkage*, which was in fact the first type of linkage to be discovered.

Sex determination

One of the general properties of the chromosome complement of a diploid organism is that it consists of pairs of chromosomes, the members of any one pair resembling each other in size, shape, and position of the centromere. This is strictly true for monoecious (hermaphrodite) organisms, but is subject to an important qualification in dioecious (unisexual) organisms. In such organisms the mating of a male and a female produces males and females in approximately equal numbers, and this 1 : 1 ratio is analogous to a cross between a heterozygote and a homozygous recessive (i.e. the monohybrid testcross). This fact would suggest that sex is a hereditary character which can be represented by genes lying on particular chromosomes. The analogy is more than superficial because it has been shown for many dioecious organisms that the two sexes differ in respect of one pair of chromosomes, the so-called *sex chromosomes*. The other chromosomes, which are of the same type in both sexes, are called *autosomes*. In most organisms where sex chromosomes occur, there is a homologous pair in the nuclei of one sex (the homogametic sex) and a dissimilar pair in the nuclei of the other sex (the heterogametic sex). The sex chromosomes which occur as a pair in the homogametic sex are called X chromosomes and, but for the existence of the heterogametic sex, could not be distinguished by their appearance from the autosomal chromosomes. The heterogametic sex, which is usually the male, has an X-chromosome and a so-called Y-chromosome which is usually much smaller than the X-chromosome. The homogametic (XX) sex produces gametes of only one type, all having one X-chromosome in addition to one member of each homologous pair of autosomes. The heterogametic (XY) sex produces equal numbers of two different types of gametes, one with an X-chromosome and one with a Y-chromosome. If the male is the heterogametic sex, an egg fertilized by an X-bearing male gamete develops into a female, and an egg fertilized by a Y-bearing male gamete develops into a male.

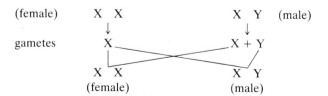

Equal numbers of males and females are therefore produced in the offspring of an organism with an X–Y method of sex determination.

Sex linkage

The transmission of the genes which are present on the sex chromosomes naturally follows the same Mendelian rules as those governing the transmission of autosomal genes. The results of crosses involving sex-linked genes can be predicted from a knowledge of these rules and of the mechanism of sex determination. The classical example of sex-linkage is concerned with the inheritance of eye colour in *Drosophila*, in which the male is the heterogametic sex. The allele *w* for 'white eye' is recessive to the wild-type allele *W* which is responsible for the normal 'red eye'. A red-eyed female crossed with a white-eyed male gives all red-eyed offspring, both male and female, in the F$_1$. In the F$_2$ the females are again all red-eyed, but the males are red and white in equal numbers. However, the reciprocal cross (i.e. the cross in which the characters of the two parents are transposed), white-eyed female by red-eyed male, gives different results. The F$_1$ females are red-eyed as before, but the males are now all white-eyed. In the F$_2$ both males and females show red and white eyes in equal numbers (Fig. 38.8).

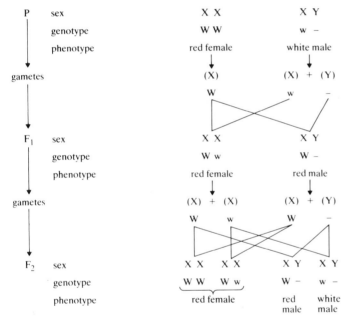

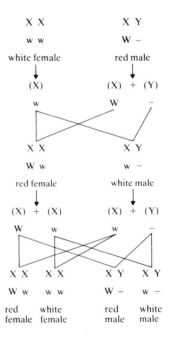

Fig. 38.8 Sex-linked inheritance of eye-colour in *Drosophila*, where the male is the heterogametic sex.

This apparently anomalous pattern of inheritance, in which the phenotypes of the offspring may depend on their sex and the proportions in which they are produced may vary according to which way round the cross is made, is seen to make sense when interpreted in terms of the sex chromosomes. From the combined results of the two reciprocal crosses, it is evident that the gene for eye colour is borne on the X-chromosome, because the recessive allele for white eye does not show in the heterozygous females. The Y-chromosome does not carry the gene so that a male has red or white eyes according to which allele is present on the X-chromosome which he received from his mother. It is also clear from these results that the male must be the heterogametic sex.

The phenomenon of sex linkage must not be confused with sex limitation. A gene, and the character determined by it, is said to be sex limited when the gene has its effect in only one sex. The gene may be present on any chromosome but, because of the influence of sex hormones or other sex-dependent factors, it is expressed in only one sex. The genes for secondary sexual characteristics, such as milk production in mammals, are sex limited. On the other hand, sex-linked genes are manifested in both sexes, but the homogametic sex, unlike the heterogametic sex, will only manifest a recessive gene if it has received it from both parents. Recessive genes will therefore be expressed more commonly in the heterogametic sex, where there is no partner to the X-chromosome, but this association of recessive sex-linked characters with the heterogametic sex is quite different from sex limitation.

Summary of the principles of Mendelian genetics

Autosomal inheritance

1. *Segregation of one character (monohybrid ratio)*

    P   *AA* × *aa*
    F_1 *Aa*
                 Mated among themselves
    F_2 3 *A*– : 1 *aa* (the 2 parental phenotypes)

Testcrosses (i.e. × *aa*)
AA – – – – 1 type; 100% *Aa*
Aa – – – – 2 types; 50% *Aa*, 50% *aa*

2. *Independent segregation of two non-linked characters (dihybrid ratio)*

    P   *AABB* × *aabb*
                    or
               AAbb × *aaBB*
    F_1 *AaBb*
                  Mated among themselves
    F_2 9 *A*–*B*– : 3 *A*–*bb* : 3 *aaB*– : 1 *aabb* (the two parental phenotypes plus the two recombinant types)

Testcrosses (i.e. × *aabb*)
AABB – – – – 1 type; 100% *AaBb*
AaBB – – – – 2 types; 50% *AaBb*, 50% *aaBb*
AABb – – – – 2 types; 50% *AaBb*, 50% *Aabb*
AaBb – – – – 4 types; 25% *AaBb*, 25% *Aabb*, 25% *aaBb*, 25% *aabb*

3. *Segregation of two linked characters exhibiting complete linkage (test-crosses)*

(a) *In coupling* (i.e. the two dominants are introduced from one parent)

    P   (*AB*) (*AB*) × (*ab*) (*ab*)
    F_1 (*AB*)(*ab*)
                  Mated with the double recessive (*ab*)(*ab*)
    F_2 50% (*AB*)(*ab*), 50% (*ab*)(*ab*) (the two parental phenotypes – cf. testcross for *one* character)
                  N.B. The dominants are 'coupled' together in the F_2 as they were in the original parents.

(b) *In repulsion* (i.e. one dominant is introduced from each parent)

P $(Ab)(Ab) \times (aB)(aB)$
F$_1$ (Ab) (aB)
         Mated with the double recessive $(ab)(ab)$
F$_2$ 50% $(Ab)(ab)$, 50% $(aB)(ab)$ (the two parental phenotypes)
         N.B. The dominants are 'repelled' (or separated) in the F$_2$ and
             thus restore the condition found in the original parents.

4. *Segregation of two linked characters exhibiting partial linkage (testcrosses)*

(a) *In coupling*

P $(AB)(AB) \times (ab)(ab)$
F$_1$ $(AB)(ab)$
         Mated with the double recessive $(ab)(ab)$
F$_2$ $50-y$% $(AB)(ab)$, $50-y$% $(ab)(ab)$, y% $(Ab)(ab)$, y% (aB)
         (ab) where y is less than 25% (the two parental phenotypes are
         in excess of the two recombinant types)

(b) *In repulsion*

P $(Ab)(Ab) \times (aB)(aB)$
F$_1$ $(Ab)(aB)$
         Mated with the double recessive $(ab)(ab)$
F$_2$ $50-y$% $(Ab)(ab)$, $50-y$% $(aB)(ab)$, y% $(AB)(ab)$, y% (ab)
         (ab) (the parental phenotypes are again in excess of the recom-
         binant types)
         N.B. (i) The ratio of the four phenotypes in the F$_2$ of these
                    crosses depends not only on the alleles but also on
                    the way they are combined in the parents.
             (ii) In these crosses $2y$ per cent is called the *recombina-
                    tion percentage* and measures the strength of the link-
                    age between two genes (i.e. the frequency of chiasma
                    formation between the two gene loci). It has a char-
                    acteristic value for any given pair of gene loci, no
                    matter whether the cross is made 'in coupling' or 'in
                    repulsion'.

Sex-linked inheritance

In sex-linked inheritance, unlike autosomal inheritance, reciprocal crosses
may give different results and, among the offspring of any given cross, one
particular phenotype may be confined to one sex. In all cases of sex-linked
inheritance, therefore, the sex as well as the phenotype of the individuals
must be specified.

   In the following examples, it is assumed that the male is the heterogametic sex and that the genes under consideration are absent from the Y-
chromosome.

5. *Segregation of one character linked to the X-chromosome*

Case 1. (Dominant character introduced from the male parent)

	Female	*Male*
P	*aa*	*AO*
F$_1$	100% *Aa*	100% *aO*
	Brother-sister mating	
F$_2$	50% *Aa*, 50% *aa*	50% *AO*, 50% *aO*

Case 2. (Dominant character introduced from female parent)

	Female	*Male*
P	*AA*	*aO*
F$_1$	100% *Aa*	100% *AO*
	Brother-sister mating	
F$_2$	100% *A–*	50% *AO*, 50% *aO*

6. *Segregation of two X-linked characters exhibiting partial linkage (testcrosses)*

(a) *In coupling*

Case 1. (Both dominant characters introduced from the male parent)

	Female	*Male*
P	*(ab)(ab)*	*(AB)O*
F$_1$	100% *(AB)(ab)*	100% *(ab)O*

Mated with the double recessive of the opposite sex

	(ab)O	*(ab)(ab)*
F$_2$	*Females*	*Females*
	50–*y*% *(AB)(ab)*	100% *(ab)(ab)*
	50–*y*% *(ab)(ab)*	
	y% *(Ab)(ab)*	
	y% *(aB)(ab)*	*Males*
	Males	100% *(ab)O*
	50–*y*% *(AB)O*	
	50–*y*% *(ab)O*	
	y% *(Ab)O*	
	y% *(aB)O*	

Case 2. (Both dominant characters introduced from the female parent)

	Female	*Male*
P	*(AB)(AB)*	*(ab)O*
F$_1$	100% *(AB)(ab)*	100% *(AB)O*

Mated with the double recessive of the opposite sex

	(ab)O	*(ab)(ab)*

	Females	*Males*	*Females*	*Males*
F$_2$	Same as Case 1		100% *(AB)(ab)*	100% *(ab)O*

N.B. No recombinants appear in the F$_2$ when an F$_1$ male is testcrossed because only one allele of each gene under consideration is present in the male.

(b) *In repulsion.* The two possible cases for this cross are similar to the above and can easily be worked out.

Some problems in genetics

Solving problems is one of the best ways of ensuring that you understand the concepts of genetics which have been outlined in this chapter and the previous one. Five problems are given below to test your powers of logical deduction. The answers to these problems are given at the end of the chapter.

1. In the flowers of the four-o'clock plant (*Mirabilis jalapa*) the allele for red flower colour, R, is incompletely dominant to an allele for white flower colour, r, the heterozygous plants being pink-flowered.

   If a red-flowered four-o'clock is crossed with a white-flowered one, what will be the flower colour of the F_1? Of the F_2? Of the offspring of a cross of the F_1 with its white parent?

2. In squash (*Cucurbita pepo*) white fruit W is dominant over yellow fruit w, and disc-shaped fruit D is dominant over spherical fruit d. The offspring of a cross between white disc and white spherical were $\frac{3}{8}$ white disc; $\frac{3}{8}$ white spherical; $\frac{1}{8}$ yellow disc; and $\frac{1}{8}$ yellow spherical. Give an explanation of this result.

3. When two true-breeding strains of maize (*Zea mays*) both having grains with a white aleurone layer were crossed, the F_1 progeny all had grains with a purple aleurone layer. When the F_1 offspring were crossed with each other the F_2 progeny consisted of 320 plants, of which 183 had purple aleurone and 137 white aleurone. Suggest a possible explanation for these results.

4. A maize plant A which was tall, pigmented (i.e. contained anthocyanin) and had ragged leaves was crossed with a triple homozygous recessive plant B which was dwarf, non-pigmented and had normal leaves. The three genes all belong to the same linkage group. The following progeny were obtained:

Tall, pigmented, ragged	146
Tall, pigmented, normal	696
Tall, non-pigmented, ragged	3
Tall, non-pigmented, normal	191
Dwarf, pigmented, ragged	221
Dwarf, pigmented, normal	5
Dwarf, non-pigmented, ragged	612
Dwarf, non-pigmented, normal	126

From the progeny, determine as far as you can the genetical constitution of the parent plant A, and the relationship of the three genes to each other. Explain your reasoning fully.

5. From two pure-breeding stocks of *Drosophila melanogaster* a female having red eyes and a yellow body was crossed with a male having white eyes and a grey body, and the F_1 all had red eyes, but the females were grey and the males were yellow. Brother–sister matings of these hybrids gave offspring as follows:

Red-eyed yellow females	76	Red-eyed yellow males	73
Red-eyed grey females	69	White-eyed grey males	68

Given that in *D. melanogaster* the male is the heterogametic sex, what can you infer about the inheritance of these characters?

Answers to problems in genetics

1. The F_1 will be pink (i.e. Rr) because all the gametes of the red-flowered parent are R and all the gametes of the white-flowered parent are r. Fer-

tilization can therefore only yield Rr (i.e. pink) plants.

The F_1 parents, both being Rr, produce ovules and pollen of R and r constitution in equal numbers. With random fertilization the F_2 progeny will have a genotypic ratio of $\frac{1}{4}RR : \frac{1}{4}Rr : \frac{1}{4}rR : \frac{1}{4}rr$, i.e. $\frac{1}{4}RR$, $\frac{1}{2}Rr$, $\frac{1}{4}rr$. Since R is incompletely dominant to r, the phenotypic ratio in this case will be the same, i.e. $\frac{1}{4}$ red, $\frac{1}{2}$ pink and $\frac{1}{4}$ white.

The F_1 parent is heterozygous (i.e. Rr) and the white parent is rr. Therefore the progeny will be 50 per cent pink and 50 per cent white, as follows:

	gametes of white parent $1r$
gametes of F_1 parent	
$\frac{1}{2}R$	$\frac{1}{2}Rr$
$\frac{1}{2}r$	$\frac{1}{2}rr$

2. Since the question involves two characters and four phenotypes, it is reasonable to assume that the problem is an example of dihybrid inheritance. Both of the parents, because they are white, must carry at least one W allele. The white disc parent must also carry at least one D allele because it has disc-shaped fruit. The white spherical parent must also be dd because it shows the fruit shape determined by the recessive allele d.

Because the offspring of this cross include plants with spherical fruit (dd), the white disc parent must be heterozygous for fruit shape, i.e. Dd. Similarly, because there are yellow fruit in the progeny, and this can result only from the fertilization of gametes of w constitution from both parents, it follows that both parents must be heterozygous, i.e. Ww, for the gene controlling fruit colour. The genotypes of the parents are therefore $Ww\ Dd$ (white disc) and $Ww\ dd$ (white spherical), and the cross can be represented as follows:

		gametes of white spherical parent	
		$\frac{1}{2}\ Wd$	$\frac{1}{2}\ wd$
	$\frac{1}{4}\ WD$	$\frac{1}{8}\ WW\ Dd$	$\frac{1}{8}\ Ww\ Dd$
gametes of	$\frac{1}{4}\ Wd$	$\frac{1}{8}\ WW\ dd$	$\frac{1}{8}\ Ww\ dd$
white disc	$\frac{1}{4}\ wD$	$\frac{1}{8}\ wW\ Dd$	$\frac{1}{8}\ ww\ Dd$
parent	$\frac{1}{4}\ wd$	$\frac{1}{8}\ wW\ dd$	$\frac{1}{8}\ ww\ dd$

The progeny of this cross will therefore occur in the following genotypic and phenotypic ratios:

Genotypic ratio	Phenotypic ratio
$\frac{1}{8}\ WW\ Dd$	
$\frac{1}{8}\ Ww\ Dd$ $\Big\}$	$\frac{3}{8}$ white disc
$\frac{1}{8}\ wW\ Dd$	
$\frac{1}{8}\ WW\ dd$	
$\frac{1}{8}\ Ww\ dd$ $\Big\}$	$\frac{3}{8}$ white spherical
$\frac{1}{8}\ wW\ dd$	
$\frac{1}{8}\ ww\ Dd$	$\frac{1}{8}$ yellow disc
$\frac{1}{8}\ ww\ dd$	$\frac{1}{8}$ yellow spherical

3. Colour of the aleurone layer is the only character under consideration. This might suggest that only one gene is involved, but the ratio of

183 : 137 in the F_2 cannot be equated with an expected 3 : 1 ratio for a monohybrid cross. Consider, therefore, the possibility that two genes are involved in producing aleurone colour. On this basis it is necessary to postulate that purple colour must depend on the presence of a dominant allele from each of the two genes, since the parents, which are both true breeding, have white aleurone. On this assumption the parents can be represented as $AA\ bb$ (white) and aa BB (white), giving an F_1 of $Aa\ Bb$ (purple). Crossing two such F_1 plants would yield an F_2 as follows:

9 $A-\ B-$:   3 $A-\ bb$: 3 $aa\ B-$: 1$aa\ bb$
9 purple   :   7 white

When applied to a total of 320 plants, this result would mean an ideal ratio of 180 purple : 140 white. The observed F_2 figures correspond so closely to this predicted result that it can be assumed that the hypothesis on which the prediction was made is correct.

4. (a) Since the question states that plant B is a triple homozygous recessive, tall (T) is dominant to dwarf (t), pigmented (P) is dominant to non-pigmented (p), and ragged (R) is dominant to normal (r). Given that the three genes all belong to the same linkage group, the two most abundant phenotypic classes (i.e. tall, pigmented, normal (696) and dwarf, non-pigmented, ragged (612)) indicate that plant A has the constitution $(TPr)(tpR)$. The fact that there are eight phenotypes must mean that crossing over has taken place between all three genes, and that the very infrequent classes presumably represent double recombinants.

(b) Gene order. Because the progeny includes double recombinants it is possible to deduce the order of the genes on the chromosome. The data for the progeny are tabulated in the order $T/t\ P/p\ R/r$. This order must be incorrect because parent A would give double recombinants of constitution $(Tpr)(tpr)$ and $(tPR)(tpr)$, whereas in fact the double recombinants are $(TpR)(tpr)$ and $(tPr)(tpr)$. Of the two other possible orders for these three genes, i.e. $(TrP)(trp)$ and $(PTr)(ptr)$, only the last will give double recombinants of constitution $(pTR)(ptr)$ and $(Ptr)(ptr)$. The order of the genes must therefore be $P/p\ T/t\ R/r$. Hence the phenotypes (with their characters now arranged in the correct order of the genes) pigmented, dwarf, ragged (221) and non-pigmented, tall, normal (191) are the result of crossing-over between the P/p and T/t genes, and similarly the phenotypes pigmented, tall, ragged (146) and non-pigmented, dwarf, normal (126) are the result of crossing-over between the T/t and R/r genes.

(c) Recombination percentages and chromosome map. Because the heterozygous parent plant A has the constitution $(PTr)(ptR)$ the percentage of the progeny in which pigmentation and dwarfness, and non-pigmentation and tallness, are combined gives the recombination percentage between the P/p and T/t genes. In the cross under consideration this is $\frac{221 + 191 + 3 + 5}{2000} \times 100 = 21.0$ per cent. The recombination between the T/t and R/r genes is given by $\frac{146 + 126 + 3 + 5}{2000} \times 100 = 14.0$ per cent. Similarly, the recombination percentage between the P/p and R/r genes is given by $\frac{221 + 191 + 146 + 126}{2000} \times 100 = 34.2$ per cent. This value is slightly less than the sum of the two previous recombination percentages because of the occurrence of double recombinants. If the cross had not yielded double recombinants, calculation of the recombination percentages between the three genes would have been the only method available for determining the gene order. Here, the values merely confirm that the gene order deduced from the phenotypes of the double recombinants is correct. On the basis of values for the three recombination percentages, a chromosome map (or linkage map) can be drawn as follows:

P/p T/t R/r

       ⟵ 21.0 ⟶ ⟵ 14.0 ⟶
       ⟵   34.2   ⟶

5. (a) The unequal distribution of the different phenotypes between the males and females of both the F_1 and F_2 generations suggests that sex linkage is involved. Because the four phenotypes in the F_2 are equally frequent, the possibility that one of the genes is autosomal and the other sex linked can be ruled out, i.e. it is safe to assume that both the genes are sex linked. The occurrence of only two phenotypes in each sex of the F_2 implies that the linkage between the two genes is complete (N.B. partial linkage would have given four phenotypes in the F_2 male, the heterogametic sex).

(b) Because all the F_1 progeny have red eyes, red-eye colour must be dominant to white-eye colour. In the knowledge that in *Drosophila melanogaster* the male is the heterogametic sex, it is also possible to infer (because the F_1 females are all grey) that grey body is dominant to yellow body.

Let A be the dominant allele for red eyes, and B the dominant allele for grey body. If the conclusions drawn above are correct, then the inheritance of the characters may be shown as follows:

		Females		*Males*
P		$(Ab)(Ab)$		$(aB)\,Y$
		red yellow	×	white grey
F_1		100% $(Ab)(aB)$		100% $(Ab)\,Y$
		red grey	×	red yellow
F_2		50% $(Ab)(Ab)$		50% $(Ab)\,Y$
		red yellow		red yellow
		+		+
		50% $(aB)(Ab)$		50% $(aB)\,Y$
		red grey		white grey

Chapter 39

Molecular aspects of genetics

The two previous chapters have shown that the hereditary characters of an organism are determined by the action of 'factors' called genes, and that genes can be considered as material particles located along the length of the chromosomes in the nucleus. However, there is a big gap between the genes on the one hand, and characters such as the shape of leaves or the colour of fruits on the other. To bridge this gap two fundamental questions need to be answered: (1) what is the nature of the genetic material? and (2) how do genes act to produce the characters whose expression they determine? A discussion of these two questions forms the subject matter of this chapter.

Nature of the gene

Chromosomes are composed of proteins (mostly histones) and deoxyribonucleic acid, DNA. Both proteins and DNA are linear molecules made up of a large number of building blocks joined end to end. This chain-like structure resembles, on a smaller scale, the linkage of genes in a linear sequence along the length of a chromosome, and so both proteins and DNA have suitable molecules to be the genetic material. Until the mid-1940s the few biologists who tried to think of the gene in chemical terms thought that proteins were more likely to be the genetic material than DNA, as proteins consist of twenty different kinds of amino-acid building blocks whereas DNA consists of only four kinds of nucleotide. It was argued that, if gene action were to be explained in terms of a 'gene language', a twenty-letter alphabet would be more likely than a four-letter one. However, DNA has been shown to be the carrier of genetic information.

The first convincing evidence that DNA is the genetic material was published in 1944 by O. T. Avery and his co-workers. They showed that the active principle involved in a phenomenon called *bacterial transformation*, which had been discovered 16 years earlier but remained unexplained, was DNA. The ability of a certain pneumonia-causing bacterium, *Diplococcus pneumoniae*, to cause infection depends on the presence of a capsule of polysaccharide material surrounding the bacterial cells. When grown on agar medium the colonies of such capsulated bacteria have a smooth surface, and the bacteria cause pneumonia when inoculated into mice. There is a genetic variant of this bacterium, the cells of which lack a polysaccharide capsule, whose agar colonies have a rough surface, and which are non-pathogenic. The smooth and rough variants are labelled S and R respectively. When mice are inoculated with either living R or heat-killed S diplococci they suffer no ill-effects, but if they are inoculated with a mixture of the two they develop

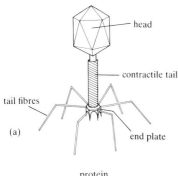

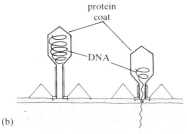

Fig. 39.1 Viral infection of bacteria. (a) Diagrammatic representation of the bacteriophage which attacks *Escherichia coli*. (b) The attachment of the bacteriophage to the surface of the bacterium (left) and the subsequent injection of its DNA into the body of the bacterium (right).

pneumonia. Living S bacterial cultures can be obtained from such infected mice, proving that some substance derived from the dead S cells must have somehow 'transformed' the living R into living S diplococci. This transformation constitutes a hereditary change because transformed cells can be propagated indefinitely as the S variant. Avery and his co-workers tested fractions of heat-killed S cells for their ability to cause this transformation, and found that only fractions containing DNA were effective. This result clearly implicated DNA as the genetic material.

Further evidence in support of this conclusion was provided in 1952 by Hershey and Chase, who used bacteriophages which attack *Escherichia coli*, a common bacterium inhabiting the human colon. A bacteriophage (literally, bacterium eater) is a virus which attacks a bacterium. Although bacteriophages vary in structure, they consist basically of an outer protein shell and an inner core of either DNA or RNA. The bacteriophage of *E. coli* is a knobkerrie-shaped structure consisting of a hexagonal head which contains DNA, a contractile cylindrical tail, and an end-plate to which six tail fibres are attached (Fig. 39.1). To attack a cell of *E. coli* a bacteriophage particle attaches itself by its end-plate to the surface of the host cell. About 20 minutes later the bacterial cell bursts and releases several hundred new viral particles which can repeat the infection process. To reach their conclusion that DNA serves as the physical basis of heredity, Hershey and Chase made use of the fact that the protein shell of the bacteriophage contains sulphur but no phosphorus, whereas the DNA in the head contains phosphorus but no sulphur. They labelled the protein of one sample of bacteriophage with ^{35}S and the DNA of another sample with ^{32}P, and demonstrated that the protein of the shell remains outside the bacterial cell but the DNA enters it through a hole made in the bacterial wall at the point of attachment. During the interval between infection and the bursting of the bacterial cell, the bacterial DNA is broken down into its component building blocks which are then resynthesized, along with other materials from the surrounding medium, into new bacteriophage particles under the direction of the invading bacteriophage DNA. When a new generation of bacteriophages has been formed inside the bacterium, the wall of the latter bursts and several hundred new infective bacteriophage particles are released. Since it is DNA and not protein which passes into the body of the bacterium, it follows that DNA must contain the code for the production of the new bacteriophage particles.

Structure of DNA

In Chapter 15 it was shown that DNA is a long chain-like molecule, the backbone of which consists of alternating sugar and phosphate residues. Attached to the sugar links in the backbone are two kinds of purines, adenine (A) and guanine (G), and two kinds of pyrimidines, cytosine (C) and thymine (T). A single purine or pyrimidine is attached to each sugar residue. The DNA molecules of all organisms conform to this general plan, but DNA extracted from different species of animals and plants has different proportions of the four possible bases. The sum of the purine bases (i.e. A + G) is always equal, within the limits of experimental error, to the sum of the pyrimidine bases (i.e. C + T) and, for any one species, the amount of A is equal to the amount of T and similarly the amount of C is equal to the amount of G. However, although the relations A = T and C = G are always valid, it is also true that A + T does not usually equal C + G (Table 39.1).

Table 39.1 Comparison of the nucleotide composition of DNA

	Per cent				$\dfrac{A}{T}$	$\dfrac{G}{C}$	$\dfrac{A+T}{G+C}$
	A	T	G	C			
Mycobacterium tuberculosis	15.1	14.6	34.9	35.4	1.03	0.98	0.42
Escherichia coli	26.1	23.9	24.9	25.1	1.09	0.99	1.00
Phage T2	32.6	32.6	18.2	16.6	1.00	1.09	1.87

The ratio of the number of A and T bases (A + T) to the number of G and C bases (G + C) is called the base ratio and varies widely between different species, ranging from under 1 to over 3.

On the basis of existing knowledge about DNA (particularly the results of X-ray crystallographic work by Wilkins) Watson and Crick proposed in 1953 their now famous model of the three-dimensional structure of DNA. They suggested that the DNA molecule, isolated from the cell, consists not of one linear chain of nucleotides (the building blocks of nucleic acids) but of two parallel chains which are linked together by hydrogen bonds between the amino (electropositive) and carbonyl (electronegative) groups of the bases on opposite chains. The bases are linked in such a way that a purine base always links with a pyrimidine base. The nucleotides in the two chains have to correspond if they are to fit together. The molecules of the purines A and G are similar in shape and differ only in certain small side groups, and those of the pyrimidines T and C are also similar to each other except for small differences in their side groups. The rule is that a nucleotide containing A can fit only with one containing T, and similarly a G nucleotide can fit only with one containing C. These specific pairs of bases are said to be *complementary* because their molecular size and structure are such that an exact fit can be formed only between the two members of a complementary pair. There are thus four possible base pairs between the two helical backbones in DNA, namely A–T, T–A, G–C, and C–G. If the order of nucleotides in one of the backbones is, for instance, T-G-C-A-T-T-C, this automatically fixes the order in the other backbone, which must be A-C-G-T-A-A-G. The two associated complementary strands then coil around each other into a helix due to the formation of hydrogen bonds, as shown in Fig. 39.2. There are two hydrogen bonds between A and T, and three between C and G. The Watson–Crick model of DNA can be compared to a ladder whose rungs are the complementary base pairs, and whose side supports are the sugar–phosphate backbones of the two molecular strands. The ladder is, however, a twisted one so that, when placed vertically, it resembles a spiral staircase because it is a double helix.

The Watson–Crick model fulfils the basic requirements of the genetic material by offering an explanation of how genes replicate their precise structure, and how they can carry genetic information in a chemical code. The model suggests a possible mechanism of precise replication of genes since the bases always pair in the same way, i.e. only complementary bases can pair. When a double helix replicates, it is postulated that the complementary strands unwind as a result of breakage of the hydrogen bonds that were holding them together, and each strand then makes a complementary copy of itself by functioning as a templet for free nucleotides in the surrounding medium (Fig. 39.3). Because of the specificity of base pairing, the new double-stranded molecules are exact duplicates of the original. Each new double

(a) DNA is constructed in the shape of a ladder

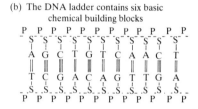

(b) The DNA ladder contains six basic chemical building blocks

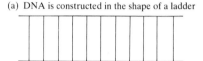

(c) The DNA ladder is coiled into a double helix

Fig. 39.2 The Watson–Crick model of DNA. P = phosphate; S = sugar (deoxyribose); A = adenine; T = thymine; G = guanine; C = cytosine.

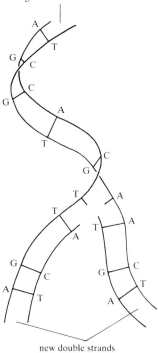

original double strand

new double strands

Fig. 39.3 Replication of DNA.

strand of DNA thus consists half of new and half of old DNA, i.e. the replication is called *semi-conservative*.

Impressive support for this model of DNA replication was provided by a classic experiment carried out in 1958 by Meselson and Stahl (Fig. 39.4). The bacterium *Escherichia coli* was cultured in a medium containing the heavy isotope of nitrogen ^{15}N as the sole nitrogen source until the DNA of all the cells was labelled with heavy nitrogen. Cells were then transferred to a medium all of whose nitrogen was the normal isotope ^{14}N. After one bacterial generation (i.e when the bacterial population had doubled), a sample of cells was removed and the DNA in them extracted.

DNA containing the heavy isotope ^{15}N (heavy DNA) can be separated from DNA containing the normal isotope ^{14}N (light DNA) by a technique known as *density gradient centrifugation*. In this technique, particles of different densities are suspended in a concentrated solution of the salt of a highly soluble heavy metal (e.g. cesium chloride), and subjected to centrifugation in an ultracentrifuge. The centrifugal field produces a density gradient in the tube of both the salt and the suspended particles. After several hours at high forces of gravity, the gradient reaches a point at which the densities of the solution and the suspended material match at a level in the tube where all the suspended particles of like density collect in a narrow band. With this technique it is possible to separate particles which are similar in size but differ in density by extremely small amounts.

When the DNA extracted from cells removed from the medium after one bacterial generation was centrifuged in the cesium chloride solution, all the DNA banded in a position exactly intermediate between that previously established for completely labelled DNA (^{15}N only) and completely unlabelled DNA (^{14}N only). This result is consistent with the hypothesis that the DNA which had replicated once in the ^{14}N medium should be 'hybrid', i.e. consist of one strand whose nitrogen is all ^{15}N ('old') and one in which the nitrogen should all be ^{14}N ('new'). After two bacterial generations the DNA extracted showed two bands in the ultracentrifuge tube, one of hybrid or intermediate density and one corresponding to the density of ^{14}N-containing DNA (light DNA). In subsequent generations, the amount of DNA in the band corresponding to light DNA increased exponentially, while the amount of hybrid DNA remained essentially constant. These results of Meselson and Stahl's experiment are thus entirely consistant with the semi-conservative mechanism of DNA replication required by the Watson–Crick model of DNA.

The experiment of Meselson and Stahl has been essentially repeated on the chromosomes of bean (*Vicia faba*) root-tip cells, using a cytological technique. The root cells were put into a solution containing tritiated (^{3}H) thymidine, and allowed to undergo mitosis so that the [^{3}H] thymidine could be incorporated into DNA. The tips were then washed and transferred to a solution containing non-radioactive thymidine. If colchicine is added at this point, spindle formation is inhibited and the chromosomes in metaphase fail to separate so that the sister chromatids remain joined together for a while by the centromere. Mitosis of the root-tip cells in the presence of radioactive thymidine for one round of duplication was shown to result in the labelling of the chromosomal DNA with the radioactive isotope. Analysis of the daughter chromosomes by autoradiography indicated that each pair was labelled with the isotope. When the daughter chromosomes were examined after a second duplication in the presence of non-radioactive thymidine,

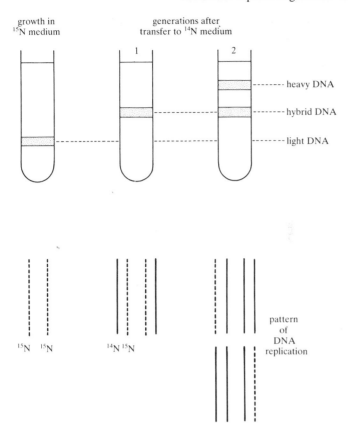

growth in
^{15}N medium

generations after
transfer to ^{14}N medium

1

2

------- heavy DNA

------- hybrid DNA

------- light DNA

^{15}N ^{15}N

^{14}N ^{15}N

pattern
of
DNA
replication

Fig. 39.4 The Meselson and Stahl experiment demonstrating the semi-conservative replication of DNA.

almost all the pairs of chromosomes which were formed from the radioactive parent contained one labelled and one unlabelled chromosome. This result indicates that the method of DNA duplication in eukaryotic organisms resembles that shown for the DNA of the bacterium *E. coli*, because it is possible to interpret it in terms of semi-conservative duplication.

Gene action

In Chapter 37 various types of gene interaction (such as epistatic and complementary gene action) which can be recognized by the standard analytic methods of Mendelian genetics were discussed. The way such interactions come about, and the way in which genes can cooperate chemically to bring about their effects, were much clarified in the early 1940s by studies on the biosynthesis of amino acids, vitamins and other substances by the bread mould *Neurospora crassa*. This fungus normally synthesizes its own amino acids and vitamins, but it cannot synthesize the vitamin biotin which it must obtain from its food. Most strains of *N. crassa* will therefore grow on a minimal growth medium containing sugar, an inorganic nitrogen source, and biotin. By subjecting the fungus either to irradiation by X-rays or to certain chemicals such as mustard gas, it is possible to produce mutant strains which will not grow on the minimal medium. Such strains are incapable of making various substances (e.g. the amino acid tryptophan) which are necessary for their growth, and these substances must be added to the minimal medium if such strains are to survive.

Among the many mutant strains of *N. crassa* that were produced, those which were incapable of tryptophan synthesis proved particularly suitable for study. Two of these strains which are unable to synthesize tryptophan have each been shown to differ from the normal type by a single gene. Both strains will grow if tryptophan, or its precursor indole, is added to the minimal medium. Evidently both strains share the ability of the normal type to convert indole into tryptophan. Furthermore, one of the strains, which will here be called the α strain, can grow if it is given anthranilic acid, which it must therefore be able to turn into indole as the normal strain does. The blockage in the synthesis of tryptophan must thus be somewhere in the chain of chemical reactions before the production of anthranilic acid. The normal strain has an allele a^1 (N.B. *Neurospora crassa* is a haploid organism, and its somatic cells have only one allele of each gene) which is able to bring about the synthesis of anthranilic acid from its precursor, whereas strain α has another allele a^2 of the same gene which is unable to carry out this reaction. In contrast to the α strain, the second strain β cannot grow when supplied with anthranilic acid, but will grow, as previously mentioned, if indole is supplied. When this strain is enabled to grow by being given indole, anthranilic acid accumulates in the medium on which the fungus is growing. Evidently this strain can make anthranilic acid, but is unable to convert it into indole. The normal strain must have an allele b^1 bringing about this conversion, whereas strain β has an allele b^2 which cannot do so. It is now possible to build up a picture of how the genes play their part in the synthesis of tryptophan.

$$\begin{array}{ll} \text{gene } a & \text{gene } b \end{array}$$

Normal strain —— $a^1 \rightarrow$ Anthranilic —— $b^1 \rightarrow$ Indole $\rightarrow$ Tryptophan
                                   acid

α Strain   —— a^2
β Strain   —— $a^1 \rightarrow$ Anthranilic —— b^2
                                   acid

Many such biochemical pathways, some of them very complex, have been analysed in *Neurospora* and other fungi. In all cases the genes have been shown to play a similar part, namely that of governing the conversion of one substance into another. Each step is a relatively simple one, such as a condensation or a methylation, so that complex synthetic pathways are achieved by the cooperation of a series of genes, each performing a specific task and each supplying the raw material for the next step in the chain of chemical reactions. The concept that a gene does its work by producing one, and only one, enzyme was expressed by Beadle and Tatum as the *one gene–one enzyme hypothesis*. However, since single genes also control the formation of non-enzymic proteins, the concept is better modified as the one gene–one protein hypothesis, or better still as the one gene–one polypeptide chain hypothesis. Although the one gene–one enzyme hypothesis has had to be modified, it represents a milestone in the understanding of how genes act.

The genetic code

The information about DNA structure considered earlier in this chapter demonstrates how DNA can replicate itself, but it does not explain how the

genetic information which determines the hereditary characters of an organism is encoded in the DNA molecule. The basic components of all organisms are proteins, of which three main groups can be recognized: (1)
enzymes, or catalytic proteins, which participate in almost all chemical
reactions that take place in living cells; (2) structural proteins which form
the building blocks for cellular components; and (3) regulatory, or control,
proteins which interact with genes or their products to control the amounts,
and presumably the times of production, of yet other protein molecules. It
is not surprising therefore, that hereditary differences between organisms are
due to differences in their protein make-up.

The three-dimensional structure of DNA is the same for all organisms, but
yet the DNA from different organisms and from different genes in the same
cell is different. It must therefore be the order of the nucleotides along the
length of a DNA molecule that determines the nature of the genetic material
at any particular locus. The discovery of the ways in which the sequence of
the nucleotides in the DNA of a chromosome is translated into the sequence
of amino acids in a protein is one of the outstanding achievements of molecular biology. This is sometimes called the transcription function of DNA,
to distinguish it from the replication function.

Protein synthesis

At the outset it is necessary to point out that in the living cell the connection
between the nucleotide sequence in the DNA and the sequence of amino
acids in a protein is not a direct one. If the DNA in the chromosomes directly
controlled the formation of proteins, then we should expect to find the proteins of the cell being produced inside the nucleus. However, experiments
with radioactive amino acids have shown that, except in very young embryonic cells in which much of the cellular machinery has probably not yet
formed, the most active sites of protein synthesis are not in the nucleus but
in the cytoplasm. The basic postulate of molecular genetics is that the information contained in DNA is first *transcribed* into an RNA molecule, from
which the information is subsequently *translated* into protein. The hypothesis
can be represented thus: DNA→RNA→ protein. It is also postulated that
genetic information is transferred only in the direction from DNA to protein;
in other words, there is no storage of information in the protein molecules
and no transcription of it back into the nucleic acids. This hypothesis is sometimes called 'the central dogma' of molecular genetics.

Protein synthesis in the cytoplasm has been shown to occur on tiny particles, invisible with a light microscope but readily detectable with an electron
microscope, called *ribosomes*. Up to a million or more of them occur in virtually all living cells. They consist of both RNA and protein. Besides ribosomal RNA, another kind of RNA called messenger RNA (mRNA) is found
in both the nucleus and the cytoplasm, and acts as a carrier of genetic information from the nucleus to the ribosomes. Messenger RNA is produced in
the nucleus but migrates into the cytoplasm, where it becomes associated
with the ribosomes. A molecule of mRNA is synthesized by using a length
(about 10^5–10^6 nucleotides long) of a single strand of DNA as a templet.
RNA consists of a single strand of nucleotides and is very similar in chemical
structure to one strand of the double helix of a DNA molecule, except that
uracil replaces thymine as the base complementary to adenine (i.e. uracil is
the only base that occurs in RNA but not in DNA) and the sugar in the

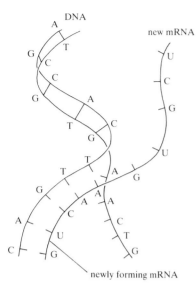

Fig. 39.5 Transcription of DNA.

backbone of the molecule is ribose instead of deoxyribose. The first step in mRNA synthesis is that a length of the double helix of a molecule of DNA unwinds by breaking the hydrogen bonds between the corresponding bases in the paired strands. The sequence of the bases A (adenine), T (thymine), C (cytosine) and G (guanine) in one of the DNA strands is copied, or transcribed, into the complementary sequence of the bases U (uracil, which replaces thymine in RNA), A, G and C in the messenger RNA. The process is shown in Fig. 39.5. Which of the two DNA strands is transcribed is obviously important because the two strands will give rise to different polynucleotide chains. The genetic information encoded in the sequence of nucleotides in the nuclear DNA has thus been transcribed by the mRNA, which then carries the information into the cytoplasm of the cell. However, this 'messenger' role of mRNA is not the only one it plays in protein synthesis; its message must be again copied, or translated, into a sequence of amino acids in a protein.

The process of translation requires the presence of a third type of RNA, called transfer RNA (tRNA). This, like mRNA, is synthesized in the nucleus and subsequently migrates into the cytoplasm. Molecules of tRNA are only seventy to eighty nucleotides long, and act as the link between mRNA and amino acids. They function as 'adaptors' for the twenty amino acids found in proteins, combining with the amino acids and then fitting them into their correct positions along the mRNA templet. Each amino acid has its own tRNA adaptor, and therefore there are at least twenty different tRNAs which correspond to the twenty amino acids. Whereas mRNAs are uncoiled molecules, tRNAs are coiled back on themselves to form approximately cross-shaped molecules with double-stranded regions of complementary base pairs. At the end of one of the 'arms' of a tRNA molecule is the site where a specific amino acid can be attached with the aid of a specific enzyme (activating enzyme). At the end of the opposite arm is a sequence of bases (actually three of them, see later in the chapter) called an *anticodon*, which pairs with a complementary sequence (also of three bases) called a *codon* on the mRNA. The anticodon on the tRNA may be compared to a 3-pin electrical plug, and the codon on the mRNA to a 3-pin socket into which the plug fits. Another sequence of bases located somewhere near the centre of the tRNA molecule is believed to interact with the ribosome, so that the tRNA molecule takes up the correct position when it is attached to the ribosome.

Although ribosomal RNA forms most of the RNA fraction of a cell, very little is known about its function. Ribosomal RNA molecules are not translated into protein but, because ribosomes (like chromosomes) consist of nucleic acid and protein, they presumably form part of the structural framework of these organelles.

The process of protein synthesis is represented diagrammatically in Fig. 39.6. One end of a molecule of mRNA attaches itself to the equatorial region of a ribosome. Lying free in the cytoplasm around the ribosome are molecules of the twenty different kinds of tRNA, each combined or *charged* with its own specific amino acid. It is postulated that the ribosome moves along the strand of mRNA, reading the sequence of nucleotides as it does so. As each codon of three bases is reached, it is recognized by the anticodon of one of the tRNAs in the surrounding cytoplasm, and the tRNA molecule then plugs into the codon of the mRNA strand. This pairing between the mRNA and the tRNA locates the amino acid attached to the 'tail' end of the

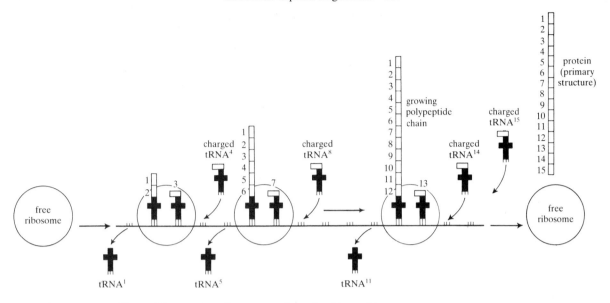

Fig. 39.6 Protein synthesis.

tRNA in a very specific position on the ribosome surface. In this position a peptide bond can be formed between this newly-arrived amino acid and a chain of amino acids attached to another tRNA already plugged into the adjacent codon of the mRNA strand. At any one time apparently two molecules of tRNA are present on the ribosome, one has already contributed its amino acid to the growing peptide chain and the other still has it attached at its tail end. When the developing peptide chain bonds to the amino acid of the newly arrived tRNA molecule, the ribosome moves along the mRNA strand so as to read the next codon in the sequence. The tRNA to which the peptide chain was previously attached, having delivered its amino acid to the peptide chain, is released into the cytoplasm. Here it can join with another molecule of its specific amino acid, ready to plug in again and deliver another amino acid molecule to the peptide chain as directed by the mRNA. Molecules of tRNA thus move in on one side of the ribosome, 'read' the mRNA codon, shift to the peptide-chain-holding site, and then move off on the other side of the ribosome. By repeated cycling of tRNAs on to and off the ribosome, the polypeptide chain grows longer and longer. When the ribosome reaches the end of the mRNA strand, it falls off. The completed protein is also released at the same time; it folds into its characteristic shape and begins its functional existence as an enzyme, or other integral part of the cell.

The language of the genetic code

All the information for making the enzymic, structural and regulatory proteins of a cell is contained in an alphabet of the four letters A, C, G and T, which are strung along the length of the DNA molecules. Somehow the order in which these four letters are arranged determines the sequences of the twenty amino-acid units that make up proteins. If a single letter coded for one amino acid, only four amino acids could be specified. If two bases were needed to specify an amino acid, this is still not enough because two letters can be selected from an alphabet of four letters to form only sixteen (i.e. 4 × 4) different two-letter words. Words of three letters are thus the smallest possible words that can code for twenty amino acids, but combinations of three letters allow sixty-four (i.e. $4 \times 4 \times 4$) different words to be constructed, which is considerably more than the necessary minimal number of twenty. Either only twenty of the sixty-four words are used, or some or all of the amino acids are coded for by more than one three-letter word. A three-letter, or *triplet*, code could be constructed in at least three different ways: (1) with words overlapping; (2) with words not overlapping but punctuated; and (3) with words not overlapping and not punctuated.

An overlapping code is composed of words that overlap one another, e.g. in a triplet code the letters of any given word may form part of one, two or three words. In the sequence C A T G T A G A G the first word is CAT, but the second is either TGT or ATG according to whether there is an overlap of one or two letters in the code. This type of code is unlikely to be the code for DNA because of the severe restrictions it imposes upon the possible sequences of amino acids in proteins. For example, if the first word of a code with three letters of which one overlaps is CAT, as in the above sequence, then the second word must begin with T, and similarly if there are two overlapping letters the second word must begin with A. Analysis of the amino-acid sequences in proteins indicates that any amino acid can follow any other amino acid, and this observed fact is incompatible with an overlapping code.

If the genetic code is a triplet non-overlapping one, then the problem of how words can be distinguished from one another remains to be solved. Assuming for the moment that the DNA code is a triplet non-overlapping one, the base sequence CAT CAT CAT . . . could be punctuated by using the fourth base G as a spacer between each CAT triplet. This would be equivalent to having a functional alphabet of only three letters, and the total number of possible triplets that could be constructed from a four-letter alphabet with such punctuation would be twenty-seven (i.e. $3 \times 3 \times 3$). However, it is known (see later) that the DNA code is degenerate, which means that some amino acids are coded for by more than one triplet. The extent of the degeneracy known to exist in the DNA code rules out the possibility of a triplet code with this type of punctuation because the number of words is too small.

The third possibility is a non-overlapping, non-punctuated code, in which the reading starts from a fixed point. This has been proved to be the type of code used by all organisms whose coding patterns have been studied. If the reading of the genetic message begins at a fixed point in a DNA strand, then the addition or removal of a single letter will alter the message from the point where the addition or deletion occurs. For example, if the sequence of three-letter words in a DNA strand is CAT GTA GAG (which would be transcribed in the RNA code as GUA CAU CUC), deletion of the second T would mean that the triplet sequence CAT GAG AG . . . is read. The first triplet remains unchanged, but all the subsequent triplets will be altered. The chemical addition of a base to a DNA sequence will similarly change the sequence of triplets beyond the point at which the base is added. A change which alters the reading of all the following triplets is called a frame shift.

Mis-sense mutations

One of the important concepts in genetics is *mutation*, the sudden change in a gene which, if it occurs in the gametes, is inherited by subsequent generations of offspring. The changed gene is allelic to the original gene, and may manifest itself as an entirely new characteristic in an individual in which it occurs. The whole framework of Mendelism depends on the occurrence of mutation in the characters whose heredity is being studied. From current knowledge of the structure of the gene, it is now possible to see how mutation can happen. Mutation results when one of the nucleotides along a DNA strand is replaced by another. When this happens, a different base is inserted in the messenger RNA and, in many cases, a different amino acid is inserted into the resulting polypeptide chain.

The acridine dyes (proflavin, acridine orange and acridine yellow, among others) are substances that bind to DNA and, in bacteriophages, cause mutations during DNA replication by becoming inserted between previously adjacent nucleotides, doubling the distance between them. This may either allow later insertion of a new nucleotide during replication, or result in the deletion of a base, thus altering the sequence. In the virus Phage T4, which infects the colon bacillus *Escherichia coli*, a large number of mutants at many different sites along its DNA molecule have been induced by treatment with acridine dyes. If a mutant resulted from the deletion of a base from the DNA strand, the addition of a base at a point beyond the first mutation would restore the reading frame of the DNA sequence, and could probably result in little or no change in the biological activity of the original protein. For example, in the sequence CAT CAT CAT CAT CAT . . . deletion of the second T causes a frame shift, and the sequence becomes CAT CAC ATC ATC AT . . . Such a mutation is described as a *mis-sense* mutation. The subsequent addition of a T (or any of the other three nucleotides) at a different position in the mis-sense sequence, say between the sixth and seventh nucleotides, causes the sequence to become CAT CAC TAT CAT CAT . . . in which the original sequence is restored after two wrong triplets. This kind of restoration-of-sense sequence clearly indicates that there is no punctuation between the codons, i.e. each codon is immediately adjacent to the next with no intervening 'spacer' bases. It is also found that, just as one frame shift results in a mis-sense mutation, so also do two, four or five frame shifts. On the other hand, just as opposing alterations (e.g. a deletion followed by an insertion), if not too far apart, restore sense, so also do three deletions or three insertions. Results such as these make sense only if the code, besides being without punctuation, is also a triplet one. The acridine-dye experiments therefore provide almost irrefutable evidence that the genetic code is a triplet code with words not overlapping and not punctuated.

The coding dictionary

Once the triplet nature of the genetic code had been established, the question of which triplets code for which amino acids was actively investigated. The code has been 'cracked' by laboratory studies on polypeptide synthesis using cellular components isolated from micro-organisms. The following three steps are involved. Step 1 – obtain ribosomes, a full complement of tRNAs, and the enzymes for attaching amino acids to them. Step 2 – obtain an mRNA with a known base sequence. Among many more, the following artificial mRNAs have been prepared:

poly U   (U–U–U–U–U–U– . . .)
poly A   (A–A–A–A–A–A– . . .)
poly C   (C–C–C–C–C–C– . . .)
poly U with single C   (C–U–U–U–U–U– . . .)

Step 3 – combine the ingredients of step 1, one of the synthetic mRNAs of step 2, and certain stock chemicals (including ATP, amino acids and magnesium). Examples of the type of result obtained are as follows:

Experiment 1

$$\left.\begin{array}{l} \text{ribosomes} \\ \text{tRNAs} \\ \text{enzymes} \end{array}\right\} + \left.\begin{array}{l} \text{amino acids} \\ \text{ATP} \\ \text{magnesium} \end{array}\right\} + \text{poly U} \rightarrow \begin{array}{l} \text{polypeptide consisting only} \\ \text{of phenylalanine} \end{array}$$

(from cells)   (stock chemicals)

Interpretation: UUU in mRNA codes for phenylalanine in protein.

Experiment 2

(as above)   +   (as above)   +   poly C   →   polypeptide consisting only of proline

Interpretation: CCC in mRNA codes for proline in protein.

Experiment 3

(as above)   +   (as above)   +   poly U with   →   polypeptide consisting of one terminal C   leucine followed by many phenylalanine residues

Interpretation: CUU in mRNA codes for leucine in protein.

Other triplets were tested for their coding abilities by synthesizing mRNAs with varying proportions of two bases. If, for example, a mixture of U and C in the proportion of 5 : 1 is synthesized into RNA, the possible triplets and their probable frequency in the artificial mRNA can be determined. The triplet UUU will be most common and will appear with the frequency ($\frac{5}{6} \times \frac{5}{6} \times \frac{5}{6}$); the triplets UUC, UCU and CUU will all appear with the frequency of $\frac{5}{6} \times \frac{5}{6} \times \frac{1}{6}$; while the triplet CCC should appear only $\frac{1}{216}$ of the time. A messenger RNA of this composition should result in the incorporation into protein of eight different amino acids. In fact, only four amino acids were present in the polypeptide produced, and this means that several of these triplets encode for the same amino acid and therefore that the code is degenerate. A knowledge of the base sequence in many artificial mRNAs and of the resulting amino-acid sequence in the polypeptides produced has revealed the code for all twenty amino acids (Table 39.2).

From even a cursory observation of Table 39.2 it is immediately apparent that there is a high degree of degeneracy in the code. For instance, the amino acid leucine is coded for by no fewer than six different codons: UUA, UUG, CUU, CUC, CUA and CUG. Other amino acids, such as methionine and tryptophan have only one codon each, AUG and UGG respectively. Most of the amino acids have either two or four codons.

In many cases it is the first two letters of a codon that define which amino acid is coded for. As mentioned above, codons with C and U as the first two letters always code for leucine, regardless of what the third letter is. Other examples are U*CX* (where *X* may be G, U, A or C) always coding for serine and GG*X* always coding for glycine. The high degree of degeneracy in certain amino acids may have the effect of buffering them against the changes of gene mutation.

The analogy of the triplet codons with words implies that the genetic code must have the equivalent of full stops and capital letters, i.e. signals for stopping and starting a sentence. Of the sixty-four possible codons listed in Table

Table 39.2 The complete genetic (mRNA) code. The three triplets with an asterisk are nonsense codons and are the code's full-stops. The amino acids are listed by their standard three-letter abbreviations, most of which may be identified by reference to Table 14.1.

SECOND LETTER

FIRST LETTER / THIRD LETTER

		U	C	A	G	
U		UUU ⎫ phe UUC ⎭ UUA ⎫ leu UUG ⎭	UCU ⎫ UCC ⎪ ser UCA ⎪ UCG ⎭	UAU ⎫ tyr UAC ⎭ UAA* UAG*	UGU ⎫ cys UGC ⎭ UGA* UGG try	U C A G
C		CUU ⎫ CUC ⎪ leu CUA ⎪ CUG ⎭	CCU ⎫ CCC ⎪ pro CCA ⎪ CCG ⎭	CAU ⎫ his CAC ⎭ CAA ⎫ gln CAG ⎭	CGU ⎫ CGC ⎪ arg CGA ⎪ CGG ⎭	U C A G
A		AUU ⎫ AUC ⎪ ile AUA ⎭ AUG met	ACU ⎫ ACC ⎪ thr ACA ⎪ ACG ⎭	AAU ⎫ asn AAC ⎭ AAA ⎫ lys AAG ⎭	AGU ⎫ ser AGC ⎭ AGA ⎫ arg AGG ⎭	U C A G
G		GUU ⎫ GUC ⎪ val GUA ⎪ GUG ⎭	GCU ⎫ GCC ⎪ ala GCA ⎪ GCG ⎭	GAU ⎫ asp GAC ⎭ GAA ⎫ glu GAG ⎭	GGU ⎫ GGC ⎪ gly GGA ⎪ GGG ⎭	U C A G

39.2, sixty-one have been shown to code for a particular amino acid. The three left over are UAA, UAG and UGA, and it is thought that these codons are nonsense codons (i.e. they do not code for any amino acid) and act as the 'full stops' on the messenger RNA which allow the completed polypeptide chain to be released so that the next polypeptide chain can be started. The mechanism for starting a polypeptide chain is not known, although the action of a special 'starter' tRNA molecule is suspected.

Regulation of gene action

All the information that is ultimately responsible for the metabolism, growth and development of cells is stored in their DNA molecules. Most of a cell's DNA is present in the chromosomes but, since some characters in certain plants are inherited cytoplasmically, there is circumstantial evidence that some genetic material is extranuclear. Chloroplasts and mitochondria contain DNA, and these organelles have a degree of autonomy consistent with the storage of some of their own genetic information. In fact, the DNA of chloroplasts is now known to code for some but not all of their proteins. Each cell receives a full complement of the primary information at cell division, and so the problem of why cells differ both structurally and functionally from one another resolves itself into the question of how genetic information is selected. At any given moment in the life cycle of a developing organism, the appropriate information must be selected by every growing and dividing cell so that each organ in all parts of the plant or animal develops in the proper way. All irrelevant or unnecessary information must be suppressed or stored so that it is not immediately accessible. Provision of the relevant information implies activating the appropriate genes to make mRNA, which will control the synthesis of the specific enzymes required at a particular site at a particular time. Development necessitates that genes will be activated in a very precise sequence, i.e. they must be 'programmed' so that each stage of development activates the next.

An easily demonstrable example of the switching on or off of gene action according to the external environment is the induced formation of the enzyme β-galactosidase by *Escherichia coli*. This enzyme catalyses the hydrolysis of lactose to galactose and glucose, and it is required by the bacterium if lactose is to be metabolized. When normal *E. coli* is grown on a medium containing lactose, it produces large amounts of β-galactosidase in order to utilize this substrate. It has been established that β-galactosidase is synthesized only in the presence of lactose, which thus acts as an enzyme inducer as well as being the substrate. Such enzymes are called *adaptive* or *inducible* enzymes to distinguish them from *constitutive* enzymes, which are synthesized continuously, whether or not the substrate is present and regardless of the cell's requirements. In contrast to the β-galactosidase of normal *E. coli*, the enzymes of this bacterium which break down glucose are constitutive enzymes.

A general mechanism whereby the regulation of gene action could be accomplished (Fig. 39.7) has been proposed by the French scientists Jacob and Monod. They proposed that a functional unit of one or several closely linked genes, called *structural genes*, produces messenger RNA for specific enzymes. This unit is kept either in an active or open state or in an inactive or closed state by the controlling influence of a single *operator gene*. The combination of a group of structural genes with an operator gene is called an *operon*. The operator gene is, in its turn, controlled in one of two ways. It may be switched off by a regulatory protein molecule, called a *repressor*, which is produced by a separate *regulator gene*. The regulator gene is not part of the operon with which it is associated, and need not even be located close to it. The repressor can be inactivated by a molecule, called an *inducer*, whose presence therefore allows the operator gene to go into the open state so that the operon becomes active. Alternatively, when an operator gene is active, another type of molecule, called a *co-repressor*, may combine with an inactive repressor so that the latter is activated, the operator gene is closed, and the operon becomes inactive. Inducer and co-repressor molecules may be simple metabolites from elsewhere in the cell or from outside the cell.

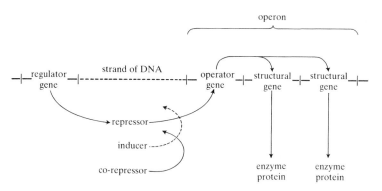

Fig. 39.7 Model of the operon proposed by Jacob and Monod.

The operon model offers a plausible explanation of how different genes may be active at different times in different tissues. It is not difficult, for example, to imagine that some metabolic activity associated with cell wall formation might produce molecules of a substance which, in addition to being an intermediate in cell wall synthesis, functions as an inducer of one of the operons responsible for the formation of messenger RNA which synthesizes cytoplasmic enzymes. These themselves might produce metabolites inducing the formation of structural components. At some stage a metabolite might also act as a co-repressor of an operon earlier in the sequence, and so terminate that sequence of reactions. Thus an orderly pattern of induction and repression could be the molecular counterpart of the orderly processes of growth and development visible to the naked eye. It must be emphasized, however, that the concept of sequences of operons represents only one possible explanation for differentiation. Specific examples of enzyme induction and repression are known, such as that involved in the formation of β-galactosidase by the bacterium *E. coli*, but the exact mechanisms by which the growth and development of flowering plants are programmed are not known. It has been suggested that certain hormones or hormone-like compounds may act to stimulate enzymes (the synthesis of hydrolases triggered off by gibberellic acid during the germination of the barley grain is described on p. 383), and also that the protein components of the chromosomes have a specific role to play in the switching on or off of genes. However, details of the nature of developmental programmes at the molecular level, and how they work, are only just beginning to be understood.

Suggestions for further reading

Ecology

Aubert de la Rue, E., Bouliere, F. and Harroy, J-P. (1957) *The tropics*. George G. Harrap.

An excellent, beautifully illustrated introduction to plant and animal life in the tropics. A background book for stimulating interest in tropical biology.

Bates, Marston. (1953) *Where winter never comes*. Gollancz.

This is an unusual but fascinating book in which a great deal of information about the tropics is presented in a most stimulating and appealing fashion. An excellent account of tropical biology in general, and of rain-forest ecology in particular, as seen through the eyes of a 'convert' rather than a native of the tropics.

Buckman, H. O. and Brady, K. C. (1969) *The nature and properties of soils*, 7th edn. Macmillan.

Probably the best elementary text on soil science.

Cloudsley-Thompson, J. L. and Chadwick, M. J. (1964) *Life in deserts*. G. T. Foulis.

A very well illustrated summary of what is known about deserts, enriched by the personal experience of the authors.

Collinson, A. S. (1977) *Introduction to world vegetation*. George Allen & Unwin.

This clearly-written, short book provides an introduction to a wide range of the topics and problems of phytogeography, especially those related to the distribution of the world's vegetation. Highly recommended.

Ewusie, J. Y. (1980) *Elements of tropical ecology*. Heinemann.

A useful introductory text written by someone who has had wide experience of both teaching and research in the tropics. This book should go a long way towards providing relevant factual information for courses on ecology.

Eyre, S. R. (1975) *Vegetation and soils, a world picture*, 2nd edn. Edward Arnold.

After an introductory section devoted to a study of how plant communities and soils develop, these two most important elements in the landscape are examined region by region. Nearly a quarter of the book deals with the tropics. Although this book is written primarily for geography students, students of botany should derive considerable benefit from reading it.

Hopkins, B. (1965) *Forest and savanna*. Heinemann.

This very readable introduction to tropical plant ecology makes special ref-

erence to West Africa, but most of the book is relevant to tropical vegetation throughout the world.

Janzen, D. H. (1975) *Ecology of plants in the tropics*. Edward Arnold.
    A concise but very informative 'evolutionary' approach to plant ecology
    in which plants are always seen as members of an ecosystem. The reader
    is soon convinced of the value of this interdisciplinary approach in such
    important ecological subjects as seed-dispersal interactions.

Kershaw, K. A. (1973) *Quantitative and dynamic plant ecology*, 2nd edn.
    Edward Arnold.
    One of the best practical guides, with good accounts of statistical methods.

Kormondy, E. J. (1969) *Concepts of ecology*. Prentice-Hall.
    An outstanding textbook of the basic concepts of modern ecology. The
    unifying theme is the structure and function of ecosystems.

Lawson, G. W. (1966) *Plant life in West Africa*. Oxford University Press.
    This is not, and does not attempt to be, a botany textbook. It succeeds
    admirably in its aim of providing a broad, mainly ecological, background
    to the botany of the region.

Odum, E. P. (1975) *Ecology*, 2nd edn. Holt, Rinehart & Winston.
    A short text in ecology from the point of view of ecosystems by a renowned
    ecologist who was one of the pioneers in developing the ecosystem con-
    cept.

Randall, R. F. (1978) *Theories and techniques in vegetation analysis*. Oxford
    University Press.
    This concise manual achieves in under 60 pages what certain larger books
    do not achieve so well in over 300 pages. Strongly recommended.

Raunkiaer, C. (1937) *Plant life forms*. Translated by H. Gilbert-Carter.
    Oxford at the Clarendon Press.
    This short book is the first part of the much larger *The life forms of plants
    and statistical plant geography*, written in 1907. There are many black-and-
    white line drawings which are a model of their kind.

Richards, P. W. (1952) *The tropical rain forest*. Cambridge University Press.
    Although somewhat dated, this scholarly book continues to be the stand-
    ard reference work in English on the subject.

Walter, H. (1972) *Ecology of tropical and subtropical vegetation*. Oliver &
    Boyd.
    A detailed account of all types of tropical and subtropical vegetation from
    rain forest to desert.

Watts, D. (1971) *Principles of biogeography*. McGraw-Hill.
    A comprehensive textbook covering the subject area where botany and
    geography overlap. Copiously referenced.

Whitmore, T. G. (1975) *Tropical rain forests of the Far East*. Oxford Uni-
    versity Press.
    A modern comprehensive text which has rapidly established itself as a
    standard work on rain forest in eastern Asia.

Genetics

Baer, A. S. (1977) *The genetic perspective*. W. B. Saunders Co.
    A stimulating book which emphasizes the practical applications of genetics
    to human welfare.

Burns, G. W. (1980) *The science of genetics*, 4th edn. Collier Macmillan.
    There are many good introductory textbooks of genetics, but this one

ranks among the best because of the very clear style in which it is written, and the logical way in which the subject matter is presented.

Kemp, R. (1970) *Cell division and heredity*. Edward Arnold.

The aim of this booklet is to show that cell division can be interpreted meaningfully only as an essential part of reproduction. Meiosis, for example, should be regarded not as an isolated process by which the number of chromosomes is halved, but as a logical sequence of events through which the genetic information of all sexually reproducing organisms must pass if they are to survive and evolve.

Strickberger, M. W. (1975) *Genetics*, 3rd edn. Macmillan.

A clearly written, up-to-date account of all aspects of genetics from the gene to the population. Deservedly a standard textbook.

Watson, J. D. (1970) *Molecular biology of the gene*, 2nd edn. Benjamin.

A concise, well-illustrated textbook suitable for the student who wants detailed information about the ultimate units of heredity.

Index

Page numbers in italics refer to figures or tables only